Molecular Biology of RNA

New Perspectives

Edited By

Masayori Inouye

Department of Biochemistry
Robert Wood Johnson Medical School at Rutgers
University of Medicine and Dentistry of New Jersey
Piscataway, New Jersey

Bernard S. Dudock

Department of Biochemistry
State University of New York at Stony Brook
Stony Brook, New York

ACADEMIC PRESS, INC.
Harcourt Brace Jovanovich, Publishers
San Diego New York Berkeley Boston
London Sydney Tokyo Toronto

ACADEMIC PRESS, INC.
1250 Sixth Avenue, San Diego, California 92101

United Kingdom Edition published by
ACADEMIC PRESS INC. (LONDON) LTD.
24–28 Oval Road, London NW1 7DX

Library of Congress Cataloging in Publication Data

Molecular biology of RNA.

Includes index.
1. Ribonucleic acid. 2. Molecular biology.
3. Molecular genetics. I. Inouye, Masayori.
II. Dudock, Bernard S. [DNLM: 1. Molecular Biology.
QU 58 M7164]
QP623.M65 1987 574.87'3283 87–11375
ISBN 0–12–372483–X (alk. paper)

PRINTED IN THE UNITED STATES OF AMERICA

87 88 89 90 9 8 7 6 5 4 3 2 1

Contents

3 Multiple Enzymatic Activities of an Intervening Sequence RNA from *Tetrahymena*

Thomas R. Cech, Arthur J. Zaug, and Michael D. Been

4 Processing and Genetic Characterization of Self-Splicing RNAs of Bacteriophage T4

Marlene Belfort, Joan Pedersen-Lane, Karen Ehrenman, Dwight H. Hall, Christine M. Povinelli, Jonatha M. Gott, and David A. Shub

II RNA Splicing

5 The Mammalian Pre-Messenger RNA Splicing Apparatus: A Ribosome in Pieces?

Joan A. Steitz

10 RNA Joining and Trypanosome Gene Expression

Nina Agabian, Karen L. Perry, and William J. Murphy

III RNA Viruses

11 The Poliovirus Genome: A Unique RNA in Structure, Gene Organization, and Replication

Steven E. Pincus, Richard J. Kuhn, Chen-Fu Yang, Haruka Toyoda, Naokazu Takeda, and Eckard Wimmer

12 Permanent Expression of Influenza Virus Genes Coding for Transcriptase Complexes: Complementation of Viral Mutants

Mark Krystal and Peter Palese

13 Molecular Mechanisms of Pathogenesis by HTLV-III

Flossie Wong-Staal and Robert C. Gallo

IV RNA in DNA Replication

V RNA: Structure, Function, and Isolation

Preface

RNA is full of surprises. An RNA acting as an enzyme, an RNA capable of self-splicing, and a branched RNA covalently linked to DNA are a few examples of new and previously unexpected roles for RNA. It is apparent that RNA is a highly complex class of molecules, with a wide variety of cellular roles, from which we have much to learn. We believe that it is appropriate at this time to take stock of recent discoveries in this rapidly developing field and to look at the approaches, strategies, and methodologies used. It is for this reason that we have undertaken, with much joy, the task of editing this book. We have been extremely fortunate in being able to present chapters written by the leaders in this field. We believe that this book will help provide new direction and insight for those already working on the subject and will serve as a useful guide to those about to start research in the molecular biology of RNA. Most of the contributing authors participated in the Fifth Stony Brook Symposium held in May 1986 on "New Perspectives on the Molecular Biology of RNA."

We wish to thank Janet Koenig for her invaluable assistance throughout this project.

Masayori Inouye
Bernard S. Dudock

I

RNA as an Enzyme

1

Cleavage of RNA by RNase P from *Escherichia coli*

SIDNEY ALTMAN, MADELINE BAER, HEIDI GOLD, CECILIA GUERRIER-TAKADA, LEIF KIRSEBOM, NATHAN LAWRENCE, NADYA LUMELSKY, AND AGUSTIN VIOQUE

Department of Biology
Yale University
New Haven, Connecticut 06511-8112

I. INTRODUCTION

RNase P performs a function in cells that is very simple in comparison to the functions performed by other ribonucleoproteins (RNPs), such as the small nuclear (sn) RNPs involved in mRNA splicing. It removes, with great accuracy, the extra nucleotides from the transcripts of tRNA genes to yield the correct 5′ terminus of the mature tRNAs (Altman *et al.*, 1982). Figure 1 shows a linear transcript of three genes: The gene in the center codes for a tRNA and, in *Escherichia coli,* the flanking genes code for other tRNAs, ribosomal RNA, or protein. In many cases in prokaryotes, the first processing event is an endonucleolytic cleavage that occurs at, or near, the 3′ end of the sequence for the mature tRNA. Simultaneously, or very shortly thereafter, a second cleavage occurs to generate the correct 5′ terminal nucleotide of the mature tRNA, and it is this reaction that will be discussed in more detail below. In eukaryotes, the order of processing events seems to be less rigidly specified (Rooney and Harding, 1986). After these two endonucleolytic cleavages, there is exonucleolytic cleavage of the extra nucleotides at the 3′ end of the tRNA and, at the same time, the nucleotide-modifying enzymes function to generate the mature tRNA molecule. The nature of the enzymatic events in eukaryotes is

Molecular Biology of RNA
New Perspectives

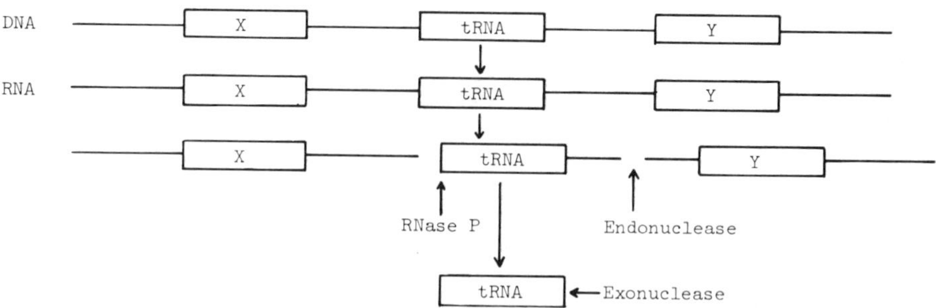

Fig. 1. A generalized scheme for the processing of tRNA gene transcripts by nucleases. X and Y denote sequences coding for protein and tRNA, rRNA and tRNA, tRNA alone, but not protein and rRNA. (Reprinted, with permission, from MIT copyright © 1981.)

similar, but most of the tRNA genes are transcribed as monomers and it is only the few extra nucleotides on either side of the mature sequence that must be removed.

Although RNase P was first identified over 15 years ago (Robertson *et al.*, 1972) in crude extracts of *E. coli,* it was only several years later that Stark *et al.* (1978) verified that the enzyme had very unusual chromatographic properties and could be inactivated by treatment with ribonucleases as well as proteases. We now know that these unusual properties were the result of the presence of an essential RNA component in the enzyme. Subsequent biochemical and genetic experiments have also demonstrated the essential nature of both RNA and protein subunits for growth of *E. coli* (Kole and Altman, 1979; Kole *et al.*, 1980).

II. ASPECTS OF THE RNase P REACTION

The reaction catalyzed by RNase P can be performed, *in vitro,* under two distinct sets of reaction conditions (Guerrier-Takada *et al.*, 1983). Most remarkably, we showed that the RNA component of RNase P can function by itself as a catalyst in the cleavage of precursors to $tRNA^{Tyr}$. Such a reaction, in the absence of protein requires the presence of 100 m*M* Mg^{2+}, although the reaction can also proceed with decreasing efficiency in the presence of lower concentrations of magnesium and added spermidine. It was shown subsequently (Gardiner *et al.*, 1985) that the RNA moiety of RNase P from *Bacillus subtilis* can act in a similar manner and we have obtained similar data for the analogous RNA from *Salmonella typhimurium* (Baer and Altman, 1985).

The RNA moiety of *E. coli* RNase P, called M1 RNA, cleaves the precursor to $tRNA^{Tyr}$ in a reaction that is second order (Guerrier-Takada *et al.*, 1986) for M1 RNA at very low concentrations of M1 (at ratios of M1 RNA to substrate molecules of approximately 1 : 10). From this result it would appear that under these conditions two molecules of M1 RNA are required to cleave a single precursor molecule. At low concentrations of M1 RNA, the reaction can be stimulated 15-fold by the addition of polyethylene glycol (PEG) and to a lesser, but still significant, extent by methylpentanediol. These effects are probably the result of enhanced local concentrations of M1 RNA that result from the excluded volume effect. If the cleavage of precursor is conducted under more standard enzymological conditions, namely in buffers containing 10 m*M* Mg^{2+}, both M1 RNA and the protein moiety C5 of the holoenzyme are required, and the reaction is first order in M1 RNA and in C5. It is possible that, in the absence of protein, the catalytic RNA molecule is unable to adopt the conformation into which it is folded in the presence of C5 protein. In the absence of protein, the appropriate folding of the catalytic RNA must be accomplished by intermolecular interactions, possible only in the presence of 100 m*M* magnesium, rather than by intramolecular interactions facilitated by the protein.

The secondary and tertiary intramolecular interactions of M1 RNA are essential for its function. Resuspended, lyophilized preparations of M1 RNA are inactive until heated or exposed to a denaturing agent and allowed to renature (Altman and Guerrier-Takada, 1986). Such renaturation is sensitive to pH and the transition from a less to more active state is most apparent between pH 7 and 7.5—a most unusual value for a pH-dependent transition of a nucleic acid, values close to the p*K* values of the bases (below pH 5 or above pH 9) being much more common. It is possible that M1 RNA is folded in such a way that the p*K* of certain ionizable groups is altered within the folded regions. The cleavage reaction itself, catalyzed by M1 RNA in the absence of protein, shows plateau values between pH 5.5 and pH 9, but the efficiency of the reaction drops precipitously on both sides of this range. In the presence of 10 m*M* manganese (the only metal ion that can substitute for magnesium M1 under these conditions), the curve is essentially identical except that it is shifted down by approximately half a pH unit, a shift that is consistent with the values for the pK_a values of these two metal ions in solution. While a metal ion–water molecule complex may be involved at the active site, the exact mechanism is not yet known and other possibilities should be entertained (Guerrier-Takada *et al.*, 1986).

The reaction catalyzed by RNase P is not a transesterification like the splicing reactions of rRNA or reactions that involve mRNA (Guerrier-Takada *et al.,* 1986). No covalent linkage is formed between the enzyme—in this case M1 RNA—and the substrate, nor are any intramolecular bonds formed transiently in the substrate during the reaction. Only magnesium and manganese can function as catalytic counterions, but magnesium, calcium, strontium, and a variety of polyamines can function as structural counterions.

Treatment of M1 RNA with RNA ligase circularizes the RNA and, surprisingly, the circles are as enzymatically active as the nonligated molecules (A. Branch and C. Guerrier-Takada, unpublished data). The circularized molecules open spontaneously on storage, and several pairs of new termini are generated, none of which correspond to the original termini of native M1 RNA. This heterogeneous group of linear molecules is active in the cleavage reaction. A second group of linear molecules can be generated by limited digestion with RNase T_1 and these have different termini again, but also retain activity. In fact, limited digestion of native M1 RNA by RNase T_1 does not destroy the activity of the catalytic RNA. That this unusual molecule can retain its catalytic activity despite sustaining a variety of phosphodieter bond cleavages suggests it maintains a high degree of structural integrity in spite of the presence of one or a few nicks. Furthermore, no particular 3′-terminal nucleotide, or hydroxyl group, is essential for the reaction.

The importance of cofactors in maintaining optimal secondary and tertiary structure of M1 RNA is emphasized by the different rates of reaction of M1 RNA with a variety of substrates in the presence or absence of C5 protein or PEG (Guerrier-Takada *et al.,* 1984, 1987). When precursors to tRNAs lack the CCA sequence that is normally present at the 3′ end of mature tRNA molecules, M1 alone is unable to cleave the precursors efficiently, in the absence of the protein cofactor. However, the same substrates can be processed as efficiently as those which contain the CCA sequence if C5 protein is added to the reaction mixtures (Guerrier-Takada *et al.,* 1984). Addition of PEG, instead of C5, is equally effective in potentiating the cleavage reaction. It can be seen from Fig. 2 that, in a typical precursor molecule, the CCA sequence is very close to the site of cleavage at the 5′ end of the mature sequence, so that the results just described suggest that there is an interaction between the CCA sequence and the active site of the catalytic RNA that is potentiated by the presence of C5 protein. In fact, there is evidence that the protein moiety of the holoenzyme actually controls the rate at which M1 RNA cleaves different substrates *in vitro* (see below).

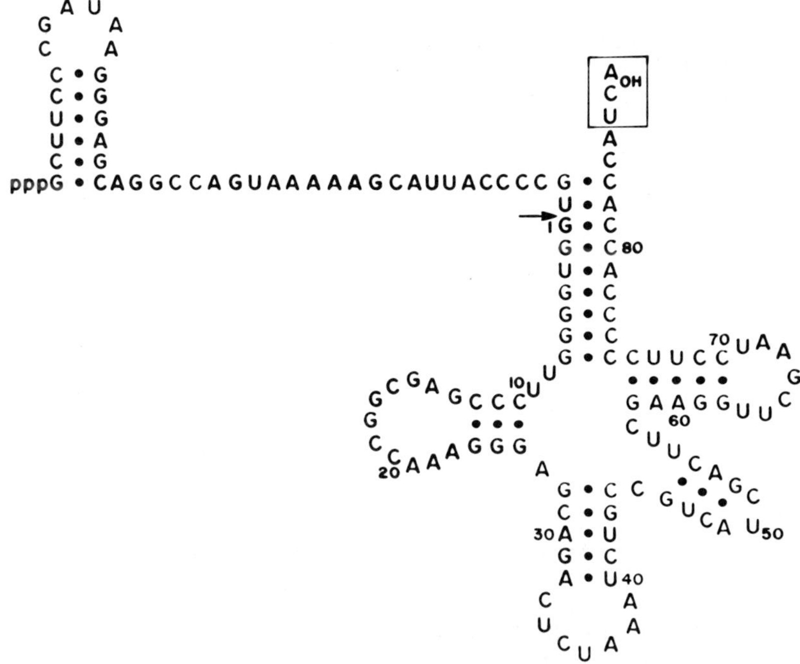

Fig. 2. Nucleotide sequence of the precursor to *E. coli* tRNA^{Tyr}su$_3$. The arrow pointing toward the sequence indicates the site of RNase P cleavage on the 5′ side of nucleotide 1 of the mature tRNA sequence. The boxed nucleotides are extra nucleotides at the 3′ terminus.

III. STUDIES OF ENZYME–SUBSTRATE INTERACTIONS

Initial studies of the reaction catalyzed by RNase P were performed with the precursor to tRNA^{Tyr} from *E. coli* (abbreviated as pTyr), whose nucleotide sequence is shown in Fig. 2. Recently, we constructed a set of altered precursor substrates with different numbers of additional nucleotides at the 5′ and 3′ ends of the sequence of the mature tRNA, as shown in Tables IA and IB (Guerrier-Takada *et al.*, 1987). These substrates were tested for their ability to serve as substrates in the reaction catalyzed by M1 RNA. If the number of extra nucleotides at the 3′ terminus of the substrates is varied systematically (as shown in Table IA), and the rate of cleavage of native precursor is taken as 1.0, the relative values for the rates of cleavage of the altered substrates derived from pGem-2/HH increase as the size of the precursor decreases. These data confirm and extend previous observations on the nature of pTyr as a substrate. How-

TABLE IA

Relative Rates of Cleavage of Different Substrates by M1 RNA[a]

Substrate	Number of extra nucleotides		Relative rate of cleavage
	5′ terminus	3′ terminus	
pTyr	43	3 (−CCACCAUCA)	1.0
pBR322/HR	43	Multiple ends	—
pGem-2/HH	72	46 (−CCACCA+46)	0.01
pGem-2/HH	72	0 (−CCACCA)	0.20
pGem-2/HH	72	20 (−CCACCA+20)	0.02
pGem-1/HR	49	25 (−CCACCA+25)	0.15
pGem-1/HB-3	49	5 (−CCACCACAUCG)	0.6
pGem-1/HB-5	49	−3(+3) (−CCAUCG)	1.33
pGem-1/HB-6	49	−2(+4) (−CCACAUCG)	0.05

TABLE IB

Effect of Polyethylene Glycol on Rate of Cleavage of Substrates with Different Termini[a]

Substrate	Rate of cleavage compared with rate of cleavage by pTyr	
	+PEG	−PEG
pTyr	1.0	1.0
pGem-1/HR	0.17	0.11
pGem-1/HB3	0.6	0.89
pGem-1/HB-3-1	1.0	1.11
pGem-1/HB-5	1.33	0.44
pGem-1/HB-5-15	1.0	0.22

[a] Reactions were carried out as described in Guerrier-Takada *et al.* (1987) in buffer that contained 100 m*M* $MgCl_2$ with or without 5% PEG. M1 RNA (12 ng) was included as the source of enzyme and approximately 3 ng of each substrate was used. The relative rates shown were calculated from the linear portion of the curve that illustrated the kinetics of cleavage of each substrate.

ever, the trend shown with the derivatives of pGem-2/HH is not adhered to if the substrate is altered within the terminal CCA sequence itself (Table IA, pGem-1 derivatives). In these experiments, pGem-1/HB-5 appears to be an even better substrate than unaltered pTyr. In separate experiments, we have also shown that pTyr lacking the terminal AUCA is cleaved almost as well as pTry itself. At present we can only speculate about the underlying reasons for these observations. It is possible that the improved rate of this latter reaction is due to the absence of a requirement for denaturation of this region during the reaction, a process that is part of the reaction of the native precursor with the holoenzyme *in vivo* and *in vitro*.

When 13 nucleotides are removed from the 5′ end of native precursor (Table IB, HB-3-1), the resultant substrate is cleaved with greater efficiency than the native substrate, but if 4 nucleotides are removed from the construct HB5, the rate of reaction decreases (Table IB, HB-5-15). Clearly, the extra nucleotides at the 5′ end of the mature sequence influence the rate of the cleavage reaction but, as yet, the nature of the interactions is unknown. Note that the presence or absence of PEG does not affect cleavage of all of these substrates in the same manner. It appears that the reaction governed by M1 RNA must be sensitive to the composition of substrates as well as to the concentration of M1 RNA in solution.

In an attempt to analyze the nature of the interactions of the RNA substrate with M1 RNA that are required for substrate recognition and cleavage, we have begun to isolate RNA–RNA enzyme–substrate complexes. Figure 3 shows a stained agarose gel after electrophoresis of precursor transcribed *in vitro* and M1 RNA. When substrate and M1 are combined, as in lanes 4 and 5, both uncleaved substrate and reaction product are visible. In lanes 2 and 3, containing M1 RNA by itself, there is a band, marked E2, that is composed of dimers of M1 RNA. In lanes 4 and 5, a faint band of such M1 RNA dimers is also visible, and there is another intense band which, on reanalysis, is found to contain both M1 RNA and substrate (E–S), in a ratio of approximately 1 : 1. A still larger band, marked X, for which the stoichiometry has not been determined, may contain M1 and substrate in the ratio of 2 : 1. This electrophoretic method for the examination of enzyme–substrate complexes will be particularly useful for analysis of the mutants of M1 RNA that are presently being constructed.

IV. STRUCTURE–FUNCTION RELATIONSHIPS IN M1 RNA

We have recently proposed a model for the secondary structure of M1 RNA from *E. coli* (Guerrier-Takada and Altman, 1984) and are able to

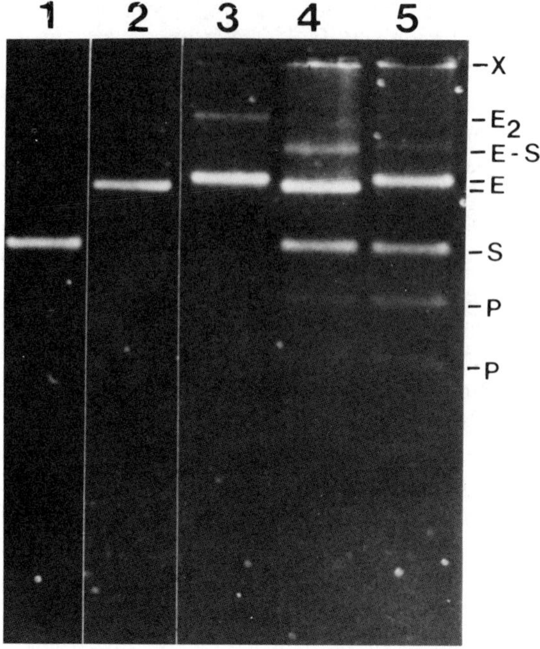

Fig. 3. Products of the cleavage reaction governed by M1 RNA analyzed in a 3% agarose gel containing ethidium bromide (Guerrier-Takada *et al.,* 1987). Products were visualized with a UV transilluminator. Lane 1, *In vitro* transcript of *E. coli* $tRNA^{Tyr}$ gene (substrate, 200 ng); lane 2, M1 RNA isolated from whole cells (260 ng); lane 3, *in vitro* transcript of gene for M1 RNA (300 ng); lane 4, M1 RNA incubated with substrate; lane 5; M1 RNA (prepared *in vitro*) incubated with substrate. P, Cleavage products of substrate; S, substrate; E, enzyme (M1 RNA prepared *in vivo* or *in vitro*); E–S, enzyme–substrate complex; E_2, dimers of M1 RNA; and X, a complex of enzyme and substrate in which E/S is greater than 1.

evaluate this model by a comparative study of analogous molecules from other organisms (Baer and Altman, 1985; Lawrence *et al.*, 1987). We have determined the nucleotide sequences of the gene for M1 RNA from *Klebsiella, Erwinia, Serratia,* and *Salmonella*. Figure 4 shows the proposed structure, based on data from *E. coli* and *Salmonella,* with the variations found in the other organisms mentioned. Of particular interest are pairs of changes that generate compensatory base pairs which, in turn, serve to maintain the structure as deduced from *E. coli* and *Salmonella*. Several such pairs of compensatory changes are marked on the figure. The existence of the compensatory base changes suggest that certain aspects of our hypothetical structure are correct and that certain hairpin structures are important for function.

In Fig. 4, nucleotides that have been altered by design are also shown (N. Lumelsky, unpublished data). A circled base indicates the position at which a change produces an altered M1 RNA that is an even better enzyme than native M1 RNA. Triangles indicate positions of changes that decrease the activity to less than 10% of native M1 RNA and boxes indicate changes that generate mutant forms of M1 RNA that function with between 10 and 100% of the efficiency of native M1 RNA. From these studies and others we hope to understand which aspects of the structure of M1 RNA affect recognition of the substrate, which aspects affect binding of the substrate, and which aspects affect the catalytic site. We have removed 120 nucleotides from the 3′ end of M1 RNA, and the altered molecule still retains between 1 and 2% of its activity (Guerrier-Takada and Altman, 1986). In this partial molecule, sequences that could hydrogen-bind with GTψCR, the longest conserved sequence in tRNA moieties, are also absent. Thus, substrate recognition by Watson–Crick pairing with this sequence is not essential for activity of the enzyme. Removal of 25 nucleotides from the 5′ end of M1 RNA results in a molecule that is active at 30% of the level of the native M1 RNA. However, if just a few nucleotides are removed simultaneously from both the 3′ and 5′ ends, activity is abolished completely. Clearly, the activity of the enzyme depends on the integrity of at least one end of the RNA.

A number of variants of M1 RNA have also been prepared by site-directed mutagenesis (Lawrence, 1986). In some cases, a certain variant may cleave its substrate at a particular rate *in vitro* in the absence of C5 protein, but the cleavage activity is fourfold less when the protein co-factor is added. In other cases, the addition of protein has a stimulatory effect on the cleavage by the variant M1 RNA. From these results, we are beginning to understand the interactions of the RNA enzyme with its protein cofactor and its substrate.

V. STUDIES OF THE PROTEIN SUBUNIT OF RNase P

The cleavage of precursors to tRNAs can be performed *in vitro* by M1 RNA in the absence of C5 protein but, as mentioned above, the protein cofactor drastically alters the rate of cleavage of different substrates by M1 RNA. C5 protein, therefore, must have important functions *in vivo*. Hansen has cloned the *rnpA* gene for C5 and determined its nucleotide sequence (Hansen *et al.*, 1985). The *rnpA* gene is part of a large cistron, in which the first gene codes for L34, a large ribosomal protein. The second gene codes for C5 and there are two more open reading frames down-stream. The gene for L34 can be amplified with ease, but it appears to be

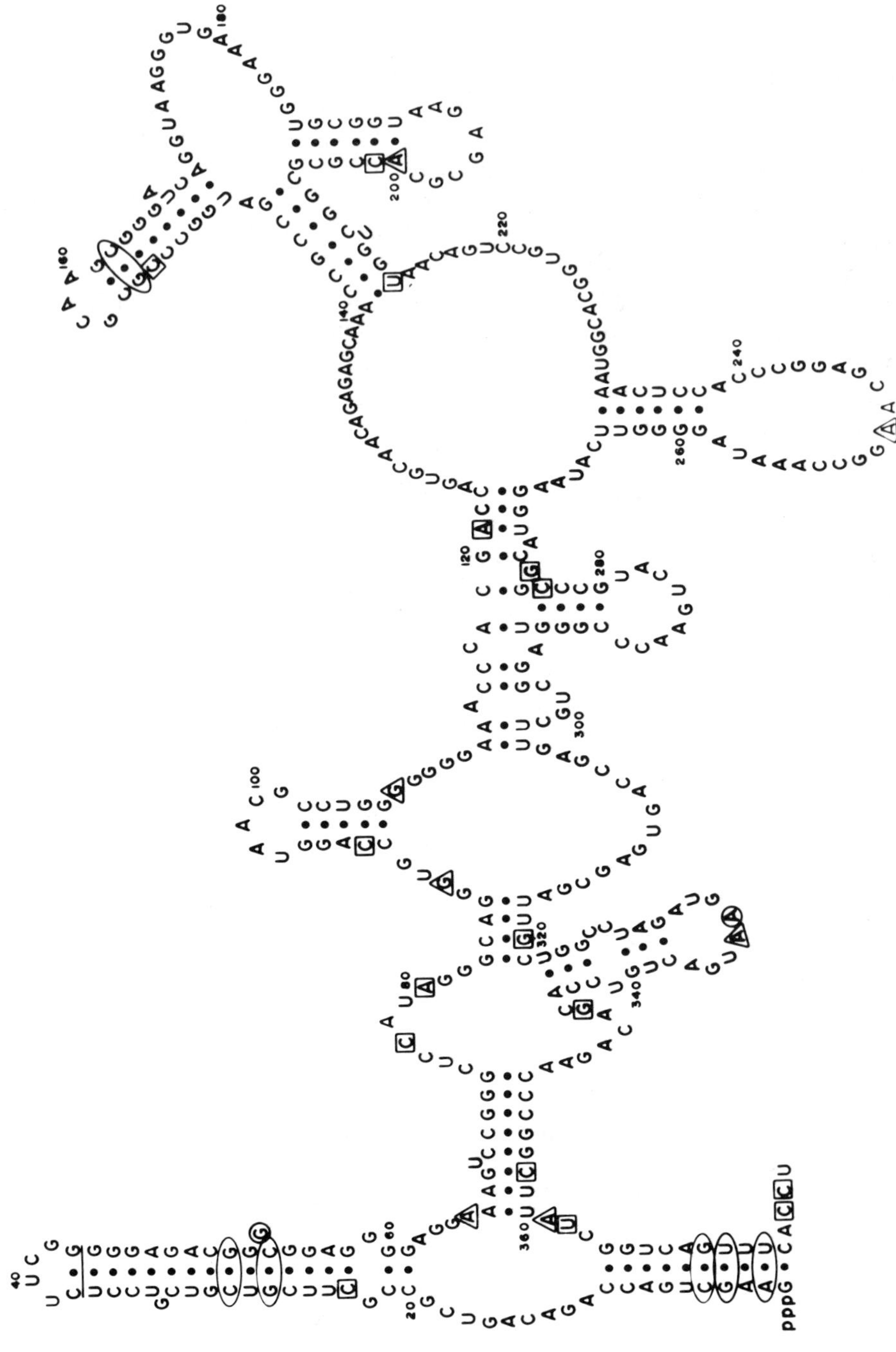

very difficult, for reasons that are as yet unclear, to amplify the expression of the gene for C5 protein. However, we have derived a plasmid that produces a sixfold amplification of the yield of C5 protein. One of the methods that we have used to purify this protein makes use of a unique affinity column (Vioque and Altman, 1986). The affinity column consists of agarose beads to which M1 RNA has been covalently linked. When crude extracts of cells that carry the plasmid just described are circulated over the column, active C5 protein, apparently pure, may be eluted by 7 *M* LiCl and 10 m*M* EDTA. The yield from this purification is approximately 15% of the total possible yield of active C5 in the crude extract.

In a footprinting study of the interactions of M1 RNA with C5, we have shown that two hairpins of the M1 RNA molecule are protected by C5 (data not shown). Whether this result means that there are two protein molecules bound to one M1 RNA molecule, and whether the two protein molecules are in contact with each other and thereby enhance the folding of the M1 RNA, remains to be proved.

The importance of C5 for the activity of RNase P *in vivo* is currently being examined in our laboratory by (L. Kirsebom, unpublished data). We have been investigating the functions of various suppressor tRNAs in the presence of *rnpA49,* a gene mutant in C5 originally isolated by Schedl and Primakoff (1973), and *rnpB709,* a gene isolated by Sakano *et al.* (1974) and mutant in M1 RNA. Suppressors SU1, SU3, and SU9 function equally well in the presence of *rnpB709* or in the wild-type background at 33°C. SU3, SU5, and SU9 function as well in the presence of *rnpA49* as in the wild-type background at 33°C. At this temperature the temperature-sensitive phenotype of A49, with respect to RNase P can just be detected. However, SU1 and SU2 function dramatically less efficiently in *rnpA49* at

---

Fig. 4. An hypothetical model for the secondary structure of M1 RNA. The model is based on data from experiments in which M1 RNA has been digested with ribonucleases under mild conditions in solution and on theoretical base-pairing schemes. The sequence shown is that for M1 RNA from *E. coli.* Circled pairs of nucleotides indicate compensatory changes found in the sequences of analogous genes for M1 RNA in certain other bacteria, as mentioned in the text (Lawrence, *et al.,* 1987). Circled individual bases indicate bases changed by mutagenesis which lead to M1 RNA molecules that have more cleavage activity *in vitro* than the parent molecule (N. Lumelsky, unpublished data). Boxed bases indicate bases changed by mutagenesis which lead to M1 RNA molecules with 10–100% the activity of the parent molecule, whereas bases in triangles indicate positions changed by mutagenesis which lead to M1 RNA molecules with less than 10% the activity of the parent molecule. The line between pairs 37–44 and 38–43 indicates the position of an insertion of two bases in the sequence of the gene for M1 RNA from *Erwinia agglomerans* which leads to the insertion of a base pair in the structure we have drawn (Lawrence, *et al.,* 1987). In *S. typhimurium,* an insertion of a G-C pair is found at this position (Baer and Altman, 1985).

33°C than in the wild-type background, at levels of activity of 40% and 10% of normal, respectively. Clearly, the mutation in the gene for C5 has a different effect on the biosynthesis of different suppressor tRNAs.

VI. HYBRID ENZYMES

We have probed the genomes of a variety of organisms with the genes for M1 RNA and C5 protein from *E. coli*. We are unable to detect any postive signal when we probe the genomes of anything other than gram-negative bacteria. In a collaborative study with Pace's group, we were unable to obtain a positive signal while probing the genome of *B. subtilis* with either gene from *E. coli* (Guerrier-Takada *et al.*, 1983). Similar negative results are observed all the way up the evolutionary ladder (Gold and Altman, 1986). However, there must be some degree of evolutionarily conserved structure, as demonstrated by the results of the following experiment. Lysed HeLa cell nuclei were passed over a DEAE column and then over an anti-Sm column, prepared with the assistance of J. Steitz' group (Gold and Altman, 1986). This crude preparation of RNase P was then separated into its protein and RNA components. When these crudely purified components were mixed with the corresponding components from *E. coli* that are necessary to reconstitute the holoenzyme, each heterologously reconstituted holoenzyme had the ability to cleave the substrate molecules. In other words, hybrid enzymes can be made with constituents from human and bacterial cells, even though the genetic sequences of the actual components have drifted to such an extent that the probes from *E. coli* cannot hybridize with the human genome. Hybrid enzymes had been made earlier with subunits from *E. coli, S. typhimurium* (Baer and Altman, 1985), and *B. subtilis* (Guerrier-Takada *et al.*, 1983). Some important aspects of tertiary structure of both RNA and protein subunits must have been preserved during evolution to allow such complementation to occur.

VII. CONCLUSION

In our studies of RNase P, we continue to investigate the role of the RNA and protein moieties of the enzyme, their interactions with each other and with their substrates. In this novel system, while the RNA plays its unusual catalytic role, the protein appears both to supply the basis for the structural integrity of M1 RNA that is necessary for the reaction to proceed *in vivo* and to control the rate of reaction with different substrates. It would appear important, therefore, if protein cofactors play similar roles in regulating the activity of other catalytic RNAs, to focus some attention on the study of these cofactors.

ACKNOWLEDGMENTS

Research in the laboratory of S. Altman is supported by grants from the National Institutes of Health and the National Science Foundation. L. Kirsebom, N. Lumelsky, and A. Vioque are postdoctoral fellows supported, respectively, by the Swedish Research Council, the National Institutes of Health, and the European Molecular Biology Organization. We thank Donna Wesolowski for excellent technical assistance.

REFERENCES

Altman, S., and Guerrier-Takada, C. (1986). *Biochemistry* **25,** 1205–1208.

Altman, S., Guerrier-Takada, C., Frankfort, H. M., and Robertson, H. D. (1982) *In* "Nucleases" (S. Linn and R. Roberts, eds.), pp. 243–274., Cold Spring Harbor Lab., Cold Spring Harbor, New York.

Baer, M., and Altman, S. (1985). *Science* **228,** 99–1002.

Gardiner, K., Marsh, T., and Pace, N. (1985). *J. Biol. Chem.* **260,** 5415–5419.

Gold, H. A., and Altman, S. (1986). *Cell (Cambridge, Mass.)* **44,** 243–249.

Guerrier-Takada, C., and Altman, S. (1984). *Biochemistry* **23,** 6327–6334.

Guerrier-Takada, C., and Altman, S. (1986). *Cell (Cambridge, Mass.)* **45,** 177–183.

Guerrier-Takada, C., Gardiner, K., Marsh, T., Pace, N., and Altman, S. (1983). *Cell (Cambridge, Mass.)* **35,** 849–857.

Guerrier-Takada, C., McClain, W. M., and Altman, S. (1984). *Cell (Cambridge, Mass.)* **38,** 219–224.

Guerrier-Takada, C., Haydock, K., Allen, L., and Altman, S. (1986). *Biochemistry* **25** 1509–1515.

Guerrier-Takada, C., Minehart, P., and Altman, S. (1987). Submitted for publication.

Hansen, F. G., Hansen, E. B., and Atlung, T. (1985). *Gene* **38,** 85–93.

Kole, R., and Altman, S. (1979). *Proc. Natl. Acad. Sci. U.S.A.* **76,** 3795–3799.

Kole, R., Baer, M., Stark, B. C., and Altman, S. (1980). *Cell (Cambridge, Mass.)* **19,** 881–887.

Lawrence, N. P. (1986). Ph.D. Thesis, Yale University, New Haven, Connecticut.

Lawrence, N. P., Richman, A., Amini, R., and Altman, S. (1987). *Proc. Natl. Acad. Sci. U.S.A.* (in press).

Robertson, H. D., Altman, S., and Smith, J. D. (1972). *J. Biol. Chem.* **247,** 5243–5251.

Rooney, R. J., and Harding, J. D. (1986). *Nucleic Acids Res.* **14,** 4849–4864.

Sakano, H., Yamada, S., Ikemura, T., Shimura, Y., and Ozeki, H. (1974). *Nucleic Acids Res.* **1,** 355–371.

Schedl, P., and Primakoff, P. (1973). *Proc. Natl. Acad. Sci. U.S.A.* **70,** 2091–2095.

Stark, B. C., Kole, R., Bowman, E. J., and Altman, S. (1978). *Proc. Natl Acad. Sci. U.S.A.* **75,** 3717–3721.

Vioque, A., and Altman, S. (1986). *Proc. Natl. Acad. Sci. U.S.A.* **83,** 5904–5908.

2

Bacillus subtilis RNase P

NORMAN R. PACE, BRYAN D. JAMES, CLAUDIA REICH, DAVID S. WAUGH, GARY J. OLSEN, AND TERRY L. MARSH[1]

Department of Biology
and Institute for Molecular and Cellular Biology
Indiana University
Bloomington, Indiana 47405

RNase P is responsible for removing the 5′-terminal, precursor-specific segments from pre-tRNA molecules during their maturation. It is a particularly interesting enzyme because its catalytic element is an RNA, not a protein (Guerrier-Takada *et al.,* 1983; Gardiner *et al.,* 1985). Although the recognition of RNase P as a catalytic RNA was preceded by the discovery of a self-splicing intron in some *Tetrahymena* 26 S rRNA precursors (Kruger *et al.,* 1982), the RNase P reaction differs in an important way: It engages in *inter*molecular reactions. In contrast, the *in vivo* self-splicing intron activity is a series of *intra*molecular rearrangements that collectively result in excision of the intron and ligation of the flanking exons (reviewed by Cech, 1985). RNase P therefore offers not only a model for RNA catalytic mechanisms, but also a system for exploring the nature of specific RNA–RNA recognition that almost certainly goes beyond the familiar Watson–Crick base-pairing interactions.

I. THE RNase P COMPONENTS

The RNase P of *Escherichia coli* was one of the first RNA-processing enzymes to be analyzed *in vitro* (Altman and Smith, 1971). It was shown

[1] Present address: Department of Biology, Hamilton College, Clinton, New York 13323.

Molecular Biology of RNA
New Perspectives

subsequently by Altman and his colleagues to consist of two components, an RNA (M1 RNA) and a protein (C5 protein) (Stark *et al.,* 1977). The ribonucleoprotein nature of RNase P has proved to have a wide phylogenetic distribution. *Bacillus subtilis,* a eubacterium distantly related to *E. coli,* also possesses an RNase P with protein and RNA elements (Gardiner and Pace, 1980). RNA subunits have been implicated in the RNase P activities of extracts from other organisms, for instance *Schizosaccharomyces pombe* (Kline *et al.*, 1981) and both the nuclei and mitochondria of human cells (Doersen *et al.,* 1985). Thus far, however, only the bacterial enzymes have been characterized in much detail.

The RNase P components from *B. subtilis* and *E. coli* are similar in their general properties. The proteins are 15–17 kDa and the RNAs are about 400 nucleotides in chain length. The RNase P proteins from each of these organisms will complement the RNA from the other, so they clearly have some homologous functions (Guerrier-Takada *et al.,* 1983). The sequences of both RNAs have been determined (below) and, evidently, the amino acid sequence of the RNase P protein from each species is known. This latter information derives from correlations (Hansen *et al.*, 1985; Ogasawara *et al.,* 1985) of structurally homologous open reading frames in DNA sequences near the origin of replication of the *E. coli* and *B. subtilis* chromosomes. A plasmid-borne version of one of the presumptive genes from *E. coli* subsequently proved able to complement a temperature-sensitive mutation in the RNase P protein. Presumably the homologous *B. subtilis* open reading frame is its RNase P protein gene.

Although RNase P undoubtedly functions *in vivo* as a ribonucleoprotein particle (RNP), it is clear that the catalytic element is the RNA moiety. This was discovered during tests of the optimum ionic environment for the enzyme reaction *in vitro* (Guerrier-Takada *et al.,* 1983). As shown in Figure 1A for the *B. subtilis* RNase P, in the presence of both protein and RNA, the cation requirements for maximum activity are moderate, about 100 m*M* NH_4^+ and 30 m*M* Mg^{2+}. In the absence of the protein (Fig. 1B), little maturation of the precursor tRNA is seen under these conditions. However, at much higher salt concentrations, the RNase P RNA alone is as effective as the holoenzyme is under moderate salt conditions. That is, high cation concentrations alleviate the requirement for the RNase P protein. Proof that the RNase P RNA is indeed the catalytic element in this reaction is rigorous: *In vitro* transcripts from the cloned *B. subtilis* (Reich *et al.,* 1986) and *E. coli* (Guerrier-Takada and Altman, 1984) RNase P RNA genes are catalytically active under the high salt, protein-independent conditions.

The most straightforward explanation of the requirement for high cation concentrations in the RNase P RNA (“RNA alone”) reaction is that

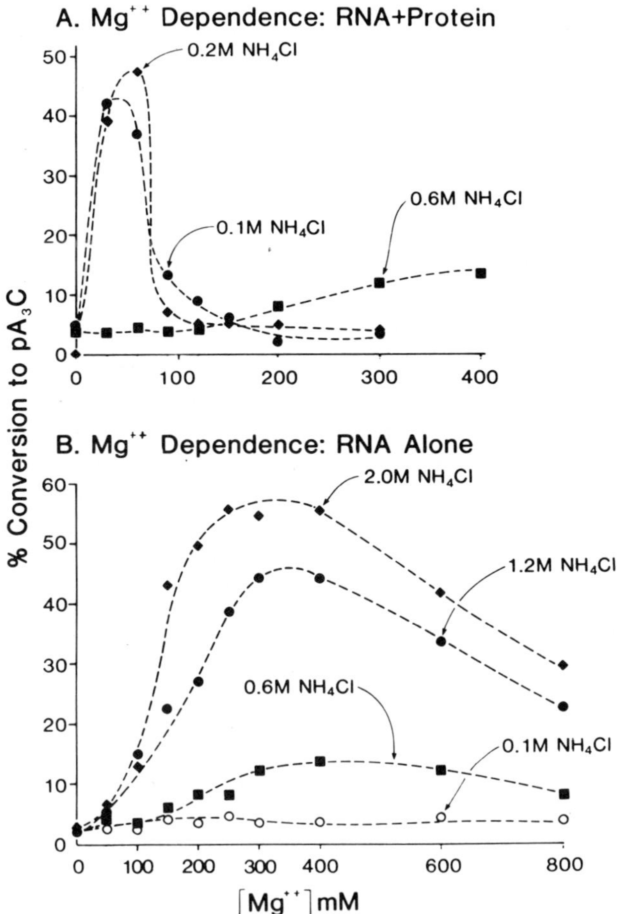

Fig. 1. Cation dependence of the RNase P reaction. As detailed previously (Gardiner *et al.*, 1985), *B. subtilis* RNase P holoenzyme (A), or RNA alone (B), was incubated with substrate at 37°C for 15 min under the indicated ionic conditions. The substrate was $tRNA^{Fmet}$ with an oligonucleotide, AAAC, added to its 5′ end. This artificial substrate was labeled at its 5′ end using [γ-^{32}P]ATP and polynucleotide kinase. The 5′-end-labeled oligonucleotide released by RNase P was separated from uncleaved substrate by thin-layer chromatography and quantitated by scintillation counting. (Reprinted, with permission, from Gardiner *et al.*, 1985.)

the cations decrease the electrostatic repulsion between the enzyme and substrate RNAs, each of which carries one negative charge per nucleotide phosphate. Without high cation concentrations, the charge repulsion between enzyme and substrate prevents the intimate contact required for catalysis. Thus, part of the role of the RNase P protein presumably is to provide charge titration. The protein is quite small relative to the RNase P RNA, so its function is likely to be localized, perhaps just to the tRNA-binding region.

It remains to be seen whether the protein participates in substrate selection by establishing specific contacts with the tRNA precursor. Both the *B. subtilis* and *E. coli* RNase P proteins are positive in net charge, so they have an intrinsic affinity for nucleic acids. However, membrane filter binding experiments suggest no selectivity for tRNA; in the absence of RNase P RNA, the *B. subtilis* protein binds 5 S ribosomal RNA as well as it binds tRNA precursors (T. L. Marsh, unpublished observations).

Although ionic screening is one interpretation of the ability of high salt concentrations to alleviate the requirement for the protein in the RNase P reaction, not all the data are completely consistent with this notion. For example, if the salt effects were simply a matter of ionic screening, it would be anticipated that alkali metal cations with small radii would be more effective at low concentrations than would larger cations. This is because the ions with smaller radii bind to nucleic acid phosphates more tightly than the larger ones, in the order $Li^+ > Na^+ > K^+ > Rb^+ > Cs^+$ (Ross and Scruggs, 1964). However, as shown in Fig. 2, the observed order of effectiveness of the ions in promoting the RNase P reaction is clearly different, in particular, Rb^+, Cs^+, K^+, $NH_4^+ > Li^+$, Na^+. It may be that only the larger cations can achieve a packing geometry with the RNA surfaces that results in activity. It is also conceivable that electrostatic shielding of phosphate residues is not the only role of cations in the RNA alone activity.

Some experimental results suggest that conformational transitions in substrate or enzyme RNAs may be involved in the RNase P RNA alone reaction (Gardiner *et al.*, 1985; Altman and Guerrier-Takada, 1986), but the findings are equally interpretable in electrostatic terms. For instance, mildly denaturing solvents (ethanol, dimethyl sulfoxide, ethylene glycol) potentiate the reaction at lower salt concentrations than required in fully aqueous media (Gardiner *et al.*, 1985). This might suggest that polynucleotide structural fluidity is important in the reaction, perhaps permitting the enzyme RNA to conform to the various pre-tRNAs and other low-molecular-weight RNA substrates. Alternatively, this effect can be explained from the electrostatic viewpoint, since the solvents reduce the dielectric constant of the reaction medium. This would, by Coulomb's

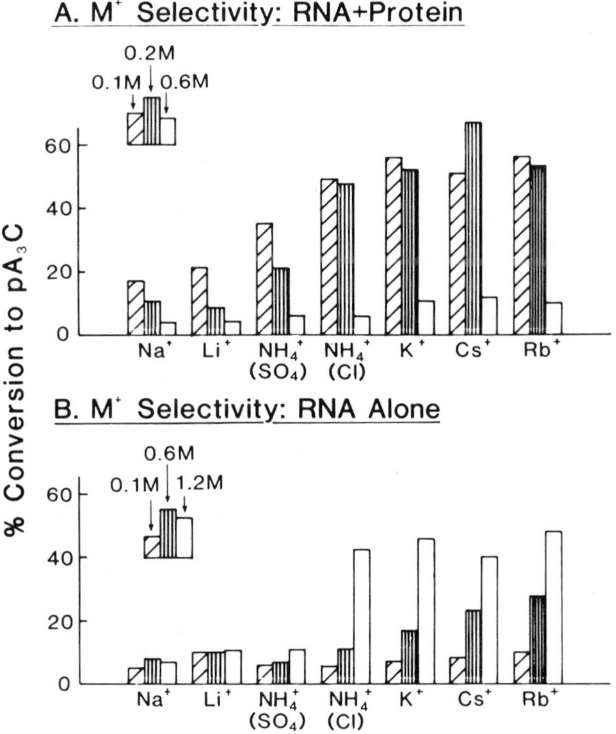

Fig. 2. Monovalent cation selectivity of RNase P. As detailed previously (Gardiner *et al.*, 1985) and in the legend to Fig. 1, RNase P holoenzyme (A), or RNA alone (B), reactions were carried out in 50 m*M* $MgCl_2$ (holoenzyme) or 0.3 *M* $MgCl_2$ (RNA alone) plus the indicated amounts and types of monovalent cations. (Reprinted, with permission, from Gardiner *et al.*, 1985.)

law, enhance the electrostatic interaction potential. Therefore, the screening cations would bind more tightly to the polynucleotides and thereby exert their influence at lower concentrations in the presence of solvents.

An important question is whether the salt effects on the RNase P reaction influence the binding process or the catalytic mechanism. This has been evaluated by inspection of the dependence of the kinetic parameters, K_m and k_{cat}, on the salt concentration of the reaction medium (C. Reich, unpublished observations). This approach presumes that the RNase P reaction rate is limited by the binding or catalytic steps, an assumption that may be incorrect (below). As shown in Fig. 3, changes in monovalent and divalent salt concentrations seem to influence the binding step (K_m);

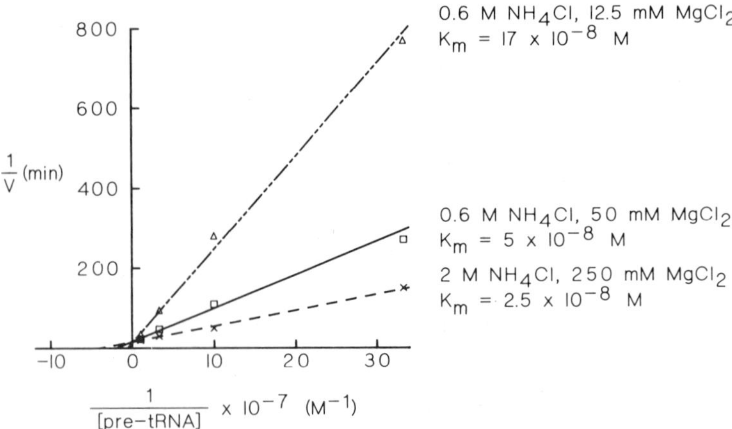

Fig. 3. Dependence of RNase P kinetic parameters on salt concentration. RNase P RNA (0.15 pmoles) was assayed for its ability to remove a nucleotide-specific sequence from precursor $tRNA^{Asp}$ at different substrate concentrations, under the indicated ionic conditions. The *in vitro*–synthesized precursor tRNA was labeled by incorporation of [α-^{32}P]UTP. Reactions were carried out for 15 min at 37°C. Reaction products were resolved on an 8% polyacrylamide gel containing 8 *M* urea. After fixing and drying the gel, products were revealed by autoradiography. The radioactive bands were cut out and their radioactivity quantitated. Reaction rates are expressed as picomoles substrate cleaved per picomole RNase P RNA per 15 min.

the catalytic rate is essentially unchanged over the salt range tested. However, the turnover number is remarkably low, about 1 mole of substrate per minute per mole of enzyme, even in the presence of saturating levels of the RNase P protein (not shown). This is far slower than the *in vivo* rate of cleavage must be. During each cell doubling, 1–2 $\times$ 10^5 tRNA molecules are generated by a relatively small number of RNase P complexes, only 20–50 per cell based on RNase P RNA recoveries. Therefore, the *in vitro* turnover rate of the enzyme seems to be about two orders of magnitude slower than the *in vivo* rate. The reason for this inefficiency in the *in vitro* reaction is not known. It may be that some factor(s) other than the two known components reside in the native RNase P complex and are required for optimal activity. However, the explanation also may be a kinetic one, namely, a slow dissociation rate of the cleaved precursor. This is suggested because, at least in the RNA alone reaction, RNase P seems not to discriminate between precursor and mature tRNA; the concentration of mature tRNA required for half-maximum inhibition (K_i) of RNase P is about the same as the K_m for a precursor tRNA substrate (T. L. Marsh, unpublished observations). It may be that the cell makes use of some mechanism for displacing the reaction product, tRNA, from the enzyme surface. The fact that RNase P does not discriminate between substrate and product also indicates that all of the information utilized by RNase P in its selection of substrates resides in the mature domain of the precursor.

II. MECHANISM OF RNase P CLEAVAGE

RNase P generates 5′-phosphate and 3′-hydroxyl termini, as do the self-splicing introns and most specific processing nucleases. The RNase P reaction is different from self-splicing, however. Both Group I and Group II self-splicing intron excisions proceed by a series of transesterifications (Cech, 1985). The reactions of Group I introns are initiated by the 3′-hydroxyl of guanosine or a guanosine-containing nucleotide. RNase P clearly does not use this mode of action. It has no requirement for a co-substrate, and periodate oxidation of the 3′ ends of both enzyme and substrate RNAs, which destroys the 2′- and 3′-hydroxyl groups, does not inactivate the enzyme (Marsh and Pace, 1985).

The initial cleavage of Group II introns occurs by a 2′-hydroxyl attack upon an intron–exon boundary, generating a covalent conjugate, an RNA "lariat" that is subsequently displaced as the exons are joined. Thus far, searches for a covalent association between RNase P RNA and its substrate, employing both the *B. subtilis* (T. L. M. and D. S. W., unpublished observations) and *E. coli* (Guerrier-Takada *et al.,* 1986) systems, have been unsuccessful. These results, albeit negative, make it likely that water, rather than a 2′-hydroxyl group, effects strand scission in the tRNA precursor. The reactions carried out by protein hydrolases are well studied and are commonly considered to be catalyzed by amino acid functional groups that donate or capture protons as needed to drive catalysis. An analogous proton-exchange scheme is shown for the RNase P reaction in Fig. 4. It is likely that the enzyme RNA activates a water molecule to participate in the reaction, rather than utilizing a free hydroxyl ion. This is suggested by the fact that the RNase P reaction rate is not directly proportional to hydroxide ion concentration over the active pH range (Marsh and Pace, 1985).

Nucleic acid groups, of course, must serve as proton sinks in the reaction depicted in Fig. 4. Both bases and phosphodiester chain elements can serve in proton transfer and catalyze general acid–general base reactions. A few examples are diagrammed in Fig. 5 along with the histidine imidazole, a commonly used group in proton-exchange reactions driven by proteins. Figure 5B is a simple base tautomerization. Figures 5C and 5D

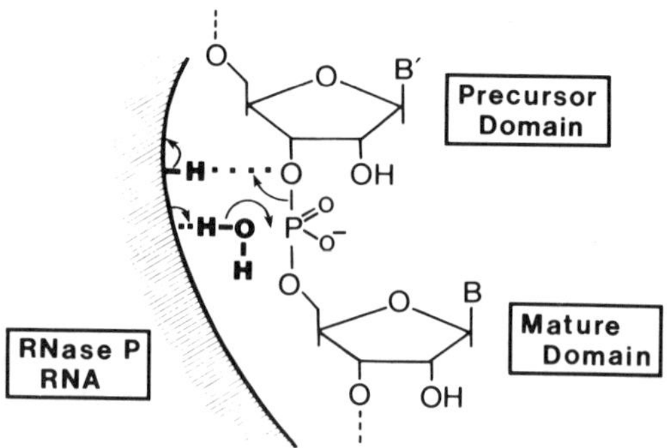

Fig. 4. Hypothetical scheme for hydrolysis of tRNA precursors by RNase P. The hatched boundary indicates the RNase P RNA surface. B and B′ are bases. The phosphodiester bond connecting the precursor and mature domains of the precursor tRNA is depicted. (Adapted, with permission, from Marsh and Pace, *Science* **229,** 79–81, 5 July 1985. Copyright 1985 by the AAAS.)

are conjectural ways that simple conformational transitions in the ribose moiety of RNA could potentiate proton transfer to overcome the energy of activation required for the hydrolytic scission of a phosphodiester bond. Figure 5C certainly does not pertain to the RNase P reaction, since it utilizes 2′- and 3′-terminal hydroxyl groups and RNase P does not require those (above). However, an internucleotide phosphate or a base exocyclic group could promote an equivalent alkoxide formation in an internal ribose (Fig. 5D).

III. TOWARD THE HIGHER-ORDER STRUCTURE OF RNase P RNA

Understanding the function of the RNase P RNA will require knowledge of the secondary and tertiary structural foldings that dictate the display of functional elements to interacting substrates. The nucleotide sequences of the RNase P RNAs from *E. coli* (Reed *et al.*, 1982; Sakamato *et al.*, 1983) and *B. subtilis* (Reich *et al.*, 1986) have been reported, but efforts to define even their secondary structures have not yet been fully successful.

Evaluating higher-order RNA structure is a process of progressive refinement. The first step attempts to predict secondary structure from nucleotide sequence, using thermodynamic estimates for helix stabilities to define possible complementary pairings. However, computational ap-

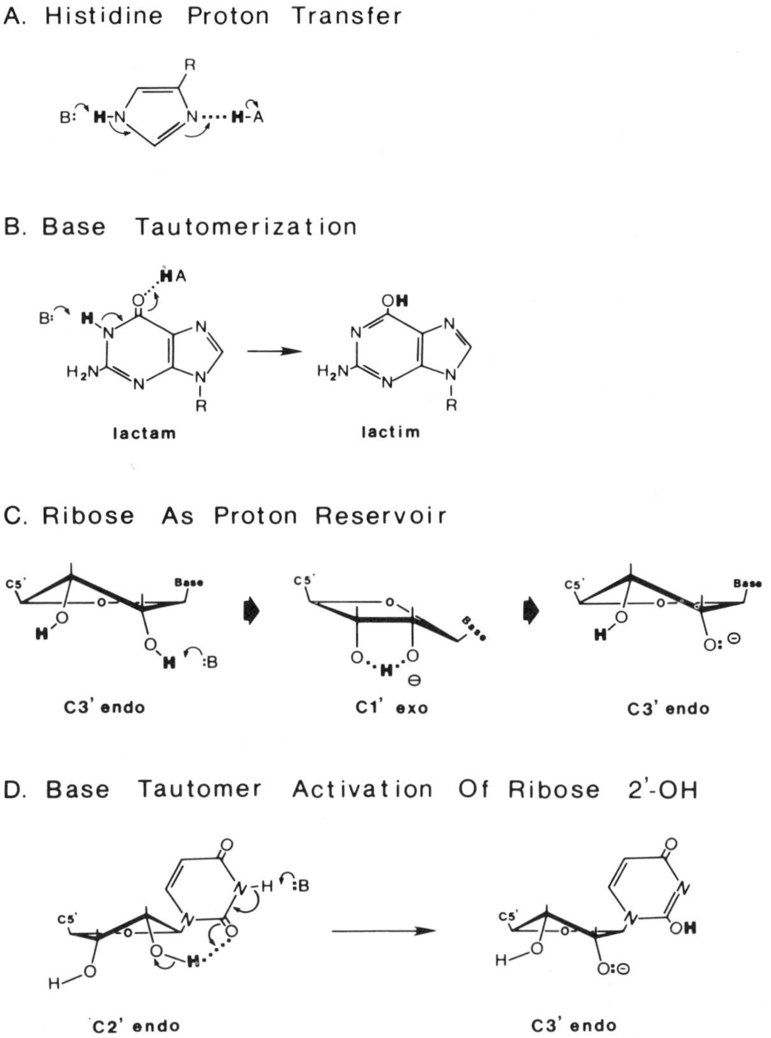

A. Histidine Proton Transfer

B. Base Tautomerization

lactam → lactim

C. Ribose As Proton Reservoir

C3' endo → C1' exo → C3' endo

D. Base Tautomer Activation Of Ribose 2'-OH

C2' endo → C3' endo

Fig. 5. Hypothetical proton transfer reactions. See text for discussion. (Reprinted, with permission, from Pace and Marsh, *Origins Life* **16,** 97–116, 1985.)

proaches to predicting helical structure from nucleotide sequences are not well developed. For instance, the structure rules summarized by Salser (1977) yield the tRNA "cloverleaf" from a tRNA nucleotide sequence in only about one-half the cases (Papanicolaou *et al.*, 1984). Similarly, only about one-half the helices predicted from *E. coli* 16 S rRNA sequence proved to be correct (Noller and Woese, 1981). There are several reasons for the inability to predict secondary structure reliably. One reason is that the available estimates of free energy values for base pairings are inexact; they are based on data from limited collections of oligonucleotides and apply to specific *in vitro* conditions. Moreover, there are no reliable estimates for the stabilities of noncanonical base pairs (e.g., A-G, G-U, etc.) or for the effects of local sequence context. Helix discontinuities, such as unpaired bases, are other common elements in natural RNAs that cannot yet be treated thermodynamically with good credibility. Finally, the existing computational methods for evaluating secondary structure from nucleotide sequence cannot account for helix-stabilizing or -destabilizing constraints imposed by tertiary structural features in a folded RNA molecule.

Although thermodynamic predictions of sequence pairings are inexact, they provide models for experimental evaluation. There are two broad approaches toward testing RNA folding models. One approach uses structure-specific enzymes and chemicals to define duplex and "single-strand" segments in the RNA under "native conditions" (Vournakis *et al.*, 1981). However, interpreting the results of such experiments requires a reliable structure model and, in any case, the results do not identify the pairing partners of duplex segments. At this time, the best *a priori* method for evaluating the secondary structures of large RNAs is the phylogenetic-comparative approach (Fox and Woese, 1975; Noller and Woese, 1981). Possible helices in an RNA, as indicated by the occurrence of complementary sequences, are tested by seeking the equivalent pairing possibilities in the homologous RNA from another organism in which the sequence varies. Helical regions are indicated by covariance in compared sequences; mutations compensate one another to maintain complementarity.

The RNase P RNAs from *E. coli* and *B. subtilis* carry out the same reaction, and the RNA and protein subunits from each organism will complement those from the other organism in the holoenzyme reaction. Therefore, the structural elements involved in catalysis and in the interaction with the RNase P proteins likely are similar in the two RNAs. Therefore, we were surprised when the nucleotide sequences of the *B. subtilis* and *E. coli* RNase P RNAs were compared (Reich *et al.*, 1986); they proved so dissimilar that homologous sequences could not be identified over most of the lengths of the molecules. This is illustrated in Fig. 6A, a "dot plot" comparison of the two sequences, showing those regions in which 9 out of 12 consecutive nucleotides are identical. For comparison, a similar analysis of the first 400 nucleotides of the 16 S rRNAs from these organisms (Fig. 6B) clearly displays much more sequence similarity. It is evident that the primary structure of RNase P RNA evolves very rapidly in comparison to that of 16 S rRNA. Our inability to unambiguously align homologous nucleotides in the two RNase P RNA sequences meant, however, that comparison of these sequences could not be used for an initial identification of helical pairings; the identification of compensatory changes that indicate helical structures demands that only homologous nucleotides are considered. Hence, it was necessary to choose organisms more closely related to *B. subtilis* than is *E. coli* for the first comparative tests of a folding model.

The choice of appropriate organisms for the phylogenetic-comparative analysis of the RNase P RNA structure requires a quantitative view of evolutionary relatedness, now available from comparative studies of rRNA sequences (reviewed by Pace *et al.*, 1986). The rRNAs are highly conservative molecules; homologous sequences are identifiable in the rRNAs of all organisms. These homologous sequences can be used to define evolutionary distances between organisms and hence to identify organisms at the appropriate phylogenetic "depth" for comparative structural analyses. As may be seen in Fig. 7, the current picture of relationships among representative organisms, *B. subtilis* and *E. coli* are fairly distant from one another. They are members of different eubacterial "phyla," of which about a dozen have so far been defined.

For analysis of the RNase P RNA structure, we chose organisms within the same phylum as *B. subtilis,* i.e., the "gram-positive and relatives" grouping, and having 5 S rRNA sequence similarities of 75–80% with *B. subtilis* and one another. These included *Bacillus megaterium, B. brevis, B. stearothermophilus, Streptococcus faecium,* and *Lactobacillus brevis*. Preliminary tests with Southern blots of restriction endonuclease digests of DNAs from these organisms showed that the *B. subtilis* RNase P RNA gene could be used as a heterologous hybridization probe to identify the corresponding gene from each of these other organisms (B. D. James, unpublished observations). In contrast, the *B. subtilis* RNase P RNA gene will not hybridize with its *E. coli* counterpart, because the sequences are so disparate (K. J. Gardiner and C. Guerrier-Takada, unpublished observations). Using the heterologous probe, the RNase P RNA genes from the close relatives of *B. subtilis* were cloned and either have been or are being

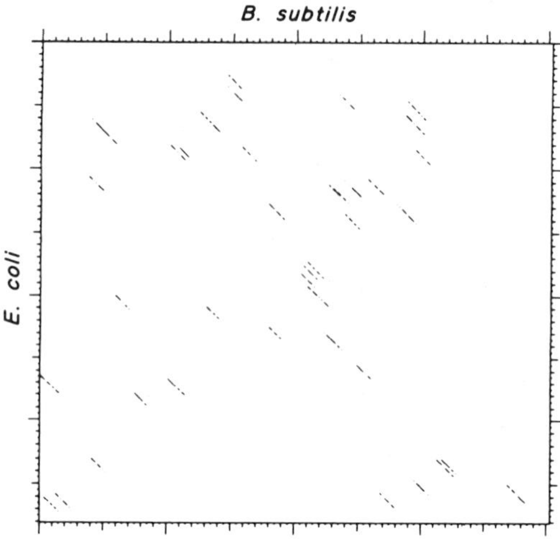

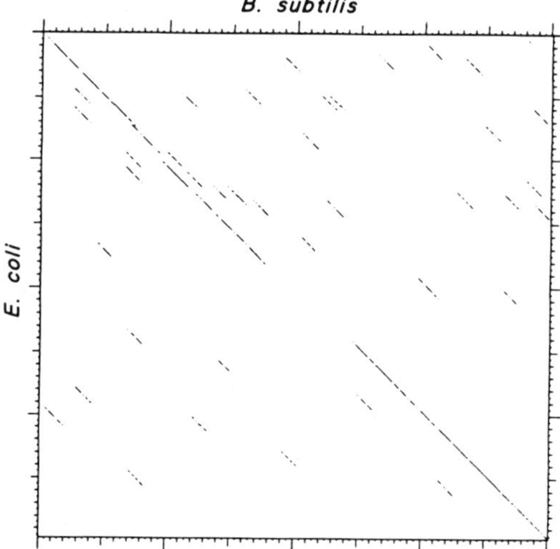

Fig. 6. Dot plot comparisons of homologous RNAs from *B. subtilis* and *E. coli*. Dots indicate pairs of nucleotides (one from each sequence) that are in regions where the two sequences are identical in at least 9 of 12 consecutive residues. Panel A compares the respective RNase P RNAs. Panel B compares the first 400 nucleotides of the *B. subtilis* 16 S rRNA with the corresponding region (396 nucleotides) of the *E. coli* 16 S rRNA. (Reprinted, with permission, from Reich *et al.*, 1986.)

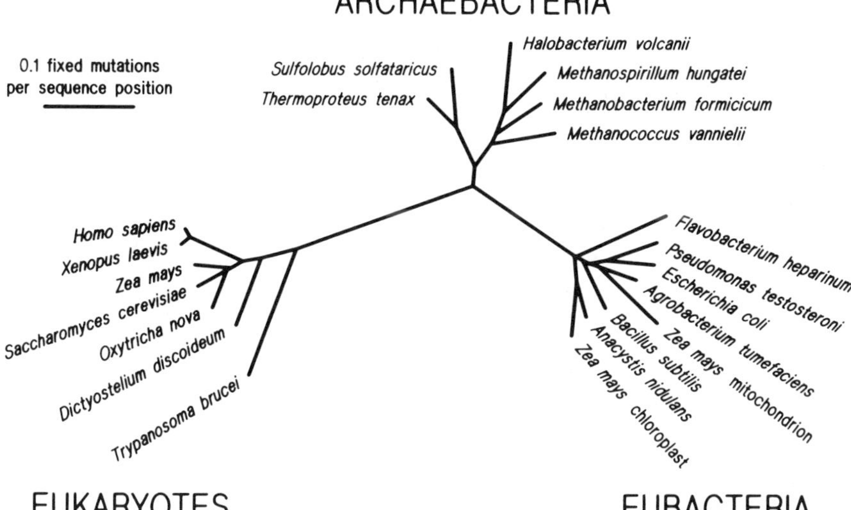

Fig. 7. Evolutionary relationships among representative organisms. An unrooted phylogenetic tree based on homologous regions of complete 16 S-like rRNA sequences from the indicated organisms was constructed as detailed (Elwood *et al.*, 1985). The scale bar corresponds to a tree branch length of 0.1 mutations fixed per sequence position. (Adapted, with permission, from Pace *et al.*, 1986. Copyright held by Cell Press.)

sequenced (B. D. James, unpublished observations). The available information permits a first test of a calculated, "minimum-energy" secondary structure of the *B. subtilis* RNase P RNA.

Figure 8 shows a computer-predicted folding of the *B. subtilis* RNase P RNA, derived using the algorithm of Zuker and his colleagues (Jacobson *et al.*, 1984) and the energy values compiled by Cech *et al.* (1983). The hatched helices are those we consider "proven" by the occurrence of at least two, independent, base-pairing changes in their sequences. Some of these changes are indicated in the figure. The phylogenetic-comparative approach also indicates incorrect helices in the minimum-energy model; possible Watson–Crick pairs in the *B. subtilis* sequence are seen as nonpairing couples in one or more other sequences. Some of these changes, and the helices they contradict, are indicated in the figure. The upshot of the analysis so far is that the thermodynamic rules are fairly successful at predicting short-range ("hairpin") foldings, but less successful at identifying "long-range" associations.

As the credibility of the *B. subtilis* RNase P RNA folding improves, we can seek homologous folding possibilities in the *E. coli* RNA. Only about

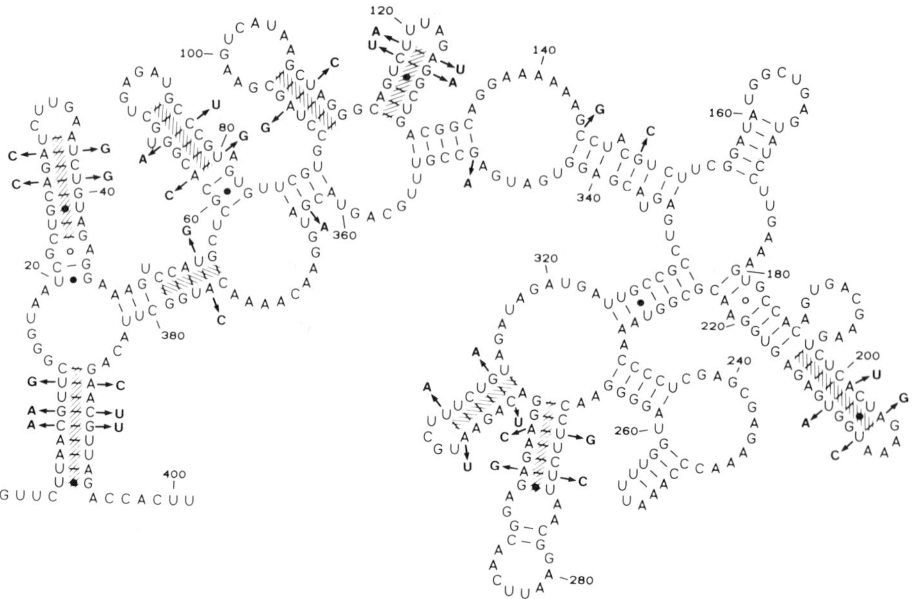

Fig. 8. A phylogenetic test of a minimum energy model for *B. subtilis* RNase P RNA. A computer-predicted folding of *B. subtilis* RNase P RNA using an algorithm by Zuker and his colleagues (Jacobson *et al.*, 1984) is shown. Hatched helices are considered "proven," i.e., they display at least two independent, compensated base-pairing changes in the helical segment. Changes in some helical segments are not compensated, thereby contradicting the minimum-energy model. Only some of the known sequence variations are indicated.

30% of the structures of the *E. coli* and *B. subtilis* RNase P RNAs can be folded in a manner clearly consistent with both sequences. The homologous structures are shown in Fig. 9. In these ordered regions the reason for the unexpected divergence of the two sequences becomes clearer. Differences between them are amplified by the presence or absence of sequence blocks as well as point changes. In the region of the molecule shown, the *E. coli* RNA has an extended helical domain. Elsewhere the *B. subtilis* RNA must have segments not present in the *E. coli* version; the *B. subtilis* RNase P RNA is 401 nucleotides in length, while that of *E. coli* is composed of only 377 nucleotides. Figure 10 includes some further examples of one of the rapidly evolving helices shown in Fig. 9. Analogous variation occurs in other regions of the molecule as well, with helical segments sometimes withering away even among the close relatives of *B. subtilis*. Does such extensive variation in the molecule suggest that the RNase P activity of the RNA may be only one of several functions that it carries out?

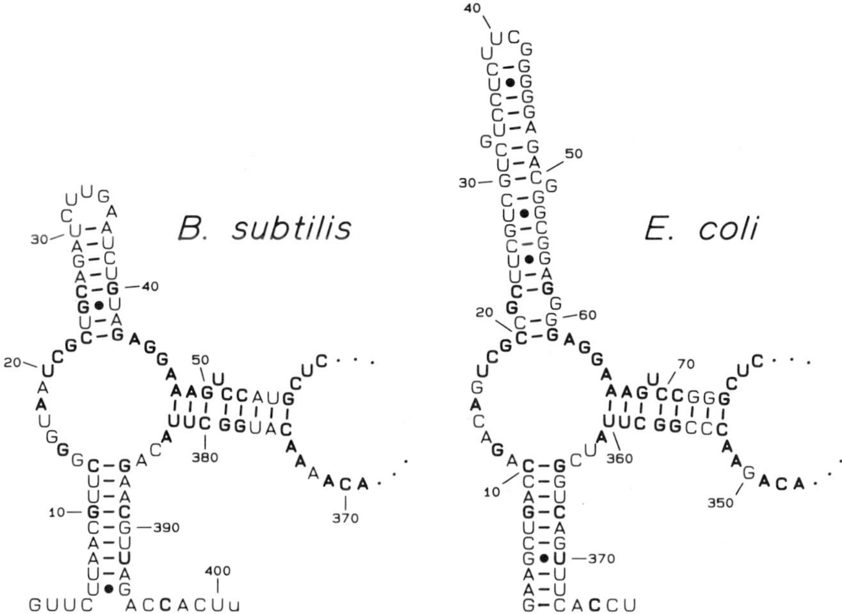

Fig. 9. Secondary structure common to the *B. subtilis* and *E. coli* RNase P RNAs. See text for discussion. Bold letters indicate the positions at which homologous residues in the two sequences are identical. (Reprinted, with permission, from Reich *et al.*, 1986.)

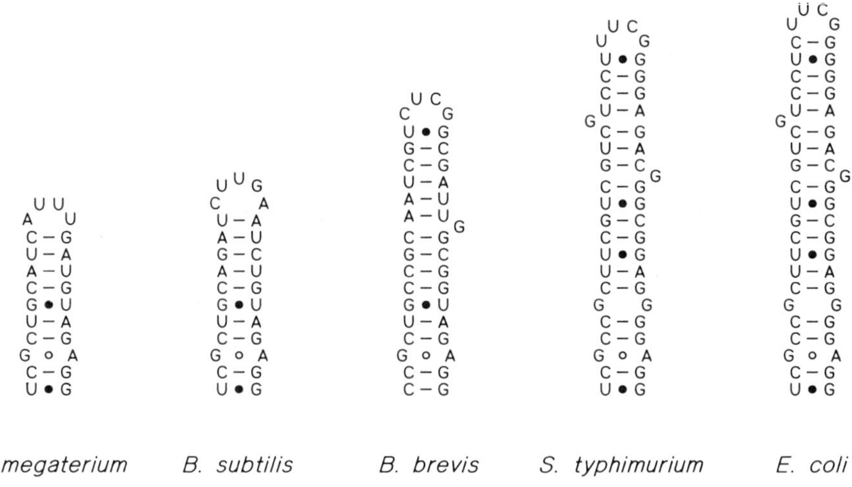

Fig. 10. Helix length variation. Homologs to the helix involving nucleotides 20–46 in the *B. subtilis* RNase P RNA are shown for *B. megaterium* (B. D. James, unpublished observations), *B. brevis* (B. D. James, unpublished observations), *E. coli* (Reed *et al.*, 1982; Sakamoto *et al.*, 1983), and *S. typhimurium* (Guerrier-Takada and Altman, 1984).

IV. STRUCTURE–FUNCTION RELATIONSHIPS IN THE RNase P RNA

One goal of our studies is to define the functional elements of the RNase P RNA. The extensive sequence dissimilarity between the RNAs of *B. subtilis* and *E. coli* suggests that some elements will prove dispensable for the processing activity, but other elements must serve roles in binding the substrate, effecting catalysis, and interacting with the RNase P protein. The extensive conservation of most of the structural organization of the domain illustrated in Fig. 9 suggests that it may have an important, albeit unknown, role in the function of the RNA. The following experimental results suggest that an internal region, about positions 240–280 in Fig. 8, is involved both in catalysis and in interaction with the RNase P protein (D. S. Waugh and C. Reich, unpublished observations)

One approach to identifying functional elements in the RNase P RNA is testing the enzymatic activities of RNAs that lack wild-type sequences. These are produced by cloning segments of the RNase P RNA gene adjacent to a phage T7 promoter, isolating *in vitro* runoff transcripts of the segments, and then annealing appropriate fragments to produce molecules lacking various sequences (Reich *et al.,* 1986; C. Reich, published observations). The enzymatic activity assays of one such series of constructs are shown in Fig. 11; the nucleotide numbers refer to Fig. 8. It is evident that annealed fragments that, when combined, include the full RNase P RNA sequence are active; specifically, the construct 1–283 + 280–401 is as active as the intact RNase P RNA isolated from cells ("*in vivo*" in Fig. 8) or prepared by transcription of the cloned gene ("*in vitro*"). However, the separate fragments are inactive in the processing reaction. This result indicates that the RNase P RNA secondary and tertiary structure, not merely a covalently contiguous nucleotide sequence, is required for activity. In contrast to the active construct, the annealing of fragments containing residues 1–239 and 280–401 does not yield an RNA that is active under the assay conditions. Moreover, inverting the sequence between residues 240 and 280, so that the complement appears in the RNA, abolishes the activity of the RNA in the assay (D. S. Waugh, unpublished observations). These results imply that sequences between residues 240 and 280 are important to the active structure or to catalysis.

This portion of the RNase P RNA also seems to be involved in the interaction with the RNase P protein. During the course of site-directed modification of the RNase P RNA gene, the nucleotide segment 236–239 (TCGA) was tandemly duplicated (D. S. Waugh, unpublished observations). As shown in Fig. 12, the transcript of this modified gene is as active

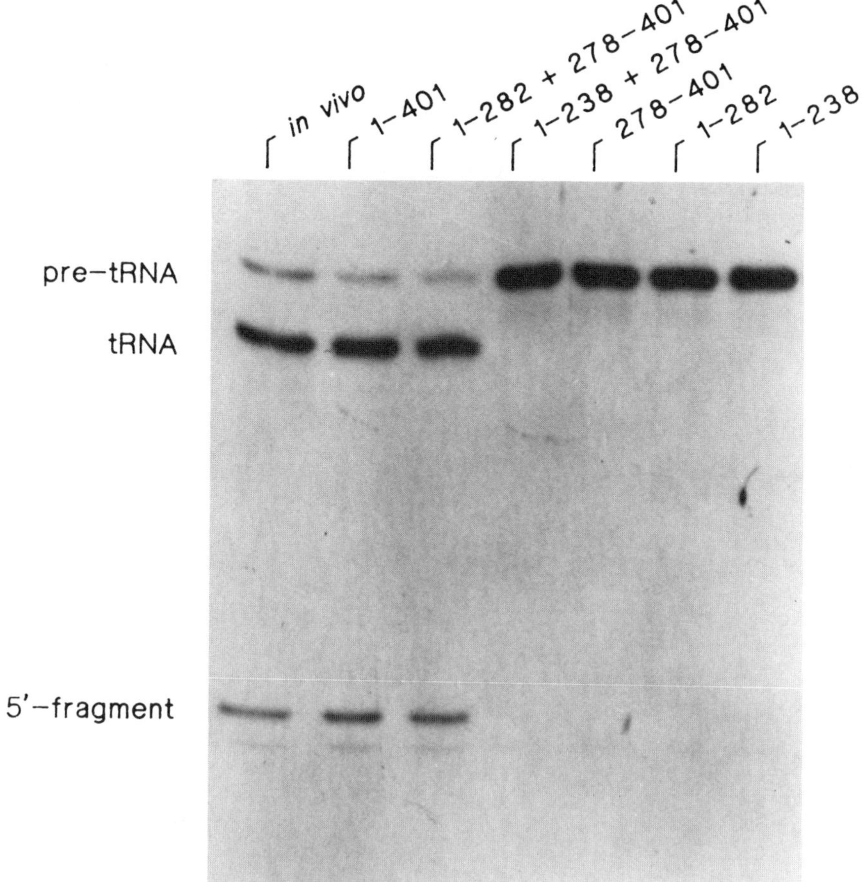

Fig. 11. Identification by fragment reconstitution of an RNase P RNA sequence important for enzymatic activity. Runoff transcripts comprising all or part of the mature RNase P RNA were assayed for their ability to cleave ^{32}P-labeled precursor tRNA^{His}. The various transcripts are designated by their first and last nucleotide position relative to the complete *B. subtilis* RNase P RNA. Reactions were carried out at 37°C for 15 min with 600 m*M* NH_4Cl and 250 m*M* $MgCl_2$. Reaction products were resolved by electrophoresis through an 8% polyacrylamide gel containing 8 *M* urea, and were visualized by autoradiography. For reactions containing more than one transcript, the fragments were "annealed" (heated to 70°C and slowly cooled to room temperature).

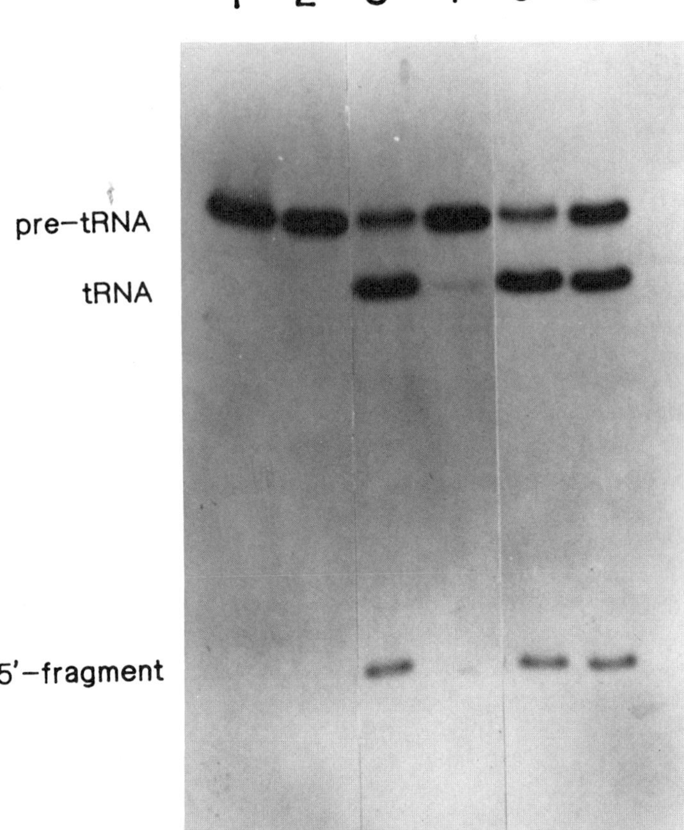

Fig. 12. Site-directed mutation of an RNase P RNA site that influences RNase P protein interaction. A small sequence (5′-TCGA-3′) was duplicated within the *Xho*I site of the *B. subtilis* RNase P RNA gene (nucleotides 236–239). T7 RNA polymerase-directed runoff transcripts of both mutant and wild-type RNase P RNAs were assayed for enzymatic activity in the absence (odd-numbered lanes), or in the presence (even-numbered lanes), of partially purified *B. subtilis* RNase P protein. Lanes 1 and 2, No RNase P RNA; lanes 3 and 4, RNase P RNA containing the tetranucleotide duplication; lanes 5 and 6, wild-type RNase P RNA.

as the native RNase P RNA in the high-salt, RNA-alone reaction; however, its activity is not stimulated by the RNase P protein under the holoenzyme reaction conditions (low salt). The most straightforward interpretation of this result is that the disrupted sequence is an important contact for the protein, although there are other possibilities. For instance, the insertion might perturb the local conformation in a manner that affects the interaction with the protein, but not catalysis.

V. WHY IS THE CATALYTIC ELEMENT OF RNase P COMPOSED OF RNA?

An interesting question to pose is: Why is the catalytic element of RNase P an RNA molecule? We know that proteins can specifically recognize RNAs and that cells generally use proteins as catalysts. Indeed, the other known, specific processing nucleases, for instance those involved in rRNA maturation, are proteins. A mechanistically trivial explanation for the fact that RNase P is an RNA would be that it is an evolutionary vestige of the very earliest life forms, before the protein-synthesizing machinery was fully established. This explanation seems unlikely, however, since the RNase P structure is so highly variable (above). A remnant of the earliest days of life on Earth might be expected to be much more conservative in its structure, at least as conservative as the rRNAs or tRNAs.

Alternatively, it must be supposed that RNase P is an RNA because it plays some role that proteins cannot so readily perform. One possibility is that the RNase P RNA provides structure to and coordinates several elements in a multienzyme tRNA processing complex analogous to the ribosome, and it happens to be a nuclease, as well, in the interests of cellular economy. All tRNA precursors must undergo multiple processing steps, for example terminal cleavages, methylations, etc. It is attractive to consider that this would occur in a multienzyme complex. Another possibility derives from the point that RNase P has a nearly unique role among enzymes—it must handle many different, yet specific, substrates: the 50 or so distinct tRNAs produced by cells. All the tRNAs have the same general form, the L-shaped tertiary structure, but they differ in their minor details, a consequence of different sequences, variable loop sizes, varying precursor-specific sequences, etc. RNase P must accommodate all of these, perhaps by molding—"induced fit"—to the substrates. Perhaps, therefore, RNase P activity is embodied in RNA because of a requirement for structural fluidity, which data suggest may be important in the reaction (above). Proteins offer a greater wealth of chemically functional groups than do polynucleotides, but RNA structure may be intrinsically more mobile than that of proteins.

ACKNOWLEDGMENT

This work was supported by National Institutes of Health Grant GM34527 to N. R. P.

REFERENCES

Altman, S., and Guerrier-Takada, C. (1986). *Biochemistry* **25,** 1205–1208.
Altman, S., and Smith, J. D. (1971). *Nature (London), New Biol.* **233,** 35.
Cech, T. R. (1985). *Int. Rev. Cytol.* **93,** 3–22.
Cech, T. R., Tanner, N. K., Tinoco, I., Weir, B. R., Zuker, M., and Perlman, P. S. (1983). *Proc. Natl. Acad. Sci. U.S.A.* **80,** 3903–3907.
Doersen, C., Guerrier-Takada, C., Altman, S., and Attardi, G. (1985). *J. Biol. Chem.* **260,** 5942–5949.
Elwood, H. J., Olsen, G. J., and Sogin, M. L. (1985). *Mol. Biol. Evol.* **2,** 399–410.
Fox, G., and Woese, C. R. (1975). *Nature (London)* **256,** 505–507.
Gardiner, K., and Pace, N. R. (1980). *J. Biol. Chem.* **255,** 7507–7509.
Gardiner, K. J., Marsh, T. L., and Pace, N. R. (1985). *J. Biol. Chem.* **260,** 5415–5419.
Guerrier-Takada, C., Haydock, K., Allen, L., and Altman, S. (1986). *Biochemistry* **25,** 1509–1515.
Guerrier-Takada, C., and Altman, S. (1984). *Science* **223,** 285–286.
Guerrier-Takada, C., Gardiner, K., Marsh, T., Pace, N., and Altman, S. (1983). *Cell (Cambridge, Mass.)* **35,** 849–857.
Hansen, F. G., Hansen, E. B., and Atlung, T. (1985). *Gene* **38,** 85–93.
Jacobson, A. B., Good, L., Simonetti, J., and Zuker, M. (1984). *Nucleic Acids Res.* **12,** 45–52.
Kline, L., Nishikawa, S., and Söll, D. (1981). *J. Biol. Chem.* **256,** 5058–5061.
Kruger, K., Grabowski, P. J., Zaug, A. J., Sands, J., Gottschling, D. E., and Cech, T. R. (1982). *Cell (Cambridge, Mass.)* **31,** 147–157.
Marsh, T. L., and Pace, N. R. (1985). *Science* **229,** 79–81.
Noller, H. F., and Woese, C. R. (1981). *Science* **212,** 402–411.
Ogasawara, N., Moriya, S., von Meyerburg, K., Hansen, F., and Yoshikawa, H. (1985). *EMBO J.* **4,** 3345–3350.
Pace, N. R., and Marsh, T. L. (1985). *Origins Life* **16,** 97–116.
Pace, N. R., Olsen, G. J., and Woese, C. R. (1986). *Cell (Cambridge, Mass.)* **45,** 325–326.
Papanicolaou, C., Gouy, M., and Ninio, J. (1984). *Nucleic Acids Res.* **12,** 31–44.
Reed, R. E., Baer, M. F., Guerrier-Takada, C., Donis-Keller, H., and Altman, S. (1982). *Cell (Cambridge, Mass.)* **30,** 627–636.
Reich, C., Gardiner, K. J., Olsen, G. J., Pace, B., Marsh, T. L., and Pace, N. R. (1986). *J. Biol. Chem.* **261,** 7888–7893.
Ross, P. D., and Scruggs, R. L. (1964). *Biopolymers* **2,** 231–236.
Sakamato, K., Kimura, N., Nagawa, F., and Shimura, Y. (1983). *Nucleic Acids Res.* **11,** 8237–8251.
Salser, W. (1977). *Cold Spring Harbor Symp. Quant. Biol.* **42,** 985–1002.
Stark, B. C., Kole, R., Bowman, E. J., and Altman, S. (1977). *Proc. Natl. Acad. Sci. U.S.A.* **75,** 3719–3721.
Vournakis, J. N., Celantano, J., Finn, M., Lockard, R. E., Mitra, T., Pavlakis, G., Troutt, A., Vandenberg, M., and Wurst, R. M. (1981). *In* "Gene Amplification and Analysis" (J. G. Chirikjian and T. S. Papas, eds.), Vol. 2, pp. 268–294. Elsevier/North-Holland, New York.

3

Multiple Enzymatic Activities of an Intervening Sequence RNA from *Tetrahymena*

THOMAS R. CECH, ARTHUR J. ZAUG, AND MICHAEL D. BEEN

Department of Chemistry and Biochemistry
University of Colorado
Boulder, Colorado 80309-0215

I. INTRODUCTION

In some species of the ciliated protozoan *Tetrahymena,* every copy of the gene for the large ribosomal RNA (rRNA) is interrupted by an intervening sequence (IVS), or intron, approximately 400 base pairs in size (Wild and Gall, 1979). As found for other IVSs, the *Tetrahymena* rRNA IVS is transcribed as part of a large precursor RNA and subsequently excised by RNA splicing (Cech and Rio, 1979; Din *et al.,* 1979; Carin *et al.,* 1980).

II. SELF-SPLICING RNA

The *Tetrahymena thermophila* pre-rRNA is self-splicing. That is, accurate splicing takes place *in vitro* in the absence of protein (Cech *et al.,* 1981; Kruger *et al.,* 1982). In this regard the splicing differs from splicing of nuclear pre-tRNAs, which requires enzymes (Greer *et al.,* 1983). The relationship to splicing of nuclear pre-mRNAs is less clear. Nuclear pre-mRNA splicing requires proteins and small nuclear ribonucleoproteins (snRNPs) but may not involve enzymes in the traditional sense (Kruger *et*

Molecular Biology of RNA
New Perspectives

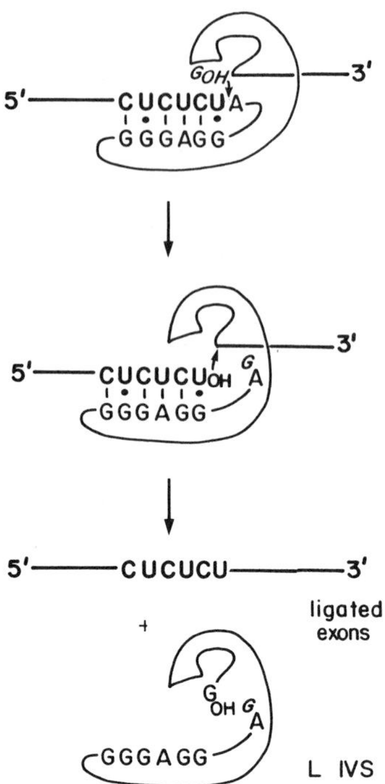

Fig. 1. Two-step transesterification mechanism for self-splicing of *Tetrahymena* pre-rRNA. Boldface letters and thick lines, exons; thin letters and lines, IVS; italicized G, free guanosine or guanosine triphosphate that adds to the 5′ end of the IVS RNA during splicing. The complex folded core structure of the IVS RNA, which is required for reactivity, is depicted as a simple curved line. The mechanism was originally proposed by Cech *et al.* (1981) and has been elaborated by Zaug *et al.* (1983), Inoue *et al.* (1985, 1986), Waring *et al.* (1986), and Been and Cech (1986).

al., 1982; Sharp, 1985; Maniatis and Reed, 1987). The RNA components of the snRNPs may be responsible for catalysis.

The mechanism of *Tetrahymena* pre-rRNA self-splicing is summarized in Fig. 1. The reaction requires Mg^{2+} (or Mn^{2+}) at a concentration of around 5 m*M*. The folded RNA molecule provides a binding site for a free guanosine nucleotide which cleaves the pre-rRNA at its 5′ splice site and becomes covalently attached to the 5′ end of the IVS (Cech *et al.*, 1981; Bass and Cech, 1984). The 3′-hydroxyl group at the end of the 5′ exon then attacks the 3′ splice site, releasing the IVS and ligating the exons. The sequence GGAGGG, originally proposed to comprise part of an internal guide sequence that aligned the exons for splicing (Davies *et al.*, 1982), has been shown to bind at least 4 of the last 6 nucleotides of 5′ exon

(Waring *et al.*, 1986; Been and Cech, 1986; Price *et al.*, 1987; L. Barfod and T. Cech, unpublished data). This base-pairing interaction is important for both steps of self-splicing. It aligns the 5′ splice site for attack by guanosine in the first step of splicing, and holds the 5′ exon into place for attack at the 3′ splice site in the second step (Inoue *et al.*, 1985; Been and Cech, 1986). Both steps in splicing occur by transesterification, an exchange of phosphate esters that requires no external energy such as is provided by ATP or GTP hydrolysis in other systems (Cech, 1983).

Following its excision from the pre-rRNA, the IVS can undergo self-catalyzed conversion to a circular form with release of the first 15 nucleotides of the IVS. Thus, cyclization is a cleavage–ligation reaction which, like RNA splicing, proceeds through a transesterification mechanism (Zaug *et al.*, 1983). Once again, the reaction requires Mg^{2+} or Mn^{2+}. Cyclization is readily reversible (Sullivan and Cech, 1985).

Self-splicing and autocyclization have been demonstrated for a number of other rRNA precursors and mRNA precursors that contain Group I and Group II IVSs. The Group I IVSs, which include the *Tetrahymena* rRNA IVS, were originally categorized by their conserved sequence elements and core secondary structure (Michel and Dujon, 1983; Waring *et al.*, 1983). Most of the Group I IVSs that have been carefully examined have been found to be self-splicing *in vitro*. Many of the self-splicing Group I RNAs are mitochondrial: the first IVS of the cytochrome *b* pre-mRNA in *Neurospora crassa* (Garriga and Lambowitz, 1984) and the IVS of the yeast mitochondrial pre-rRNA (Van der Horst and Tabak, 1985). Additional examples occur in pre-mRNAs transcribed from the thymidylate synthase and ribonucleotide reductase genes of bacteriophage T4 (Chu *et al.*, 1986; Gott *et al.*, 1986). In all of these cases, splicing takes place by the same guanosine-dependent transesterification mechanism described for the *Tetrahymena* IVS.

The mitochondrial Group II IVSs share a different set of conserved sequence elements and structural features (Michel and Dujon, 1983). Two of these have been found to be self-splicing *in vitro* (Peebles *et al.*, 1986; Van der Veen *et al.*, 1986; Schmelzer and Schweyen, 1986). In these cases, however, self-splicing occurs by a nucleotide-independent transesterification mechanism involving lariat formation. The lariats are very similar to those formed in nuclear pre-mRNA splicing (Cech, 1986a).

III. THE IVS RNA ENZYME

After its excision from *Tetrahymena* pre-rRNA, the linear IVS RNA undergoes a series of cyclization and self-cleavage reactions (Zaug *et al.*, 1984). The final product, the L-19 IVS RNA, is missing the first 19 nucleo-

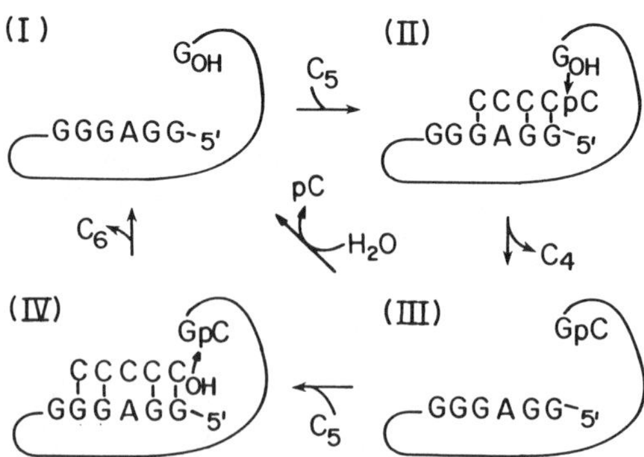

Fig. 2. A shortened form of the *Tetrahymena* IVS RNA acts as an enzyme. With pentacytidylic acid (C_5) as a substrate, the L-19 IVS RNA (I) acts as a nucleotidyltransferase [poly(C) polymerase, oligo(C) dismutase] by the reaction sequence I → II → III → IV → I. At pH 9, the same ribozyme also has substantial activity as a ribonuclease by the reaction sequence I → II → III → I. (Adapted from Zaug and Cech, 1986a. Copyright 1986 by the AAAS.)

tides of the IVS that contain the major and minor sites of cyclization. Therefore it cannot undergo intramolecular reactions. The L-19 IVS RNA still retains activity, however, and can catalyze cleavage–ligation reactions on other RNA molecules in the presence of 5–40 mM $MgCl_2$ (Zaug and Cech, 1986a). The L-19 IVS RNA is regenerated in each reaction cycle. Thus, it is acting as a true enzyme.

One of the enzymatic activities of the L-19 IVS RNA resembles that of RNA polymerase (Fig. 2). With C_5 as a substrate, the L-19 IVS RNA catalyzes the reaction $2C_5 \rightarrow C_4 + C_6$. One C_5 is used as a "monomer" unit to donate a pC moiety, and the other C_5 acts as a primer to be extended. The C_4 is equivalent to the pyrophosphate produced during the normal RNA polymerase reaction. Products as large as C_{30} are produced after 1 hr. Thus the L-19 IVS RNA can act as a poly(C) polymerase (Zaug and Cech, 1986a).

The reaction mechanism (Fig. 2) involves steps that are intermolecular versions of pre-rRNA self-splicing and other self-processing reactions. More specifically, formation of the covalent intermediate (II → III) is analogous to IVS RNA autocyclization, and resolution of the intermediate (IV → I) is analogous to exon ligation or the reversal of cyclization.

The K_m for C_5 is 42 μM, and the turnover number is approximately 100/hr (Zaug and Cech, 1986a). The reaction is specific for ribonucleotides, no reaction taking place with d-pC_5 or d-pA_5. Among the oligoribonucleotides, pU_6 is a much poorer substrate than pC_5 or pC_6, and pA_6 gives no reaction. The substrate specificity of this ribozyme has been altered by site-specific mutagenesis of the sequence GGAGGG in the active site; changing active-site G residues to As changes the substrate from oligo(C) to oligo(U) (Been and Cech, 1986). Thus, the ribozyme is not restricted to being a poly(C) polymerase.

One of the major differences between the ribozyme and RNA polymerase is that the former enzyme uses an internal template to direct the polymerization reaction, while the latter enzyme is dependent on an external template. It might be possible physically to remove the template portion of the ribozyme from the catalytic portion and get it to function with an external template. If so, the ribozyme could conceivably act as a primordial RNA replicase (Zaug and Cech, 1986a; Cech, 1986b). The work reported recently by Szostak (1986) could be another step in this direction.

At pH 9, the L-19 IVS RNA has activity as a ribonuclease (Fig. 2). The covalent ribozyme–substrate intermediate (III) undergoes hydrolysis, releasing the nucleotide or oligonucleotide attached to its 3′ end (in this case pC) and regenerating the ribozyme (I). The lability of the phosphodiester bond following the 3′-terminal guanosine (G^{414}) of the IVS had been previously noted in the site-specific hydrolysis of the circular IVS RNA (Zaug *et al.*, 1984) and of the 3′ splice site of pre-rRNA (Inoue *et al.*, 1986). The pH dependence of the hydrolysis reactions is indicative of attack by hydroxide ion (specific base catalysis; Zaug *et al.*, 1985). The ribonuclease activity of the L-19 IVS RNA shown in Fig. 2 does not provide a useful tool for sequence-specific cleavage of RNA, because hydrolysis cannot be separated cleanly from the transesterification activity. Thus, at pH 9 the ribozyme not only cleaves RNA, it also can rearrange RNA sequences.

When we used a substrate with a 3′-terminal phosphate (C_5p) we expected that the RNA enzyme would be blocked. Instead we found that the L-19 IVS RNA dephosphorylates the substrate (Zaug and Cech, 1986b). The substrate 3′-phosphate is transferred to the 3′-terminal guanosine of the enzyme. The pH dependence of the reaction (optimum at pH 5) indicates that the enzyme has activity toward the dianion and much greater activity toward the monoanion form of the 3′-phosphate of the substrate. Phosphorylation of the enzyme is reversible by C_5-OH and other oligo(pyrimidines) such as UCU-OH. Thus, the RNA enzyme acts as a phosphotransferase, transferring the 3′-terminal phosphate of C_5p to UCU-OH with multiple turnover (Fig. 3). At pH 4 and 5, the phosphoenzyme undergoes slow hydrolysis to yield inorganic phosphate (Fig. 3). Thus, the enzyme has acid phosphatase activity (Zaug and Cech, 1986b).

We have recently found that the *Tetrahymena* ribozyme can also act as

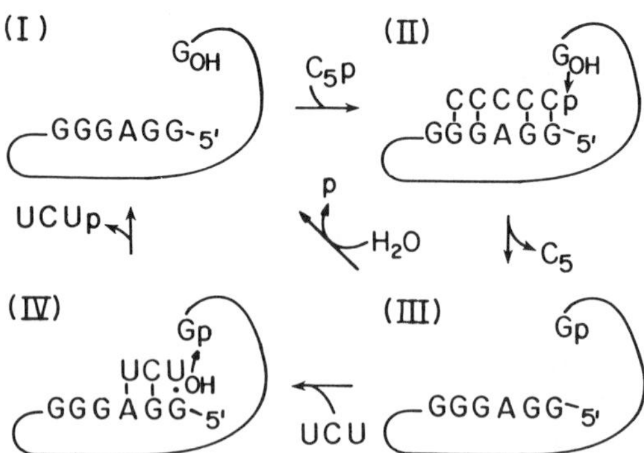

Fig. 3. The L-19 IVS RNA also has enzymatic activity toward phosphate monoester substrates. With C_5p and UCU_{OH} as co-substrates, the L-19 IVS RNA (I) acts as a phosphotransferase by the pathway I → II → III → IV → I. At pH 4–5, the phosphoenzyme (III) undergoes slow hydrolysis with release of inorganic phosphate. Thus, the ribozyme can act as an acid phosphatase (pathway I → II → III → I). (For details, see Zaug and Cech, 1986b.)

a sequence-specific endoribonuclease by a guanylyl transfer mechanism (Zaug *et al.*, 1986). It cleaves other RNA molecules at sequences that resemble the 5′ splice site of the rRNA precursor. Cleavage is concomitant with addition of a free guanosine nucleotide to the 5′ end of the downstream RNA fragment; thus, one product can be readily end-labeled by inclusion of [^{32}P]GTP in the reaction. The ribozyme has a K_m of 44 μM for GTP, a K_m of 0.8 μM for the RNA substrate, and a turnover number of approximately 8/hr for an oligonucleotide substrate GGCCCUCUAAAAA.

The endoribonuclease reaction is analogous to the first step of pre-rRNA self-splicing (compare right-hand structure in Fig. 4 to top structure in Fig. 1). Unlike the reactions shown in Figs. 2 and 3, this reaction does not involve formation of a covalent intermediate between the ribozyme and its substrate. Thus L-19 IVS RNA that has had G^{414} removed by periodate oxidation and β-elimination (L-19 IVS_β in Fig. 4) is as effective a catalyst as intact L-19 IVS RNA.

The ribozyme has high specificity for cleavage after the nucleotide sequence CUCU (Zaug *et al.*, 1986). Its specificity for cleavage of single-stranded RNA is greater than that of known ribonucleases, and may approach that of the DNA restriction endonucleases. Site-specific mutations in the sequence GGAGGG in the active site of the IVS RNA (the same sequence that serves as the 5′ exon-binding site for self-splicing; see Fig. 1) alter the sequence specificity of the ribozyme in a predictable manner (Zaug *et al.*, 1986). Ribozymes that cleave after the sequences

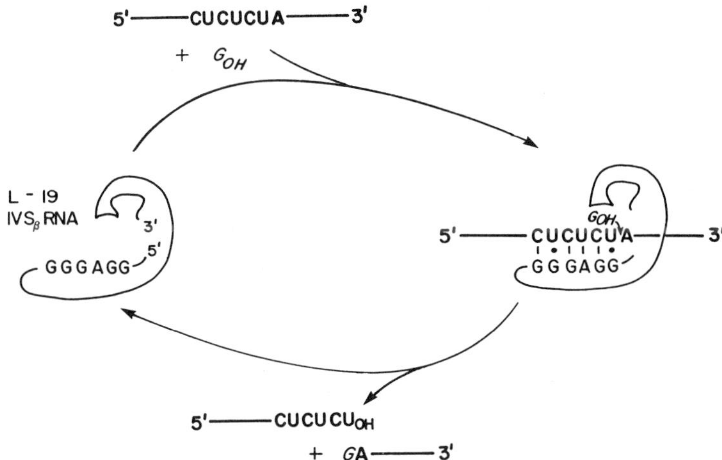

Fig. 4. The *Tetrahymena* ribosome acts like an RNA restriction endonuclease (Zaug *et al.,* 1986). The L-19 IVS RNA is subjected to periodate oxidation and β-elimination to remove its 3'-terminal guanosine (G^{414}). The resulting ribozyme (L-19 IVS_{β} RNA) binds an RNA substrate (boldface letters and thick lines) and a free guanosine nucleoside or nucleotide (G in italics). It then catalyzes the cleavage of the substrate by transesterification.

CUGU, CGCU, and GUCU have been synthesized and tested. Thus, we have taken the first few steps toward creating a whole set of sequence-specific endoribonucleases that may be useful as tools for studying the molecular biology of RNA.

REFERENCES

Bass, B. L., and Cech, T. R. (1984). *Nature (London)* **308,** 820–826.

Been, M. D., and Cech, T. R. (1986). *Cell (Cambridge, Mass.)* **47,** 207–216.

Carin, M., Jensen, B. F., Jentsch, K. D., Leer, J. C., Nielsen, O. F., and Westergaard, O. (1980). *Nucleic Acids Res.* **8,** 5551–5566.

Cech, T. R. (1983). *Cell (Cambridge, Mass.)* **34,** 713–716.

Cech, T. R. (1986a). *Cell (Cambridge, Mass.)* **44,** 207–210.

Cech, T. R. (1986b). *Proc. Natl. Acad. Sci. U.S.A.* **83,** 4360–4363.

Cech, T. R., and Rio, D. C. (1979). *Proc. Natl. Acad. Sci. U.S.A.* **76,** 5051–5055.

Cech, T. R., Zaug, A. J., and Grabowski, P. J. (1981). *Cell (Cambridge, Mass.)* **27,** 487–496.

Chu, F. K., Maley, G. F., West, D. K., Belfort, M., and Maley, F. (1986). *Cell (Cambridge, Mass.)* **45,** 157–166.

Davies, R. W., Waring, R. B., Ray, J. A., Brown, T. A., and Scazzocchio, C. (1982). *Nature (London)* **300,** 719–724.

Din, N., Engberg, J., Kaffenberger, W., and Eckert, W. (1979). *Cell (Cambridge, Mass.)* **18,** 525–532.

Garriga, G., and Lambowitz, A. M. (1984). *Cell (Cambridge, Mass.)* **39,** 631–641.
Gott, J., Schub, D., and Belfort, M. (1986). *Cell (Cambridge, Mass.)* **47,** 81–87.
Greer, C. L., Peebles, C. L., Gegenheimer, P., and Abelson, J. (1983). *Cell (Cambridge, Mass.)* **32,** 537–546.
Inoue, T., Sullivan, F. X., and Cech, T. R. (1985). *Cell (Cambridge, Mass.)* **43,** 431–437.
Inoue, T., Sullivan, F. X., and Cech, T. R. (1986). *J. Mol. Biol.* **189,** 143–165.
Kruger, K., Grabowski, P. J., Zaug, A. J., Sands, J., Gottschling, D. E., and Cech, T. R. (1982). *Cell (Cambridge, Mass.)* **31,** 147–157.
Maniatis, T., and Reed, R. (1987). *Nature (London)* **325,** 673.
Michel, F., and Dujon, B. (1983). *EMBO J.* **2,** 33–38.
Peebles, C. L., Perlman, P. S., Mecklenburg, K. L., Petrillo, M. L., Tabor, J. H., Jarrell, K. A., and Cheng, H.-L. (1986). *Cell (Cambridge, Mass.)* **44,** 213–223.
Price, J. V., Engberg, J., and Cech, T. R. (1987). *J. Mol. Biol.* (in press).
Schmelzer, C., and Schweyen, R. J. (1986). *Cell (Cambridge, Mass.)* **46,** 557–565.
Sharp, P. A. (1985). *Cell (Cambridge, Mass.)* **42,** 397–400.
Sullivan, F. X., and Cech, T. R. (1985). *Cell (Cambridge, Mass.)* **42,** 639–648.
Szostak, J. W. (1986). *Nature (London)* **322,** 83–86.
Van der Horst, G., and Tabak, H. F. (1985). *Cell (Cambridge, Mass.)* **40,** 759–766.
Van der Veen, R., Arnberg, A. C., Van der Horst, G., Bonen, L., Tabak, H. F., and Grivell, L. A. (1986). *Cell (Cambridge, Mass.)* **44,** 225–234.
Waring, R. B., Scazzocchio, C., Brown, T. A., and Davies, R. W. (1983). *J. Mol. Biol.* **167,** 595–605.
Waring, R. B., Towner, P., Minter, S. J., and Davies, R. W. (1986). *Nature (London)* **321,** 133–139.
Wild, M. A., and Gall, J. G. (1979). *Cell (Cambridge, Mass.)* **16,** 565–573.
Zaug, A. J., and Cech, T. R. (1986a). *Science* **231,** 470–475.
Zaug, A. J., and Cech, T. R. (1986b). *Biochemistry* **25,** 4478–4482.
Zaug, A. J., Grabowski, P. J., and Cech, T. R. (1983). *Nature (London)* **301,** 578–583.
Zaug, A. J., Kent, J. R., and Cech, T. R. (1984). *Science* **224,** 574–578.
Zaug, A. J., Kent, J. R., and Cech, T. R. (1985). *Biochemistry* **24,** 6211–6218.
Zaug, A. J., Been, M. D., and Cech, T. R. (1986). *Nature (London)* **324,** 429–433.

4

Processing and Genetic Characterization of Self-Splicing RNAs of Bacteriophage T4

MARLENE BELFORT,* JOAN PEDERSEN-LANE,* KAREN EHRENMAN,*† DWIGHT H. HALL,‡ CHRISTINE M. POVINELLI,‡ JONATHA M. GOTT,§ AND DAVID A. SHUB§

* *Wadsworth Center for Laboratories and Research*
New York State Department of Health
Albany, New York 12201
† *Department of Microbiology and Immunology*
Albany Medical College
Albany, New York 12208
‡ *School of Applied Biology*
Georgia Institute of Technology
Atlanta, Georgia 30332
§ *Department of Biological Sciences*
State University of New York
Albany, New York 12222

I. INTRODUCTION

The interruption of a protein-coding sequence in a prokaryote was first reported for the thymidylate synthase (*td*) gene of bacteriophage T4, which was shown to contain a 1017-nucleotide intron (Chu *et al.*, 1984). Subsequently the *td* gene was shown to be expressed via an RNA splicing mechanism (Belfort *et al.*, 1985, 1986) that could proceed autocatalytically *in vitro* (Chu *et al.*, 1985, 1986). Conserved sequence elements in and around the *td* intron (Chu *et al.*, 1984, 1986) suggested it to be a group

I intervening sequence (IVS), similar to those found in genes of fungal mitochondria and plant chloroplasts and in nuclear genes encoding ribosomal RNAs (rRNAs) of protozoa and slime molds (reviewed by Michel *et al.,* 1982; Waring and Davies, 1984; Cech, 1986).

There are three major implications associated with these findings: First, if indeed the *td* intron is a group I intron, the splicing reaction should generate characteristic group I splice products. Second, the use of prokaryotic genetic strategies and simple phenotypic screens should delineate the regions of the *td* gene that direct the splicing pathway. Third, since autocatalytically processed group I introns can be labeled *in vitro* in the absence of proteins, this reaction should provide the means to identify other intron-containing genes in T4. An exploration of these implications provides the basis for this review.

II. GROUP I SPLICING MECHANISM FOR T4 *td* RNA

While invariant nucleotides associated with group I introns are limited to a U immediately upstream of the intron, and a G at its 3′ end (Fig. 1), conserved sequences also exist within the intron. Indeed, the group I eukaryotic introns were first typed into a distinct class based on these conserved sequence elements, which are believed to guide folding of the intron into a characteristic core structure (Davies *et al.,* 1982; Michel *et al.,* 1982). As part of this core structure, pairings between an internal guide sequence (IGS) of the intron and sequences at the 3′ end of the upstream exon (and perhaps also at the 5′ end of the downstream exon) are believed to play a major role in aligning the RNA into an active splicing conformation and in targeting the 5′ splice site (Davies *et al.,* 1982; Waring *et al.,* 1986; Been and Cech, 1986). Not only can conserved sequence elements be identified within the *td* intron, but the IVS also possesses the invariant U and G residues as well as potential IGS–exon pairings (Chu *et al.,* 1986). The question therefore arises as to whether the *td* splicing mechanism also conforms to the group I processing pathway.

Self-splicing was first demonstrated for *Tetrahymena thermophila* nuclear large rRNA and subsequently for several fungal mitochondrial group I introns (Garriga and Lambowitz, 1984; Van der Horst and Tabak, 1985). The first step in the RNA-catalyzed splicing pathway for *Tetrahymena* rRNA is cleavage at the 5′ splice site via a transesterification reaction in which a guanosine cofactor attacks the phosphodiester bond 3′ to the conserved U and displaces the upstream exon (Fig. 1, step 1) (Cech *et al.,* 1981). Via a second transesterification, now at the 3′ splice site, the exons are joined and the intron excised (step 2). Covalent cyclization of

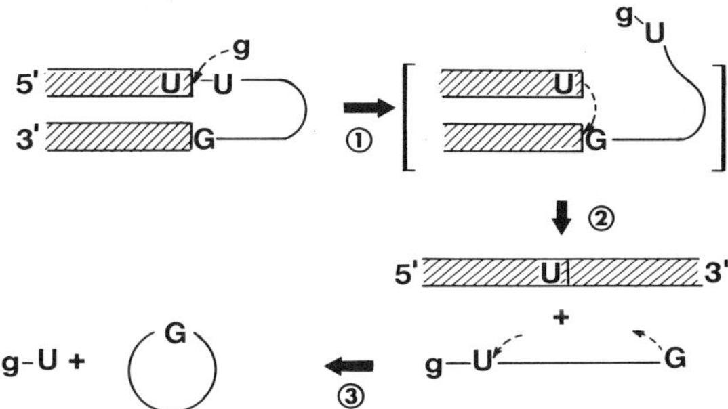

Fig. 1. The processing pathway of group I introns. This scheme was proposed for *Tetrahymena* rRNA splicing (reviewed by Cech, 1986) and shows many elements in common with other group I introns of eukaryotes and with the *td* intron of phage T4. The conserved nature of the U residue 5′ to the bond that is susceptible to G attack in both steps 1 and 3 is discussed in the text. G represents the invariant nucleotide at the 3′ end of the intron, and g the guanosine cofactor. The oligoribonucleotide released from the 5′ end of the intron upon cyclization is designated g-U.

the intron may then occur as its conserved 3′ terminal G adds to a bond near the 5′ end of the intron, liberating a 5′-terminal intron oligomer (Step 3) (Zaug *et al.*, 1983). The first distinguishing feature of the reaction is addition of the G to the 5′ end of the intron during step 1. Second, the intron product formed in the third transesterification reaction is a circle with a 3′–5′ linkage at the cyclization junction. Therefore, to determine mechanistic parallels between the *td* intron and its eukaryotic group I counterparts, we characterized both linear and circular forms of the *td* intervening sequence.

A. A Noncoded G at the 5′ End of the Intron

To determine whether splicing of *td* RNA may proceed via a G-initiated transesterification reaction, we incubated deproteinized RNAs from *Escherichia coli* transformants with [α-^{32}P]GTP. Figure 2 shows the presence of a strongly labeled band at ~1 kb when RNA from *E. coli* containing the transcriptionally induced *td* gene on plasmid pKTd2 was incubated with [^{32}P]GTP under self-splicing conditions. This band, corresponding in size to that of labeled intron (lane 3), was absent when RNA from cells carrying the vector (pKC30) alone was incubated with [α-^{32}P]GTP (lane 2). Furthermore, the label was removed from the RNA by treatment with alkaline phosphatase (lane 4) as predicted for an intron terminally labeled with [^{32}P]GTP. Additionally, incubation of RNA from clone pKK223-3-

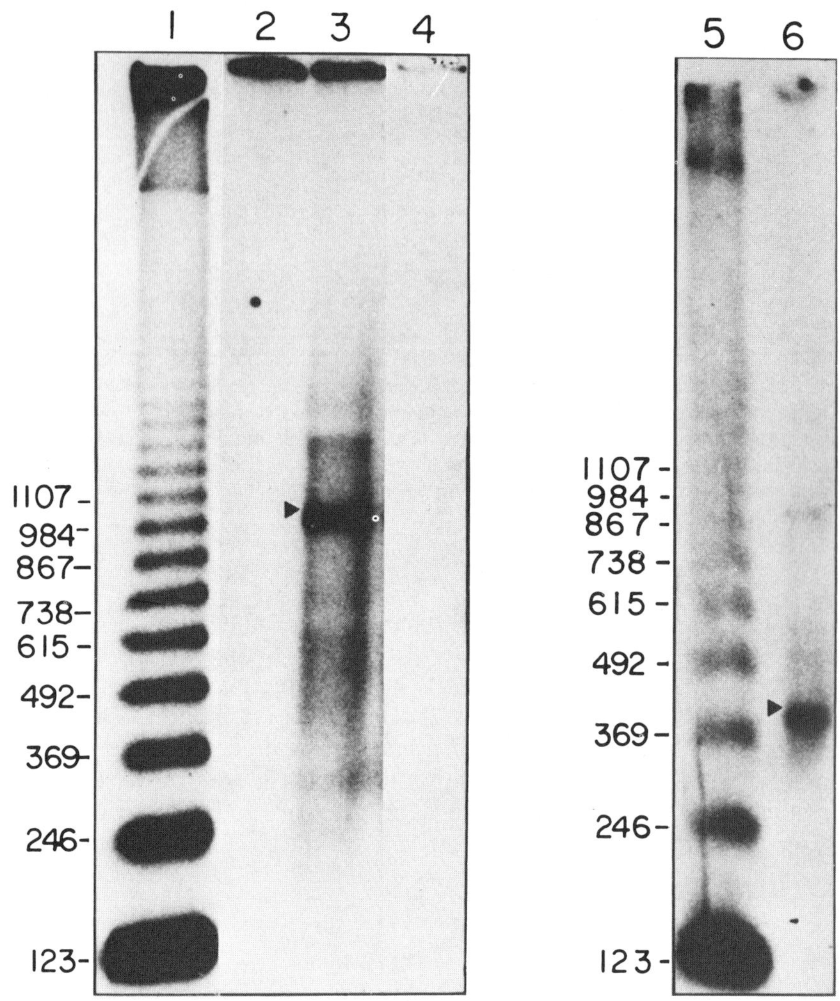

Fig. 2. Labeling of intron RNA with [α-^{32}P]GTP. Total deproteinized cellular RNAs (10 μg) from transcriptionally induced pKC30 vector (lane 2) or pKTd2 (Belfort *et al.*, 1983) (lanes 3 and 4), or from pKK223-3-*td*Δ1-3 (Belfort *et al.*, 1987) (lane 6) were labeled by incubation with [α-^{32}P]GTP under self-splicing conditions (Ehrenman *et al.*, 1986). The RNA in lane 4 was treated with alkaline phosphatase after labeling. Size markers (lanes 1 and 5) are ^{32}P-labeled 123 nucleotide DNA ladders. Separation was on a 1.5% agarose–2.2 *M* formaldehyde gel.

*td*Δ1-3, that is splicing proficient despite a 631-nucleotide deletion within the intron, yielded a labeled band at ~0.4 kb, corresponding in size to the partially deleted intron (lane 6). Digestion of purified [^{32}P]GTP-labeled intron RNA with nucleases P1, T2, or T1 yielded labeled species, the sizes of which were consistent with covalent G addition, via a 3′–5′ phosphodiester linkage, to the 5′ end of the linear intron (Ehrenman *et al.*, 1986).

The G-addition reaction seems not to be confined to splicing *in vitro,* as demonstrated by chemical (Maxam–Gilbert) sequencing of the reverse transcript of the linear intron generated *in vivo*. A cytosine was apparent at the 3′ end of the cDNA, corresponding to the G adduct at the 5′ end of the intron (data not shown). Thus, covalent addition of a G to the 5′ end of the intron, a reaction that distinguishes the eukaryotic group I intervening sequences (IVSs), is also a feature of the *td* intron.

B. The Intron Cyclization Junction

The *td* intron can undergo a second reaction to form a covalently closed structure, as demonstrated by the following experiments: (1) Aberrant migration in polyacrylamide gels of this intron product, synthesized either *in vitro* or *in vivo* (Chu *et al.*, 1986; Ehrenman *et al.*, 1986); (2) mobilization of the spuriously migrating circular species to a linear form by treatment with an intron-specific oligonucleolide and ribonuclease H (Chu *et al.*, 1986; Ehrenman *et al.*, 1986); (3) circular intron structures of appropriate size identified by heteroduplex analysis and electron microscopy (Ehrenman *et al.*, 1986). Since group II self-splicing introns are distinguished by forming branched circles with 2′–5′ junctions (Peebles *et al.*, 1986; Van der Veen *et al.*, 1986), it was of importance to define the cyclization junction of the *td* intron to substantiate its status as a group I IVS.

The first indication that the *td* intron is a true circle with a 3′–5′ linkage, rather than a lariat with a 2′–5′ junction, came from incubation of the *td* intron in the presence of active HeLa cell debranching extract (Ruskin and Green, 1985). The 2′–5′ phosphodiesterase activity present in the extract had no effect on the anomalous mobility of the *td* intron (data not shown). Second, the primer extension analysis shown in Fig. 3 argues firmly against a branched structure, which would impose a reverse transcriptase block at the 5′ intron boundary (Ruskin *et al.*, 1984).

Purified linear and circular intron species synthesized *in vivo* were analyzed by dideoxy sequencing, using a primer that anneals to the 5′ region of the intron (Fig. 3). The sequence found at the 5′ end of the linear species is that predicted based on G addition, with a reverse transcriptase

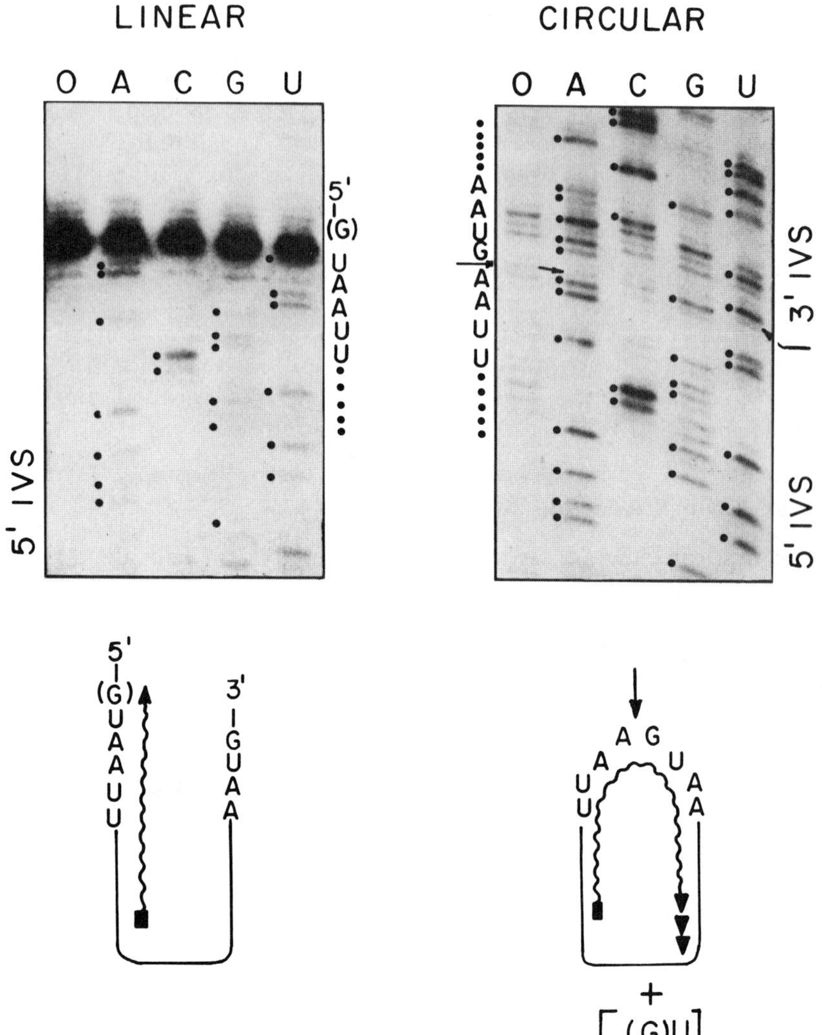

Fig. 3. (Top) Primer extension dideoxy sequence analysis of purified intron RNAs. Linear and circular intron isolated from cells expressing the cloned *td* gene from pKTd2 were analyzed by reverse transcription in the absence (lane 0) or presence of chain terminators ddT (lane A), ddG (lane C), ddC (lane G), or ddA (lane U). The sequences at the 5′ end of the linear intron and at the cyclization junction (arrows) of the circular intron are shown at the right and left margins of the respective autoradiograms. (Bottom) The corresponding reverse transcription products (wavy arrows) from the primer (■) are shown below. The noncoded guanosine (G) at the 5′ end of the intron is inferred from the results of Ehrenman *et al.* (1986). [(G)U] is a postulated product released upon cyclization when the 3′G attacks the bond between residues 2 (U) and 3 (A) of the intron.

stop in all the lanes 1 residue beyond the U at the known 5′ end of the coded intron sequence (Chu *et al.,* 1984, 1986). In contrast, when cDNA synthesis is primed from the circular intron, the sequence reads continuously from the 5′ directly into the 3′ end of the IVS. Notable is the loss of the two 5′ intron residues, namely the noncoded G and the U to which it was covalently joined.

Thus, the *td* splice products extend the analogies between this prokaryotic intron and that of the eukaryotic group I introns. The autocatalytic covalent addition of a G to the 5′ end of the intron, as well as the cyclization event described above, suggest a similar splice mechanism to that described for *Tetrahymena* large rRNA (Zaug *et al.,* 1983) and the *Saccharomyces cerevisiae* mitochondrial large rRNA (Tabak *et al.,* 1984).

Interestingly, while the four known cyclization sites (the major and minor sites of *Tetrahymena* rRNA intron, as well as the *S. cerevisiae* mitochondrial rRNA and T4 *td* cyclization sites) vary in their distance from the 5′ intron terminus (from 2 to 19 residues), they have one striking feature in common: They are all preceded by a U residue. This observation suggests that this U may be a major determinant in cyclization site recognition, much as the conserved U at the 3′ end of the upstream exon is important in targeting the 5′ splice site. This U in the first exon is thought to pair with a critical G residue in the IGS in effecting 5′ splice site selection (Waring and Davies, 1984; Waring *et al.,* 1986). The conserved U at the intron cyclization site may have a similar role and pair with the same G to impart specificity to the circularization reaction. The similarity of sequences upstream of the respective 5′ splice site and cyclization site(s) for each of the three group I introns under discussion (Tabak *et al.,* 1984; Sullivan and Cech, 1985; Ehrenman *et al.,* 1986) is consistent with the IGS first pairing with the 3′ end of the upstream exon, to target the 5′ splice site, and after cleavage at this site, pairing with residues preceding the cyclization site. While this latter interaction may be thought of as guiding cyclization site selection, as proposed originally by Cech and colleagues (reviewed by Cech and Bass, 1986), thorough mutational analysis will be helpful in rigorously testing this notion.

III. NONDIRECTED MUTAGENESIS AND DELINEATION OF TWO FUNCTIONAL DOMAINS FOR SPLICING IN THE *td* INTRON

Some structure–function relationships for group I introns have been inferred from cis-dominant intron mutations that disrupt splicing of *S. cerevisiae* mitochondrial RNAs (Anziano *et al.,* 1982; De La Salle *et al.,*

1982; Weiss-Brummer *et al.*, 1983) and from studies involving second-site revertants (Holl *et al.*, 1985). Additionally, such relationships are being probed by site-directed mutagenesis targeted to conserved sequence elements (Burke *et al.*, 1986; Waring *et al.*, 1986). Furthermore, artificial substrates are being used to ascribe functional roles to specific residues within the catalytically active RNA "ribozyme" of *Tetrahymena* and *Neurospora* (Sullivan and Cech, 1985; Garriga *et al.*, 1986). More random genetic strategies would be extremely useful, but are somewhat limited for the available eukaryotic group I introns, in part by the lack of facile transformation systems for fungal mitochondria and *Tetrahymena*.

In contrast, the *td* gene of T4 is extremely amenable to genetic analysis. The success of the approach is based primarily on screening methods for the thymidylate synthase-defective phenotype (Simon and Tessman, 1963; Hall, 1967), which allowed the isolation of a large collection of T4 *td* mutants following random mutagenesis targeted against the intact phage. Furthermore, analysis of these nondirected *td* mutations has been facilitated by the promiscuous recombination system of T4 (Broker and Doermann, 1975). Thereby, as described below, 81 independently isolated mutations were mapped by marker rescue (Mattson *et al.*, 1977) in cells containing cloned *td* subfragments. Of these mutations, over one-fourth were localized to the IVS, delineating two intron domains that are functional in splicing. A mutation from each domain, one from the 5′ region of the intron and one from the 3′ region, were selected for further study.

A. Intron Domains Defined by Genetic Mapping

The T4 *td* mutants were isolated on the basis of plaque morphology after *in vitro* nitrous acid or hydroxylamine mutagenesis. A T4 *td* mutant forms a small plaque on a host lacking thymidylate synthase when thymidine in the media is limiting, and a large wild-type-sized plaque when thymidine is in excess (Hall, 1967). The mutations were mapped by marker rescue in cells containing defined portions of the *td* gene (Fig. 4 and Hall *et al.*, 1987). This mapping is based on the fact that a given mutant will only show positive marker rescue (i.e., produce wild-type progeny) by recombination with plasmid subclones that contain the wild-type DNA sequence corresponding to that altered in the mutant (Mattson *et al.*, 1977). The distribution of mutations in different regions of the *td* gene is shown in Fig. 4.

The major results from this mapping study are: (1) All 81 mutations were rescued by the *Eco*RI fragment shown (Fig. 4). This result suggests that all the mutations map to the *td* region, rather than to other loci on the T4 genome. (2) At least 23 of the mutations were localized to the intron,

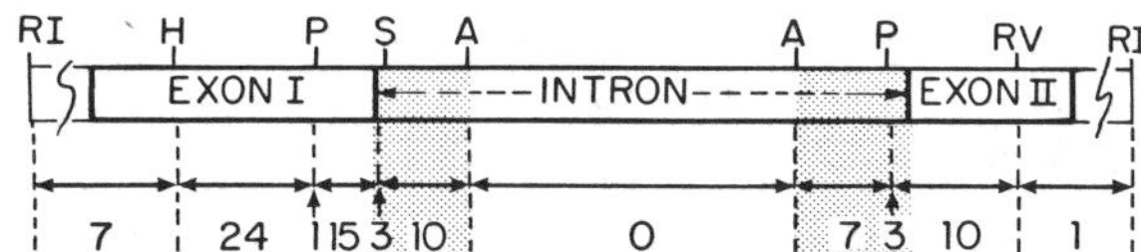

Fig. 4. Marker rescue mapping of *td* mutations. The physical map of the *td* gene is shown above, with the restriction sites (RI, *Eco*RI; H, *Hin*dIII; P, *Pvu*II; S, *Stu*I; A, *Aha*III; RV, *Eco*RV) that define the ends of cloned fragments used to establish eight deletion intervals. Numbers immediately under the bidirectional arrows, demarcating the eight deletion intervals, show the distribution of 81 independent *td* mutations. Upward-directed arrows indicate mutations that map to a particular restriction site. All 81 mutations were rescued by the full-length (2.85 kb) *Eco*RI fragment. The exons and intron are drawn to scale, spanning a total of 1.87 kb.

and 25 more map to fragments that overlap the intron–exon boundaries. (3) None of the mutations is in the middle of the intron, as defined by a deletion of 631 base pairs. This result corroborates the splicing proficiency of the cloned *td* construct that is deleted for these central 631 intron residues (Fig. 2, Ehrenman *et al.*, 1986; West *et al.*, 1986; Belfort *et al.*, 1987). (4) Of the known intron mutations 13 cluster to the 5′ region of the intron, and 10 map to the 3′ region. The mapping therefore defines two functional domains of approximately 200 residues each at the ends of the intron.

Like many eukaryotic group I introns, the *td* intervening sequence contains an open reading frame (ORF). The *td* intron ORF of 735 nucleotides (Chu *et al.*, 1986) spans the central nonessential domain. The ORF sequences of other group I introns are not usually highly conserved and are looped out of secondary structure models for RNA folding (Michel *et al.*, 1982; Waring and Davies, 1984). Similarly, for the *td* intron these genetic studies imply that the bulk of the intron ORF is not essential for formation of the active splicing conformation.

B. Map Location of Two Selected Mutations Is Corroborated by Sequence Change

To verify the accuracy of the mapping techniques, and to determine the structural basis for the mutant phenotypes, T4*td*N57 and T4*td*N47 were selected for sequencing. Recombination frequency suggested that *td*N57 mapped very close to the *Stu*I site at the 5′ end of the intron, while *td*N47 was located very close to the *Pvu*II site at its 3′ end (Fig. 4). Since the mutant genes have yet to be cloned, we determined the sequence change directly at the RNA level. This was achieved by using synthetic oligodeoxynucleotides to prime cDNA synthesis from the RNA of infected cells in the presence of each of the four ddNTPs. Thus, the sequence change of T4*td*N57 was shown to be a C → T transition directly within the *Stu*I site, 10 residues from the 5′ splice site, while T4*td*N47 was shown to be a G → A transition directly within the *Pvu*II site, 49 nucleotides from the 3′ splice site.

C. Splicing Defects of Two Selected Mutations

As a preliminary screening method for splicing-defective mutants, we developed an RNA dot blot assay with RNA from phage-infected cells. Synthetic oligodeoxynucleotides complementary to strategic regions of the primary transcript or RNA splice products were used as probes (Fig. 5). The intron-specific probe (probe 1) that hybridizes to premessage and excised intron gave a signal of roughly equal intensity with wild-type and mutant RNAs. On the other hand, probes 2 and 3, specific to the 5′ and 3′ premessage splice sites, respectively, gave more intense signals with the mutant RNAs than with RNA from the wild-type infection. This result

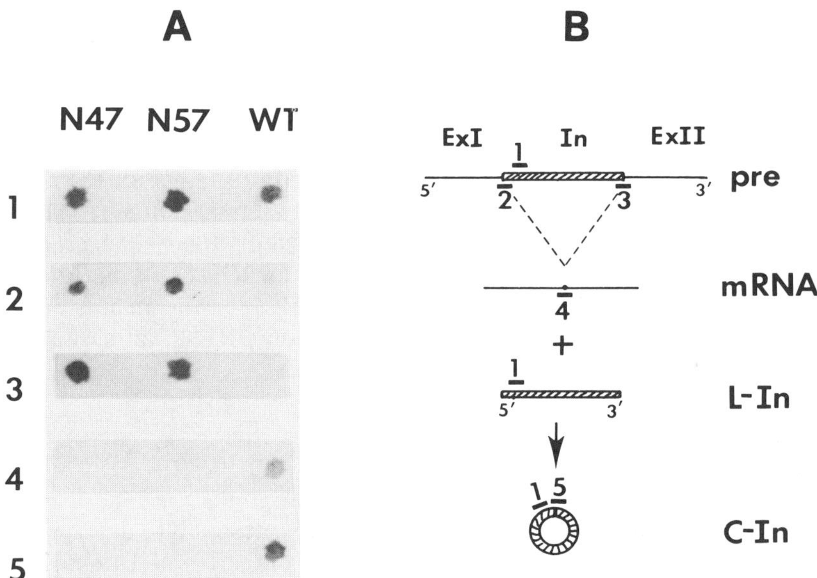

Fig. 5. Dot blot splicing assay. Deproteinized RNA (100 ng) from *E. coli* B infected with wild-type (WT) T4 or with the indicated T4 *td* mutants was spotted onto nitrocellulose. The filters were probed with the 5′ end-labeled oligodeoxynucleotides. The numbering in A corresponds to the hybridization probe used (see B) (Hall *et al.*, 1987). Abbreviations are: ExI, exon I; In, intron; ExII, exon II; pre, premessage; L-In, linear intron; and C-In, circular intron. (From Hall *et al.*, 1987. Copyright held by Cell Press.)

suggests a build-up of the precursor transcript in the mutants. In contrast is the behavior of the probes directed against mature splice products. The oligonucleotides complementary to the mRNA splice junction (probe 4) and the cyclization junction (probe 5) yielded strong signals with wild-type RNA, but showed no hybridization to the mutant RNAs. These results suggest that both *td*N57 and *td*N47 direct the synthesis of a precursor transcript that is defective in yielding the mature splice products.

This type of assay should have wide application in screening for splicing-defective mutants in other systems. It has particular utility in a prokaryotic system, where transcripts are short-lived, and in which an RNA stability mutant mapping to the intron might initially masquerade as splicing defective. It is therefore very important to establish early on in an analysis a measure of splicing efficiency, which is best reflected in the mRNA : pre-mRNA ratio.

Hybridization analysis with RNAs separated on agarose-formaldehyde gels corroborated the buildup of pre-mRNA and the dramatic reduction in both mature mRNA and excised 1.0 kb intron in the mutants relative to wild type (Hall *et al.,* 1987). Interestingly, in the case of the mutant *td*N47, the small amount of excised intron was truncated at its 3′ end, to a species of ~0.85 kb (Hall *et al.,* 1987).

Primer extension analysis was also used to map the mutant and wild-type transcripts. Using a 5′ intron-specific oligodeoxynucleotide (oligomer 1 in Fig. 5) in the presence of ddNTPs, reduction in cleavage at the 5′ splice site was demonstrated for both T4*td*N57 and T4*td*N47 (Hall *et al.,* 1987). Furthermore, with an exon II primer, a molecule corresponding in size to an intron–exon II intermediate was apparent for both mutants, but not for the wild type (Hall *et al.,* 1987). These results suggest that while the primary defect for both T4*td*N57 and T4*td*N47 is an impairment of 5′ splice site recognition, cleavage at the 3′ splice site is also defective for both mutants. Whereas the trace amount of free intron was of apparently normal size for T4*td*N57, the excised intron of T4*td*N47 appeared to be truncated suggesting a loss of specificity in addition to a reduction of cleavage activity for this mutant.

D. Relating Sequence Change to Splicing Defect

Interestingly, the C → U transition of *td*N57 in the 5′ intron domain occurs within the putative IGS of the intron (Chu *et al.,* 1986), disrupting a G-C pairing that is thought to stabilize the interaction with exon I sequences (Fig. 6). Such a pairing is supported by mutation *td*H6, a G → A change in the second-to-last residue of exon I, which results in a splicing-defective phenotype (data not shown). The mutated G in *td*H6 is

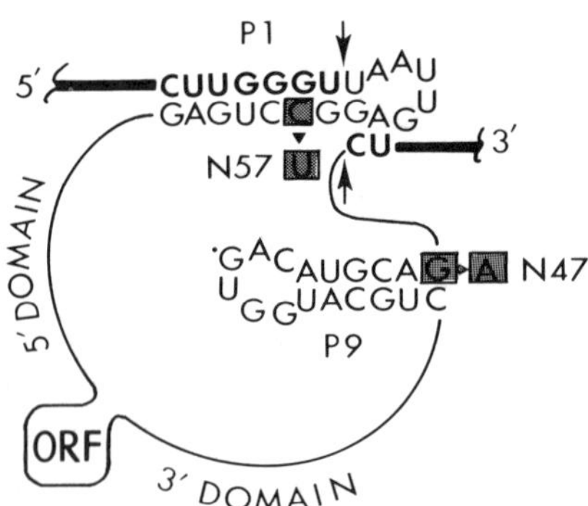

Fig. 6. Potential secondary structures disrupted by *td*N57 and *td*N47. The exons are shown in bold lettering and heavy lines and the 5′ and 3′ splice sites are demarcated by downward- and upward-directed arrows, respectively. The 5′ and 3′ intron domains, in which mutations cluster, are separated by the intron open reading frame (ORF) sequence, which is looped out (not to scale). The site of the N57 mutation, the C → U change at residue 10 of the intron (boxed), is shown within the internal guide sequence. This mutation is believed to disrupt the P1 pairing (Davies *et al.*, 1982) in the 5′ domain. The site of N47, the G → A change at residue 969 of the intron (boxed), is believed to destabilize the P9 pairing in the 3′ domain. The P9 pairing is identical for the *nrdB* intron with a single residue difference in the associated loop. The G indicated by a dot is an A in *nrdB* (Sjöberg *et al.*, 1986; M. Belfort, J. Pedersen-Lane, J. M. Gott, and D. A. Shub, unpublished). (From Hall *et al.*, 1987. Copyright held by Cell Press.)

precisely that residue that is thought to pair with the mutated C in *td*N57. These results support the role of the exon I–IGS pairing (P1 pairing of Davies *et al.*, 1982) in targeting the 5′ splice site. A second role for the mutated residue in T4*td*N57, at the 3′ splice site, is suggested by our data. This role is consistent with the proposal of Waring *et al.* (1986) that nucleotides making up the 3′ attack site are a subset of the IGS.

A possible secondary structure involving the *td*N47 mutation, 49 residues from the 3′ splice site, is also shown in Fig. 6. This is believed to correspond to the phylogenetically conserved P9 pairing of the core structure of Davies *et al.* (1982). Remarkably, 17 of the 18 residues associated with the P9 structure shown in Fig. 6 are identical in the *nrdB* intron, which is the second intron to be described in T4 (Gott *et al.*, 1986; Sjöberg *et al.*, 1986; Section III below). A disruption of the indicated G-C pairing would destabilize the stem (Fig. 6). This structural disturbance seems to alter intron interactions in such a way as to reduce cleavage at the 5′ splice site as well as to cause accumulation of a 3′ truncated intron, resulting perhaps from reduced specificity at the 3′ splice site. How this comes about is still a matter of speculation. A gross alteration of the core intron structure is one possibility, the disruption of tertiary interactions is another. RNA structure mapping and the isolation of second-site revertants will be instrumental in addressing such issues.

The facile methods described for isolating and mapping *td* mutations under physiological conditions, as well as the assay developed to screen splicing-defective mutants, places us in a position to analyze exhaustively libraries of *td* mutations. This saturation mutagenesis approach should facilitate a division of domains for splicing into functional subdomains and should eventually yield a complete genetically authenticated secondary structure model for the *td* ribozyme. Furthermore, clear-cut splicing phenotypes of the mutants will be instrumental in assigning specific roles to sequences and structures involved in apposition of reactive residues that bring about RNA catalysis.

IV. MULTIPLE SELF-SPLICING INTRONS IN T4

The discovery of the intron in the *td* gene of T4 raised the question of the generality of introns and RNA splicing in prokaryotes. Although introns have been reported to exist in archaebacterial tRNAs (Kaine *et al.*, 1983; Daniels *et al.*, 1985) and rRNA (Kjems and Garrett, 1985), the phylogenetic status of these prokaryotes is unusual. Archaebacteria have more characteristics in common with eukaryotes than do eubacteria and may form a third primary kingdom (Fox *et al.*, 1980). It was therefore of interest to determine whether RNA splicing is a common feature of gene expression in eubacteria, from the standpoint of the evolutionary occurrence of introns and to investigate novel modes of genetic regulation in prokaryotes.

The G-initiated splicing reaction has been used successfully to detect self-splicing RNAs in total mitochondrial RNA of *Neurospora* (Garriga and Lambowitz, 1984), *Saccharomyces* (Van der Horst and Tabak, 1985), and *Podospora* (Michel and Cummings, 1985), and we reasoned that it could also be used to detect and map previously unknown introns in T4. Indeed, when deproteinized RNA from T4-infected cells was incubated with [α-^{32}P]GTP, multiple end-labeled species were evident (Fig. 7). These results indicate that T4 genes other than *td* may be expressed by a G-initiated RNA-processing mechanism, and that the T4 genome contains multiple introns. Interesting questions are therefore raised regarding the genetic loci involved and the possible regulatory roles for RNA splicing in T4.

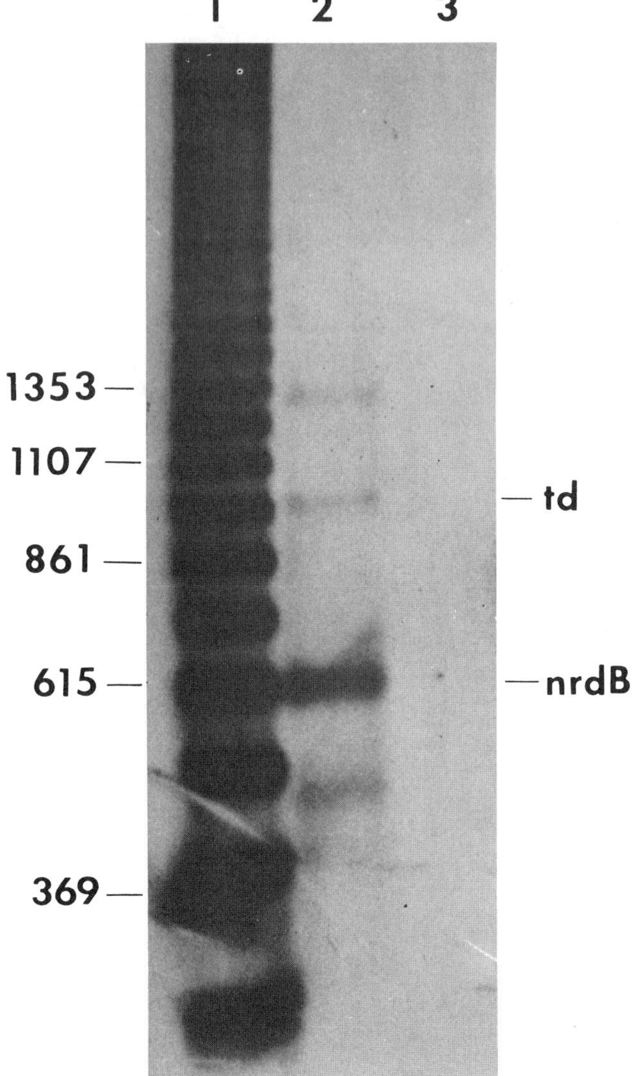

Fig. 7. Labeling of T4 RNA with [α-^{32}P]GTP. RNA extracted from *E. coli* 7 min after T4 infection was labeled with [α-^{32}P]GTP under self-splicing conditions for 1 hr at 38°C (Ehrenman *et al.*, 1986). RNA was separated directly on an agarose/formaldehyde gel (lane 2) or after treatment with alkaline phosphatase (lane 3). Size markers in lane 1 are a 123-bp DNA ladder. The *td* intron was identified by comigration with the cloned gene product, and identification of the *nrdB* intron is discussed in the text.

A. Identification and Cloning of Loci Encoding Group I Introns

The T4 RNA species labeled with [α-^{32}P]GTP under self-splicing conditions had several properties in common with the [α-^{32}P]GTP-labeled *td* intron (Gott *et al.*, 1986): (1) The label was sensitive to alkaline phosphatase treatment (Fig. 7, lane 3); (2) the temperature-dependence of the labeling reaction was similar, with most incorporation occurring at temperatures between 38°C and 48°C; and (3) the labeling of each band was most intense in RNA extracted 7 min postinfection. These results suggest that there is a family of T4 genes expressed in concert via a group I splicing reaction.

The strategy for mapping of these loci is shown in Fig. 8. [α-^{32}P]GTP-labeled T4 RNA was used as hybrization probe against restricted T4 DNA

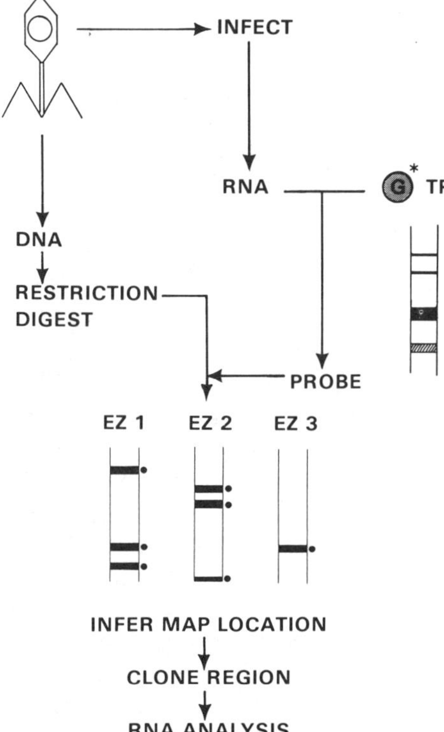

Fig. 8. Identification scheme for multiple introns in phage T4. The use of the mixed [α-^{32}P]GTP-labeled T4 RNA probe to map and guide the cloning of group I intron-containing genes in T4 is described in the text. EZ, Restriction endonuclease.

fragments that had been electrophoretically separated. Choice of enzymes for the analysis was guided by the T4 restriction map (Rueger and Kutter, 1984). By combining the mapping data for a number of enzymes (*Bgl*II, *Hae*II, *Xba*I, *Eco*RV, *Sph*I, and *Nde*I) two regions of the T4 genome distinct from the *td* locus were identified as hybridizing to the labeled RNA population (Gott *et al.*, 1986). These loci have been mapped to the ribonucleoside diphosphate (RDP) reductase subunit B (*nrdB*) gene and to a distant locus on the T4 map in the general vicinity of, but not within, *nrdC* (Fig. 9). Subsequent cloning of both loci and RNA analysis provided evidence for group I splicing.

We shall confine ourselves here to a description of the *nrdB* intron. The intensely labeling RNA at 0.6 kb (Fig. 7) was found to hybridize to DNA around map position 137 kb of the T4 genome, within the *nrdB* gene (Gott *et al.*, 1986). Whereas we inferred an intron in *nrdB* through the use of a mixed [^{32}P]GTP-labeled T4 RNA probe that traced T4 genes encoding group I splice products, Sjöberg *et al.* (1986) arrived at a similar conclusion from an entirely different perspective. By sequence analysis of T4 *nrdB* and comparison of the deduced amino acid sequence with that of the B subunit of the *E. coli* RDP reductase, an interruption in coding sequence of ~0.6 kb was inferred. Various DNA fragments spanning this region were cloned (see Fig. 9) and used to characterize the RNA products.

B. Self-Splicing of *nrd*B-Encoded RNA

Processing of RNA transcribed both *in vivo* and *in vitro* from cloned *nrdB* sequences yields products characteristic of a group I splicing mechanism. When RNA transcribed *in vivo* from an *nrdB* clone with an intact intron was incubated *in vitro* with [α-^{32}P]GTP under self-splicing conditions, a 0.6-kb RNA was labeled (Gott *et al.*, 1986).

In vitro transcription products, separated on a 4% acrylamide/8 *M* urea gel are shown in Fig. 10. When templates pJBK1, pJHK1 and pJSK6 (Fig. 9) were linearized within the intron, a single RNA species of the size expected for the runoff transcript was seen for each template (Fig. 10A, lanes 1–3). However, templates cleaved downstream of the intron yielded four RNA species, as expected for processing-proficient transcripts. In addition to runoff transcripts of a size predicted based on the template, an RNA species approximately 600 nucleotides shorter than the runoff transcript was seen for each template (Fig. 10A, lanes 4–6). Similarly, *in vitro* transcription of pJSE17 (Fig. 9), which is an extended clone including exon II sequences, yielded multiple products when cleaved at various positions in the downstream exon. In each case one of these was 0.6 kb

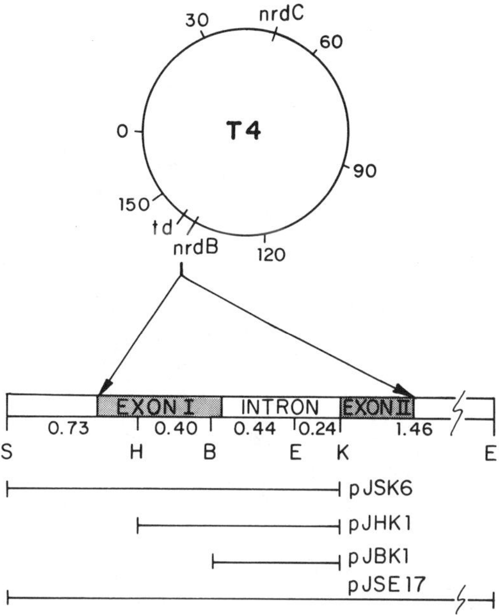

Fig. 9. Map of intron-containing regions of T4. The circular T4 map (166 kb) shows the relative locations of the *td* and *nrdB* intron-containing genes and the *nrdC* gene, which is in the neighborhood of another intron-containing locus. The physical map of the *nrdB* gene is shown below (Sjöberg *et al.*, 1986; Gott *et al.*, 1986) relative to the constructs used in this work. S, *Sph*I; H, *Hin*dIII; B, *Bgl*II; E, *Eco*RI; and K, *Kpn*I. The 3′ splice site lies within the *Kpn*I site (data not shown). The numbering below the linear map and around the circular map is in kilobases.

shorter than the runoff transcript. We have tentatively identified these latter species as the exon ligation products. The remaining two RNA species were identical in mobility for each of the three templates utilized in Fig. 10A, lanes 4–6. One of these bands was the same size (0.6 kb) as the major band labeled when T4 RNA (Fig. 7) or RNA from cells containing pJSK6 was incubated under self-splicing conditions in the presence of [α-^{32}P]GTP (Gott *et al.*, 1986). These data suggest that this 0.6-kb band is linear autoexcised intron. That the second uniform species, of lower mobility, is likely to be the circular form of the intron is supported by the anomalous mobility of this RNA, which has migrated off the diagonal in the two-dimensional gel shown in Fig. 10B. (Circular intron was seen only with *Kpn*I-cleaved templates, for reasons discussed by D. A. Shub, J. M. Gott, and M. Belfort, manuscript in preparation.) The existence of these RNA products strongly suggests that indeed the *nrdB* gene of T4 is interrupted by a group I intron.

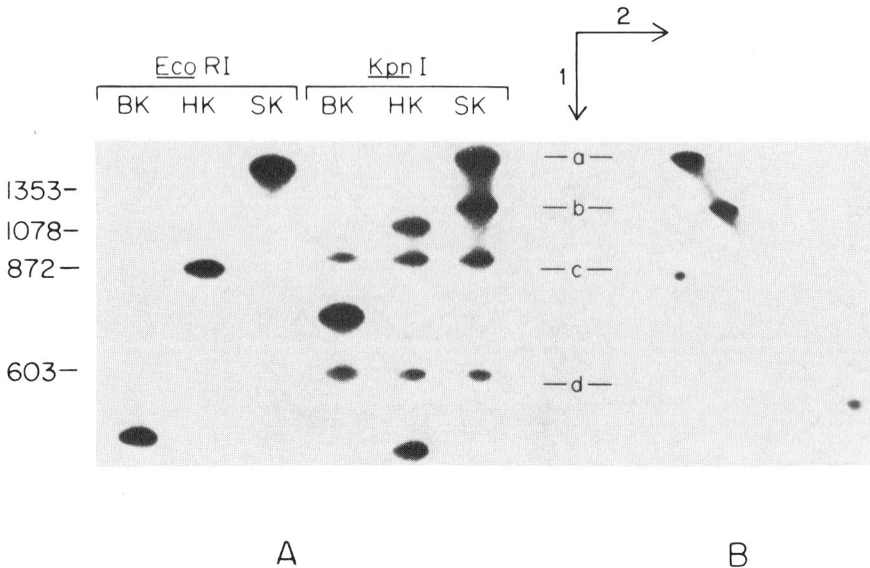

Fig. 10. *In vitro nrdB* transcripts. pJBK1 (BK), pJHK1 (HK), and pJSK6 (SK) (Fig. 9) linearized with *Eco*RI or *Kpn*I were separated on a 4% acrylamide/8 *M* urea gel (A). Size markers on the left are from *Hae*III-digested ϕX174 DNA. In B, the RNA transcribed from *Kpn*I-digested pJSK6 was separated in the second dimension on a 6% acrylamide/8 *M* urea gel, with bands labeled between A and B corresponding to runoff transcript (a), exon product (b), circular intron (c), and linear intron (d). (From Gott *et al.*, 1986. Copyright held by Cell Press.)

C. A Regulatory Role for Introns in T4?

Splicing provides the means for expression of multiple gene products by the same gene. For the *td* gene, there is an exon I product in addition to the synthase encoded by the ligated exons (Chu *et al.*, 1985; Belfort *et al.*, 1986). Our preliminary splice junction data (M. Belfort, J. Pedersen-Lane, J. M. Gott, and D. A. Shub, unpublished) in conjunction with the *nrdB* sequence data (Sjöberg *et al.*, 1986) predict that *nrdB* is also capable of encoding two products. Furthermore, splicing may result in the differential expression of such products from the same gene, as has been demonstrated for *td* (Belfort *et al.*, 1986). These interrupted genes therefore have the potential for splicing-dependent regulation at the single gene level.

One might also consider a global regulatory potential for group I introns in T4. Might control of the group I splicing reaction regulate this entire family of interrupted genes, which appear to be expressed in concert after T4 infection? Since guanosine is an essential cofactor in the group I reaction, it or one of its 5′ phosphates (Cech *et al.*, 1981; Bass and Cech, 1984) may be a positive regulator of the reaction. A consideration of the functions of the two known split genes of T4 suggests possible regulatory circuits. Both thymidylate synthase and RDP reductase are pivotal enzymes in DNA precursor metabolism, providing the replication machinery with dNTPs. Might an increase in the rNTP pools (including GTP, which is in equilibrium with guanosine and its other ribonucleotide derivatives) provide a positive splicing signal, which would result in RDP reductase and thymidylate synthase production to deliver dNTPs for timely DNA synthesis? Alternatively, might other guanosine metabolites, like dGTP, a product of this pathway that is known to inhibit group I splicing (Bass and Cech, 1986), or an alarmone such as guanosine tetraphosphate, act as negative regulators? Might these signal dNTP excess or physiological distress, respectively, and halt futile channeling of rNTPs into the dNTP pool? Definitive answers to these questions must await the identification of the third splicing locus and studies that define the effects of the concentrations of various guanosine nucleotides on the splicing reaction.

V. CONCLUSIONS

1. We have demonstrated that the products of the prokaryotic *td* splicing reaction conform to those of the group I introns of eukaryotes. These include a linear intron with a noncoded G at its 5′ end, and a circular form of the IVS that has lost two 5′ intron residues and that is closed by a 3′–5′ phosphodiester linkage. A conserved U residue 5′ to the four characterized cyclization sites of group I introns (including *td*) suggests that this residue may be important in cyclization site selection.

2. The utility of prokaryotic genetics in defining functional domains of the intron was demonstrated using nondirected mutagenesis of the *td* gene in the intact phage. Thus, clustering of mutations toward the intron boundaries (~200 nucleotides at each end of the 1017-residue IVS) implicated these regions as its functional splicing elements. Two selected splicing-defective mutations were shown to fall within phylogenetically conserved pairings. The first is thought to disrupt the exon I–IGS folding in the 5′ domain, while the second is predicted to impair the P9 helix in the 3′ domain. This saturation mutagenesis approach promises to provide a useful adjunct to structure–function analyses of group I introns.

3. Exploitation of the self-splicing G-addition reaction has enabled us to identify and investigate two other group I introns in T4. One maps to the *td*-linked *nrdB* gene, while the other is near the distant *nrdC* gene. By cloning the *nrdB* gene we have shown that its RNA product is matured via

the group I splicing pathway. We speculated on regulatory roles for introns in T4, both at the single-gene level (allowing multiple products to be differentially expressed from the same gene) and as providing a guanosine sensor that may co-regulate all intron-containing genes. This scheme would control channeling of precursors between the rNTP and dNTP pools.

It has been argued that RNA was the primordial nucleic acid and that the RNA-catalyzed reaction represents the ancestral splicing process (reviewed in Cech, 1985; Sharp, 1985; Darnell and Doolittle, 1986; Gilbert, 1986). Introns do not, however, appear to be a common feature of prokaryotes, and it has been proposed that prokaryotes have lost their intervening sequences under pressure to streamline their genomes (Doolittle, 1978). Certainly, the remarkable conservation of both structural features and splicing mechanism between the T4 introns and the eukaryotic group I introns reflects their common ancestry. While the existence of introns in prokaryotes is consistent with the early evolutionary occurrence of self-splicing (Darnell and Doolittle, 1986), one must bear in mind that some group I introns are mobile elements (Dujon *et al.,* 1986). This leads to the possibility that their occurrence in phage T4 might be the result of a relatively recent genetic exchange (Michel and Dujon, 1986). In either event, the selective pressure for the maintenance of introns in specific T4 genes results presumably from the replicative advantage they confer to the phage. This may be a consequence of the regulatory potential that introns afford in terms of gene expression and/or physiological control.

ACKNOWLEDGMENTS

We are thankful to the colleagues within our laboratories and outside who helped by sharing their insights and their expertise. This work was supported by grants from the National Institutes of Health (GM33314 to M. B. and GM16306, GM24455, GM36714 to D. H. H.) and from the National Science Foundation (DMB-8502961 and DMB-8505527 to M. B. and DMB-8609066 to D. A. S.).

REFERENCES

Anziano, P. Q., Hanson, D. K., Mahler, H. R., and Perlman, P. S. (1982). *Cell (Cambridge, Mass.)* **30,** 925–932.

Bass, B. L., and Cech, T. R. (1984). *Nature (London)* **308,** 820–826.

Bass, B. L., and Cech, T. R. (1986). *Biochemistry* **25,** 4473–4477.

Been, M. D., and Cech, T. R. (1986). *Cell (Cambridge, Mass.)* **47,** 207–216.

Belfort, M., Moelleken, A., Maley, G. F., and Maley, F. (1983). *J. Biol. Chem.* **258,** 2045–2051.

Belfort, M., Pedersen-Lane, J., West, D., Ehrenman, K., Maley, G., Chu, F., and Maley, F. (1985). *Cell (Cambridge, Mass.)* **41,** 375–382.

Belfort, M., Pedersen-Lane, J., Ehrenman, K., Chu, F. K., Maley G. F., Maley, F., McPheeters, D. S., and Gold, L. (1986). *Gene* **41,** 93–102.

Belfort, M., Chandry, P. S., and Pedersen-Lane, J. (1987). *Cold Spring Harbor Symp. Quant. Biol.* **52,** in press.

Broker, T. R., and Doermann, A. H. (1975). *Annu. Rev. Genet.* **9,** 213–244.

Burke, J. M., Irvine, K. D., Kaneko, K. J., Kerker, B. J., Oettgen, A. B., Tierney, W. M., Williamson, C. L., Zaug, A. J., and Cech, T. R. (1986). *Cell (Cambridge, Mass.)* **45,** 167–176.

Cech, T. R. (1985). *Int. Rev. Cytol.* **93,** 3–22.

Cech, T. R. (1986). *Cell (Cambridge, Mass.)* **44,** 207–210.

Cech, T. R., and Bass, B. L. (1986). *Annu. Rev. Biochem.* **55,** 599–629.

Cech, T. R., Zaug, A. J., and Grabowski, P. J. (1981). *Cell (Cambridge, Mass.)* **27,** 487–496.

Chu, F. K., Maley, G. F., Maley, F., and Belfort, M. (1984). *Proc. Natl. Acad. Sci. U.S.A.* **81,** 3149–3153.

Chu, F. K., Maley, G. F., Belfort, M., and Maley, F. (1985). *J. Biol. Chem.* **260,** 10680–10688.

Chu, F. K., Maley, G. F., West, D. K., Belfort, M., and Maley, F. (1986). *Cell (Cambridge, Mass.)* **45,** 157–166.

Daniels, C. J., Gupta, R., and Doolittle, W. F. (1985). *J. Biol. Chem.* **260,** 3132–3134.

Darnell, L. E., and Doolittle, W. F. (1986). *Proc. Natl. Acad. Sci. U.S.A.* **83,** 1271–1275.

Davies, R. W., Waring, R. B., Ray, J. A., Brown, T. A., and Scazzocchio, C. (1982). *Nature (London)* **300,** 719–724.

De La Salle, H., Jacq, C., and Slonimski, P. P. (1982). *Cell (Cambridge, Mass.)* **28,** 721–732.

Doolittle, W. F. (1978). *Nature (London)* **272,** 581–582.

Dujon, B., Colleaux, L., Jacquier, A., Michel, F., and Monteilhet, C. (1986). *In* "Extrachromosomal Elements in Lower Eukaryotes" (R. B. Wickner, A. Hinnebusch, A. M. Lambowitz, I. C. Gunsalus, and A. Hollaender, eds.), pp. 5–27. Plenum, New York.

Ehrenman, K., Pedersen-Lane, J., West, D., Herman, R., Maley, F., and Belfort, M. (1986). *Proc. Natl. Acad. Sci. U.S.A.* **83,** 5875–5879.

Fox, G. E., Stackebrandt, E., Hespell, R. B., Gibson, J., Maniloff, J., Dyer, T. A., Wolfe, R. B., Balche, W. E., Tanner, R. S., Nagrum, L. J., Zalden, L. B., Blakemore, R., Gupta, R., Lewis, B. J., Stahl, D. A., Luehrsen, K. R., Chen, K. N., and Woese, C. R. (1980). *Science* **209,** 457–463.

Garriga, G., and Lambowitz, A. M. (1984). *Cell (Cambridge, Mass.)* **39,** 631–641.

Garriga, G., Lambowitz, A. M., Inoue, T., and Cech, T. R. (1986). *Nature (London)* **322,** 86–89.

Gilbert, W. (1986). *Nature (London)* **319,** 618.

Gott, J. M., Shub, D. A., and Belfort, M. (1986). *Cell (Cambridge, Mass.)* **47,** 81–87.

Hall, D. H. (1967). *Proc. Natl. Acad. Sci. U.S.A.* **58,** 584–591.

Hall, D. H., Povinelli, C. M., Ehrenman, K., Pedersen-Lane, J., Chu, F., and Belfort, M. (1987). *Cell (Cambridge, Mass.)* **48,** 63–71.

Holl, J., Rodel, G., and Schweyen, R. J. (1985). *EMBO J.* **4,** 2081–2085.

Kaine, B. P., Gupta, R., and Woese, C. R. (1983). *Proc. Natl. Acad. Sci. U.S.A.* **80,** 3309–3312.

Kjems, J., and Garrett, R. A. (1985). *Nature (London)* **318,** 675–677.

Mattson, T., Van Houwe, G., Bolle, A., Selzer, G., and Epstein, R. (1977). *Mol. Gen. Genet.* **154,** 319–326.

Michel, F., and Cummings, D. J. (1985). *Curr. Genet.* **10,** 69–79.

Michel, F., and Dujon, B. (1986). *Cell (Cambridge, Mass.)* **46,** 323.

Michel, F., Jacquier, A., and Dujon, B. (1982). *Biochimie* **64,** 867–881.

Peebles, C. L., Perlman, P. S., Mecklenburg, K. L., Petrillo, M. L., Tabor, J. H., Jarrell, K. A., and Cheng, H.-L.(1986). *Cell* (*Cambridge, Mass.*) **44,** 213–223.

Rueger, W., and Kutter, E. (1984). *In* "Genetic Maps" (S. J. O'Brian, ed.), Vol. 3, pp. 27–34. Cold Spring Harbor Lab., Cold Spring Harbor, New York.

Ruskin, B., and Green, M. R. (1985). *Science* **229,** 135–140.

Ruskin, B., Krainer, A. R., Maniatis, T., and Green, M. R. (1984). *Cell* (*Cambridge, Mass.*) **38,** 317–331.

Sharp, P. A. (1985). *Cell* (*Cambridge, Mass.*) **42,** 397–400.

Simon, E. H., and Tessman, I. (1963). *Proc. Natl. Acad. Sci. U.S.A.* **50,** 526–532.

Sjöberg, B.-M., Hahne, S., Mathews, C. Z., Mathews, C. K., Rand K. N., and Gait, M. J. (1986). *EMBO J.* **5,** 2031–2036.

Sullivan, F. X., and Cech, T. R. (1985). *Cell* (*Cambridge, Mass.*) **42,** 639–648.

Tabak, H. F., Van der Horst, G., Osinga, K. A., and Arnberg, A. C. (1984). *Cell* (*Cambridge, Mass.*) **39,** 623–629.

Van der Horst, G., and Tabak, H. F. (1985). *Cell* (*Cambridge, Mass.*) **40,** 759–766.

Van der Veen, R., Arnberg, A. C., Van der Horst, G., Bonen, L., Tabak, H. F., and Grivell, L. A. (1986). *Cell* (*Cambridge, Mass.*) **44,** 225–234.

Waring, R. B., and Davies, R. W. (1984). *Gene* **28,** 277–291.

Waring, R. B., Towner, P., Minter, S. J., and Davies, R. W. (1986). *Nature* (*London*) **231,** 133–139.

Weiss-Brummer, B., Holl, J. Schweyen, R. J., Rodel, G., and Kaudewitz, F. (1983). *Cell* (*Cambridge, Mass.*) **33,** 195–202.

West, D. K., Belfort, M., Maley, G. F., and Maley, F. (1986). *J. Biol. Chem.* **261,** 13446–13450.

Zaug, A. J., Grabowski, P., and Cech, T. R. (1983). *Nature* (*London*) **301,** 578–583.

II

RNA Splicing

5

The Mammalian Pre-Messenger RNA Splicing Apparatus: A Ribosome in Pieces?

JOAN A. STEITZ

Howard Hughes Medical Institute
Department of Molecular Biophysics and Biochemistry
Yale University School of Medicine
New Haven, Connecticut 06510

I. INTRODUCTION

Only within the past year has evidence finally been amassed to implicate all of the most abundant small ribonucleoproteins (RNPs) of mammalian cell nuclei in the splicing of pre-messenger RNA (pre-mRNA). Small RNPs can be simply defined as one or more proteins tightly complexed with a small RNA molecule (having a chain length up to about 300 nucleotides). They fall into two major categories: those that are specifically localized in the nucleus (small nuclear RNPs, abbreviated snRNPs) and those that reside in the cytoplasm (small cytoplasmic RNPs, abbreviated scRNPs). Both their high abundance and the conservation of their protein and RNA components across higher eukaryotic species suggest that small RNPs play central roles in cellular metabolism. All whose functions are currently known participate in some aspect of gene expression, very likely using their RNA components to make specific contacts with other nucleic acid molecules.

As potent probes for deciphering the structure and function of small RNPs, we have been using autoantibodies present in the sera of some patients suffering from rheumatic diseases such as systemic lupus erythematosis (SLE). In most cases, patient autoantibodies react with the pro-

Molecular Biology of RNA
New Perspectives

tein component(s) of a small RNP and are therefore useful in determining which particles are related to others by virtue of common protein constituents. Several classes of small RNPs have been distinguished in this manner.

The small RNPs that are involved in splicing belong to the Sm class of snRNPs. There are four abundant particles (present in about 10^6 copies per mammalian cell nucleus) and an unknown number of less abundant particles in this class. All appear to share several small proteins that possess determinants recognized by anti-Sm antibodies; also present on each particle are one or more unique proteins that are likewise sometimes recognized by patient autoantibodies. The RNAs contained in the four highly abundant Sm snRNP particles are U1 (165 nucleotides), U2 (185 nt), U5 (116 nt) and lastly a particle containing two RNAs, U4 (145 nt) and U6 (106 nt). These are stable nuclear RNA molecules first described in the late 1960s by the laboratories of Harris Busch (Muramatsu *et al.,* 1966) and Sheldon Penman (Weinberg and Penman, 1968). All except U6 contain an unusual $m_3^{2,2,7}G$ cap structure at their 5′ ends (Busch *et al.,* 1982).

II. THE DISCOVERY OF snRNPs

Our involvement with snRNPs began serendipitously in early 1978. The year prior to that I had attempted, during a sabbatical leave, to raise antibodies directed against the protein components of hnRNP (heterogeneous nuclear ribonucleoprotein). At that time, hnRNP were the best characterized nonribosomal RNA–protein complexes in cells. It was thought that the RNA component of these structures was the nuclear precursor to mRNA, although the long length of hnRNA molecules was not understood until after the discovery of split genes (late 1977). It had been shown that a relatively small number of nuclear proteins packaged the rapidly turning-over hnRNA molecules into a beads-on-a-string structure. Mild ribonuclease treatment resulted in the conversion of these structures to surprisingly homogeneous monoparticles that sedimented at 30–40 S and contained about 80% protein plus an approximately 600-nt-long fragment of hnRNA. Because hnRNP proteins (like snRNP proteins) are highly conserved across species, they are poorly immunogenic. Hence, my attempts to raise antibodies in 1977 failed completely. Subsequently, these valuable tools for investigating hnRNP structure and function have been obtained (Dreyfus *et al.,* 1984).

After my return to Yale, Michael Lerner, an MD/PhD student joined my laboratory. One day while glancing at a newly arrived issue of *Nature* we noticed an article describing "anti-RNP" antibodies present in patient sera (Alarcon-Segovia *et al.*, 1978). Although the article focused on antibody uptake into cells, it and several other papers that were referenced informed us that "RNP" was an abundant and highly conserved nuclear antigen of unknown molecular identity, apparently consisting of both RNA and protein. It seemed possible that these spontaneously occurring autoantibodies might be the very ones that I had tried (and failed) to produce a year earlier. Michael, fresh from his medical school courses, knew both that anti-RNP was a fairly frequently occurring autoantibody and who the local rheumatologists were. Within hours, he had obtained patient sera containing anti-RNP (and the related anti-Sm) antibodies and had begun to work with them. He soon learned that neither autoantibody reacted with the 30 S hnRNP, but rather with a smaller antigen sedimenting at 10–12 S. However, both because of Sm and RNP's abundance and because of the recent evidence from Sidney Altman's laboratory that an *E. coli* tRNA processing enzyme, RNase P, consisted of RNA as well as protein (Stark *et al.*, 1978), we decided that study of these nuclear antigens was a worthwhile endeavor.

The RNP antigen turned out to be the U1-containing snRNP. But it took more than a year of frustrating experiments to determine this. Purification of the RNP antigen according to protocols described in the medical literature began with tissue, such as rat liver. As Michael Lerner purified, ribonuclease chewed away and eventually destroyed the antigen, which he was assaying by the relatively insensitive Ouchterlony immunodiffusion technique. The turning point came after Joan Brugge described in a seminar the use of a new reagent called Pansorbin. Pansorbin is simply a fixed preparation of *Staphyloccus aureus* bacteria, but the Protein A contained in their cell walls specifically binds the constant region of most mammalian IgG molecules.

The use of ^{32}P-labeled cultured cells (rather than liver) as the antigen source and of Pansorbin to select out antigen–antibody complexes immediately revealed that the Sm and RNP antigens are associated with discrete small RNA molecules. Both the rapidity with which the assay could be done and the use of cultured cells (instead of tissues, which are bathed in RNase-containing serum) circumvented destruction of the RNA during analysis. Moreover, the sensitivity for detecting an RNA molecule bound to an antigenic protein was vastly improved. Using this simple technique, we screened hundreds of serum samples from patients with various rheumatic diseases and identified over a dozen different small RNA–protein complexes as targets of autoimmunity. These are listed in Table I.

We found that anti-Sm antibodies specifically recognize snRNPs containing two previously identified RNA species (U1 and U2) and three new RNAs, which we named U4, U5, and U6 (Lerner and Steitz, 1979). Anti-

TABLE I

Small RNPs Recognized by Autoantibodies

RNP class[a]	Antibody	RNA components	Antigenic proteins[b,c] (kDa)
Sm snRNPs	Anti-Sm	U4	26, 18, 13
		U6	
		U5	
	Anti-(U2)RNP	U2	33, 26
	Anti-(U1)RNP	U1	68, 34, 22
Ro scRNPs	Anti-Ro	Mouse Y1-Y2	60
		Human Y1-Y5	
La snRNPs	Anti-La	Many cellular RNAs:	50
		Rat 4.5 S_I	
		Mouse or hamster 4.5 S	
		tRNA precursors	
		pre-5 S rRNA	
		Pre-7SL	
		Viral RNAs:	
		VAI	
		VAII	
		EBER 1	
		EBER 2	
Jo scRNP	Anti-Jo	$tRNA^{His}$	65
scRNPs	Anti-LL, anti-SU	Other tRNAs	?
snoRNP[d]	Anti-5 S RNP	5 S rRNA	35
snoRNP	Anti-Th	7-2	?
snoRNP	Anti-(U3)RNP	U3	36
scRNP	Anti-SRP	7SL	54

[a] Many of these specificities have been discovered and studied in the laboratory of our colleague, Dr. John Hardin, Section of Rheumatology, Department of Internal Medicine, Yale University School of Medicine.

[b] Usually proteins other than the antigenic one(s) are also present in the RNP.

[c] Values are taken from the literature or our unpublished work.

[d] snoRNP stands for *s*mall *n*ucleolar *RNP*.

RNP sera precipitated only U1-containing complexes and was therefore renamed anti-(U1)RNP. U3 snRNA was already at that time known to be nucleolar whereas anti-Sm and anti-RNP antibodies gave nucleoplasmic fluorescence; thus it was not so surprising that U3 RNA might not be included in the Sm group of particles. Experiments using ^{35}S- instead of ^{32}P-labeled cell extracts showed that seven small nuclear proteins (in the size range of 11–34 kDa) were immunoprecipitated by both anti-Sm and anti-(U1)RNP. These proteins are very tightly bound to the RNAs and could be shown to be distinct from histones and hnRNP proteins. Rather, they comprise a discrete set of abundant nuclear proteins.

III. THE snRNPs-AND-SPLICING HYPOTHESIS

Even before we obtained definitive evidence that anti-Sm and anti-RNP antibodies reacted with the U1 snRNP, we had realized that the sequence adjacent to the m^3G cap of U1 RNA could theoretically base pair with sequences at the ends of introns in eukaryotic pre-messenger RNAs. By 1979, enough gene sequences had been analyzed so that short consensus sequences had been developed for 5′ and 3′ splice sites. The striking complementarity between these and U1 prompted both Rogers and Wall (1980) and our group (Lerner *et al.,* 1980) to propose that U1 RNA might serve to align the two ends of an intron during the splicing process. Initially, this suggestion was supported only by circumstantial evidence: (1) the extreme conservation both of splice site sequences and of U1 RNA and its associated antigenic proteins across a wide range of higher eukaryotic species, and (2) the observed association of U1 (and other abundant Sm) snRNPs with hnRNP (containing the pre-mRNA) in nuclear extracts.

Direct evidence for the involvement of U1 snRNPs in splicing emerged only slowly over the next several years in parallel with the development of *in vitro* splicing systems. Initially, we received a number of requests for antibodies from people who tried to inject them into cells to inhibit the expression (splicing) of one or another transcript. Only one such collaboration was fruitful. Jane Flint and her associates at Princeton had shown that nuclei isolated from adenovirus-infected cells could carry out the splicing of transcripts from the adenoviral early regions 1 and 2. Addition of anti-(U1)RNP or anti-Sm sera (but not normal serum or antibodies directed against other classes of small RNPs) to the nuclei *in vitro* inhibited the production of spliced transcripts (Yang *et al.,* 1981). This was the first experimental evidence that snRNPs participate in RNA splicing.

IV. snRNP COMPONENTS AND STRUCTURE

Meanwhile, two postdoctoral fellows, Monique Hinterberger and Ingvar Pettersson, worked to develop a biochemical purification procedure for the Sm snRNPs starting with HeLa cell nuclei. We reasoned that only in this way could we ascertain whether different snRNPs truly have different protein compositions. Moreover, such fractions might provide snRNPs for functional studies. Their procedure (Hinterberger *et al.,* 1983) yielded fairly clean U1 snRNPs and a mixture of U2-, U4-, U5-, and U6-containing particles. Analyses of these fractions combined with the results of immunoblots (Pettersson *et al.*, 1984) established that U1 snRNPs have three unique proteins (68, 34, 22 kDa) that possess determinants recognized by anti-(U1)RNP sera. Sm determinants are carried by the two or three largest of the smaller proteins (26, 18, 13, 12, 11 kDa) that seem to reside on all Sm snRNPs. A monoclonal anti-Sm antibody (Lerner *et al.*, 1981) reacted with both the 26-kDa and 18-kDa proteins, indicating that these two snRNP polypeptides share epitopes. Fractionation studies also revealed that Sm snRNPs are very stable RNA–protein complexes; they defy denaturation conditions ordinarily used to take ribosomes apart.

Tsuneyo Mimori, an MD from Tokyo, joined our lab in 1983, bringing a collection of some 300 Japanese patient sera. Upon screening these, he identified a rare specificity, anti-(U2)RNP. This antibody (Mimori *et al.*, 1984) allowed us to conclude that the U2 snRNP (like the U1 snRNP) possesses unique proteins (here 33 and 26 kDa) as well as the common smaller proteins.

Following up various hints, Carl Hashimoto (a graduate student) was able to show, as did Bringmann *et al.* (1984), that U4 and U6 RNAs reside in the same snRNP particle (Hashimoto and Steitz, 1984). This explained why antibodies directed against the distinctive m^3G cap precipitated U6 RNA from cell extracts or snRNP preparations but not from a mixture of naked RNAs. The U4/U6 particle is therefore unique in containing two different snRNAs compared to the U1, U2, and U5 single-RNA particles.

V. IS EUKARYOTIC RNase P AN Sm snRNP?

During the development of a biochemical fractionation procedure for Sm snRNPs, we decided also to assay fractions for RNase P activity. This was done in collaboration with Sidney Altman's lab, which had demonstrated an activity in mammalian cell extracts that would cleave *Escherichia coli* tRNA precursors to generate the same precise 5′ ends that are made by *E. coli* RNase P (Koski *et al.*, 1976). Indeed, activity could be detected and during purification it seemed to comigrate through five or six steps precisely with the U2 snRNP! These observations caused a great deal of excitement; perhaps the U2 particle was eukaryotic RNase P. The critical experiment was then performed. Anti-Sm antibody bound (via protein A) to Sepharose beads was found to remove over 95% of the Sm snRNPs (as assessed by RNA profile) from a highly purified fraction. Yet, over 95% of the RNase P activity remained unbound in the supernatant. Thus, despite the tantalizing cofraction results, we concluded that RNase P is not an Sm snRNP.

VI. U1 snRNPs BIND 5′ SPLICE SITES

Purified snRNP fractions also provided material with which to attempt functional studies of snRNPs. Thus, even before the emergence of extracts active in pre-mRNA splicing, Steve Mount (a graduate student) tried asking whether purified U1 snRNPs might bind specifically to splice sites in a model pre-mRNA substrate. The substrate was a labeled RNA transcribed from a phage T7 promoter cloned upstream from the first intron and flanking exon sequences of the mouse β-globin gene. This long-shot experiment yielded surprisingly pleasing results (Mount *et al.*, 1983). The U1 snRNP bound and protected a 15- to 17-nucleotide region spanning the 5′ splice site. No 3′ splice site sequences were observed among protected fragments, consistent with the earlier realization that the region of U1 RNA originally proposed to pair with the 3′ splice site was not as highly conserved as that which is complementary to 5′ splice sites (Mount and Steitz, 1981). Integrity of both the RNA and protein components of the U1 snRNP were found to be essential for specific and efficient binding to the 5′ splice site. Thus, the U1 snRNP resembles a ribosome in that both RNA and proteins contribute importantly to biological function.

When true *in vitro* splicing systems first became available, two more types of experimental evidence for U1 snRNP involvement emerged. First, both anti-(U1)RNP and anti-Sm antibodies were shown to inhibit the reaction (Padgett *et al.*, 1983). Curiously, anti-(U2)RNP did not inhibit (but see below). Second, RNase H and deoxyoligonucleotides complementary to the 5′ end of U1 RNA were used to decapitate U1 snRNPs in a splicing extract (Kramer *et al.*, 1984). This clever trick abolished splicing, showing that not only the U1 snRNP particle but also (more specifically) the 5′ end of U1 RNA is essential for splicing.

VII. U2, U5, AND U4/U6 snRNPs ALSO PARTICIPATE IN SPLICING

In 1984, analysis of intermediates in the splicing process led to the exciting discovery of branched RNA structures (Ruskin *et al.*, 1984; Padgett *et al.*, 1984). In the first step of splicing, cleavage at the 5′ splice site is apparently coupled to formation of a 2′, 5′-phosphodiester bond between the 5′ end of the intron and an A residue located about 30 nucleotides upstream of the 3′ splice site. In the second step, the two exons are ligated with release of the intron still in the form of a lariat. The intron branch site coincides with a highly conserved sequence previously shown to be essential for splicing of yeast pre-mRNAs. In mammalian pre-mRNAs the branch point sequence is much less rigidly defined; if the sequence normally utilized is deleted, the splicing apparatus can usually find another acceptable A residue with which to form the branch. Several laboratories (Brody and Abelson, 1985; Grabowski *et al.,* 1985; Frendeway and Keller, 1985) also ran splicing extracts on gradients and discovered that splicing intermediates sediment very fast (30 S in yeast, 50 S in mammalian systems). These active splicing complexes have been dubbed "spliceosomes."

The discovery of the intron branch point suggested that a third pre-mRNA region, in addition to the 5′ and 3′ splice sites, might need to be recognized during pre-mRNA splicing. Hints that this splicing component might be the U2 snRNP (Keller and Noon, 1984) were followed up using the RNase H/oligonucleotide degradation technique both in our lab by Doug Black (a graduate student) and in Tom Maniatis' lab at Harvard by Adrain Krainer. Both found that targeted degradation of U2 RNA in a nuclear extract abolished splicing activity (Black *et al.,* 1985; Krainer and Maniatis, 1985). Meanwhile, Benoit Chabot (a graduate student) probed *in vitro* splicing reactions with anti-(U2)RNP antibodies (Black *et al.,* 1985). Whereas no interaction with the pre-mRNA substrate had been observed in prior attempts using purified U2 snRNPs, after incubation of an active splicing extract in the presence of ATP, nuclease protection of a relatively large region of the intron (about 40 nt) that included the branch site was observed with anti-(U2)RNP antibodies. Thus, U2 snRNPs were shown to associate with the branch site.

When RNase protection experiments were performed using anti-Sm antibodies, binding at the 3′ splice site (in addition to the 5′ splice site and branch point) were observed. Benoit Chabot (Chabot *et al.,* 1985) went on to demonstrate that the 3′ splice site binding component had a m^3G cap moiety and responded in expected ways to various mutations in the 3′ splice site region. The fortuitous observation that the binding component is highly resistant to micrococcal nuclease led to its probable identification as the U5 snRNP. Of the abundant RNAs in Sm snRNP particles, only U5 RNA was insensitive to nuclease degradation.

Most recently, the remaining abundant Sm snRNP, containing U4 and U6 RNAs, was shown by Doug Black (Black and Steitz, 1986) to be required for splicing by oligonucleotide-targeted RNase H digestion of either U4 or U6 RNA in splicing extracts. It is not yet clear whether this particle interacts directly with the pre-mRNA or serves to arrange other snRNPs and splicing components in the proper conformation for splicing to occur. Mixing experiments performed using extracts in which U1, U2, or U4/U6 snRNPs had been separately destroyed indicated that each of these particles contributes independently to the splicing reaction. Moreover, no splicing intermediates were formed unless all these snRNPs are intact, suggesting their involvement at an early stage of the reaction.

VIII. THE SPLICEOSOME–RIBOSOME ANALOGY

With the realization that four different snRNPs are involved in intron removal from pre-mRNA molecules, the similarity of splicing to translation becomes striking. Both of these essential cellular processes involve as basic components RNA- and protein-containing subunits (snRNPs or ribosomal subunits) that assemble onto an RNA substrate together with other components to form a catalytic complex. Apparently, more pieces (subunits) are needed in the spliceosome than the ribosome because of the requirement to recognize several distant sites on the substrate before assembly. The large size and multiplicity of components in the spliceosome or active ribosome (factors, in addition to RNP subunits) probably reflects the extreme versatility of both the pre-mRNA splicing and translation machinery: An almost infinite variety of substrates can be acted upon with high fidelity.

Another aspect of pre-mRNA splicing that is elucidated by our current understanding of how individual snRNPs interact with the pre-mRNA (see Fig. 1) is the issue of minimum intron size. The lower size limit that is

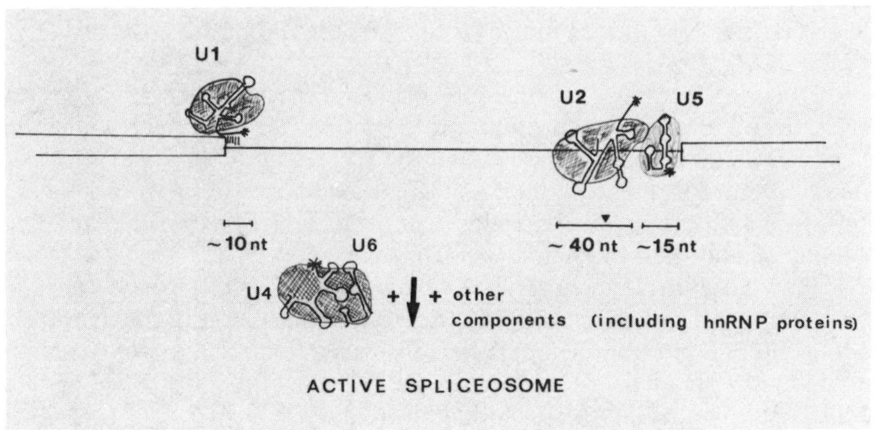

Fig. 1. Participation of RNP components in pre-mRNA splicing. Binding sites for specific snRNPs on the intron undergoing excision are indicated. All evidence is discussed in the text, except for participation of hnRNP proteins (Choi *et al.*, 1986).

observed for most naturally occurring introns, or when deletions have reduced an intron to its minimum functional length, is about 60 nucleotides (Wieringa *et al.*, 1984; Ruskin *et al.*, 1985; Rautman and Breathnach, 1985). This correlates nicely with the sum of the sizes of the (intron) regions protected from RNase by the U1, U2, and U5 snRNPs. If an intron is shortened below this length, the splicing apparatus reaches back into the 5′ exon to find a suitable "cryptic 5′ splice site." Thus, it appears that minimum intron length is determined quite simply by the geometry of the initial binding of snRNPs to specific regions of the substrate.

Many intriguing questions for future analysis are also raised by the spliceosome–ribosome analogy. Some of these are listed below.

1. Is there an absolutely defined assembly pathway for snRNPs and other components to form a spliceosome? Or, depending on the particular intron being spliced, can components assemble in various orders?
2. Do snRNPs or auxiliary splicing factors possess catalytic activity for one or both of the splicing reactions? Or, is the spliceosome like a ribosome in that it aligns an already activated substrate into a favorable geometry for "self-splicing"? The recent discovery that certain yeast mitochondrial introns self-splice via lariat intermediates (Peebles *et al.*, 1986; Van der Veen *et al.*, 1986) suggests that mammalian pre-mRNA splicing might also be essentially a self-splicing process.
3. How is the ATP known to be required for both steps of splicing utilized? Is ATP hydrolysis used as it is in the initiation of protein synthesis to scan the pre-mRNA to achieve correct splice site pairing? Is it used, again as in protein synthesis, to release from the spliceosome snRNPs or other factors in order to proceed to the next step of the process? There are hints that snRNPs may bind ATP and undergo conformational change (Black *et al.*, 1985).
4. Do snRNPs interact with each other and the pre-mRNA via RNA–RNA, RNA–protein, or protein–protein contacts? Precisely where do these interactions map?
5. Is there a step in splicing analogous to the ribosomal proofreading step, in which energy is expended to reject incorrectly formed splicing complexes and thereby maintain fidelity?
6. How is alternative splicing, as documented for different tissues or developmental stages, achieved? According to the protein synthesis model, several possibilities must be considered: modification of snRNPs, modification of other splicing components, or modulation of splice site availability by specific RNA-binding proteins.
7. How did these two RNA-centered machines evolve? Is their RNA content simply an evolutionary relic, or has RNA survived because it can better accomplish the required manipulations than protein?

Despite certain progress since the discovery of snRNPs, 7 years ago, much remains to be learned about their involvement in one of the critical steps in mammalian gene expression.

ACKNOWLEDGMENTS

I am indebted to numerous younger colleagues—those mentioned, as well as others—whose enthusiasm and dedication to science produced the results and insights summarized here. The work of my laboratory has been generously supported by grants from the National Institutes of Health and National Science Foundation and most recently by the Howard Hughes Medical Institute.

REFERENCES

Alarcon-Segovia, D., Ruiz-Arguelles, A., and Fishbein, E. (1978). *Nature (London)* **271,** 67–69.

Black, D. L., and Steitz, J. A. (1986). *Cell (Cambridge, Mass.)* **46,** 679–704.

Black, D. L., Chabot, B., and Steitz, J. A. (1985). *Cell (Cambridge, Mass.)* **42,** 737–750.

Bringmann, P., Appel, B., Rinke, J., Rewter, R., Thiessen, H., and Luhrmann, R. (1984). *EMBO J.* **3,** 1357–1363.

Brody, E., and Abelson, J. (1985). *Science* **228,** 963–967.

Busch, H., Reddy, R., Rothblum, L., and Choi, Y. C. (1982). *Annu. Rev. Biochem.* **51,** 617–654.

Chabot, B., Black, D. L., LeMaster, D. M., and Steitz, J. A. (1985). *Science* **230,** 1344–1349.

Choi, Y. D., Grabowski, P. J., Sharp, P. A., and Dreyfuss, G. (1986). *Science* **231,** 1534–1539.

Dreyfuss, G., Choi, Y. D., and Adam, S. A. (1984). *Mol. Cell. Biol.* **4,** 1104–1112.

Frendeway, D., and Keller, W. (1985). *Cell (Cambridge, Mass.)* **42,** 355–367.

Grabowski, P. J., Seiler, S. R., and Sharp, P. A. (1985). *Cell (Cambridge, Mass.)* **42,** 345–353.

Hashimoto, C., and Steitz, J. A. (1984). *Nucleic Acids Res.* **12,** 3283–3293.

Hinterberger, M., Pettersson, I., and Steitz, J. A. (1983). *J. Biol. Chem.* **258,** 2604–2613.

Keller, E. B., and Noon, W. A. (1984). *Proc. Natl. Acad. Sci. U.S.A.* **81,** 7417–7420.

Koski, R., Bothwell, A., and Altman, S. (1976). *Cell (Cambridge, Mass.)* **9,** 101–116.

Krainer, A. R., and Maniatis, T. (1985). *Cell (Cambridge, Mass.)* **42,** 725–736.

Kramer, A., Keller, W., Appel, B., and Luhrmann, R. (1984). *Cell (Cambridge, Mass.)* **38,** 299–307.

Lerner, E. A., Lerner, M. R., Janeway, C. A., Jr., and Steitz, J. A. (1981). *Proc. Natl. Acad. Sci. U.S.A.* **78,** 2737–2741 (1981).

Lerner, M. R., and Steitz, J. A. (1979). *Proc. Natl. Acad. Sci. U.S.A.* **76,** 5495–5499.

Lerner, M. R., Boyle, J. A., Mount, S. M., Wolin, S. L., and Steitz, J. A. (1980). *Nature (London)* **283,** 220–224.

Mimori, T., Hinterberger, M., Pettersson, I., and Steitz, J. A. (1984). *J. Biol. Chem.* **259,** 560–565.

Mount, S. M., and Steitz, J. A. (1981). *Nucleic Acids Res.* **23,** 6351–6368.

Mount, S. M., Pettersson, I., Hinterberger, M., Karmas, A., and Steitz, J. A. (1983). *Cell (Cambridge, Mass.)* **33,** 509–518.

Muramatsu, M., Hodnett, J. L., and Busch, H. (1966). *J. Biol. Chem.* **241,** 1544–1547.

Padgett, R. A., Mount, S. M., Steitz, J. A., and Sharp, P. A. (1983). *Cell (Cambridge, Mass.)* **35,** 101–107.

Padgett, R. A., Konarska, M. M., Grabowski, P. J., Hardy, S. F., and Sharp, P. A. (1984). *Science* **225,** 898–903.

Peebles, C. L., Perlman, P. S., Mecklenburg, K. L., Petrillo, M. L., Tabor, J. H., Jarnell, K. A., and Cheng, H.-L. (1986). *Cell (Cambridge, Mass.)* **44,** 213–223.

Pettersson, I., Hinterberger, M., Mimori, T., Gottlieb, E., and Steitz, J. A. (1984). *J. Biol. Chem.* **259,** 5907–5914.

Rautman, G., and Breathnach, R. (1985). *Cell (Cambridge, Mass.)* **41,** 95–105.

Rogers, J., and Wall, R. (1980). *Proc. Natl. Acad. Sci. U.S.A.* **77,** 1877–1879.

Ruskin, B., Krainer, A. R., Maniatis, T., and Green, M. R. (1984). *Cell (Cambridge, Mass.)* **38,** 317–331.

Ruskin, B., Greene, J. M., and Green, M. R. (1985). *Cell (Cambridge, Mass.)* **41,** 833–844.

Stark, B. C., Kole, R., Bowman, E. J., and Altman, S. (1978). *Proc. Natl. Acad. Sci. U.S.A.* **75,** 3717–3721.

Van der Veen, R., Arnberg, A. C., Van der Horst, G., Bonen, L., Tabak, H. F., and Grivell, L. A. (1986). *Cell (Cambridge, Mass.)* **44,** 225–234.

Weinberg, R. A., and Penman, S. (1968). *J. Mol. Biol.* **38,** 289–304.

Wieringa, B., Hofer, E., and Weissmann, C. (1984). *Cell (Cambridge, Mass.)* **37,** 915–925.

Yang, V. W., Lerner, M. R., Steitz, J. A., and Flint, S. J. (1981). *Proc. Natl. Acad. Sci. U.S.A.* **78,** 1371–1376.

6

Exon Sequences and Splice Site Proximity Play a Role in Splice Site Selection

ROBIN REED AND TOM MANIATIS

Department of Biochemistry and Molecular Biology
Harvard University
Cambridge, Massachusetts 02138

I. INTRODUCTION

Most genes in higher eukaryotes are interrupted by at least one, and often several, introns which range in size from 30 up to many thousands of nucleotides. In addition, some transcripts are alternatively spliced, generating two or more different spliced mRNAs from the same pre-mRNA (see Padgett *et at.,* 1986 for review). Although a number of studies have been directed toward elucidating the mechanism by which the correct pairs of splice sites are chosen during splicing, the mechanism of splice site selection is still largely unknown. The importance of the conserved sequence elements at the 5′ and 3′ splice junctions of each intron was established through analyses of both naturally occurring and *in vitro*–generated mutations in these elements (Padgett *et al.,* 1986). The mutations either reduce or abolish splicing from the affected junction, and in most cases, one or more cryptic splice sites are used (Treisman *et al.,* 1983; Wieringa *et al.,* 1983). Interestingly, cryptic splice sites, which closely resemble normal splice sites, are efficiently used in the absence of the normal sites but are never used in their presence.

The development of efficient *in vitro* splicing systems (Krainer et at., 1984) led to the elucidation of a two-step splicing pathway (Krainer *et al.,* 1984; Ruskin *et al.,* 1984; Grabowski *et al.,* 1984) which implicated a role

Molecular Biology of RNA
New Perspectives

for an additional sequence element in splicing. The first step of the splicing pathway is cleavage at the 5′ splice junction and formation of the lariat intermediate, an RNA species in which the 5′ terminal nucleotide of the intron is linked via a 2′-5′ phosphodiester bond to a specific adenine residue located near the 3′ splice site. The second step of the splicing reaction entails cleavage at the 3′ splice site and ligation of the exons. Based on the observation that lariat formation occurs at a specific adenine residue within the intron, sequence comparisons of a number of different introns were performed. These analyses led to the identification of a weakly conserved sequence element at the branch site (Ruskin *et al.*, 1984; Zeitlin and Efstratiadis, 1984; Keller and Noon, 1984). As discussed below, a number of subsequent experiments established that this element plays a role in splicing both *in vivo* and *in vitro*.

A role for the branch point sequence during *in vitro* splicing was demonstrated by mapping the site of lariat formation for a number of different introns (Reed and Maniatis, 1985). Included among these were α- and β-like globin introns, a rat insulin intron, and the *Drosophila ftz* intron. In all of these cases, the predominant site of lariat formation occurs within the weakly conserved sequence element located between 18 and 40 nucleotides from the 3′ splice site. A further demonstration of a role for the branch point sequence was provided by Rautmann and Breathnach (1985). After finding that some pre-mRNAs which contain synthetic 5′ and 3′ splice sites separated by random sequence were very inefficiently spliced *in vivo* (Rautmann *et al.*, 1984), these investigators inserted the branch point sequence from human β-globin into the artificial intron near the 3′ splice and observed efficient splicing.

As was mentioned above, mutation of the 5′ or 3′ splice sites usually results in the efficient use of cryptic splice sites. A similar phenomenon is also observed when the branch point sequence is mutated or absent. This was demonstrated by analyzing the *in vitro* splicing products generated from a pre-mRNA containing an artificial intron (obtained from Rautmann *et al.*, 1984; Reed and Maniatis, 1985). All of the characteristic splicing products and intermediates were generated, indicating that the synthetic intron was spliced via the normal two-step splicing pathway. When the site of lariat formation was mapped in this intron, branch formation was localized to a region about 20 nucelotides from the 3′ splice site. Thus, in the absence of a normal branch point sequence, lariat formation occurs within a cryptic sequence located the conserved distance from the 3′ splice site. Mutation of the normal branch point sequence also results in the activation of one or more cryptic branch point sequences both *in vivo* and *in vitro* (Ruskin *et al.*, 1985).

The observation that both cryptic splice sites and cryptic branch point sequences are often efficiently used in the absence of their normal counterparts suggests that the sequence requirements for splicing are nonstringent in higher eukaryotes. By contrast, the 5′ and 3′ splice sites, as well as the branch point sequence, are stringently conserved in yeast (Langford and Gallwitz, 1983; Pikielny *et al.*, 1983). In addition, mutation of any of these elements almost always abolishes splicing in yeast, and cryptic sequences are usually not activated (Gallwitz, 1982; Langford and Gallwitz, 1983; Pikielny *et al.*, 1983; Langford *et al.*, 1984; Newman *et al.*, 1985).

Although three sequence elements, the branch point sequence, the 5′ splice site and the 3′ splice site, are required for efficient splicing, these elements alone are not sufficient to explain the mechanism of splice-site selection in higher eukaryotes. This mechanism must ensure that real splice sites are distinguished from many very similar cryptic splice sites which are located throughout both introns and exons. In addition, the mechanism of splice-site selection must ensure that only the correct pairs of 5′ and 3′ splice sites are used in pre-mRNAs which contain multiple introns. To determine whether other sequence elements within pre-mRNA play a role in splice-site selection, we employed a cis-competition assay in which both the wild-type sequences and their mutant counterparts are present on the same precursor (Reed and Maniatis, 1986). The cis-competition assay was employed as a means of circumventing the difficulties associated with identifying important sequences when alternative cryptic sequences can function in the absence of their normal counterparts.

The cis-competition experiments described below led to two new conclusions regarding splice site selection. First, the data show that sequences within exons play an important role in the use of the adjacent 5′ or 3′ splice sites. Second, the data indicate that the proximity of the 5′ and 3′ splice sites to one another is an important determinant in splice site selection. The relevance of this data to splice selection in both normal and alternatively spliced pre-mRNAs is discussed.

II. EXON SEQUENCES AND SPLICE SITE PROXIMITY PLAY A ROLE IN SPLICE SITE SELECTION

The cis-competition experiments were carried out by analyzing the *in vitro* splicing products generated from ^{32}P-labeled SP6 pre-mRNA substrates (Green *et at.*, 1983; Krainer *et al.*, 1984; Ruskin *et al.*, 1984). The precursors examined contain tandem duplications of either the 5′ or the 3′ splice site of human β-globin intervening sequence (IVS) 1. Similar precursors originally were used to test scanning models for splice site selection (see Discussion, Lang and Spritz, 1983; Kuhne *et al.,* 1983). In the experiments described here, the duplicated splice sites were introduced near the middle of the intron so that each duplicated site is flanked by about half of the normal intron (see Fig. 1). Thus, all of the 3′ splice site duplications contain the normal branchpoint sequence.

As shown in Fig. 1, a series of precursors containing tandem duplications of the 3′ splice site was examined. The full-length exon 2 flanks the external 3′ splice site in each precursor while the internal splice site lies adjacent to portions of exon 2 ranging in size from 14 nucleotides to full length. *In vitro* splicing of precursors containing the 14- or 55-nucleotide portion of the exon adjacent to the internal splice site (3′D-14, 3′-55) yields RNA derived from exclusive utilization of the external splice site. By contrast, with the 115- or 120-nucleotide portion of the exon in this position (3′D-115, 3′D-120), both splice sites are used, and the internal site is preferred. Finally, when the full-length exon is adjacent to both the external and internal splice site (3′D-205), only the internal site is used. These experiments suggest that either changes in the distance between the splice sites or the changes in sequence could account for the observed differences in the pattern of splice-site selection.

To distinguish between these possibilities, a number of precursors containing sequences substituted for the internal exon 2 were examined (Fig. 2). These sequences were derived from introns, exons, and random sequences. For example, the normal human exon 3 (3′S-220A), a 500-nucleotide fragment from an IVS (3′S-500), and a sequence from the 3′-flanking region of the human β-globin gene (3′S-96A) were used for the substitutions. Interestingly, most of the sequences substituted for the normal exon did not allow use of the adjacent 3′ splice site (Fig. 2A). Instead, the external 3′ splice site adjacent to the full-length, normal exon 2 was used efficiently. Thus, these experiments show that a normal 3′ splice site can be inactivated completely by the sequences that are adjacent to it. As shown in Fig. 2B, a few of the sequences substituted for the second exon did allow efficient use of the internal 3′ splice site. However, no obvious sequence element or secondary structure was discerned in either the functional or nonfunctional substitution sequences.

The data with all of the 3′ splice site duplication precursors described above indicate that the sequences within the exon play an important role in the use of the adjacent 3′ splice site. Most sequences substituted for the normal exon, whether other exons or random sequences, prevent use of the 3′ splice site. In addition, these data suggest that the proximity of the splice sites is an important criterion for splice site selection in these precursors since only the internal splice site is used when both duplicated sites are flanked by the full-length exon.

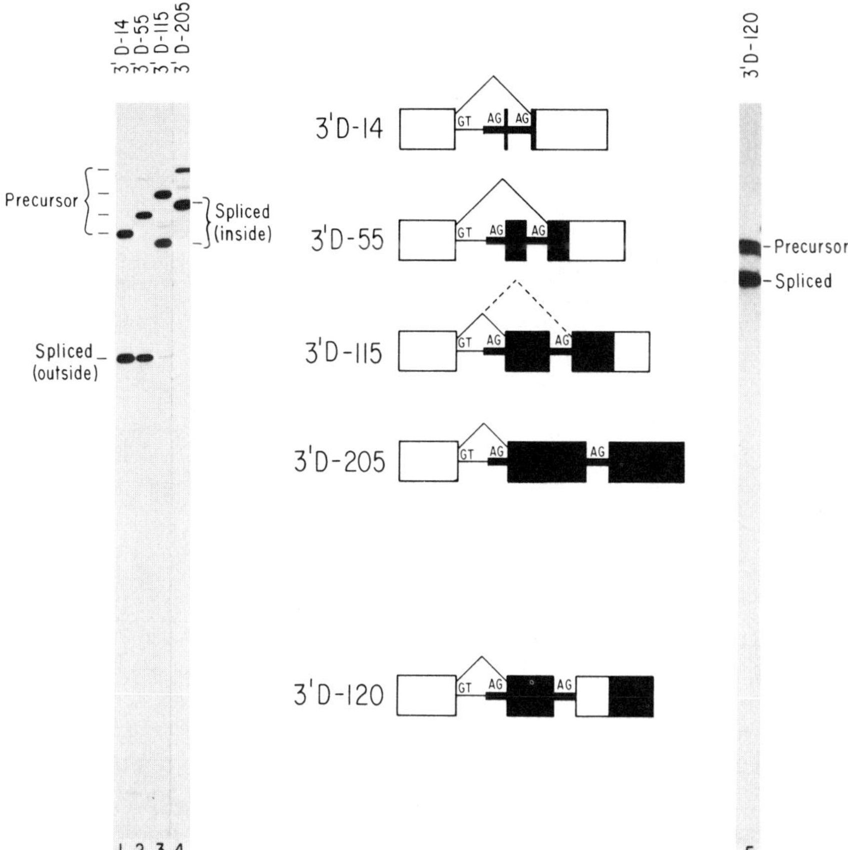

Fig. 1. The effect of exon 2 truncations on the relative use of duplicated 3′ splice sites. The schematic diagrams depict the structures of the SP6 RNA precursors that were spliced *in vitro* and then analyzed by electrophoresis on a 5% denaturing polyacrylamide gel. Exon sequences are denoted by boxes and intron sequences by the horizontal line. The shaded regions within the boxes indicate the duplicated portion of the exon while the heavy line denotes the duplicated part of the intron. The use of the 5′ (GT) and 3′ (AG) splice sites is shown in the figure; the solid lines indicate the predominant splice sites used, and the dashed line represents the minor splice sites used. Lanes 1–4, 3′ truncations of the internal exon 2 sequences; lane 5, 5′ truncation of the internal exon 2 sequences. The precursor RNA and the spliced RNA resulting from the use of the internal or external splice sites are indicated adjacent to the autoradiograms. The other splicing intermediates and products were identified on longer film exposures and account for the faint bands seen in each lane.

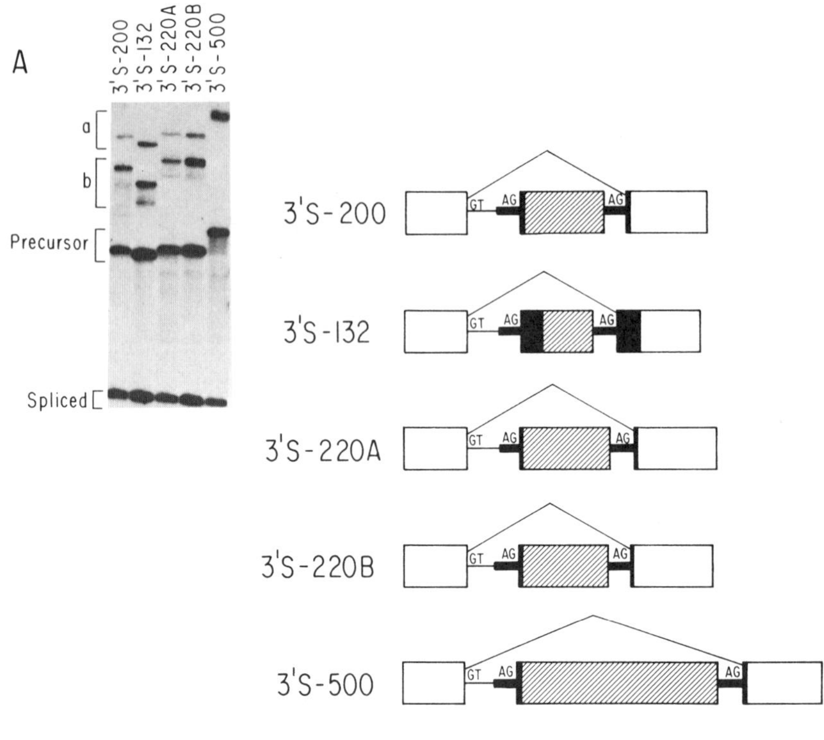

Fig. 2. The effect of internal second exon substitutions on the relative use of duplicated 3′ splice sites. The schematic diagrams depict the structure of the SP6 precursors that were spliced *in vitro* and then analyzed by electrophoresis on a 5% denaturing polyacrylamide gel. The boxes, lines, and shaded regions refer to the structures described in Fig. 1 legend. The cross-hatching designates the sequences that were substituted for the normal exon sequences. The precursor and spliced RNAs are indicated next to the autoradiogram. (A) Exon substitutions that inactivate the adjacent 3′ splice site. a, Intron–exon lariat; b, intron lariat; c, exon 1. (In lane 5, the intron–exon lariat and the intron lariat are not resolved from each other.) (B) Exon substitutions that result in the utilization of the adjacent 3′ splice site. a, Exon 1; b, excised intron. In each lane the exon–intron lariat intermediate is not resolved from the precursor band. On an 8% gel, this RNA species is detected as a band migrating above the precursor.

An analogous set of experiments was carried out with the 5′ splice site of human β-globin IVS 1 (Reed and Maniatis, 1986). The precursors analyzed in this set of experiments contain tandem duplications of the 5′ splice site. The external 5′ splice site was flanked by the full-length exon 1, whereas the internal splice site was flanked by portions of exon 1 ranging in size from 16 nucleotides to full length. Similar to the results

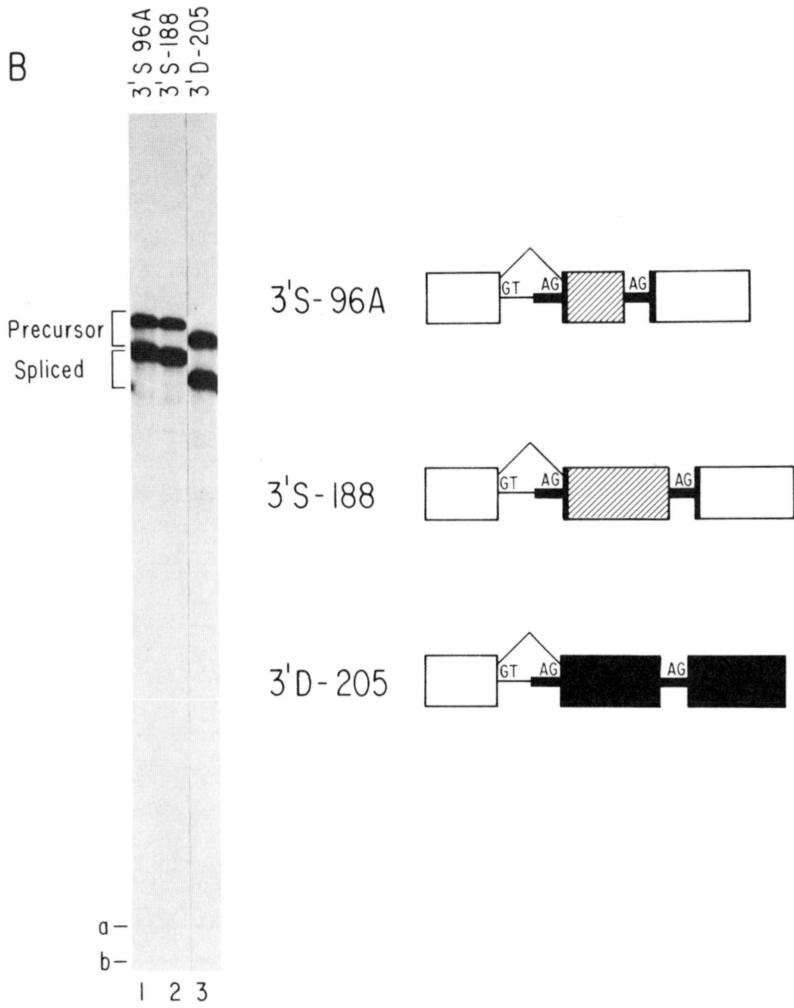

Fig. 2. (*Continued*)

obtained with the 3′ splice site duplication precursors, when the small portion of exon 1 is adjacent to the internal splice site, the external splice site is used exclusively, whereas the internal splice site is used when the large portions of the exon or the full-length exon is adjacent to the internal splice site.

To determine whether sequence or distance accounts for the altered patterns of splice-site selection seen with the 5′ duplication precursors, a number of sequences were substituted for the exon adjacent to the internal 5′ splice site. Similar to the results obtained for the 3′ substitution precursors, most of the sequence substitutions result in complete inactivation of the adjacent 5′ splice site, and all spliced RNA is derived from efficient use of the external splice site adjacent to the normal, full-length exon 1. Partial utilization of the internal 5′ splice site was observed with one of the sequences used for substitution, although this sequence does not function as well as the normal exon 1. The data with the 5′ duplication precursors reveal that the sequences within the exon also plays an important role in the utilization of the adjacent 5′ splice sites. In addition, proximity of the 5′ splice site is an important criterion in the choice between splice sites, since only the internal site is used when the full-length exon is adjacent to both duplicated sites.

Precursors containing duplications of both the 5′ and 3′ splice sites on the same precursor were also examined (Reed and Maniatis, 1986). These data revealed that the same splice sites that are used in the precursors containing one duplicated site are also used in the double duplication precursors. In addition, the importance of proximity in the selection of splice sites is seen with the double duplication precursor containing both duplicated 5′ and 3′ splice sites flanked by the full-length exon. In this precursor, only the internal pair of splice sites is used.

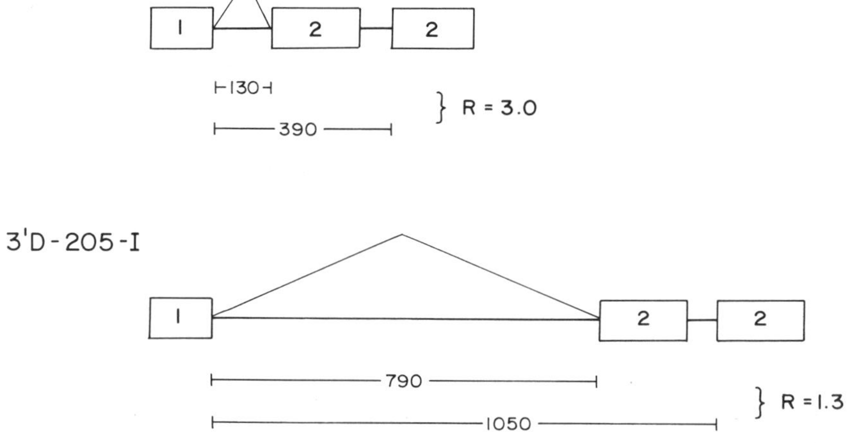

Fig. 3. The effect of distance on the relative use of duplicated 3′ splice sites. The schematic diagrams depict the SP6 precursors that were spliced *in vitro* and fractionated on a 5% denaturing polyacrylamide gel. The boxes and horizontal lines designate exons and introns, respectively. The splice sites used are indicated in the figure. R refers to the ratio of the distance between the outside 3′ splice site to the 5′ splice site and the inside 3′ splice site to the 5′ splice site.

To investigate the mechanism by which the most proximal splice sites are used, an experiment was performed to determine whether the relative use of duplicated 3′ splice sites would change if their relative distance from the 5′ splice site was more similar. This experiment was carried out by inserting a 790-nucleotide fragment into the middle of IVS 1 in precursor 3′D-205 (Fig. 3). Consequently, in 3′D-205-I the external 3′ splice site is about three times further away from the 5′ splice site than is the internal splice site, whereas in 3′D-205-I the external 3′ splice site is only about 1.3 times further away from the 5′ splice site. As shown schematically in Fig. 3, when these two precursors were spliced *in vitro,* the internal splice site is used to the same extent in both precursors. This result is consistent with the idea that the most proximal sites in pre-mRNA may be brought together by a mechanism involving one-dimensional tracking along the RNA. This mechanism is distinguished from previously proposed scanning mechanisms by requiring that tracking occur after splicing factors are bound at the splice sites (see Section V).

III. THE PATTERN OF SPLICE SITE SELECTION IS ALTERED IN DIFFERENT EXTRACT PREPARATIONS AND IN DILUTED EXTRACTS

Further evidence that exon sequences play a role in splice site selection was obtained when the duplication precursors were examined in different preparations of HeLa cell nuclear extracts and, subsequently, in dilutions of the same extract preparation (Reed and Maniatis, 1986). The experiments with different preparations of extracts revealed different patterns of splice site selection with some of the 3′ duplication precursors. Specifically, an increase in the use of the external splice site was observed with some of the precursors shown in Fig. 2B, which contain substitution sequences adjacent to the internal splice site. By contrast, the relative use of the internal and external splice sites remained the same in the 3′ duplication precursor containing both duplicated sites flanked by the full-length exon (3′D-205, Fig. 1). These data suggested the possibility that a factor(s) may be present at different concentrations in the different extract preparations. If the factor is limiting, the splice site adjacent to the normal exon may be able to compete more effectively than splice sites adjacent to any of the substitution sequences. In factor excess, both the substitution exon and the normal exon (or adjacent splice sites) have access to the factor, and then the most proximal splice site is used for splicing. This hypothesis predicts that dilution of the splicing extract will result in a relative increase in the use of the external splice site in the precursors containing the substitution sequences. Examination of a number of the substitution precursors in diluted extracts revealed that this is the case (Reed and Maniatis, 1986).

IV. SPLICE SITE SELECTION CAN BE ALTERED BY COMPETITION IN TRANS

In an attempt to discern the nature of the factor(s) responsible for the altered patterns of splice site selection, trans-competition experiments were performed (data not shown). The competing RNA used in these experiments contains the human β-globin IVS 1 3′ splice site, the branch point sequence, and full-length exon 2. These data revealed that increasing competitor RNA results in a relative increase in the use of the external splice site adjacent to the normal exon when the internal splice site is adjacent to a substitution sequence. By contrast, the same concentrations of competitor did not alter the pattern of splice site selection in the precursor containing the full-length exon adjacent to both duplicated splice sites. In subsequent experiments, evidence was obtained indicating that the competitor RNAs form stable complexes with factors in the extract. This is based on the observation that preincubation of the extract with the competitor alters the pattern of splice site selection whereas the same amount of competitor added to the reaction with the pre-mRNA has no effect. To determine whether the factor responsible for the altered patterns of splice site selection is dependent on the presence of the 3′ splice site in the competitor RNA, a similar competitor RNA lacking a 3′ splice site was tested. Although the pattern of splice site selection was not altered with the latter competitor, these experiments are difficult to interpret because the competitor lacking the splice site is degraded in the extract at a much faster rate than the competitor containing an intact splice site. Additional studies to determine the nature of factors involved in splice site selection are underway.

A further illustration of the importance of exon sequences in determining the relative use of adjacent splice sites was observed with a human β-globin IVS 1 3′ duplication precursor containing the mouse β-globin exon 2 adjacent to the internal splice site and the normal human exon 2 adjacent to the external splice site (Reed and Maniatis, 1986). When this precursor is spliced in diluted extracts, both the external and internal splice sites are used about equally. This result suggests that the splice site adjacent to the normal human exon can compete more effectively than the site adjacent to the mouse exon even though these exons are nearly 90% identical.

These data indicate that even subtle changes in exon sequences can effect the use of adjacent splice sites.

V. DISCUSSION

The experiments described in this report demonstrate that the sequences within exons play an important role in the use of adjacent 5′ or 3′ splice sites. In most cases, sequences that were substituted for normal exons result in decreased use of the adjacent splice site. Moreover, subtle changes in exon sequences were found to effect splice site use. The proximity of splice sites to one another also appears to play a role in determining the splice site chosen. The effect of exon deletions and substitutions on splice site selection may result either from removal of sequences that facilitate splicing or the insertion of sequences that interfere with splicing. Regardless of the mechanism, the observed effects are not a general property of exons since replacement of exon 1 or 2 with the human β-globin exon 3 and other exon sequences results in the inactivation of the adjacent splice sites. The nature of the sequences within the normal flanking exons that result in efficient splice site selection is not known. A particular sequence element or structure may be contained within these exons, or an interaction between exons 1 and 2 may be required for maximal efficiencies of splicing. Although we cannot distinguish between these and other possibilities, the data indicate that the observed effects of exon sequences are localized to the adjacent splice site. Based on this observation, and the effect of extract dilution on splice site selection, it is possible that exon sequences participate directly in splice site recognition by determining the affinity of adjacent splice sites for small nuclear ribonucleoproteins (snRNPs) or other splicing components. The facilitation of splice site recognition by exon sequences could be the consequence of RNA secondary structure which brings the adjacent splice sites into a more accessible configuration, or the exon sequences might lead to a more favorable RNP structure.

A role for exon sequences in splicing has also been identified in viral pre-mRNAs *in vivo*. Mutations within exon sequences have been shown to affect the use of alternate splice sites in SV40 pre-mRNAs (Somasekhar and Mertz, 1985). Also consistent with the observation that splice site selection can be affected by sequences outside of the intron is the finding that the human β-globin IVS 1 is not spliced *in vivo* when it is placed within the 3′ untranslated region of the herpes virus thymidine kinase gene, a gene that normally contains no introns (Greenspan and

Weissman, 1985). Similarly, the *in vivo* splicing efficiency of a yeast actin intron depends upon its location within a gene that normally contains no introns (Klinz and Gallwitz, 1985).

A. Implications on Scanning Mechanisms for Splice Site Selection

A simple linear scanning mechanism for splice site selection involves recognition of a 5′ or 3′ splice site by the splicing machinery followed by a linear scan to locate the nearest matching splice site (Lewin, 1980; Sharp, 1981). To test this model two groups studied splice site use in human γ-globin (Lang and Spritz, 1983) or rabbit β-globin (Kuhne *et al.*, 1983) RNA precursors containing duplicated 5′ or 3′ splice sites. Conflicting results were obtained with precursors containing 3′ splice site duplications. Exclusive use of the internal 3′ splice site was observed with the γ-globin precursor, while the external 3′ splice site was used in the β-globin precursor. The latter result rules out the simple 5′ → 3′ scanning mechanism stated above. The discrepancy between these studies can now be reconciled in light of the role of exon sequences in splice site selection. In the rabbit β-globin precursor the internal 3′ splice site is adjacent to only 43 nucleotides of exon 3 whereas almost all of exon 3 is similarly located in the γ-globin precursor. Thus, consistent with the data obtained with the analogous human β-globin IVS 1 duplications, preferential use of the internal splice site is observed if a large portion of the exon is adjacent to this site whereas the external site is preferred if small exon truncations are located internally.

Also consistent with our data is the observation that the external 5′ splice site was used with both rabbit and γ-globin precursors containing a small exon truncation adjacent to the internal duplicated 5′ splice site (Kuhne *et al.*, 1983; Lang and Spritz, 1983). On the basis of these observations, the most straightforward interpretion of the previous studies directed toward testing the scanning model is that the pair of strongest 5′ and 3′ splice sites in closest proximity are used. A model in which scanning occurs after the recognition of both the 5′ and 3′ splice sites is suggested by the experiments presented in Fig. 3. These experiments revealed that the inside 3′ splice site is still preferentially used when the duplicated splice sites are located about the same relative distance from the 5′ splice site (3′D-205-I). Alternatively, RNA secondary structure could be responsible for the pattern of splice site selection observed.

B. Possible Mechanisms for Splice Site Selection

The patterns of RNA splicing observed with the precursors containing duplicated 5′ or 3′ splice sites provide insights into the two fundamental problems in splice site selection in normal pre-mRNAs. First, these data suggest that exon sequences may play a key role in distinguishing between normal splice sites and the many identical or nearly identical cryptic splice sites located throughout pre-mRNAs. Second, the observation that only the nearest two 5′ and 3′ splice sites are used when the duplicated splice sites are flanked by identical exon sequences may be relevant to the mechanism of correct pairing of 5′ and 3′ splice sites in normal pre-mRNAs. Below we discuss possible mechanisms for splice site selection that incorporate these observations, are consistent with studies of β-thalassemia splicing mutations, and which take into account the formation of a multicomponent splicing complex.

Discrimination between normal and cryptic splice sites may result from differences in the affinities of splicing components for these two types of sites. Additional specificity in the splicing reaction may be achieved during the formation of a multicomponent splicing complex. At least five different snRNPs, possibly a number of different heterogeneous nuclear ribonucleoproteins (hnRNP) proteins, as well as other factors appear to comprise the splicing complex (see Padgett *et al.*, 1986, for review). The assembly of specific multicomponent nucleoprotein complexes has also been proposed to play a role in other processes requiring high levels of accuracy, such as site-specific recombination and initiation of DNA replication (Echols *et al.*, 1984).

In the case of splicing, a variety of RNA protection analyses have revealed specific interactions between cellular factors and the 5′ and 3′ splice sites and the branch point sequence (Chabot *et al.*, 1985; Black *et al.*, 1985; Ruskin and Green, 1985). The factors bound to these sites may interact with each other to form a specific three-dimensional structure which could be required for optimal splicing efficiencies. It is possible that exon sequences play an important role in the formation of such a structure. Cryptic splice sites would then be excluded in the presence of normal splice sites by virtue of the relatively inefficient complex formation with the cryptic sites. It is also possible that the relative stability of splicing complexes formed with cryptic or normal splice sites plays a role in splice site selection. For example, if there is a time lag between the formation of the splicing complex and the first step in the splicing reaction, complexes formed with cryptic splice sites may dissociate before splicing can be initiated. The discrimination between normal and cryptic splice sites may therefore occur at several steps along the splicing pathway, including splice site recognition, complex formation, and complex stability.

The second aspect of splice site selection, pairing the correct 5′ and 3′ splice sites, could be achieved by a balance between differential splice site strength and the proximity of the 5′ and 3′ splice sites to one another. (Splice site strength is defined here as the affinity of a splice site for splicing factors and/or the ability of a splice site to participate in splicing complex formation.) The splicing factors bound to the 5′ and 3′ splice sites may be brought together for complex formation by a one-dimensional diffusion mechanism, by random collision, or as a result of RNA folding. Thus, both the proximity of the sites and their strength would determine the pairs used. This hypothesis predicts that perturbations of either parameter, the strength of a splice site or its location, could alter the pattern of splice site selection. The β-thalassemia splicing mutants provide interesting examples of such perturbations (Treisman *et al.*, 1983). For example, exon 2 skipping is never observed in the normal human β-globin gene, but when a single base mutation in the 5′ splice site of IVS 2 results in the activation of a weak cryptic 5′ splice site, exon 1 is joined to exon 3 in approximately 10% of the spliced RNA (Treisman *et al.*, 1982). Another striking example of exon jumping is provided by a naturally occuring splicing mutation in the human phenylalanine hydroxylast (PAH) gene which contains 13 exons (S. Woo, personal communication). Inactivation of the 5′ splice site of intron 12 results in skipping of exon 12, and remarkably, only the 5′ splice site of IVS 11 is used for skipping, even though 10 other normal 5′ splice sites are present in the precursor. These data provide further evidence of the importance of splice site proximity in splice site selection.

REFERENCES

Black, D. L., Chabot, B., and Steitz, J. A. (1985). *Cell (Cambridge, Mass.)* **42,** 737–750.

Chabot, B., Black, D. L., LeMaster, D. M., and Steitz, J. A. (1985). *Science* **230,** 1344–1349.

Echols, H., Dodson, M., Better, M., Roberts, J. D., and McMacken, R. (1984). *Cold Spring Harbor Symp. Quant. Biol.* **49,** 727–733.

Gallwitz, D. (1982). *Proc. Natl. Acad. Sci. U.S.A.* **79,** 3493–3497.

Grabowski, P. J., Padgett, R. A., and Sharp, P. A. (1984). *Cell (Cambridge, Mass.)* **37,** 415–417.

Green, M. R., Maniatis, T., and Melton, D. A. (1983). *Cell (Cambridge, Mass.)* **32,** 681–694.

Greenspan, D. S., and Weissman, S. M. (1985). *Mol. Cell Biol.* **5,** 1894–1900.

Keller, E. B., and Noon, W. A. (1984), *Proc. Natl. Acad. Sci. U.S.A.*, **81,** 7417–7420.

Klinz, F., and Gallwitz, D. (1985), *Nucleic Acids Res.* **13,** 3791–3804.

Krainer, A. R., Maniatis, T., Ruskin, B., and Green, M. R. (1984), *Cell (Cambridge, Mass.)* **36,** 993–1005.

Kuhne, T., Wieringa, B., Reiser, J., and Weissman, C. (1983). *EMBO J.* **2,** 727–733.

Lang, K. M., and Spritz, R. A. (1983). *Science* **220,** 1351–1355.

Langford, C. J., and Gallwitz, D. (1983). *Cell (Cambridge, Mass.)* **33,** 519–527.

Langford, C. J., Klinz, F. J., Donath, C., and Gallwitz, D. (1984). *Cell (Cambridge, Mass.)* **36,** 645–653.

Lewin, B. (1980). *Cell (Cambridge, Mass.)* **22,** 324–326.

Newman, A. J., Lin, R.-J., Cheng, S.-C., and Abelson, J. (1985). *Cell (Cambridge, Mass.)* **42,** 335–344.

Padgett, R. A., Grabowski, P. J., Konarska, M. M., Seiler, S., and Sharp, P. A. (1986). *Annu. Rev. Biochem.* **55,** 1119–1150.

Pikielny, C. S., Teem, J. L., and Rosbash, M. (1983). *Cell (Cambridge, Mass.)* **34,** 395–403.

Rautmann, G., and Breathnach, R. (1985). *Nature (London)* **315,** 430–432.

Rautmann, G., Matthes, H. W. D., Gait, M. J., and Breathnach, R. (1984). *EMBO J.* **3,** 2021–2028.

Reed, R., and Maniatis, T. (1985). *Cell (Cambridge, Mass.)* **41,** 95–105.

Reed, R., and Maniatis, T. (1986). *Cell (Cambridge, Mass.)* **46,** 681–690.

Ruskin, B., and Green, M. R. (1985). *Cell (Cambridge, Mass.)* **43,** 131–142.

Ruskin, B., Krainer, A. R., Maniatis, T., and Green, M. R. (1984). *Cell (Cambridge, Mass.)* **38,** 317–331.

Ruskin, B., Greene, J. M., and Green, M. R. (1985). *Cell (Cambridge, Mass.)* **41,** 833–844.

Sharp, P. (1981). *Cell (Cambridge, Mass.)* **23,** 643–646.

Somasekhar, M. B., and Mertz, J. E. (1985). *Nucleic Acids Res.* **13,** 5591–5609.

Treisman, R., Proudfoot, N., Shander, M., and Maniatis, T. (1982). *Cell (Cambridge, Mass.)* **29,** 903–911.

Treisman, R., Orkin, S., and Maniatis, T. (1983). *Nature (London)* **302,** 591–596.

Wieringa, B., Meyer, F., Reiser, J., and Weissman, C. (1983). *Nature (London)* **301,** 38–43.

Zeitlin, S., and Efstratiadis, A. (1984). *Cell (Cambridge, Mass.)* **39,** 589–602.

7

Factors That Influence Alternative Splice Site Selection *in Vitro*

JAMES L. MANLEY, JONATHAN C. S. NOBLE, XIN-YUAN FU, AND HUI GE

Department of Biological Sciences
Columbia University
New York, New York 10027

I. INTRODUCTION

Alternative splicing of pre-mRNAs presents a potential mechanism for regulating gene expression. Although considerable information concerning the pathway and mechanism of pre-mRNA splicing has accumulated in recent years, very little is known about factors that influence alternative splice site selection (see reviews by Padgett *et al.*, 1986; Leff *et al.*, 1986). One possibility is that alternative splicing primarily reflects the random utilization of splice sites, with the ratio determined by the relative "strengths" of the competing splice sites. By this model, splice site utilization would be controlled by factors such as homology with established consensus sequences (Mount, 1982) and the position of splice sites within the pre-mRNA. Studies that correlate changes in alternative splicing with differences in the pre-mRNA 5′ or 3′ ends, which could change the secondary or tertiary structure of the RNA, or in some other way influence splice site selection, are consistent with this model. Examples of such a correlation include the myosin light-chain gene (Nabeshima *et al.*, 1984), the mouse α-amylase gene (Young *et al.*, 1981; Schibler *et al.*, 1983), and the *Drosophila* alcohol dehydrogenase gene (Benyajati *et al.*, 1983). Different patterns of alternative splicing could also result from differential

Molecular Biology of RNA
New Perspectives

use of poly(A) addition sites, such as in the mouse immunoglobin μ and δ genes (Early *et al.*, 1980; Maki *et al.*, 1981) and the calcitonin gene (Amara *et al.*, 1982; Rosenfeld *et al.*, 1984).

The early region of the DNA tumor virus SV40 offers an excellent model to begin studying the mechanism and regulation of alternative splicing. SV40 early pre-mRNA can be spliced *in vivo* to produce large T and small t mRNAs, which result from utilization of either of two alternative 5′ splice sites and a shared 3′ splice site (for review, see Tooze, 1981). The amount of large T mRNA that accumulates in HeLa cells and a number of other mammalian cell lines, including infected and transformed cells, is three- to fivefold greater than the amount of small t mRNA (e.g., Manley *et al.*, 1986). Somewhat surprisingly, when the products of SV40 early pre-mRNA splicing *in vitro* were analyzed, small t splicing was not detected, even though large T splicing occurred with a reasonable efficiency (Noble *et al.*, 1986). This defect, which apparently reflects a lack of cleavage at the small t 5′ splice site, is not an inherent property of the pre-mRNA used, because the same precursor was efficiently spliced to small t mRNA when microinjected into *Xenopus laevis* oocytes (Manley *et al.*, 1986).

One explanation for our failure to detect significant levels of small t mRNA splicing *in vitro* is that a 5′ → 3′ "scanning" mechanism is the major determinant of 5′ splice site selection *in vitro* [see Sharp (1981) for a discussion of scanning as a mechanism of splice site selection]. By this model, a component (or components) of the splicing apparatus would recognize the 5′ end of the pre-mRNA and scan, or diffuse, along the RNA until encountering a 5′ splice site, which would be utilized in conjunction with the appropriate 3′ splice site. Such a model would explain the almost exclusive utilization *in vitro* of the SV40 large T 5′ splice site.

Evidence in other systems both for and against the existence of a scanning mechanism of RNA splicing has been documented. Lang and Spritz (1983) analyzed the splicing pattern *in vivo* of a series of 5′ and 3′ splice site duplications in the human Gγ-globin second intron and found that the 5′-most splice site was always utilized, consistent with a 5′ → 3′ scanning model. Kuhne *et al.* (1983) analyzed a similar set of mutations involving the rabbit β-globin large intron, and found that, for both 5′ and 3′ splice sites, the site further removed from the intron was used exclusively. These results argue against a 5′ → 3′ scanning model for selection of the 3′ splice site, although they are consistent with such a model for selection of the 5′ splice site. Finally, Reed and Maniatis (1986) proposed, based on *in vitro* splicing experiments using mutants containing splice site duplications within the human β-globin gene, that the length of intact exon sequences adjacent to a splice site is an important factor. These experiments argue against a scanning mechanism and suggest that sequences within exons influence selection of the correct splice sites in pre-mRNA.

To determine whether a $5' \rightarrow 3'$ scanning mechanism is a major determinant in SV40 early pre-mRNA splicing *in vitro,* we constructed a number of recombinant plasmids containing mutations affecting splicing signals in the SV40 early region, and analyzed the effects of these mutations on splicing in a HeLa cell nuclear extract.

II. MATERIALS AND METHODS

The details of the construction and structure of the plasmids used here have all been described previously (Noble *et al.,* 1986; Fu and Manley, 1987), with the exception of $pYSf^{\Delta}$. This plasmid was constructed by inserting an *Sfa*NI–*Hin*dIII (SV40 nucleotides 4893–4002) SV40 DNA fragment into the *Escherichia coli* RNA polymerase expression vector used previously (Noble *et al.,* 1986, 1987). pRSP-ΔIVS1, which contains the first intron and flanking exons from the adenovirus late transcription unit fused to a bacteriophage SP6 promoter (Konarska *et al.,* 1984), was obtained from R. Padgett and P. Sharp.

Preparation of HeLa nuclear extracts, *in vitro* splicing reactions, and analysis of RNA samples were all performed as described previously (Noble *et al.,* 1986, 1987) and in the figure legends.

III. RESULTS

Capped SV40 early pre-mRNAs were synthesized *in vitro* from DNA templates containing either bacteriophage SP6 or *E. coli* RNA polymerase promoters. A 1169-base pair *Hin*dIII fragment of SV40 DNA, containing the early introns and several hundred basepairs of 5′- and 3′-flanking sequences, was used as the wild-type template, and all mutations used in this study were constructed with this fragment (see Fig. 1). The choice of promoter did not influence in any way the subsequent processing of the pre-mRNAs.

The RNA products obtained following incubation of an [α-^{32}P]GTP labeled SV40 pre-mRNA in a HeLa nuclear extract (Dignam *et al.,* 1983; Krainer *et al.,* 1984) are shown in Fig. 2, in which purified RNAs were denatured by glyoxal and resolved by electrophoresis through a 1.4% agarose gel. Two products, corresponding to spliced large T RNA and the released large T intron, are readily detected. Between 10 and 25% of the precursor is converted to large T-related spliced products, depending on

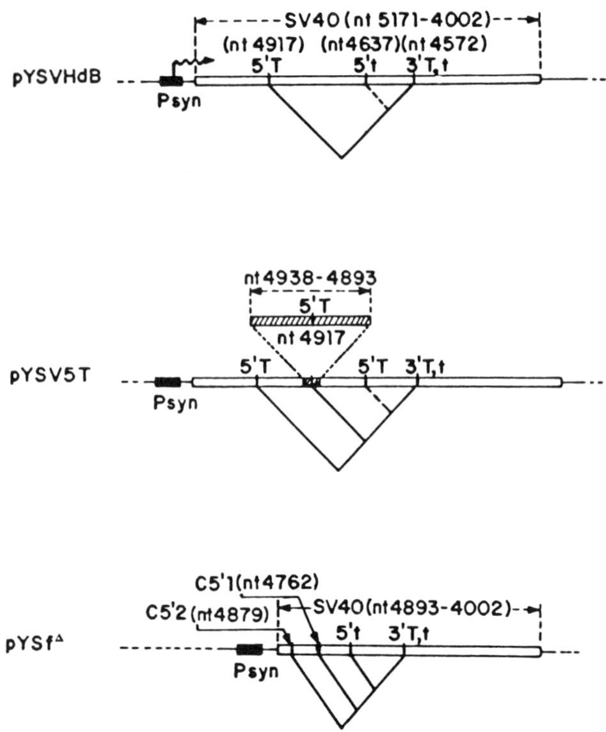

Fig. 1. Wild-type and mutant constructs. The structures of wild-type and mutant constructs used in the preparation of precursor RNAs are illustrated schematically. The locations of normal large T and small t 5′ splice sites and T/t 3′ splice site are indicated for pYSVHdB. The size of the fragment containing the inserted large T 5′ splice site in pYSV5T is given, together with the approximate position of the insertion within the large T intron (nucleotide 4529). The locations of cryptic 5′ splice sites c1 and c2 in pYSF$^{\Delta}$ are also indicated. The synthetic *E. coli* RNA polymerase consensus promoter, which defines the 5′ end of precursors generated for both wild-type and mutant templates, is designated P_{Syn}. Coordinates are SV40 nucleotide numbers (Tooze, 1981).

the extract preparation used. In comparison, 60–75% of a precursor containing a deleted form of the first intron from the adenovirus late transcription unit (L1–L2) is converted to spliced products with the extracts we have used in these experiments, similar to results obtained by others with this precursor (Konarska *et al.*, 1984).

Notable in their absence in Fig. 2 are products reflecting small t mRNA splicing. Although small t-related RNAs have never been detected by the type of analysis shown, more sensitive analyses reveal that very low levels of small t splicing do occur when the wild-type precursor is pro-

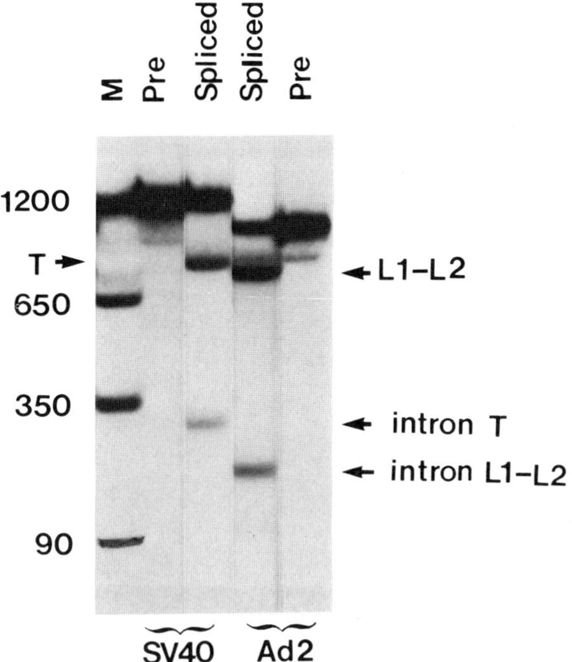

Fig. 2. Gel analysis of wild-type SV40 early and Ad2 IVS 1 pre-RNAs processed *in vitro*. pYSVHdB (SV40) and pRSPΔIVS 1 (Ad2) precursor RNAs were synthesized, processed *in vitro*, and analyzed after glyoxalation on a 1.4% agarose gel (McMasters and Carmichael, 1977). Pre in each case indicates unprocessed precursor RNA. SV40, Species corresponding to large T spliced and lariat large T intron RNAs are indicated; Ad2, Products representing spliced L1–L2 and lariat IVS1 are indicated; M, ^{32}P-labeled RNA markers of sizes 1200, 650, 350, and 90 nucleotides.

cessed *in vitro*. Figure 3 shows the results of such an analysis. Following incubation in nuclear extract, RNA was purified and used as a template for reverse transcriptase with a 5′ ^{32}P-labeled 32-nucleotide synthetic oligonucleotide complementary to sequences just downstream of the 3′ splice site as primer (see Noble *et al.*, 1986). cDNA products were resolved by denaturing polyacrylamide gel electrophoresis and cDNAs of a size corresponding to spliced small t RNA were eluted. An aliquot of this 5′ end-labeled cDNA was hybridized to a complementary, unlabeled RNA transcribed *in vitro* with *E. coli* RNA polymerase from a *Hin*dIII fragment isolated from a small t cDNA clone (obtained from M. Botchan, Berkeley), and which was recloned in an *E. coli* promoter-containing plasmid. Hybrids were treated with S1 nuclease and resolved in a dena-

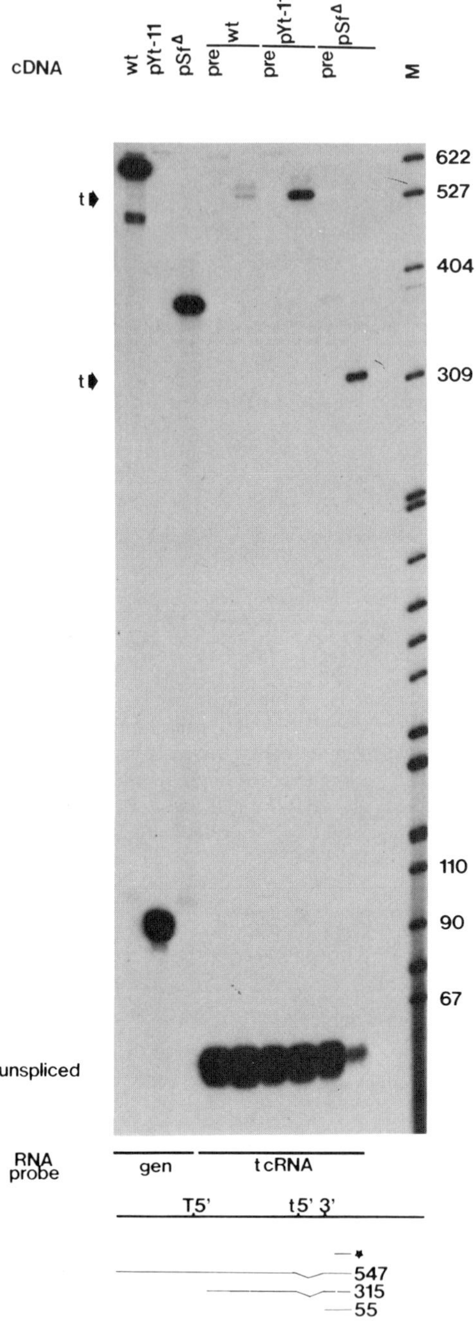
cDNA
wt
pYt-11
pSf$^{\Delta}$
pre
wt
pre
pYt-11
pre
pSf$^{\Delta}$
M
t
t
622
527
404
309
110
90
67
unspliced
RNA probe
gen
tcRNA
T5'
t5' 3'
547
315
55

turing polyacrylamide gel (Fig. 3). By this analysis, only an elongated DNA produced from RNA spliced to small t RNA *in vitro* should be resistant to S1 nuclease. A faint band of the expected size (550 nucleotides) can be observed in the lane labeled "wt." This band is not present in the lanes in which precursor RNA was substituted for RNA from a processing reaction, nor when an RNA probe transcribed from genomic DNA (gen) was substituted for the cRNA in the hybridization analysis, supporting the contention that the band detected represents small t spliced RNA. By estimating the fraction of the primer extension products that were the size of small t cDNA, and the fraction of this cDNA that actually represented spliced small t RNA (Fig. 3), we estimate that approximately 0.1% of the precursor was spliced *in vitro* to small t. Thus, wild-type SV40 early pre-RNA is spliced to small t 0.5–1% as frequently as it is spliced to large T RNA. If we assume that the ratio of small t to large T mRNA detected in transfected HeLa cells (e.g., Manley *et al.*, 1986; Fu and Manley, 1987) reflects the relative use of splice sites, then SV40 early pre-RNA is spliced to small t *in vivo* 20% as frequently as it is spliced to large T mRNA. The relative efficiency of small t splicing *in vitro* is then lower by a factor of 20–40 than it is *in vivo*.

Figure 3 also shows the results obtained when two mutant pre-RNAs were analyzed for small t splicing. One of these, designated pYt-11, contained an insertion of 11 nucleotides within the small t intron (between SV40 nucleotides 4611 and 4612). Approximately fivefold more spliced

---

Fig. 3. S1 nuclease analysis of small t splicing using wild-type and mutant pre-RNAs. PYSVHdB (wt), PYSf$^{\Delta}$ (pSf$^{\Delta}$), and pYt-11 precursor RNAs were synthesized, processed *in vitro*, and analyzed by primer extension using a 5′ end-labeled 32-nucleotide synthetic oligonucleotide complementary to a region immediately downstream of the 3′ splice junction (nucleotide 4546–4547). In each case cDNAs corresponding in size to unspliced precursor RNA and that expected for spliced small t RNA were eluted and analyzed further by S1 mapping. The two unlabeled probes used in this analysis (gen and t cRNA) were prepared by *in vitro* transcription of genomic and small t cDNA early region templates, respectively. gen, cDNAs corresponding to full-length precursor RNA were analyzed by S1 mapping using a genomic RNA probe. wt and pSf$^{\Delta}$ yield protected products of 660 and 390 nucleotides. pYt-11 gives a product of 90 nucleotides which maps a region of divergence between the cDNA and genomic probe due to insertion of an *Eco*RI linker in the small t intron of pYt-11. t cRNA, cDNAs corresponding to spliced small t RNA were analyzed by S1 mapping using a small t cRNA probe. wt and pYt-11 yield protected products of 547 and 55 nucleotides representing spliced small tRNA and unspliced RNA, respectively. pSf$^{\Delta}$ gives products of 315 and 55 nucleotides corresponding once again to small t spliced and unspliced RNA, respectively. Note that equivalent amounts of processed RNA were analyzed for wt and pYt-11, while one-fifth the amount was used for pSf$^{\Delta}$. M, ^{32}P-labeled *Hpa*II digest of pBR322. The sizes and identities of expected products are reproduced schematically at the base of the figure.

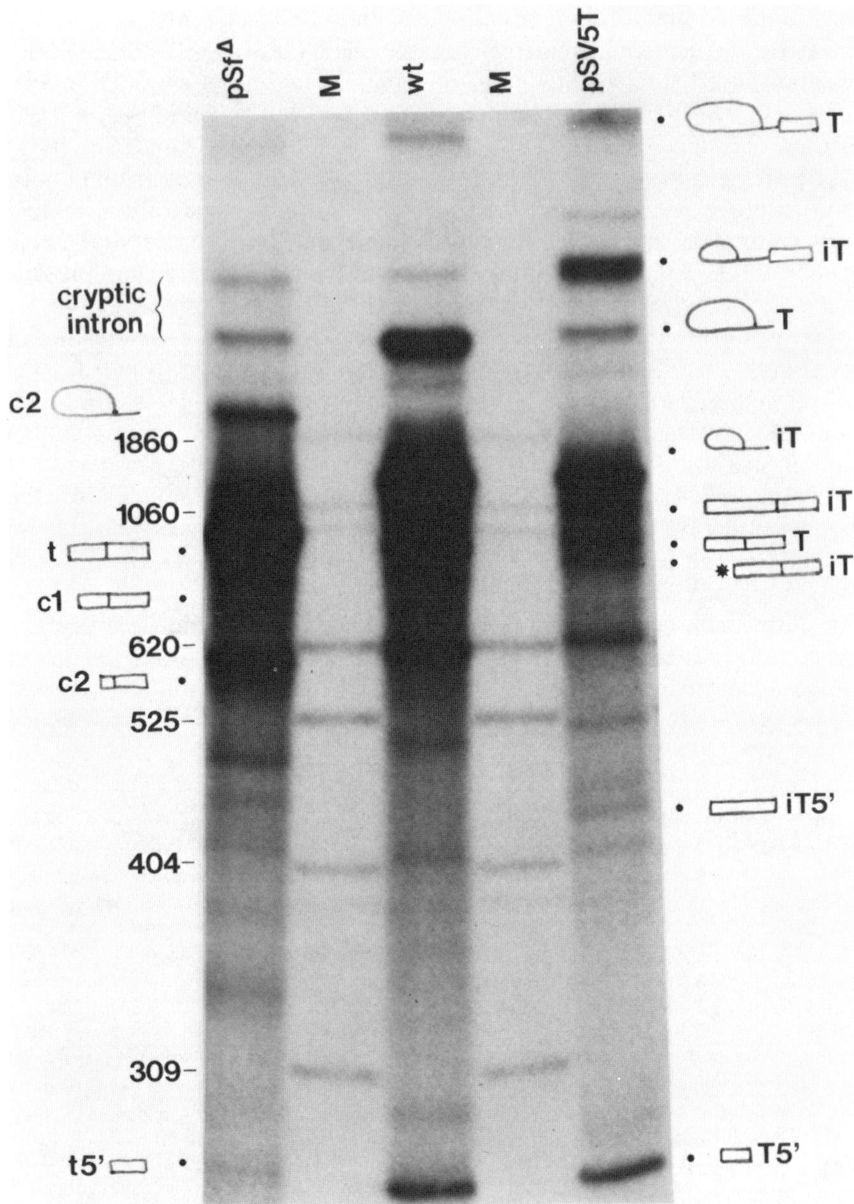

Fig. 4. Gel analysis of wild-type and mutant pre-RNAs processed *in vitro*. PYSVHdB (wt), pYSf$^{\Delta}$ (pSf$^{\Delta}$), and pYSV5T (pSV5T) precursor RNAs were synthesized, processed *in vitro,* and the products fractioned on a 5% acrylamide–8 *M* urea gel. RNA species are labeled using the following symbols, intermediates; 5′ ▭, 5′ exon; lariat intron–3′ exon, products; ▭▭, ligated exons, lariat intron. The identity of each species was

small t RNA was produced from this mutant than from wild-type pre-RNA (Fig. 3). Similar enhancement of small t splicing was observed when this mutation was transferred to a plasmid containing an intact SV40 early region and the RNAs produced *in vivo* following transfection of HeLa cells were analyzed (Fu and Manley, 1987). The enhanced efficiency of small t splicing, which has also been observed with other similar insertion mutants, appears to be due to the increase in the size of this intron (from 66 to 77 nucleotides). Specifically, we believe that a steric constraint on the efficient binding of splicing factors, i.e., snRNPs, has been eliminated by these mutations (see Fu and Manley, 1987, for a discussion).

The other mutant analyzed in the experiment shown in Fig. 3, designated pSf$^{\Delta}$ (pYSf$^{\Delta}$ in Fig. 1), resulted in a more substantial (greater than 10-fold) enhancement in small t splicing, represented by a product of 315 nucleotides. (The amount of pSf$^{\Delta}$ RNA used in the analysis was 20% the amount of wild-type and pYt-11 RNA used.) pSf$^{\Delta}$ is a deletion mutant that lacks the large T 5′ exon and the first 21 nucleotides of the large T intron. Several explanations can be put forth to explain the observed activation of small t splicing with this mutant. One of these is that a 5′ → 3′ scanning model does indeed dictate the pattern of SV40 early splice site selection *in vitro*. However, additional analysis of the products of pSf$^{\Delta}$ splicing suggests that this is not the case. Figure 4 displays directly the RNA products obtained from the *in vitro* splicing of pSf$^{\Delta}$ pre-RNA. A number of species are apparent, and the identity of some of these are indicated in the figure. Most notably, three distinct spliced species resulted from splicing of this precursor *in vitro*. The structures of the RNAs were deduced by Maxam–Gilbert sequencing of the cDNAs produced from the primer extension analysis referred to above (Fig. 5). All three species utilized the natural 3′ splice site. One, as expected, was produced by splicing to the small t 5′

---

confirmed by a combination of primer extension and S1 nuclease analysis of purified RNAs. pSf$^{\Delta}$, Small t spliced (835 nucleotides) and 5′ exon species (265 nucleotides) are indicated. Spliced products utilizing cryptic 5′ splice sites c1 (710 nucleotides) and c2 (590 nucleotides) are also indicated, together with the lariat intron resulting from the latter splice. pSV5T, 5′ Exon species resulting from utilization of either normal (245 nucleotides) or inserted (435 nucleotides) large T 5′ splice site are indicated, as are the lariat intron–3′ exon intermediates in each case. Spliced products corresponding to both normal (820 nucleotides) and inserted (100 nucleotides) large T 5′ splice sites are indicated, as are the lariat intron species. In addition a product of the inserted large T splice labeled with an asterisk (*) is identified. This species is a protection product similar to those described previously (Noble *et al.*, 1986), corresponding to partial degradation of RNA spliced utilizing the inserted large T 5′ splice site. Two anomalously migrating RNAs bracketed and designated "cryptic intron" have been identified as the intermediate and product of a cryptic splice downstream of the T/t 3′ splice junction. The mechanism of activation of this cryptic splice is to be described elsewhere (Noble *et al.*, 1987).

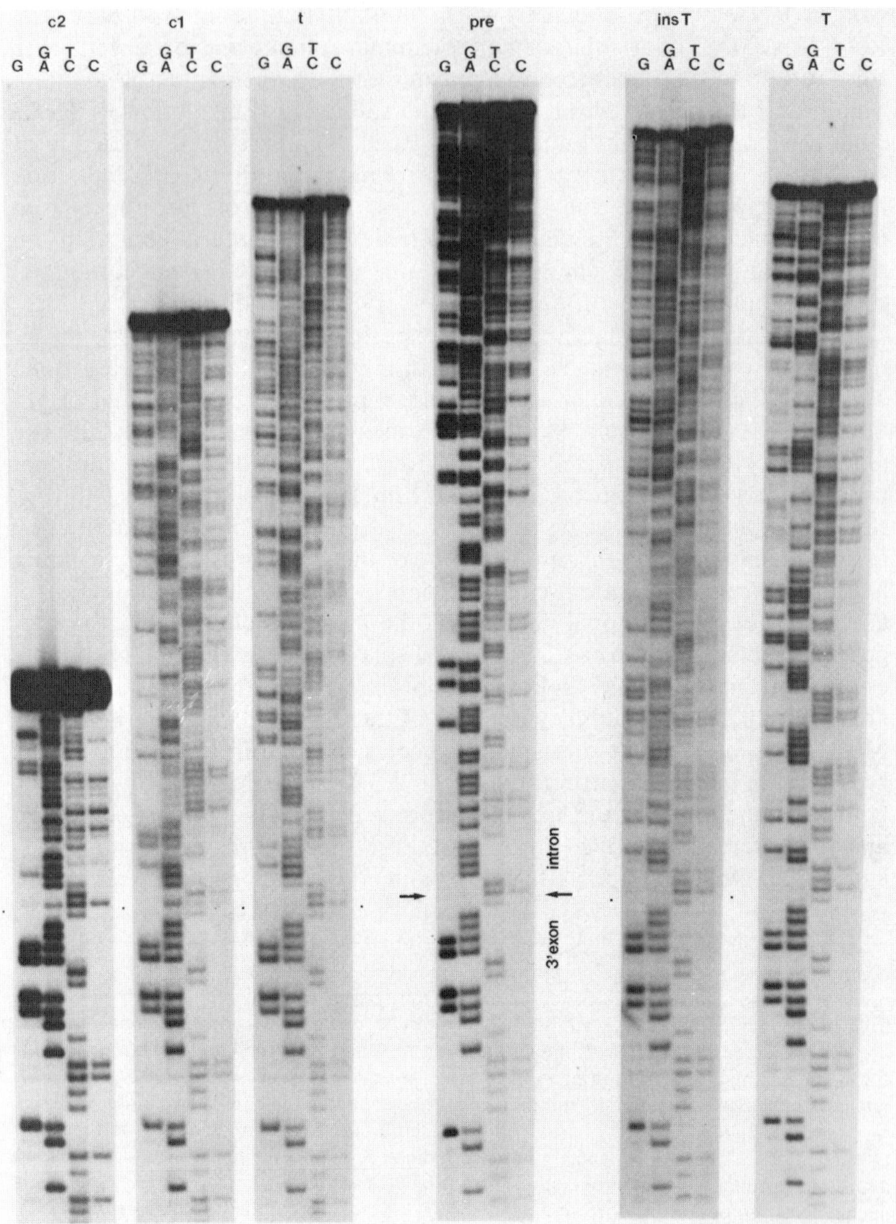

Fig. 5. Nucleotide sequence analysis of cDNAs corresponding to spliced products of wild-type and mutant pre-RNAs. pYSVHdB, pYSf$^{\Delta}$, and pYSV5T precursor RNAs were synthesized, processed *in vitro,* and the products analyzed by primer extension using the

TABLE I

Cryptic 5′ Splice Sites Utilized in Splicing of pSf$^{\Delta}$ Pre-RNA[a]

5′ Splice site consensus	$^{C}_{A}$AG/GTRAGT	Match
5′ Splice site		
Large T	GAG/GTATTT	6/9
Small t	AAG/GTAAAT	8/9
Cryptic c1	TGT/GTGGGT	5/9
Cryptic c2	AAT/GTACTG	5/9

[a] The nucleotide sequences of normal and cryptic 5′ splice sites utilized in splicing of pSf$^{\Delta}$ pre-RNA are shown and compared to the 5′ splice site consensus sequence. R, Purine. The location of the 5′ splice junction in each case is represented by a /. Homology to the consensus is indicated by underlining and described quantitatively in terms of the number of matching nucleotides.

splice site. The other two products, designated c1 and c2 in Fig. 5 (see also Fig. 1), were derived from utilization of upstream, cryptic 5′ splice sites that are not used in wild-type pre-mRNA splicing. Thus, the deletion of the large T 5′ splice site in pSf$^{\Delta}$ results in activation of three downstream 5′ splice sites. The splice sites utilized in this mutant pre-mRNA are reproduced in Table I.

The utilization of multiple 5′ splice sites in splicing of pSf$^{\Delta}$ pre-RNA is not consistent with the strictest form of 5′ → 3′ scanning mechanism, which would predict utilization of only the most upstream 5′ splice site. However, a variation of this model, that incorporates an element of splice site "strength", i.e., degree of homology with the 5′ splice site consensus sequence (Mount, 1982), would be consistent with the data. The two

3′exon primer described in Fig. 3. The resulting cDNAs were fractionated on a 5% acrylamide–8 *M* urea gel and individual species with sizes approximating those expected for the spliced products were eluted and sequenced using the Maxam–Gilbert technique (see Noble *et al.*, 1986). Pre, The nucleotide sequence of a cDNA corresponding to unspliced pYSV5T precursor is given for comparison. The boundary between 3′ exon and T/t intron sequences is indicated. pYSf$^{\Delta}$, Nucleotide sequences of cDNAs representing spliced products utilizing the normal 3′ splice site and small t, c1, and c2 cryptic 5′ splice sites are shown and designated t, c1, and c2, respectively. pYSV5T, Nucleotide sequences of cDNAs corresponding to splices using the normal and inserted large T 5′ splice sites are shown, designated T and ins T, respectively.

cryptic 5′ splice sites share only a limited degree of homology with the consensus sequence (see Table I), and might not effectively "capture" a hypothetical scanning splicing factor. Therefore, to provide an additional test of the scanning model, we constructed and analyzed the duplication mutant pSV5T which contains a small DNA fragment encompassing the large T 5′ splice site inserted into a unique restriction endonuclease cleavage site located near the center of the large T intron [see Fig. 1 and Fu and Manley (1987) for details]. Because the authentic large T 5′ splice site is intact in these mutants, the scanning model predicts that this upstream splice site would be greatly preferred in *in vitro* splicing reactions. However, the results of *in vitro* splicing (Fig. 4) indicate that both splice sites were used with approximately equal efficiency. The identities of the spliced products indicated were confirmed by Maxam–Gilbert sequencing of primer extension products produced from these RNAs (Fig. 5). These findings argue strongly against the notion that a scanning mechanism plays a dominant role in SV40 early splice site selection *in vitro*.

IV. DISCUSSION

The experiments presented here provide a number of insights into factors that influence the alternative splicing pathway displayed by SV40 early pre-mRNA *in vitro*. Although not detected previously among the products of wild-type pre-mRNA processing (Noble *et al.*, 1986), we have shown here that the small t splice does occur *in vitro,* albeit at an extremely low efficiency. Only about 0.1% of the precursor was spliced to small t mRNA, which is over two orders of magnitude less than the efficiency of large T mRNA splicing.

What elements are responsible for the extremely low levels of small t mRNA splicing detected *in vitro?* One possibility is that a different, or additional, trans-acting factor, not required for large T splicing, is required for small t splicing. Although we have presented evidence previously consistent with the notion that differences in trans-acting factors can influence the pattern of SV40 early alternative splicing *in vitro* (Fradin *et al.*, 1984; Fu and Manley, 1987), the experiments described here do not address this possibility, and we will not discuss it further.

The data presented indicate that the inefficiency of small t splicing *in vitro* cannot be attributed solely to a scanning mechanism of splice site selection *in vitro*. Although deletion of the upstream large T 5′ splice site did result in enhancement of small t splicing, it also activated, with approximately equal efficiencies, two upstream cryptic splice sites. However, when the large T (or small t) 5′ splice site was duplicated at a site within the large T intron, the original large T and duplicated 5′ splices sites were used with nearly equal efficiencies, arguing that a scanning model, if valid, is at best only one element in a complexity of factors influencing splice site selection in SV40 early pre-RNA *in vitro*.

What then does control utilization of SV40 early splice sites *in vitro*, and why is the small t splice site used so inefficiently? Two findings described here provide insights into these questions.

A mutation that increased the size of the small t intron from 66 to 77 nucleotides led to an approximately fivefold enhancement of small t splicing *in vitro* (pYt-11, see Fig. 3). Similar results (not shown) were obtained when other mutants containing insertions at the same site were analyzed. These findings suggest that the size of the small t intron is a limiting factor for small t splicing *in vitro*. However, because a very similar increase was detected *in vivo* (Fu and Manley, 1987), we believe that the size constraint on small t splicing is not the primary cause of the extremely low efficiency of small t splicing *in vitro*. Rather, we believe that the size of this intron is an important factor controlling the ratio of small t to large T mRNA splicing *in vivo* (see Fu and Manley, 1987, for discussion).

We have also presented data here that deletion of the large T 5′ splice site resulted in a substantial increase in small t splicing *in vitro* (pSf$^{\Delta}$; Fig. 3). This mutation also resulted in activation of two cryptic 5′ splice sites at positions upstream of the small t splice site (see Fig. 1 and Table I for a summary). When *in vitro* splicing of additional deletion mutants that remove these cryptic splice sites as well as the large t 5′ splice site was analyzed, an even more significant enhancement of small t mRNA splicing was detected (Noble *et al.*, 1986). A comparison of the amount of spliced RNA produced using the small t splice site by these deletion mutants with the amount produced by wild-type pre mRNA reveals an increase of greater than 20-fold. Interestingly, when this efficiency of small t splicing *in vitro* is compared with the efficiency of large T splicing from a wild-type precursor the ration is very similar to that found *in vivo* (see above).

At least two explanations for this data, not mutually exclusive, suggest themselves. First, the deletions may alter the secondary or tertiary structure of the RNA such that the small t 5′ splice site is more amenable to the splicing apparatus. Two lines of evidence, however, suggest that if higher-order RNA structure does block small t 5′ splice site utilization *in vitro*, it is not the sole factor. The first derives from our analysis of other "wild-type" precursors, which contain more or less SV40 sequences at their 5′ or 3′ ends, or which contain alterations within the large T intron, but leave other known splicing signals intact. No enhancement of small t splicing has been detected with any of these mutants (results not shown), consistent with the idea that the important feature of mutations that activate small t splicing may be the removal of competing 5′ splice sites. Also relevant to the question of whether higher-order RNA structure prevents small t splicing *in vitro* are previous studies that analyzed RNA species arising from protection of uncapped wild-type precursor from degradation by an endogenous 5′ → 3′ exonuclease active in splicing extracts (Noble *et al.,* 1986). One protection product with a 5′ end mapping to a position approximately 10 nucleotides upstream of the small t 5′ splice site appeared to represent association of a splicing component, probably U1 small nuclear ribonucleoprotein (snRNP) (Mount *et al.,* 1983), with the small t 5′ splice site. The existence of this RNA, which was detected with relatively good yield, suggests that at least the initial interaction of splicing components with this splice site can occur *in vitro*. Thus, if RNA structure does prevent small t splicing, it is not because the splice sites are inaccessible to splicing factors.

A second explanation for the inactivity of the small t 5′ splice site *in vitro* might be suggested by the primary sequence of the splice site. Inspection of Table I suggests that, if the extent of homology to the consensus 5′ splice site sequence is an indication of splice site strength, then the small t splice site should be the most, not the least, efficiently used in SV40 early pre-mRNA. Indeed, since naturally occurring perfect matches with the 9-nucleotide consensus have not been noted (Mount, 1982; S. Mount, personal communication), the small t 5′ splice site displays the best match with the consensus sequence of any known 5′ splice site. In light of the observed inactivity of the small t splice site *in vitro,* could it be that this match may, in fact, be too good? According to this view, it may be imagined that U1 snRNP, which recognizes and binds to 5′ splice sites (Mount *et al.,* 1983), might interact so strongly with the small t 5′ splice site that it actually impedes small t splicing *in vitro*. Consistent with this possibility are recent experiments suggesting that U1 snRNPs may interact only transiently with the pre-mRNA, and may not form part of the splicing complex (Grabowski and Sharp, 1986). Perhaps release of U1 snRNP from the small t 5′ splice site does not occur efficiently *in vitro*. Arguing against such a "tight binding inhibition" model in its simplest form, however, are our observations that the small t 5′ splice site is used efficiently both *in vivo* (Fu and Manley, 1987) and *in vitro* (unpublished data) when moved further upstream in the pre-mRNA.

In summary, we have shown that although SV40 small t mRNA splicing is not readily detected *in vitro,* it does occur at a very low efficiency. This inefficiency does not appear to be due solely to the operation of 5′ → 3′ scanning mechanism of splice site selection *in vitro*. Rather, it is likely that several factors contribute to the inactivity of the small t 5′ splice site.

These most likely include the small size of the intron, the higher-order structure of the RNA, the nucleotide sequence at the 5′ splice site, and/or a requirement for different trans-acting components. While the studies here have suggested a number of factors that may influence the selection of alternative splice sites in pre-mRNA, the nature of the role played by each requires further studies.

ACKNOWLEDGMENTS

We thank M. Kopczynski for technical assistance, R. Padgett and P. Sharp for the plasmid pRSP-1-ΔIVS, and C. Prives and S. Mount for helpful discussions. This work was supported by National Institutes of Health Grant CA33620 and National Science Foundation Grant PCM-82-16798.

REFERENCES

Amara, S. G., Jonas, V., Rosenfeld, M. G., Ong, E. S., and Evans, R. M. (1982). *Nature (London)* **298,** 240–244.

Benyajati, C., Spoerel, N., Naymerle, H., and Ashburner, M. (1983). *Cell (Cambridge, Mass.)* **20,** 293–301.

Dignam, J. D., Lebovitz, R. M., and Roeder, R. G. (1983). *Nucleic Acids Res.* **11,** 1475–1489.

Early, P., Rogers, J., Davis, M., Calame, K., Bond, M., Wall, R., and Hood, L. (1980). *Cell (Cambridge, Mass.)* **20,** 313–319.

Fradin, A., Jove, R., Hemenway, C., Keiser, H. D., Manley, J. L., and Prives, C. (1984). *Cell (Cambridge, Mass.)* **37,** 927–936.

Fu, X.-Y., and Manley, J. L. (1987). *Mol. Cell. Biol.* **7,** 738–748.

Grabowski, P. J., and Sharp, P. A. (1986). *Science* **233,** 1294–1299.

Konarska, M. M., Padgett, R. A., and Sharp, P. A. (1984). *Cell (Cambridge, Mass)* **38,** 731–736.

Krainer, A. R., Maniatis, T., Ruskin, B., and Green, M. R. (1984). *Cell (Cambridge, Mass)* **36,** 993–1005.

Kuhne, T., Wieringa, B., and Weissmann, C. (1983). *EMBO J.* **2,** 727–733.

Lang, K. M., and Spritz, R. A. (1983). *Science* **220,** 1351–1355.

Leff, S. E., Rosenfeld, M. G., and Evans, R. M. (1986). *Annu. Rev. Biochem.* **55,** 1091–1117.

Maki, R., Roeder, W., Traunecker, A., Sidman, C., Wable, M., Raschke, W., and Tonegawa, S. (1981). *Cell (Cambridge, Mass.)* **24,** 353–365.

McMaster, G. K., and Carmichael, G. C. (1977). *Proc. Natl. Acad. Sci. U.S.A.* **74,** 4835–4838.

Manley, J. L., Noble, J. C. S., Chandhuri, M., Fu, X.-Y., Michaeli, T., Ryner, L., and Prives, C. (1986). "Cancer Cells," Vol. 4., pp. 259–265. Cold Spring Harbor Lab., Cold Spring Harbor, New York.

Mount, S. M. (1982). *Nucleic Acids Res.* **10,** 459–472.

Mount, S. M., Petersson, I., Hinterberger, M., Karmas, A., and Steitz, J. A. (1983). *Cell (Cambridge, Mass.)* **33,** 509–518.

Nabeshima, Y., Kurijama-Fujii, Y., Muramatsu, M., and Ogata, K. (1984). *Nature (London)* **308,** 333–338.

Noble, J. C. S., Prives, C., and Manley, J. L. (1986). *Nucleic Acids Res.* **14,** 1219–1235.

Noble, J. C. S., Pan, Z. Q., Prives, C., and Manley, J. L. (1987). *Cell (Cambridge, Mass.)* **50,** in press.

Padgett, R. A., Grabowski, P. J., Konarska, M. M., Seiler, S., and Sharp, P. A. (1986). *Annu. Rev. Biochem.* **55,** 1119–1150.

Reed, R., and Maniatis, T. (1986). *Cell (Cambridge, Mass.)* **46,** 681–690.

Rosenfeld, M. G., Amara, S. G., and Evans, R. M. (1984). *Science* **225,** 1315–1320.

Schibler, U., Hagenbuchle, O., Wellauer, P. K., and Pittet, A. C. (1983). *Cell (Cambridge, Mass.)* **33,** 501–508.

Sharp, P. A. (1981). *Cell (Cambridge, Mass.)* **23,** 643–646.

Tooze, J. (1981). "DNA Tumor Viruses," 2nd ed. Cold Spring Harbor Lab., Cold Spring Harbor, New York.

Young, R. A., Hagenbuchle, O., and Schibler, U. (1981). *Cell (Cambridge, Mass.)* **23,** 451–458.

8

Messenger RNA Splicing in Yeast

ARTHUR J. LUSTIG,[1] REN-JANG LIN,[2] AND JOHN ABELSON

Division of Biology
California Institute of Technology
Pasadena, California 91125

I. AN OVERVIEW OF NUCLEAR mRNA SPLICING

The process of nuclear mRNA splicing, the efficient and precise deletion of intervening sequences (IVS, or introns) from precursor mRNAs, is a process common to most, if not all, eukarytotic organisms. Recent studies in numerous laboratories have resulted in a significant advancement in our understanding of the mechanism of this process. In particular, the development and characterization of *in vitro* splicing systems (Hernandez and Keller, 1983; Grabowski *et al.*, 1984; Padgett *et al.*, 1984; Ruskin *et al.*, 1984; Lin *et al.*, 1985; Newman *et al.*, 1985), as well as *in vivo* characterization of splicing intermediates (Domdey *et al.*, 1984; Rodriguez *et al.*, 1984; Zeitlin and Efstratiadis, 1984), have led to the elucidation of a two-step mechanism for nuclear mRNA splicing (Fig. 1).

In the first step of the reaction, a cleavage at the 5′ end of the intron is accompanied by the formation of a 2′-5′ phosphodiester bond between the 5′ nucleotide of the intron and an internal conserved sequence (see below). Both the first exon (exon 1) and an unusual lariat intermediate, containing the intron and exon 2, are formed as intermediates. This reac-

[1] Present address: Program in Molecular Biology, Memorial Sloan-Kettering Cancer Center, New York, New York 10021.

[2] Present address: Department of Microbiology, University of Texas, Austin, Texas 78712.

Molecular Biology of RNA
New Perspectives

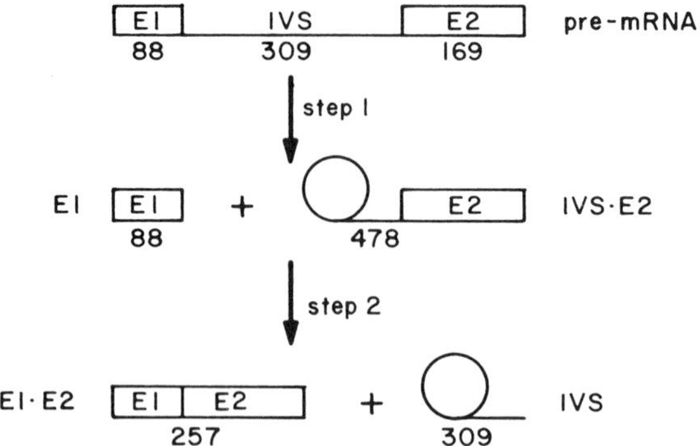

Fig. 1. Mechanism of mRNA splicing *in vitro*. This figure shows the synthetic SP6 actin pre-mRNA, utilized in many of the *in vitro* assays described in the text, and its processing in an *in vitro* splicing system. The SP6 promoter–actin fusion was constructed as described previously (Lin *et al.*, 1985). The labeled pre-mRNA produced *in vitro* by SP6 polymerase is processed in a two-step reaction, described in the text, giving rise to the intermediates, an 88-nucleotide exon 1 (E1) and a 478-nucleotide lariat (IVS·E2), and the products, a 257-nucleotide mature mRNA (E1·E2) and a 309-nucleotide lariat intervening sequence (IVS). (Reproduced from Lustig *et al.*, 1986. Copyright held by Cell Press.)

tion is followed by a cleavage at the 3′ end of the intron and ligation of the two exons. The final products are, therefore, the spliced mRNAs and an intron in a lariat configuration. This two-step mechanism has been demonstrated by the analysis of *in vitro* splicing systems, as well as through biochemical and microscopic characterization of *in vivo* intermediates and products.

A number of sequences critical to the splicing process have been defined by sequence conservation (Fig. 2) and *in vitro* mutagenesis studies

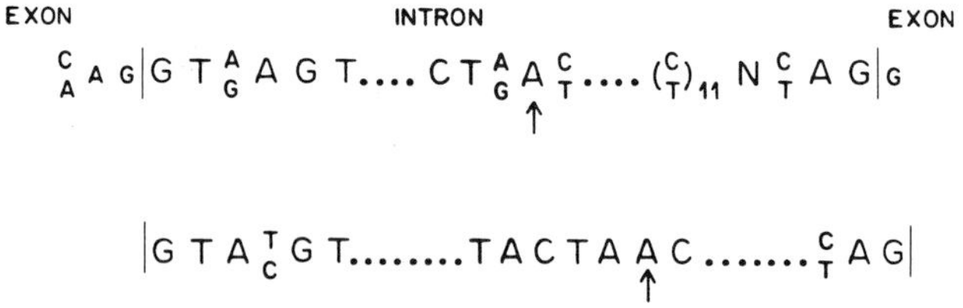

EXON INTRON EXON

$^{C}_{A}$ A G | G T $^{A}_{G}$ A G T.... C T $^{A}_{G}$ A $^{C}_{T}$.... $(^{C}_{T})_{11}$ N $^{C}_{T}$ A G | G
↑ (under branch A)

| G T A $^{T}_{C}$ G T........T A C T A A C.......$^{C}_{T}$ A G |
↑ (under branch A)

Fig. 2. Consensus splicing sequences in yeast and higher eukaryotes. Shown here is a comparison of the sequences suggested to be important in splicing through sequence conservation studies. The top of the figure (Mount, 1982; Keller and Noon, 1984) demonstrates the sequences that are likely to be involved in higher eukaryotic splicing, while the bottom of the figure (Teem *et al.*, 1984) shows the consensus sequences present in yeast. Arrows indicate the site of branch formation. The left and right solid lines demarcate the 5′ and 3′ ends of the intron, respectively.

(reviewed in Padgett *et al.,* 1986). In yeast, the 5′ splice site occurs immediately upstream of the highly conserved GTATGT sequence. This 5′G forms the 2′-5′ bond with the penultimate nucleotide within the internal strictly conserved TACTAAC box. Furthermore, the sequence immediately preceding the 3′ splice site is invariably an AG dinucleotide. In contrast, in higher eurkaryotes the sequences diverge considerably from those present in yeast. Nonetheless, a GT dinucleotide at the 5′ end of the intron, an AG dinucleotide at the 3′ end of the intron, and a loosely conserved TACTAAC-like sequence also appear to be present, the latter being the site of branch formation.

The manner in which the essential splicing sequences are brought together to allow the various steps of the reaction to proceed is not yet well understood. However, some recent data may shed some light on this problem. In both yeast and higher eukaryotes, at least a part of the splicing process appears to take place on a large multimeric complex of 40–60 S (Brody and Abelson, 1985; Frendewey and Keller, 1985; Grabowski *et al.,* 1985). Such a complex can be identified by mixing radiolabeled precursor mRNA to a splicing extract in the presence of ATP (see Fig. 3 below), followed by glycerol gradient sedimentation. The formation of this complex depends both upon the presence of ATP and sequences essential for splicing. However, until recently, there has been very little evidence that this complex is directly involved in splicing. The spliceosome, which is approximately the size of a ribosomal subunit, may consist of as many as 30 proteins, as well as numerous RNA components. The nature of the constituents of the spliceosome is only beginning to be understood.

Small nuclear ribonucleoprotein particles (snRNPs), composed of a set of highly conserved proteins tightly associated with a small nuclear RNA (snRNA), are also intrinsically involved in this process. In higher eukaryotes at least four snRNP particles (U1, U2, U4/U6, U5) are essential components of the mRNA splicing machinery (Yang *et al.,* 1981; Padgett *et al.,* 1983; Bozzoni *et al.,* 1984; Fradin *et al.,* 1984; Kramer *et al.,* 1984; Black *et al.,* 1985; Chabot *et al.,* 1985; Krainer and Maniatis, 1985; Berget and Robberson, 1986; Black and Steitz, 1986). These snRNPs are capable of recognizing and interacting with specific intronic sequences (Mount *et al.,* 1983; Tatai *et al.,* 1984; Black *et al.,* 1985; Chabot *et al.,* 1985) and at least some have been demonstrated to be present in the spliceosome (Grabowski *et al.,* 1985; Grabowski and Sharp, 1986). In yeast, no snRNP has been clearly identified. However, recent studies have shown that snRNAs are indeed involved in splicing (Cheng and Abelson, 1986) and, similar to their mammalian counterparts, are constituents of the spliceosome (Pikielny and Rosbash, 1986; J. Abelson, personal communication; C. Guthrie, personal communication).

Furthermore, at least one protein (the C protein) present in heterogeneous nuclear (hn) RNPs appears to be an important component of the splicing process and is likely to be present within the spliceosome (Choi *et al.,* 1986).

The functions of the spliceosome, and the sn and hnRNPs, remain unknown. However, the complexity of this system suggests that these complexes may serve as a scaffold upon which the reactive sequences are brought into close proximity. Alternatively, some of these RNP complexes may be the catalytic components responsible for endonucleolytic and ligation activities (see Section V).

We have been investigating the splicing process in the yeast *Saccharomyces cerevisiae*. In this chapter, we will summarize studies conducted in our laboratory over the past several years that have utilized a combination of biochemical and genetic approaches, uniquely available in yeast, to begin to dissect the components involved in mRNA splicing and to ascertain their function.

II. PRELIMINARY *IN VITRO* AND *IN VIVO* CHARACTERIZATION OF YEAST mRNA SPLICING

The development of a yeast *in vitro* system has been critical for the understanding of the splicing process in this organism (Lin *et al.,* 1985; Newman *et al.,* 1985). In this system, ^{32}P-labeled RNA is transcribed from a truncated actin or *CYH* ribosomal protein gene under the control of the highly efficient bacteriophage SP6 promoter. The resulting synthetic intron-containing mRNA precursor is processed within a yeast whole-cell extract to form splicing intermediates and products in an ATP-dependent reaction. A number of criteria were used to establish the identity of the *in vitro* species. First, nucleotides spanning the exon1/exon2 junction were hybridized to the products of the *in vitro* system and subsequently digested with the single-stranded nuclease S1. A fragment predicted to be protected by the oligonucleotide in the mature mRNA, but not in precusor RNA, was identified. Characterization of the mRNA product by ribonuclease digestion confirmed the accuracy of the splicing event. Furthermore, the exon 1 intermediate and the mRNA product migrated at the expected position on denaturing polyacrylamide gels. However, both the intron-containing intermediate and product displayed anomolous behavior on these gel systems, suggesting that these molecules may have a circular topology. Further analysis of these species by ribonuclease digestion revealed the presence of a branch point resulting from a unique 2′-5′ linkage. These data indicated that the intron intermediate and product were in a lariat configuration. In this way, the splicing process in yeast was demonstrated to conform to the model shown in Fig. 1. As discussed above, other investigators have demonstrated that a similar mechanism is present in higher eukaryotes. Parallel characterization of labeled splicing intermediates and products demonstrated that the lariat molecules are formed *in vivo,* as well as *in vitro* (Domdey *et al.,* 1984). Thus, by these criteria, the *in vitro* system is likely to reflect the mechanism present *in vivo*.

Utilizing this *in vitro* system, our laboratory has demonstrated that splicing *in vitro* involves a large multimeric complex, which we have termed the spliceosome (Fig. 3) (Brody and Abelson, 1985). When the *in vitro* splicing reaction is carried out in the presence of ATP and ^{32}P-labeled precursor, a 40 S complex is observed on glycerol gradients. This 40 S complex binds the precursor as well as the intermediates, the intron lariat product, and some mature mRNA. The formation of the complex is dependent upon the presence of essential intronic sequences (Brody and Abelson, 1985; Vijayraghavan *et al.,* 1986). Some mutations introduced at the 5′ splice site and alterations of the branch point sequence lead to an inability to form the spliceosome as well as to the complete loss of reactivity. In contrast, a mutation at the 3′ splice site allows the spliceosome to form and intermediates to accumulate. These results demonstrate the importance of sequences present at the branch point and the 5′ splice junction for the formation of the yeast spliceosome. In addition, a 20 S complex has also been identified that contains predominantly the end-products of the splicing reaction (C. Guthrie, personal communication). Several recent studies have indicated that the 40 S and 20 S complexes are associated with snRNAs that are likely to be essential components of the splicing process (Cheng and Abelson, 1986; Pikielny and Rosbash, 1986; Hostomsky *et al.,* 1987; Riedel *et al.,* 1987). The functions of these RNAs remain unknown.

III. CHARACTERIZATION OF MUTATIONS IN THE SPLICING PROCESS

A. The Yeast *rna* Mutations

The development of the *in vitro* system described above has yielded a great deal of information regarding the basic steps involved in mRNA splicing. In addition, it has also afforded us the opportunity to screen yeast mutants potentially defective in the splicing process. The identification and characterization of such mutations could help to define the steps

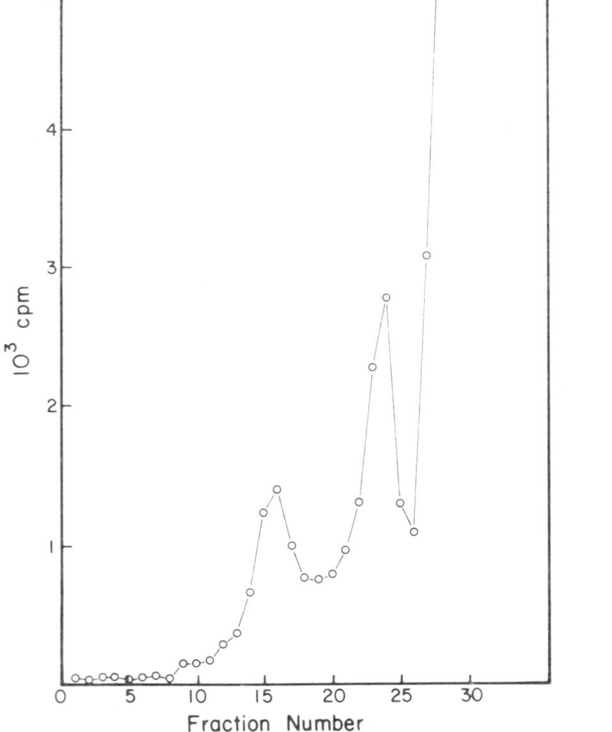

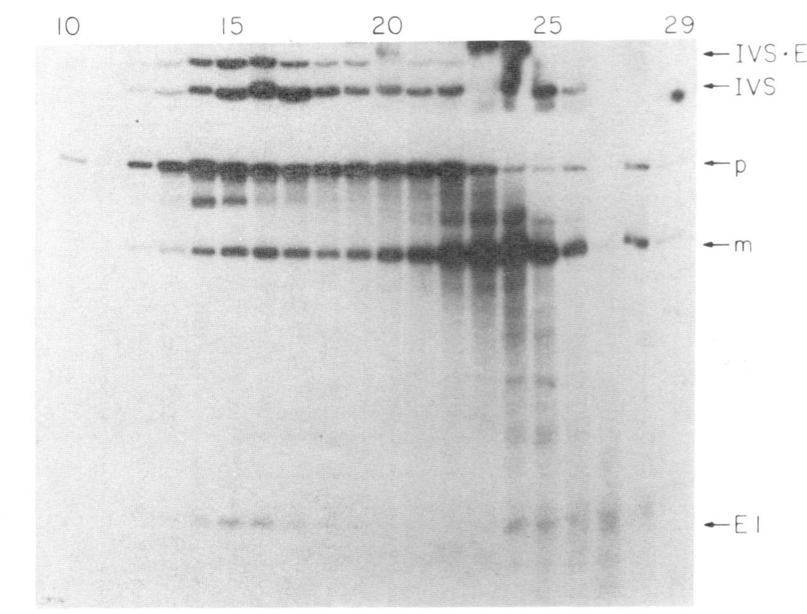

Fig. 3. Spliceosome formation in *vitro*. (Left) This figure demonstrates the sedimentation properties of the splicesome formed *in vitro*. ^{32}P-Labeled synthetic SP6 action transcript (see Fig. 1) was added to a wild-type whole cell extract in the presence of ATP under splicing conditions (Lin *et al.*, 1985), and the reaction sedimented through a glycerol gradient. Fractions were collected from the gradient and the ^{32}P-labeled transcript and splicing products counted. The right side of the gradient represents the top of the gradient. Two complexes were observed. The first peak migrates at 40 S and corresponds to the spliceosome complex. The second peak migrates at 20 S and represents the "product" peak. The label at the top of the gradient is predominantly degraded precursor mRNA. (Right) Fractions collected from the gradient were deproteinized, denatured, and fractionated on a denaturing polyacrylamide gel. The fraction numbers corresponding to the gradient are listed on top. The arrows on the right hand side indicate the position of the lariat intermediate (IVS·E2), intron lariat (IVS), precursor (p), mature mRNA (m), and exon 1 (E1) described in Fig. 1. All species comigrate with the 40 S peak [illegible] contained within the 20 S peak. (This figure courtesy of Dr. S.-C. Cheng.)

and components participating in mRNA splicing. Similar approaches utilized in other systems (e.g., bacterial DNA replication) have been critical for the dissection of those processes.

The mutants that we have been analyzing are the yeast *rna* mutants. These mutants were initially isolated by Hartwell (1967) by nitrosoguanidine mutagenesis. The resulting temperature sensitive (ts) mutations were screened for their inability to accumulate RNA, but not protein, at the restrictive temperature 36°C. The mutants isolated fell into 11 complementation groups, termed *rna*1–*rna*11.

One of these mutations, *rna*1 (Hutchinson *et al.*, 1969), results in pleiotropic effects, including the accumulation of tRNA precursors (Hopper *et al.*, 1978; Knapp *et al.*, 1978; O'Farrell *et al.*, 1978; Etcheverry *et al.*, 1979) and a defect in some aspect of RNA transport (Hutchinson *et al.*, 1969; Shiokawa and Pogo, 1974). Due to its complexity, the nature of this defect has remained unknown, but may involve a general defect in RNA transport from nucleus to cytoplasm.

The *rna*2–*rna*11 mutations, on the other hand, cause a rapid cessation of mature rRNA formation at the restrictive temperature (Hartwell *et al.*, 1970). It was initially thought that these mutations were directly involved in the processing of the 35 S rRNA precursor to its mature products, since the RNA precursor accumulates at the restrictive temperature (Warner and Udem, 1972). However, it was subsequently determined that ribosomal protein synthesis was diminished at 36°C in these mutants due to a decrease in mature mRNA coding for the ribosomal proteins (Gorenstein and Warner, 1976; Warner and Gorenstein, 1977). Presumably, the inability of ribosomal proteins to assemble into a pre-ribosome indirectly leads to the loss of rRNA precursor processing.

In recent years, it has been found that, unlike many eukaryotes, yeast contains introns in only a subset of genes, including most of the ribosomal protein genes (Rosbash *et al.*, 1981; Bollen *et al.*, 1982; Leer *et al.*, 1982; Kaufer *et al.*, 1983; Mitra and Warner, 1984), the actin (Gallwitz and Sures, 1980; Ng and Abelson, 1980) and tubulin (Schatz *et al.*, 1986) genes, the *MAT***a1** gene (Miller, 1984), and the gene coding for one of the nuclear-coded subunits of cytochrome oxidase (subunit V) (M. Cumsky, personal communication). This discovery led to the suspicion that the *rna*2–*rna*11 mutations may be defects in some aspect of mRNA splicing. Indeed, two lines of evidence pointed in that direction. First, all precursors containing introns accumulate at the restrictive temperature in all of those mutants tested (Fried *et al.*, 1981; Rosbash *et al.*, 1981; Larkin and Woolford, 1983; Teem *et al.*, 1983; Miller, 1984). Second, precursor accumulation in an *rna*2-containing strain is intron dependent. When the bacterial β-galactosidase gene, under the control of a yeast promoter, is interrupted by a ribosomal protein intron, its expression in a yeast strain carrying the *rna*2 mutation becomes temperature sensitive (Teem and Rosbash, 1983).

Although these data suggest a role of the *RNA* gene products in mRNA processing and splicing, the nature of this involvement has remained unclear. Concern over the direct impact of these mutations on splicing was accentuated by the case of the *rna*1 mutation, in which the effect on tRNA processing appears to be secondary to a defect in RNA transport (Hutchinson *et al.,* 1969; Shiokawa and Pogo, 1974). The presence of mutations that suppress both the *rna*1 and *rna*2–*rna*11 mutations further confused this issue (Pearson *et al.,* 1982).

Clearly, if the *rna* mutations do result in intrinsic defects in splicing, they can be utilized both in the isolation and characterization of the splicing proteins and snRNPs, and in the dissection of the spliceosome. Therefore we attempted to ascertain the nature of the *RNA* gene products through the utilization of *in vitro* approaches to this problem. Details of the work described below have been presented elsewhere (Lustig *et al.,* 1986).

B. An *in Vitro* Heat Sensitivity Assay

To ascertain the nature of the *rna* mutant defects, we sought to develop an *in vitro* assay that mimics the *in vivo* temperature-dependent accumlation of intron-containing precursors. To this end, we utilized the *in vitro* system described above. For these studies, we have used, as substrate, a transcript produced from a truncated actin gene under the control of the bacteriophage SP6 promoter (see Fig. 1).

To analyze the impact of the *rna* mutations on splicing, we have assayed the lability of extracts isolated from the *rna*2–*rna*11 mutants to normally permissive *in vitro* temperatures. To this end, we have developed an *in vitro* heat-inactivation assay. The extracts were incubated at 30–32°C under a number of different conditions, and the splicing reaction carried out in the presence of ATP and precursor at an *in vitro* permissive temperature of 15°C. Details of this assay have been described elsewhere (Lustig *et al.,* 1986). An example of such a heat-inactivation assay is shown in Fig. 4 (left). This figure demonstrates that splicing activity in extracts isolated from *rna*2, *rna*5, and *rna*11 mutant strains is eliminated after 30 min of incubation. Extracts isolated from most wild-type strains, on the other hand, were stable under these conditions. Both intermediates and products were eliminated, suggesting that these defects are present in early stages of the splicing reaction.

All of the *rna* mutants have been assayed under a number of different

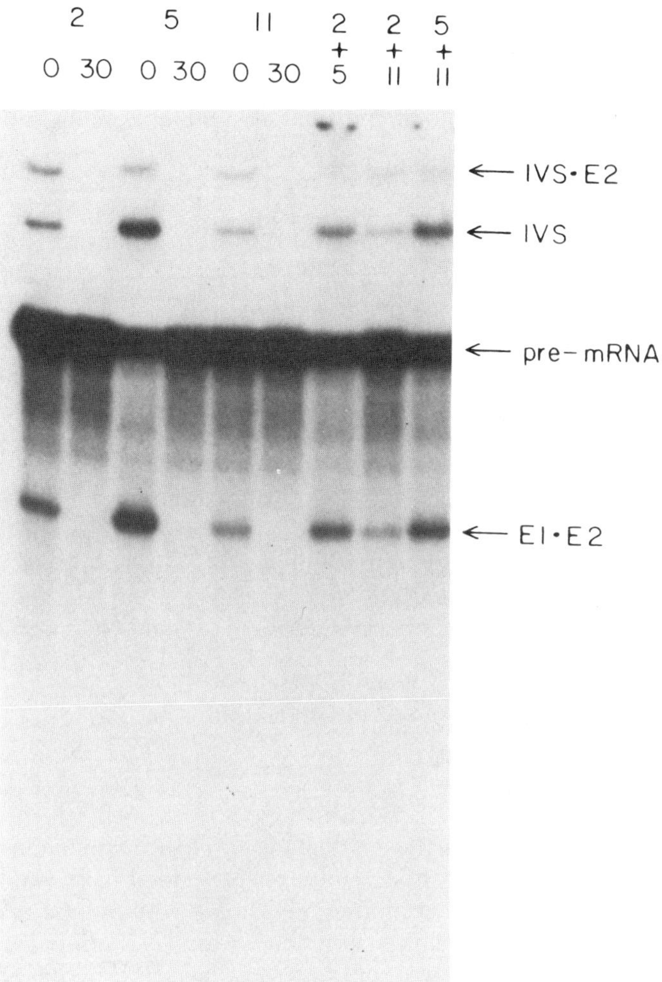

Fig. 4. *In vitro* complementation of *rna* mutant extracts. *rna*2, *rna*5, and *rna*11 extracts (2, 5, 11) were heat-inactivated for the times indicated (in minutes) and were assayed either separately or in pairwise combination (2 + 5, 2 + 11, 5 + 11) following this treatment. Heat inactivation was performed as described (Lustig *et al.*, 1986). The splicing reaction was carried out using the synthetic SP6 actin transcript, shown in Fig. 1, and the products analyzed by denaturing polyacrylamide gel electrophoresis. Arrows indicate the position of the precursor, intermediates, and products shown in Fig. 1. All combinations of these extracts restored splicing activity. (Reproduced from Lustig *et al.*, 1986. Copyright held by Cell Press.)

conditions in which the pH, potassium concentration, and ATP levels were varied. This analysis indicated that extracts isolated from all of the *rna* mutants, with the exception of *rna*6 and *rna*9, can be heat-inactivated *in vitro*. In all cases intermediates do not accumulate, indicating that the defects affect an early stage of the process.

The ability of most of the *rna* mutant extracts to demonstrate heat sensitivity *in vitro* suggested that the *RNA* gene products may be directly involved and essential for the splicing process. It should be pointed out that the inability to obtain an *in vitro* phenotype for *rna*6 and *rna*9 mutations may be the consequence of the assay conditions which demand an irreversible denaturation of splicing components. Thus, the *RNA*6 and *RNA*9 gene products may be splicing components capable of renaturation following heat inactivation. Alternatively, the proper inactivation conditions may not yet have been identified for these mutants. It is also possible that these gene products do not directly affect splicing, but rather serve as modulators or regulators of the splicing process.

The demonstration of heat inactivation is of course not sufficient to demonstrate a role of these gene products in splicing. Nonspecific effects such as proteolysis and nucleolytic digestion during the heat inactivation could give rise to similar effects. We therefore sought to determine both the specificity of the inactivation and the relationship of the *in vitro* defects to the *rna* mutations.

C. Specificity of the Heat-Inactivation Assay

One approach to address these issues is *in vitro* complementation. If extracts isolated from different *rna* mutants are added pairwise following heat inactivation, then most combinations should result in the restoration of activity (or complementation) if the defects are (1) in different splicing components and (2) the components are exchangeable. Nonspecific effects, on the other hand, would not be expected to result in a consistent pattern of complementation.

Extracts were mixed after heat inactivation, the splicing reaction initiated, and the products analyzed. The results of such an analysis are shown in Fig. 4. As discussed above, when present alone, extracts isolated from *rna*2, *rna*5, and *rna*11 strains are heat sensitive *in vitro*. However, when combinations of heat-inactivated *rna*2, *rna*5, and *rna*11 extracts were added following heat inactivation, partial or complete reactivity was restored. This assay was extended to all of the heat-sensitive *rna* mutants. Most combinations of extracts were able to complement. In contrast, all pairs of extracts isolated from cells carrying the same mutation in different genetic backgrounds failed to demonstrate *in vitro* complementation.

Several conclusions can be drawn from these data. First of all, the presence of a large number of extract conbinations that complement indicates that (1) many of the defects are in components of the splicing process and (2) these components are exchangeable. Second, the absence of complementation in like combinations argues for the specificity of this process and the linkage of the *in vitro* defects to the *rna* mutations (see below). Finally, the absence of complementation in some unexpected cases may be due to nonspecific effects, or alternatively, to tightly bound components of a single particle (e.g., a snRNP) that cannot exchange under the assay conditions. These data therefore argue that many of the effects observed *in vitro* are defects in the splicing process in an early step of the reaction.

D. Linkage of Heat-Inactivation Behavior to the *rna* Mutations

These data demonstrated that heat inactivation is, in many cases, due to the specific loss of an essential splicing component. However, we still needed to demonstrate that the splicing defects are the consequence of the *rna* mutations. We took several approaches to address this issue. For these studies, we concentrated on the analysis of the *rna*2, *rna*5, and *rna*11 mutations.

First of all, revertants of the *rna*2 mutation were isolated by their ability to grow at the restrictive temperature. These revertants were demonstrated genetically to be within the *RNA2* gene. Most of these revertants (3/4) also reverted the *in vitro* phenotype, indicating the linkage of the splicing defect with the *rna*2 mutation.

We have also approached this problem by standard yeast tetrad analysis. In yeast, sporulation of a diploid heterozygous for a nuclear mutation yields a tetrad, a product of a single meiotic event, containing two wild-type spores and two mutant spores. Thus, in a diploid heterozygous for an *rna* mutation, the *in vitro* defect should segregate 2 : 2 and cosegregate with the *ts* gene. Diploids heterozygous for either the *rna*2, *rna*5, or *rna*11 mutation were constructed and sporulated, and the cosegregation of the ts marker and the *in vitro* heat sensitivity was analyzed. As expected, the ts marker segregated 2 : 2 in most tetrads. In each of two tetrads obtained from diploids heterozygous for *rna*2, *rna*5, and *rna*11, extracts isolated from all spores carrying an *rna* mutation were heat sensitive *in vitro*. In contrast, extracts isolated from 9 out of 12 wild-type spores demonstrated resistance to heat inactivation.

In addition, *in vitro* complementation of extracts isolated from most mutant spores was carried out in all possible combinations. These experiments have shown that only extracts isolated from spores containing different mutations complement. Complementation is never observed in like combinations. The data further demonstrate that heat sensitivity of *rna*2, *rna*5, and *rna*11 extracts is due to the presence of an *rna* mutation. As described above, a less extensive *in vitro* complementation analysis has been carried out for the rest of the *rna* mutants with similar results. Thus, it is likely that, in most of these cases, the heat sensitivity of the extract is the consequence of an *rna* mutation.

In summary, these data indicate that (1) many of the *RNA* gene products are essential for an early stage of mRNA splicing, (2) these gene products are directly involved in this process, and (3) these gene products are exchangeable under the assay conditions. The data also strongly suggest that the *RNA* gene products are themselves splicing components. However, the formal possibility that they are modifying enzymes or regulators of the splicing processes cannot be ruled out. For simplicity, however, in the remainder of this paper, we will refer to the defective splicing components as the *RNA* gene products.

It should be pointed out that the nature of the complementing activity is unknown. Although all of the *RNA* genes examined to data code for proteins (Last, 1986; Lee *et al.*, 1986; J. Beggs, personal communication; R. Sternglanz, personal communication), the complementing unit may be a free protein, a protein complex or a protein–RNA complex, such as an snRNP or hnRNP.

The development of the *in vitro* complementation assay described above will also permit the biochemical purification of the individual complementing activities. Indeed, recent experiments have demonstrated that certain fractions of the wild-type *in vitro* splicing extract are capable of complementing the heat-inactivated extracts from the *rna* mutants (Lustig *et al.*, 1986). Thus, we feel that this assay will be a powerful tool for the isolation and purification of the components of the yeast mRNA splicing machinery.

IV. THE *RNA* GENE PRODUCTS AND THE SPLICEOSOME

The identification of mutations directly involved in splicing has permitted a closer analysis of the mechanism of yeast mRNA splicing. We were particularly interested in the possible role of the *RNA* gene products in spliceosome formation. To test this, we heat inactivated the *rna* mutant

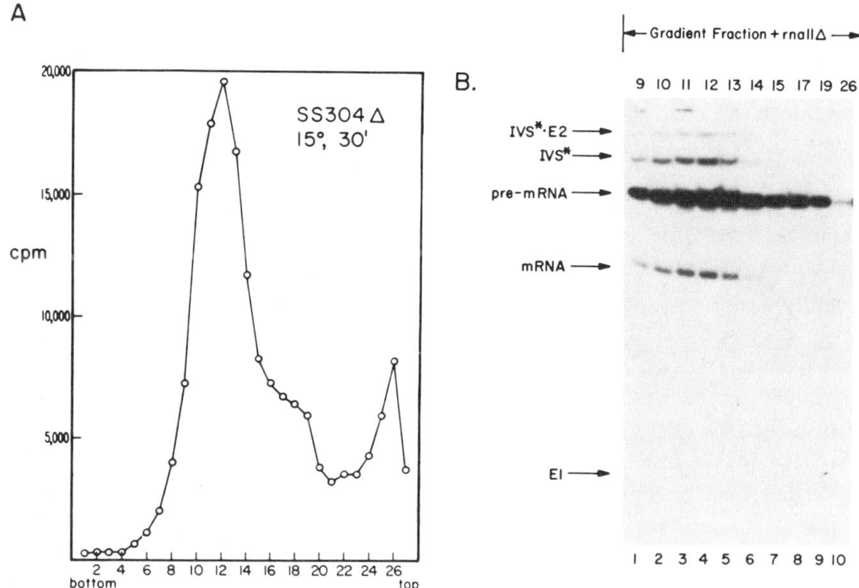

Fig. 5. Chasing of the pre-mRNA in the heat-inactivated (Δ) *rna*2 spliceosome with an *rna*11 Δ extract. (A) A heat-inactivated (Δ) *rna*2 extract (derived from strain SS304) was incubated with ^{32}P-labeled synthetic SP6 actin pre-mRNA under splicing conditions at 15°C for 30 min. The reaction mixture was subsequently fractionated on a glycerol gradient, and the ^{32}P-labeled precursor counted. The peak observed migrates at the position of the 40 S complex. Degraded RNA is present at the top of the gradient. (B) Shown here is an autoradiogram of a gel assay of the chasing reaction products. Specific gradient fractions were mixed with a heat-inactivated (Δ) *rna*11 extract under splicing conditions. The reaction products were subsequently fractionated on a denaturing polyacrylamide gel. The fraction numbers of the gradient (shown on the left) are indicated on the top of the autoradiogram. The arrows on the left indicate the position of the lariat intermediate (IVS*E2), intron lariat (IVS*), pre-mRNA, mature mRNA, and exon 1 (E1), corresponding to the products listed in Fig. 1. These data show that all of the complementing activity is contained within the 40 S peak. (Reproduced from Lin *et al.*, 1987.)

extracts and assayed their ability to form a 40 S complex after addition of ATP and precursor (Lin *et al.* 1987). Fig. 5 (left) shows the result of such an analysis for the *rna*2 mutation. *rna*2 extracts can form the spliceosome both before and after heat inactivation. In the absence of ATP, a 30 S complex is present, the function of which has remained unknown. In contrast, the other heat-sensitive *rna* mutants (*rna*3, *rna*4, *rna*5, *rna*7, *rna*8, and *rna* 11) are incapable of forming the 40 S complex after heat inactivation.

These data suggest that many of the *rna* mutants are defective in spliceosome formation. In contrast, the finding that heat-inactivated *rna*2 extracts can form the 40 S complex containing unspliced precursor indicates that the *RNA*2 gene product is essential for a step in splicing following 40 S complex formation, but before the production of splicing intermediates.

The previous study indicating that heat-inactivated *rna*2 extracts can carry out the formation of the complex allowed a direct determination of the functionality of the 40 S complex in *in vitro* splicing (Lin *et al.*, 1987). Complexes were isolated from heat-inactivated *rna*2 extracts containing unspliced precursor mRNA. These complexes were then added to heat-inactivated extracts isolated from many of the other *rna* mutants, and the products of the reaction analyzed. The result of one such experiment is shown in Fig. 5. On the left is shown a glycerol gradient profile of *rna*2 extracts following heat inactivation and initiation of the splicing assay. Each fraction was added to a heat-inactivated *rna*11 extract and the splicing activity assayed. As shown in Fig. 5 (right), conversion of the bound precursor mRNA into mature products was observed. Furthermore, this activity cosedimented with the 40 S complex. Reinitiation did not take place, as a second exogenous precursor RNA of different size could not be spliced under these conditions. Similar results were obtained when heat-inactivated *rna*3, *rna*4, *rna*5, *rna*7, *rna*8, and *rna*11 extracts were added to the spliceosome fractions. This chasing reaction was also dependent upon the addition of ATP, indicating the ATP is required both for spliceosome formation and for a subsequent step of the reaction.

Three conclusions can be drawn from these data. First of all, the spliceosome isolated from the *rna*2 heat-inactivated extract, although not necessarily identical to the wild-type spliceosome, can act as a functional intermediate in the splicing process. Second, the *RNA*3, *RNA*4, *RNA*5, *RNA*7, *RNA*8, and *RNA*11 gene products appear to be essential for spliceosome formation. The complementation data further indicate that these same gene products are either part of the 40 S complex or, alternatively, are necessary for the assembly of the complex but not for subsequent steps. Finally, the *RNA*2 gene product must be involved in a step after spliceosome formation, but before the production of the intermediates.

V. SPECULATIONS

On the basis of these data, as well as other data from our laboratory, we have derived a working model for mRNA splicing in yeast (Fig. 6) (Lin *et al.*, 1987). The splicing reaction has been divided into three steps. Step A, spliceosome formation, involves the products of the *RNA*3, *RNA*4,

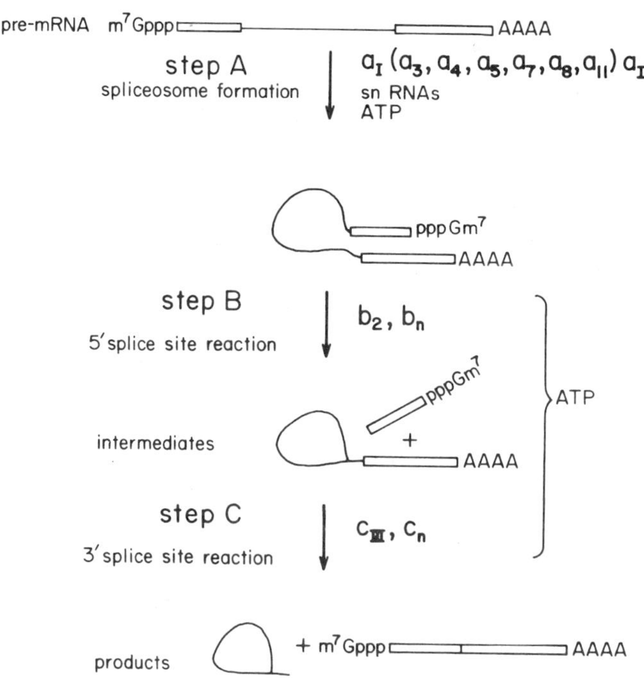

Fig. 6. A model for nuclear mRNA splicing in yeast. Shown here is a representation of the model for mRNA splicing described in the text. Lower-case letters (a, b, and c) represent factors that are involved in the corresponding step. The arabic number in subscript indicates the product of the corresponding RNA gene (e.g., b_2). Presented here also are factors, denoted in roman numerals (e.g., a_I), uniquely present in fractionated wild-type extracts (Cheng and Abelson, 1986) that act at a specific stage of the process. (Reproduced from Lin *et al.*, 1987.)

*RNA*5, *RNA*7, *RNA*8, and *RNA*11 genes (a_3–a_{11}), each of which may be intrinsic components of the spliceosome. This step requires ATP and involves the participation of snRNAs. Step B, the formation of the intermediates, is mediated by the *RNA*2 gene product (b_2). In addition, on the basis of experiments not described here, a nonspliceosomal factor, b_n, also participates in this step. Step C, the formation of the products, involves one spliceosomal factor (c_{III}) and one nonspliceosomal factor, c_n. ATP is required for one or both of steps B and C. Of course, any product participating in an early step of the reaction may also be involved in a subsequent step of splicing.

The actual functions of the *RNA* gene products in splicing are still a subject for speculation. Two general possibilities exist. One of these arises from an analysis of the behavior of yeast mitochondrial introns. At least one yeast mitochondrial gene contains an intron that is processed in a fashion similar to the yeast nuclear genes, forming a lariat intermediate and product (Peebles *et al.,* 1986; Van der Veen *et al.,* 1986). Remarkably, although present in the same organism, this reaction can be carried out in the absence of any exogenous macromolecules *in vitro* (but probably not *in vivo*), apparently proceeding through a transesterification mechanism as in *Tetrahymena* rRNA (reviewed in Cech and Bass, 1986). These data led to the proposal that the complexity of the nuclear processing machinery may be the result of functions, once a part of the precursor RNA molecule, which have now have been relegated to other RNAs or to proteins (Cech, 1986). In this model, the *RNA* gene products may represent such proteins. It is indeed possible that such a separation of functions led to a more efficient splicing reaction.

If, however, the mitochondrial and nuclear reactions are only superficially similar, then the nuclear system may proceed through more conventional endonuclease and ligase enzymatic activities. The complexity found in the mRNA splicing process may then be the consequence of the requirement for a scaffold, which may be necessary to bring reactive sequences in close proximity. Such a matrix may greatly increase the efficiency of the reaction. In this case, the *RNA* gene products may serve either a structural or enzymatic role in splicing.

Several approaches are possible in the yeast system to define more clearly the roles of the *RNA* gene products in mRNA splicing. First of all, the *in vitro* complementation assay should serve as an excellent assay for the purification of the gene products by conventional biochemical procedures. Second, once such products are isolated, it will be possible to evaluate the intrinsic catalytic activity of the these components as well as their role in spliceosome formation. Clearly, a gene product that is an intrinsic component of the spliceosome can also serve as a marker for spliceosome purification and assembly.

In addition, the isolation of second-site suppressors of some of the *rna* mutations may define important interactions taking place between *RNA* gene products and other factors. The analysis of allele specificity *in vitro* may also shed light on the activities present in different domains of proteins participating in splicing. Finally, the isolation of additional mutations is likely to aid further in the identification of intrinsic components and regulators of splicing activity and spliceosome assembly. Therefore we expect that the combination of genetic and biochemical techniques available in the yeast system will continue to lead to a better understanding of the process of eukaryotic nuclear mRNA splicing.

ACKNOWLEDGMENTS

We would like to thank Drs. J. Woolford, J. Beggs, L. H. Hartwell, and A. Hopper for providing many of their *rna* strains. This work was supported in part by a National Institutes of Health postdoctoral fellowship (I F32 GM 10087-01A1) to R.-J. Lin and by a grant from the National Institutes of Health (GM 32637).

REFERENCES

Berget, S. M., and Robberson, B. L. (1986). *Cell (Cambridge, Mass.)* **46,** 691–696.

Black, D. L., and Steitz, J. A. (1986). *Cell (Cambridge, Mass.)* **46,** 697–704.

Black, D. L., Chabot, B., and Steitz, J. A. (1985). *Cell (Cambridge, Mass.)* **42,** 737–750.

Bollen, G., Molenaar, C., Cohne, L. H., van Raamsdonk-Duin, M., Mager, W. H., and Planta, R. J., (1982). *Gene* **18,** 29–37.

Bozzoni, I., Annesi, F., Beccari, E., Fragapane, P., Pierandrei-Amaldi, P., and Amaldi, F. (1984). *J. Mol. Biol.* **180,** 1173–1178.

Brody, E., and Abelson, J. (1985). *Science* **228,** 963–967.

Cech, T. R. (1986). *Cell (Cambridge, Mass.)* **44,** 207–210.

Cech, T. R., and Bass, B. L. (1986). *Annu. Rev. Biochem.* **55,** 599–630.

Chabot, B., Black, D. L., LeMaster, D. M., and Steitz, J. A. (1985). *Science* **230,** 1344–1349.

Cheng, S.-C., and Abelson, J. (1986). *Proc. Natl. Acad. Sci. U.S.A.* **83,** 2387–2391.

Choi, Y. D., Grabowski, P. J., Sharp, P. A., and Dreyfuss, G. (1986). *Science* **231,** 1534–1539.

Domdey, H., Apostol, B., Lin, R.-J., Newman, A., Brody, E., and Abelson, J. (1984). *Cell (Cambridge, Mass.)* **39,** 611–621.

Etcheverry, T., Colby, D., and Guthrie, C. (1979). *Cell (Cambridge, Mass.)* **18,** 11–26.

Fradin, A., Jove, R., Hemenway, C., Keiser, H. D., Manley, J. W., and Prives, C. (1984). *Cell (Cambridge, Mass.)* **37,** 927–936.

Frendeway, D., and Keller, W. (1985). *Cell (Cambridge, Mass.)* **42,** 355–367.

Fried, H. M., Pearson, N. J., Kim, C. H., and Warner, J. R. (1981). *J. Biol. Chem.* **256,** 10176–10183.

Gallwitz, D., and Sures, I. (1980). *Proc. Natl. Acad. Sci. U.S.A* **77,** 2546–2550.

Gorenstein, C., and Warner, J. R. (1976). *Proc. Natl. Acad. Sci. U.S.A.* **73,** 1547–1551.

Grabowski, P. J., and Sharp, P. A. (1986). *Science* **223,** 1294–1299.

Grabowski, P. J., Padgett, R. A., and Sharp, P. A. (1984) *Cell (Cambridge, Mass.)* **37,** 415–427.

Grabowski, P. J., Seiler, S. R., and Sharp, P. A. (1985). *Cell (Cambridge, Mass.)* **42,** 345–353.

Hartwell, L. H. (1967). *J. Bacteriol.* **93,** 1662–1670.

Hartwell, L. H., McLaughlin, C. S., and Warner, J. R. (1970). *Mol. Gen. Genet.* **109,** 42–56.

Hernandez, N., and Keller, W. (1983). *Cell (Cambridge, Mass.)* **35,** 89–99.

Hopper, A., Banks, K., and Evangelidis, V. (1978). *Cell (Cambridge, Mass.)* **14,** 211–219.

Hutchinson, H. T., Hartwell, L. H., and McLaughlin, C. S. (1969). *J. Bacteriol.* **99,** 807–814.

Kaufer, N. F., Fried, H. M., Schwindinger, W. F., Jasin, M., and Warner, J. R. (1983). *Nucleic Acids Res.* **11,** 3123–3135.

Keller, E. B., and Noon, W. A. (1984). *Proc. Natl. Acad. Sci. U.S.A.* **81,** 7417–7420.
Knapp, G., Beckmann, J. S., Johnson, P. F., Fuhrman, S. A., and Abelson, J. (1978). *Cell (Cambridge, Mass.)* **14,** 221–236.
Krainer, A. R., and Maniatis, T. (1985). *Cell (Cambridge, Mass.)* **42,** 725–736.
Kramer, A. R., Keller, W., Appel, B., and Luhrmann, R. (1984). *Cell (Cambridge, Mass.)* **38,** 299–307.
Larkin, J. C., and Woolford, J. L. (1983). *Nucleic Acids Res.* **11,** 403–420.
Last, R. L. (1986). Ph.D Thesis, Carnegie-Mellon University, Pittsburgh, Pennsylvania.
Lee, M. G., Lane, D. P., and Beggs, J. D. (1986). *Yeast* **2,** 59–67.
Leer, R. J., van Raamsdonk-Duin, M., Molenaar, C., Cohen, L. H., Mager, W. H., and Planta, R. J., (1982). *Nucleic Acids Res.* **10,** 5869–5878.
Lin, R.-J., Newman, A. J., Cheng, S.-C., and Abelson, J. (1985). *J. Biol. Chem.* **260,** 14780–14792.
Lin, R.-J., Lustig, A. J., and Abelson, J. (1987). *Genes Dev.* **1,** 7–18.
Lustig, A. J., Lin, R.-J., and Abelson, J. (1986). *Cell (Cambridge, Mass.)* **47,** 953–963.
Miller, A. M. (1984). *EMBO J.* **3,** 1061–1065.
Mitra, G., and Warner, J. R. (1984). *J. Biol. Chem.* **259,** 9218–9224.
Mount, S. M. (1982). *Nucleic Acids Res.* **10,** 459–472.
Mount, S. M., Pettersson, I., Hinterberger, M., Karmas, A., and Steitz, J. A. (1983). *Cell (Cambridge, Mass.)* **33,** 509–518.
Newman, A. J., Lin, R.-J., Cheng, S.-C., and Abelson, J. (1985). *Cell (Cambridge, Mass.)* **42,** 335–344.
Ng, R., and Abelson, J. (1980). *Proc. Natl. Acad. Sci. U.S.A.* **77,** 3912–3916.
O'Farrell, P. Z., Cordell, B., Valenzuela, P., Rutter, W. J., and Goodman, H. M. (1978). *Nature (London)* **274,** 438–445.
Padgett, R. A., Mount, S. M., Steitz, J. A., and Sharp, P. A. (1983). *Cell (Cambridge, Mass.)* **35,** 101–107.
Padgett, R. A., Konarska, M. M., Grabowski, P. J., Hardy, S. F., and Sharp, P. A. (1984). *Science* **225,** 898–903.
Padgett, R. A., Grabowski, P. J., Konarska, M. M., Seler, S., and Sharp, P. A. (1986). *Annu. Rev. Biochem.* **55,** 1119–1150.
Pearson, N. J., Thornburn, P. C., and Haber, J. E. (1982). *Mol. Cell. Biol.* **2,** 571–577.
Peebles, C. L., Perlman, P. S., Mecklenburg, K. L., Petrillo, M. L., Tabor, J. H., Jarrell, K. A., and Cheng, H.-L. (1986). *Cell (Cambridge, Mass.)* **44,** 213–223.
Pikielny, C. W., and Rosbash, M. (1986). *Cell (Cambridge, Mass.)* **45,** 869–877.
Rodriguez, J. R., Pikielny, C. W., and Rosbash, M. (1984). *Cell (Cambridge, Mass.)* **39,** 603–610.
Rosbash, M., Harris, P. K. W., Woolford, J. L., and Teem, J. L. (1981). *Cell (Cambridge, Mass.)* **24,** 679–686.
Ruskin, B., Krainer, A. R., Maniatis, T., and Green, M. R. (1984). *Cell (Cambridge, Mass.)* **38,** 317–331.
Schatz, P., Pillus, L., Grisafi, P., Solomon, F., and Botstein, D. (1986). *Mol. Cell. Biol.* **6,** 3711–3721.
Shiokawa, K., and Pogo, A. O. (1974). *Proc. Natl. Acad. Sci. U.S.A.* **71,** 2658–2662.
Tatai, K., Takemura, K., Mayeda, A., Fujiwara, Y., Tanaka, H., Ishihama, A., and Ohshima, Y. (1984). *Proc. Natl. Acad. Sci. U.S.A.* **81,** 6281–6285.
Teem, J. L., and Rosbash, M. (1983), *Proc. Natl. Acad. Sci. U.S.A.* **80,** 4403–4407.
Teem, J. L., Rodriguez, J. R., Tung, L., and Rosbash, M. (1983). *Mol. Gen. Genet.* **192,** 101–103.

Teem, J. L., Abovich, N., Kauler, N. F., Schwindinger, W. F., Warner, J. R., Levy, A., Woolford, J., Leer, R. J., van Raamsdonk-Duin, M., Mager, W., Planta, R. J., Schultz, L. Friesen, J., Fried, H., and Rosbash, M. (1984). *Nucleic Acid Res.* **12,** 8295–8328.

Van der Veen, R., Arnberg, A. C., Van der Horst, G., Bonen, L., Tabak, H. F., and Grivell, L. A. (1986). *Cell (Cambridge, Mass.)* **44,** 225–234.

Vijayraghavan, U., Parker, R., Tamm, J., Iimura, Y., Rossi, J., Abelson, J., and Guthrie, C. (1986). *EMBO J.* **5,** 1683–1695.

Warner, J. R., and Gorenstein, C. (1977). *Cell (Cambridge, Mass.)* **11,** 201–212.

Warner, J. R., and Udem, S. A. (1972). *J. Mol. Biol.* **65,** 243–257.

Yang, V. W., Lerner, M. R., Steitz, J. A., and Flint, S. J. (1981). *Proc. Natl. Acad. Sci. U.S.A.* **78,** 1371–1375.

Zeitlin, S., and Efstratiadis, A. (1984). *Cell (Cambridge, Mass.)* **39,** 589–602.

9

Architecture of Fungal Introns: Implications for Spliceosome Assembly

ROY PARKER

Department of Biology
University of California at San Diego
La Jolla, California 92093

BRUCE PATTERSON

Department of Biochemistry and Biophysics
University of California at San Francisco
San Francisco, California 94143

I. INTRODUCTION

The removal of intervening sequences (IVS) by the process of RNA splicing is an elementary step in gene expression. Nuclear mRNA splicing in mammals and in yeast is mediated by a group of small nuclear RNAs complexed with protein (snRNPs) (for reviews, see Maniatis and Reed, 1987; Green, 1986). One role of snRNPs in the splicing process appears to be to assemble with the precursor into a large complex, termed the spliceosome, which is capable of the splicing process. The discovery of self-splicing in mitochondrial Group II introns (in which splicing proceeds via the lariat intermediate seen in nuclear mRNA splicing) (Peebles *et al.*, 1986) has suggested the possibility that the assembly of the splicing complex by snRNPs may function primarily to bring the precursor into the proper geometry for a reaction which can then be viewed as being fundamentally autocatalytic (see Cech, 1986).

Molecular Biology of RNA
New Perspectives

Two general paradigms have emerged for the mechanism by which snRNPs participate in the assembly of the splicing complex. Considerable evidence supports the conclusion that specific snRNPs recognize and bind conserved sequences within the intron. In particular, the U1 RNA containing snRNP (Mount *et al.*, 1983), the U2 snRNP (Black *et al.*, 1985), and the U5 snRNP (Chabot *et al.*, 1985; Gerke and Steitz, 1986; Tazi *et al.*, 1986) bind to the 5′ junction, the branch site, and the 3′ splice site, respectively. In addition, the snRNPs are thought to organize the spliceosome by interactions with one another, though there are no data that bear directly on this point. Thus, spliceosome assembly, and hence splicing, would appear to be dependent on a variety of both intron–snRNP and snRNP–snRNP interactions.

In principle, the structure of the intron or the spatial arrangement of the conserved intron sequences may promote or influence splicing in two ways. First, if the reaction is fundamentally autocatalytic, an intron with a high degree of the proper secondary and tertiary structure may position the nucleotides involved in bond cleavage and ligation in the proper geometry for the splicing process to occur. Indeed, if the degree of structure is sufficient, such introns may be able partially or wholly to dispense with some trans interactions. Note that, while some mitochondrial Group II introns can self-splice *in vitro,* genetic evidence suggests that other members of this class of intron require trans-acting factors (Bertrand *et al.*, 1982; McGraw and Tzagoloff, 1983; Piller *et al.*, 1983; Simon and Faye, 1984). Alternatively, and what seems more likely from our current understanding of nuclear pre-mRNA splicing, the spatial arrangement of the conserved sequence elements may position the snRNPs in the proper geometry to interact with each other, and therefore to assemble into a functional spliceosome. If this was true, we would expect to find the spatial relationship between splicing signals to be nonrandom and, on average, to illustrate the optimal arrangement between any such pair of signals. For example, branch points in mammalian introns appear to be a relatively constant distance from the 3′ splice site, thus suggesting an interaction between factors, presumably snRNPs, at each of these sites (Ruskin *et al.*, 1985; Reed and Maniatis, 1985).

In this light we have examined the spacing of the conserved intron sequences found at the 5′ splice site, the branch site, and the 3′ splice junction in introns from *Saccharomyces cerevisiae* and other fungi. In that the primary sequences of the 5′ junction sequence and the branch point, or TACTAAC box in *Saccharomyces cerevisiae,* are more strictly conserved in fungal introns (see Guthrie *et al.*, 1986), possible secondary structures or spatial arrangements may also be highly conserved. Based on our analyses, we conclude that there are conserved spatial relationships between both the 5′ splice site and the branch point, as well as between the branch point and the 3′ splice site. We hypothesize that these spacings are optimal to promote interactions of the snRNPs at each of these sites. In cases where the spacing between elements is not optimal, as based on linear distance, we propose that there are additional intron features that restore the optimal geometry. In the case of the branch point to 3′ splice site, we suggest this to occur by the interaction of a trans-acting factor with an additional conserved sequence element, a polyuridine tract. Alternatively, as in the 5′ junction to branch point case, we can identify cis-acting secondary structures that restore the optimal spacing.

II. BRANCH SITE–3′ SPLICE JUNCTION RELATIONSHIP

As our first step in the examination of the relationships of conserved intron sequences, we gathered a data base of a number of fungal introns. The introns we have used for the analyses in this paper include 21 introns from *Saccharomyces cerevisiae,* five introns from *Schizosaccharomyces pombe,* three introns from *Schizophyllum commune,* five introns from *Neurospora crassa,* five introns from *Asperigillus niger,* and four introns from *Trichoderma reesei.* These introns are listed and the references given in the legend to Fig. 1.

A. Two Types of 3′ Splice Sites in Fungi

We have examined the structure of the 3′ splice site by comparing the distance between the branch point and the AG at the 3′ junction. The results of this analysis, illustrated in the histogram in Fig. 1, indicate that the distance between these two sequence elements is not random, and moreover suggest that there are two types of spacings. In introns from fungi other then *S. cerevisiae* and in two of the *S. cerevisiae* introns (found in the ***MATa1*** gene), the branch site is quite close to the AG at the 3′ junction, the distance varying between 5 and 15 nucleotides. We refer to introns containing this type of 3′ splice site as being type 3′S introns (for short 3′ splice site). In contrast, in the remainder of *S. cerevisiae* introns the spacing between the branch site and the 3′ junction is significantly larger (22–137 nucleotides) with a distinct preference for a spacing of approximately 40 nucleotides. We refer to this type of intron as being a type of 3′L intron (for long 3′ splice site).

Having discerned two different classes of 3′ splice site regions in fungal introns, we searched for further similarities within the introns in each class. Type 3′S introns do not appear to possess any conserved sequences

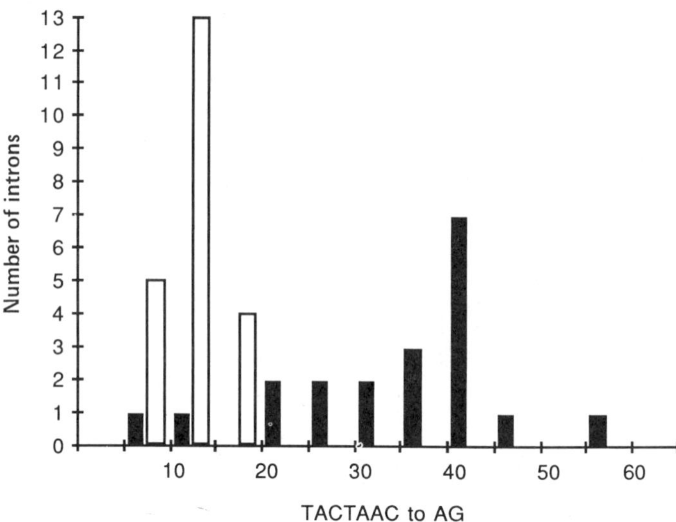

Fig. 1. Branch point to AG spacings in fungal introns. The figure shows the number of nucleotides between the branch point and the 3′ splice site in a number of fungal introns. The nucleotides are counted from the nucleotide 3′ of the branch point to the nucleotide immediately preceding the AG. The number of introns within a given spacing interval are represented for *Saccharomyces cerevisiae* (dark bars) and several other fungi (open bars). The introns represented from *S. cerevisiae* are: actin (Ng and Abelson, 1980), *cyh2* (L29, Kaufer *et al.,* 1983), *kin28* (Simon *et al.,* 1986), L17A and L25 (Leer *et al.,* 1984), L34A (Schaap *et al.,* 1984), *MATa1*-1 and *MATa1*-2 (Miller, 1984), rp16A, rp16B, rp28A, and rp28B (Molenaar *et al.,* 1984), rp29 (Mitra and Warner, 1984), rp51A (Teem and Rosbash, 1983), rp51B (Abovich and Rosbash, 1984), rp59 (Teem *et al.,* 1984), rp73 (J. Warner, personal communication), S10A (Leer *et al.,* 1982), S10B (Leer *et al.,* 1985), *TUB1* and *TUB3* (not shown but with a spacing of 137 nucleotides) (Schatz *et al.,* 1986). The other fungal introns are: from *Schizosaccharomyces pombe, NDA2* (Toda *et al.,* 1984), *NDA3* introns 1–5 (Hiraoka *et al.,* 1984); from *Schizophyllum commune, IG2* introns 1–3 (Dona *et al.,* 1984); from *Neurospora crassa,* histone H3 and histone H4 introns 1 and 2 (Woudt *et al.,* 1983), *am1* and *am2* (Kinnaird and Fincham, 1983); from *Aspergillus niger,* glucoamylase introns 1–5 (Boel *et al.,* 1984); and from *Trichoderma reesei, CBH* introns 1 and 2 (Shoemaker *et al.,* 1983), the *EG1* introns 1 and 2 (M. Innis, personal communication).

other than the absolutely conserved AG dinucleotide found at the 3′ splice sites of all nuclear mRNA introns (data not shown). Even the nucleotide preceding the AG, which is a pyrimidine in all mammalian introns and all fungal 3′L introns, is sometimes a purine in 3′S introns.

In 3′L introns, on the other hand, we find significant sequence conservation in the region upstream of the AG dinucleotide. As illustrated in Fig. 2, we observe both a 3′ splice site sequence of N(U or A)APyAG, as well as a general enrichment for uridines extending at least to position

−20 (as measured from the site of 3′ cleavage). This enrichment is especially striking from nucleotides −7 to −13, and includes a 100% conserved uridine at position −9. The relative enrichment of uridines is demonstrated in Fig. 2B, which shows the expected (based on the base composition of the 21 *S. cerevisiae* introns discussed in this work) and observed frequencies of the consensus nucleotides at each position.

Extensive sequence analyses in both mammals (Mount, 1982) and plants (Brown *et al.*, 1986) have been performed, and it is interesting to compare the consensus sequences derived with the one we suggest for 3′L introns of *S. cerevisiae*. Mammalian introns terminate in the sequence NPyAG, those from plants in GPyAG, and those in yeast in N(U or A)APyAG (in all cases, we arbitrarily define the 5′ end of these subsequences as the first position which is not a pyrimidine). Thus, in each case we observe a conserved sequence motif, although the sequences share only the PyAG as a common signal. In all cases, the region preceding this sequence is pyrimidine rich. In mammalian introns, the enrichment is especially striking, positions −5 to −15 displaying 70–93% occurrence of pyrimidines, with U being the preferred one (Mount, 1982). Both yeast and plant introns show an even greater preference for uridine; indeed, at least in the case of yeast, the proportion of cytosines is close to that expected if there were no specific enrichment for them. Though the net pyrimidine enrichment in both yeast and plant introns is less than that observed in mammals, the prevalence of uridine residues in yeast 3′L introns, particularly from positions −7 to −13, is striking (each position being from 58 to 100% U for 19 introns).

On the basis of the above observations we conclude that there are two distinct types of 3′ splice site geometries in fungal introns. One class, which we designate 3′S, contains introns in which the branch point is quite close to the 3′ splice site (5–15 nucleotides) and which display no apparent conserved sequence element 3′ of the branch point other than the AG dinucleotide. The second class of introns, 3′L, is distinguished by a significantly greater branch point to 3′ splice site spacing, clustering about 35–45 nucleotides. In these introns we observe both an enrichment for uridines in positions −13 to −7 and a preferred 3′ splice site sequence of (U or A)APyAG. These introns appear to be analogous to introns identified thus far in mammalian and plant systems.

B. Implications of 3′ Splice Site Variations

The observation that there are two different types of 3′ splice sites in fungal introns raises three important questions. (1) Are these introns spliced by the same splicing machinery? (2) How is the 3′ splice site, in

A

U	G	U	C	U	A	U	A	U	U	A	U	A	U	G	U	U	U	A	G	actin
A	C	A	A	U	A	U	U	U	U	U	U	U	U	G	U	A	C	A	G	cyh2 (L29)
U	G	A	U	U	A	A	U	G	A	U	U	C	A	U	C	G	C	A	G	kin28
U	U	G	A	G	A	U	C	U	U	U	U	U	U	A	A	C	U	A	G	L17A
U	G	A	A	U	U	U	A	C	U	U	U	U	U	G	U	A	U	A	G	L25
U	A	C	C	A	U	U	U	U	A	U	U	U	U	U	A	A	U	A	G	L34
U	U	U	G	A	U	U	U	U	G	U	U	U	U	C	U	A	C	A	G	rp16A
U	U	G	U	G	C	A	U	U	U	U	U	U	C	A	A	U	U	A	G	rp16B
U	U	U	U	C	A	U	U	U	U	U	U	U	U	U	U	U	U	A	G	rp28A
U	U	U	U	C	U	U	G	U	C	U	U	A	A	U	C	A	C	A	G	RP28B
A	A	A	A	A	C	G	U	G	G	A	U	U	A	A	U	A	U	A	G	rp29
U	U	U	U	G	U	A	U	U	G	C	U	U	U	U	A	A	U	A	G	rp51A
G	A	U	U	A	U	U	G	C	U	A	U	U	U	U	U	A	U	A	G	rp51B
G	A	U	U	U	A	C	U	A	U	U	U	C	C	A	U	U	U	A	G	rp59
U	A	U	C	G	U	U	U	A	C	A	U	U	U	C	A	A	C	A	G	rp73
A	A	U	G	G	U	A	U	U	A	U	U	U	A	U	A	A	C	A	G	S10A
C	U	G	U	U	G	A	A	A	A	U	U	U	A	A	A	A	U	A	G	S10B
U	U	U	U	U	G	A	U	U	U	C	U	C	U	U	U	A	C	A	G	TUB1
A	U	U	U	U	U	C	U	U	U	U	U	U	A	C	A	A	C	A	G	TUB3
-20	-19	-18	-17	-16	-15	-14	-13	-12	-11	-10	-9	-8	-7	-6	-5	-4	-3	-2	-1	position
4	6	4	4	4	6	6	3	3	4	4	0	2	6	5	8	13	0	19	0	A (6.2)
2	3	3	2	5	2	1	2	2	3	0	0	0	0	3	0	1	0	0	19	G (3.3)
1	1	1	3	2	2	2	1	2	2	2	0	3	2	3	2	1	8	0	0	C (3.0)
12	9	11	10	8	9	10	13	12	10	13	19	14	11	8	9	4	11	0	0	U (6.4)
21	32	21	21	21	32	32	16	16	21	21	0	11	32	26	42	68	0	100	0	%A (33)
11	16	16	11	26	11	5	11	11	16	0	0	0	0	16	0	5	0	0	100	%G (18)
5	5	5	16	11	11	11	5	11	11	11	0	16	11	16	11	5	42	0	0	%C (16)
63	47	58	53	42	47	53	68	63	53	68	100	74	58	42	47	21	58	0	0	%U (34)
U	U	U	U	U	U	U	U	U	U	U	U	U	U	N	U/A	A	U/C	A	G	Consensus

B

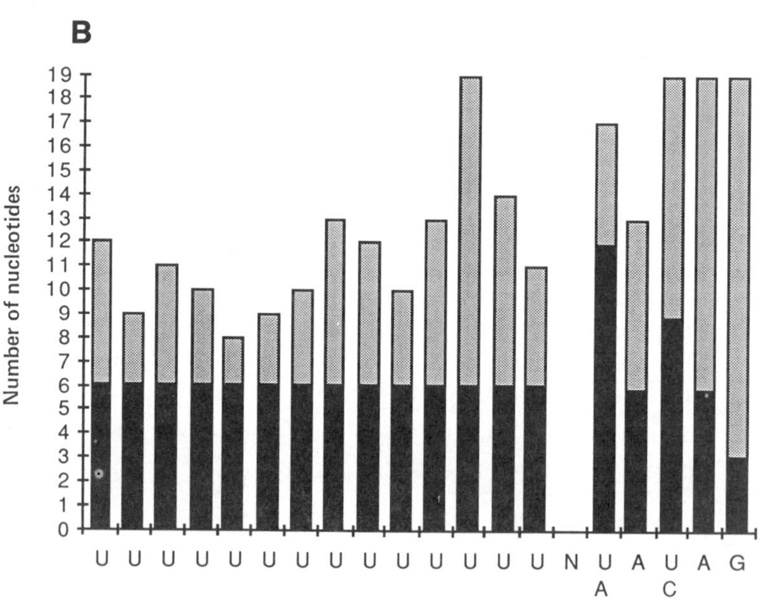

particular the AG dinucleotide, recognized in each case? (3) What is the function of the conserved polyuridine sequence found in 3′L introns? In the following sections we address these questions.

1. Splicing of Both Types of Introns Involves Common Factors

Several observations from *S. cerevisiae* suggest that, in general, the same splicing machinery interacts with both 3′S and 3′L introns. Several genes that have been identified as being required for the splicing of 3′L introns, are also required for the splicing of 3′S introns. These include both protein components of the spliceosome (the *RNA2* gene product) (Miller, 1984; Lustig *et al.,* 1986), and an snRNA (snR7) required for splicing *in vivo* (Patterson and Guthrie, 1987). In addition, 3′S introns contain all known splicing signals other than the polyuridine tract. Furthermore, a deletion of the actin intron (3′L type) which brings the AG within 7 nucleotides of the TACTAAC box (thus creating a type 3′S intron) is spliced only marginally less efficiently than wild type (Fouser and Friesen, 1987). From these arguments, we conclude that 3′L and 3′S introns are fundamentally similar, and that the 3′L introns represent a slightly more complex intron than the 3′S type.

2. Recognition of the 3′ Splice Site

How is the 3′ splice site recognized in these two classes of introns? To answer this question, it is useful to review the process of 3′ splice site recognition in mammalian introns, which are similar to 3′L introns. In this case, the recognition of the 3′ splice site appears to be initiated by the binding of a factor, probably the U5 snRNP, to the polypyrimidine stretch (Frendeway and Keller, 1985; Ruskin and Green, 1985; Chabot *et al.,* 1985). It is important to note that the actual interaction with the substrate appears to be mediated by a protein component of the U5 snRNP (Tazi *et al.,* 1986; Gerke and Steitz, 1986). This is followed by the binding of the

Fig. 2. Analysis of conserved sequences in 3′ splice sites in 3′L introns. (A) Sequence alignment and analysis: the 3′ terminal 20 nucleotides (including the AG) of 19 *Saccharomyces cerevisiae* introns have been aligned. The absolute number of nucleotides observed at each position (total of 19) are shown, as well as the percentage of each nucleotide at each position. The expected numbers of and percentages for each nucleotide (calculated from the base composition of the *S. cerevisiae* introns listed in the legend to Fig. 1) are shown in parentheses on the right. The consensus sequence we suggest is shown at the bottom. (B) Relative frequencies of consensus nucleotides at each position. Solid bars, the expected number of occurrences of the consensus nucleotide(s) at each position; stippled bars, occurrence of the consensus nucleotide(s) above the level expected. Introns represented are the same as those shown in Fig. 2A.

U2 snRNP to the branch point in a manner that may require both U2–U5 interactions as well as U2–intron interactions. In contrast, in yeast the first detectable intermediate in the assembly pathway involves the binding of the yeast U2 snRNP (Ares, 1986), snR20 (also called LSR1), to the branch point in a manner dependent on recognition of the 5′ splice junction (Pikielny *et al.*, 1986). This observation suggests that in yeast the TACTAAC box is the primary signal in 3′ intron recognition.

Though the order of snRNP addition may be different, we suggest that recognition of the 3′ splice site in yeast 3′L introns occurs in a manner analogous to mammalian introns. We have recently identified a yeast snRNA, snR7, with striking homology to mammalian U5 (Patterson and Guthrie, 1987). We suggest that the snR7 snRNP is involved in the recognition of the polyuridine tract and 3′ splice junction in yeast in a manner analogous to the interaction of the U5 snRNP with the 3′ splice site in mammalian introns. Furthermore, the preferred TACTAAC-to-AG spacing we observe in yeast (about 40 nucleotides) is strikingly reminiscent of that seen for branch point-to-AG spacing in mammals, raising the possibility that in yeast the U2 (snR20) and U5 (snR7) snRNPs involved in the recognition of sequences in the 3′ portion of the intron interact as they are thought to do in mammals.

In this light it is interesting to consider the recognition of the 3′ splice site in 3′S introns. In these introns, the AG dinucleotide must be recognized in the absence of a polyuridine tract. This is a notable problem given that the U5-associated protein which binds the polypyrimidine tract is dependent on the presence of the AG for efficient binding (Tazi *et al.*, 1986; Gerke and Steitz, 1986). If recognition of the polyuridine-AG sequence in yeast is analogous to that in mammals, then recognition of the AG in 3′S introns presumptively occurs in one of the following ways: (1) The protein that recognizes the AG in 3′L introns also recognizes it in 3′S introns, but in a manner independent of the polyuridine tract; or (2) Alternatively, recognition of the AG in 3′S introns occurs by a mechanism independent of the protein-mediated U5 snRNP–intron interaction identified to date. Note that this recognition could be mediated by an RNA–RNA interaction. While such a recognition process could be unique to 3′S introns, it seems more likely that it would also occur in 3′L introns, concurrent with or subsequent to, the recognition of the polyuridine tract. In either case, because 3′L introns require additional interactions, we predict that there should be specific mutations that inhibit the splicing of 3′L introns but do not affect the splicing of 3′S introns.

3. *Function of the Polyuridine Tract*

As we have just argued, it seems likely that the polyuridine tract is recognized by the yeast equivalent of U5, snR7. Is there any evidence which indicates that the polyuridine tract plays a role similar to the mammalian polypyrimidine tract? Several reports have described the deletion of sequences between the TACTAAC box and the AG. In general, the conclusion is that these sequences are not required for the first (Rymond and Rosbash, 1985; Fouser and Friesen, 1987) or second step (Fouser and Friesen, 1987) in the splicing process. However, in that deletions in this region would, in many cases, convert a 3′L intron into a 3′S intron such experiments do not address the role of these sequences in their original configuration. Similarly, in many instances, the deletions reported either fail to remove the conserved sequences, or replace them with sequences similar to the consensus. This leads us to conclude that, while these experiments demonstrate significant differences between splicing in yeast and in mammals [where the deletion of the normally used polypyrimidine stretch completely blocks splicing (Fukamaki *et al.*, 1982; Ruskin and Green, 1985)], the role of the conserved nucleotides near the 3′ splice site in 3′L introns has not been tested rigorously and requires further experiments.

Given that 3′S introns are spliced without the contribution of a polyuridine stretch, its presence in 3′L introns must be explained. We suggest that its role is to adapt the greater spacing seen in 3′L introns such that it resembles that seen in 3′S introns. In other words, since 3′S introns are spliced efficiently with a branch point to AG spacing of about 10 nucleotides, this may represent a preferred or optimal spatial configuration for splicing. Recognition of the polyuridine tract in 3′L introns may allow a *trans*-acting factor to bind, perhaps snR7, and through interactions with other snRNPs, position the AG dinucleotide into the proper geometry for cleavage and ligation.

Is there any evidence that the polyuridine tract plays a role in the positioning of the AG dinucleotide? The only data we are aware of is indirect. The observation that a mutation of the AG dinucleotide in the actin intron to AC (C303/305, Vijayraghavan *et al.*, 1986), is still spliced, though inefficiently, to the proper site, suggests to us that the position to the 3′ splice site is marked in a manner independent of the AG dinucleotide. We believe it is unlikely that the observed splicing represents a low-level ability of the splicing machinery to recognize the dinucleotide AC, because (1) a mutation that alters the AG to GG appears to show the same behavior (Fouser and Friesen, 1987) and (2) in many yeast introns, an AC immediately precedes the AG, thus erroneous splicing to this dinucleotide would result in nonfunctional proteins.

Further evidence that a polyuridine-AG motif might provide a positioning function comes from the comparison of two introns with similar branch point-to-AG spacings but different nucleotide compositions. In the case of the *TUB3* intron, the branch point-to-AG spacing is 137 nucleotides and the region upstream of the AG is an excellent match to the consensus shown in Fig. 2. By Northern analysis, no precursor or lariat is detected *in vivo* (Schatz *et al.*, 1986). We compare this to a synthetic intron constructed by the insertion of pBR322 sequences just 3′ of the TACTAAC box in the yeast actin intron, generating a TACTAAC-to-AG spacing of 120 nucleotides. Here, the sequence upstream of the AG is devoid of uridines in the −7 to −13 region. *In vivo,* the primary product of this gene is the lariat intermediate (Cellini *et al.*, 1986). Although other explanations are also possible, this observation is consistent with the polyuridine tract playing a role in positioning the 3′ splice site over large distances.

In summary, we propose that the function of the polyuridine tract is to interact with a component of the splicing machinery, probably the snR7 snRNP. Whether or not this interaction contributes to the efficiency of the assembly of the spliceosome requires further experiments. In addition, we suggest the possibility that the polyuridine tract functions to position the 3′ splice junction in the appropriate geometry for cleavage and ligation. A prediction of this model is that mutation of the polyuridine tract in an intron in which the AG has been previously mutated, for example, the C303/305 mutation discussed above, should prevent the low-level accurate splicing which still occurs in this mutant.

III. BRANCH-SITE–5′ SPLICE JUNCTION RELATIONSHIP

A second feature of the intron that would be anticipated to affect the efficiency of the splicing process is the spatial relationship of the 5′ splice junction to the branch site. While an optimal relationship may in itself provide certain aspects of the proper intron geometry for splicing, it is more likely that conserved relationships between these intron sequences represent features that promote interactions between the U1 snRNP bound at the 5′ splice site and the U2 snRNP bound at the branch site. Interaction between these two snRNPs has been proposed previously (Black *et al.*, 1985), and the coimmunoprecipation of U1 snRNPs and U2 snRNPs by monospecific antisera directed against either of these snRNPs is suggestive of their direct interactions (Mattaj *et al.*, 1986).

As described above, a functional yeast analog of U2, snR20, (also called LSR1), has been identified previously. Pikielny and his co-workers (1986) have observed that the first detectable intermediate in the assembly pathway for a yeast intron involves the binding of this U2 snRNP to the intron in a manner dependent on the presence of the 5′ splice junction. This suggests that interactions between the U1 snRNP and the U2 snRNP

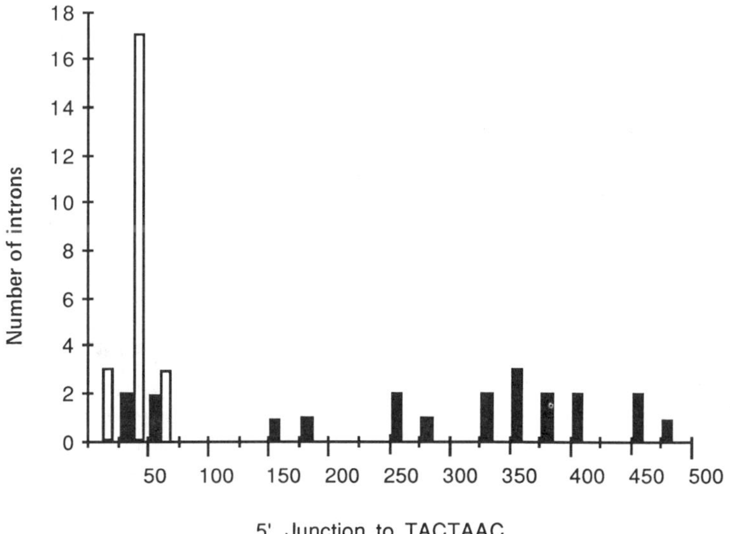

Fig. 3. Analysis of 5′ junction to branch point spacings in fungal introns. Introns are grouped by the number of nucleotides between the branch point and 5′ junction, excluding the conserved nucleotides that make up each signal (6 nucleotides at the 5′ junction and the 7 nucleotides equivalent to the TACTAAC box at the branch point). The number of introns within a given spacing interval are shown for *S. cerevisiae* (solid bars) as well as other fungi (open bars). The introns represented are the same group shown in Fig. 1.

may initiate, and therefore dominate, the assembly process in yeast. (Although, no U1 analog has yet been identified in *Saccharomyces cerevisiae,* the strong similarity of the splicing process in many aspects, including the presence of a conserved sequence at the 5′ junction similar to that which interacts with the U1 snRNP in mammalian introns, argues strongly that one exists and will soon be described.) Since the initial events that can be detected in the assembly of mammalian spliceosomes occur at the 3′ splice site, interactions between the U1 and U2 snRNPs may be more important to the assembly process in yeast and other fungi as compared to mammals. This suggests that features of the intron that influence interactions between these two snRNPs may be more highly conserved then in mammalian introns.

As shown in Fig. 3, we have examined the relationship between the 5′ splice junction and branch site and again have found two distinct distance relationships. In all introns from fungi other than *S. cerevisiae* and four of the *S. cerevisiae* introns, the spacing between these two elements is quite small (about 40 nucleotides). We refer to introns containing this spacing between the 5′ splice site and the branch point as being a 5′S intron. It seems likely that this small spacing could facilitate direct interaction between the U1 and the U2 snRNPs. The size of the region bound be each of these snRNPs in mammalian extracts [approximately 15 nucleotides for U1 (Mount *et al.*, 1983), about 35 nucleotides for U2 (Black *et al.*, 1985)] is consistent with this notion. In contrast, in the remainder of *S. cerevisiae* introns the spacing is significantly larger (200–450 nucleotides). We refer to introns containing this spacing between the 5′ splice site and the branch point as being a 5′L intron.

How do the U1 and U2 snRNPs interact in introns with a large spacing between the 5′ junction and branch site? One possibility is that secondary structure features of the intron could bring the 5′ splice junction and the branch site closer together in three-dimensional space. Interestingly, we have noted that most of the *S. cerevisiae* introns with a large spacing between the 5′ junction and the branch site can be folded into secondary structures which position the 5′ junction and the branch site in a similar linear arrangement to that of the small introns just discussed. Some of these structures are shown in Fig. 4 and the remainder are listed in Table I. A common feature of these structures is that, while the exact position of the helix varies, the number of nucleotides between the 5′ splice site and the TACTAAC box is relatively constant (after subtracting the nucleotides in the stem and loop), around 45 nucleotides (see Table I), and is approximately the same as in the small introns.

Are there any data that suggest these helices may actually play a role in the splicing of the large yeast introns? First, it is worthwhile pointing out that the trans-splicing experiments of Solnick (1985) and Konarska and her colleagues (1985) demonstrate that the splicing machinery has the ability to splice across such helical structures. In addition, the observation that the conserved sequences required for splicing in yeast can be found in other transcripts which are not spliced (Parker, 1985) suggest that there are additional features to yeast introns that have not yet been identified. Similarly, the observation that yeast does not in general use cryptic 5′ splice junctions cannot be due solely to the strong sequence conservation of yeast splicing signals. The yeast splicing machinery can clearly use a mutant 5′ splice site in the right place [note the use of many mutant junctions (Parker and Guthrie, 1985; Fouser and Friesen, 1986; Vijayraghavan *et al.*, 1986; Jacquier *et al.*, 1985)], though a perfect 5′ consensus sequence is not used when inserted a significant distance into the intron, even in the absence of the normal 5′ junction (A. Newman, personal communication). This argues that the optimal position of the 5′ junction is marked in some way.

In addition, two experimental observations are consistent with these

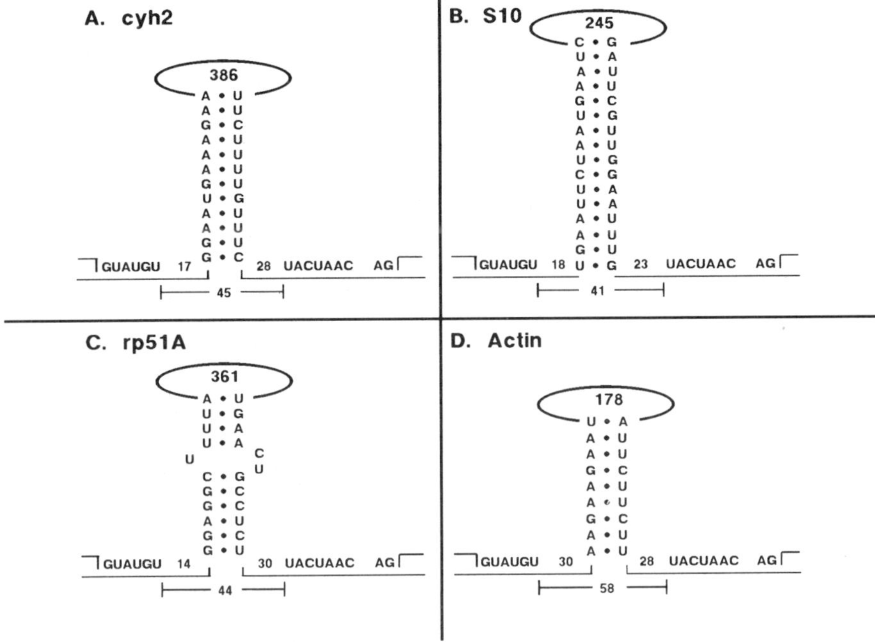

Fig. 4. Possible helical structures in *Saccharomyces cerevisiae* introns. The figure shows possible secondary structures for four introns from *S. cerevisiae*. Exon sequences are shown as open boxes, introns sequences as a solid line. The conserved sequences found at the 5′ splice junction, the branch site, and the AG at the 3′ splice junction are also shown. Nucleotide distances between the 5′ junction sequence and the base of the helix are shown with the first base counted as the base following the 3′ uridine of the 5′ consensus. The distance between the helix and the TACTAAC box is counted from the first unpaired nucleotide 3′ of the helix to the nucleotide preceding the first U of the UACUAAC sequence. The distance between the 5′ junction and the branch point across the helix is the sum of these two. The number of nucleotides in the large loop is shown above each helix. The S10 intron shown is from the S10A gene.

helices playing a significant, though perhaps nonessential role in splicing. As noted by Pikielny and Rosbash (1985), there is a small region near the 5′ splice site in the intron of the rp51A gene which is required for optimal splicing. The region they defined includes the majority of the helix we predict for the rp51A intron (the bottom 6 base pairs of the helix shown in Fig. 4). Similarly, when small deletions are made which remove the 3′ portion of this stem in both the actin and *cyh2* introns, splicing is inhibited both *in vivo* and *in vitro* (A. Newman, personal communication). Interestingly, at least in some cases, the splicing of these introns can be restored by much larger deletions, consistent with a role of the helix in bringing the

TABLE I

Possible Helical Structures in *Saccharomyces cerevisiae* Introns[a]

Gene	5′ Portion	3′ Portion	Spacing across helix
L17A	UGGUUAAAUUGA	UCAAGAUUAACCA	58
L25	AGCACGAGAAAAUUGAGAGGAAGA	UUUUCUUCUAUUUGGUGCU	44
rp16A	AGUAAUUGGUAAUGCAAAGU	GCUUUGUAUAUCAUUAUU	48
rp16B	UAGAGAAUGAAUGGAU	AUUCCAUAUUCUAUUUUUA	47
rp28B	AAGAGCUAU	AUAGGACUUCUU	46
rp51B	GGAUGAUGAUGGA	UCUAUCUUAUCC	45
rp59	ACAUAGUGAACAUUUUU	AAAGAUACUAUGU	35
rp73	UAUUUCGCUUUAUUCUUUCAUGGGA	UCUCCAUGACUGUGAAAUAAAAGUGAGAUA	40

[a] The table shows the 5′ and 3′ portions of large helical structures formed within some *S. cerevisiae* introns. In each case, the nucleotides forming the stem are shown in a 5′ → 3′ direction. Nucleotides which are bulged are underlined. No strong helical region could be found for the rp28A intron which gave a similar spacing to the others. The introns from the L34, S10B, and rp29 genes were not examined for this structure. The spacing across the helix is calculated as described in the legend to Fig. 4.

5′ junction and the TACTAAC box into close proximity (A. Newman, personal communication).

On the basis of the above observations, we can make a model in which the interaction between U1 at the 5′ junction and U2 at the branch site is optimized in yeast introns by one of two ways. First, the distance between the two signals may be such that it promotes this interaction directly as in the small introns. Alternatively, larger introns may fold into secondary structures which bring these two elements into closer proximity.

IV. PERSPECTIVES

To this point, we have discussed two different types of 3′ intron organization as well as two distinct manners in which the 5′ splice junction and the branch point are related to each other. Interestingly, the majority of introns with a short 5′ splice site to branch point distance (5′S) also have a short branch point to 3′ splice site spacing (3′S). Similarly, introns with a large spacing between the 5′ junction and the branch site (5′L) always contain a splice site distinguished by a large branch site to 3′ splice site distance and a conserved polyuridine tract (3′L). Though the significance of this nonrandom pairing of organizational features of the intron is not understood, the observation that mixed introns exist, either naturally (for example, the *kin28* intron is both a 5′S and a 3′L intron) or can be created (as discussed in Section III,B) suggest that these different types of intron organization are not absolutely required together.

This raises the interesting possibility that the different types of introns we see reflect the evolution of pre-mRNA introns. The simplest, and perhaps progenitor, pre-mRNA intron is represented by the fungal introns distinguished by a short spacing between the 5′ junction and the branch point as well as a 3′S splice site. By our models, this type of intron is optimized for snRNP–snRNP interactions and as such requires none of the additional features we have discussed. The introns found in the ***MATa1*** gene in *S. cerevisiae* are a good example of this class of intron.

If our view of 5′S and 3′S introns as being optimized for spliceosome assembly is correct, why have more complex introns evolved and been maintained? One possibility is that the additional requirements for the splicing of more complex introns represent opportunities for the regulation of gene expression. For example, in 5′L introns factors that destabilize the helix (perhaps by stabilizing a mutually exclusive structure) would be expected to decrease the efficiency of splicing. This notion provides a basis for why mammalian genomes, with their greater requirement for regulatory networks, encode, in general, introns of the more complex type.

ACKNOWLEDGMENTS

We gratefully acknowledge Dr. Christine Guthrie for her intellectual and financial support. We thank L. Fouser, J. Friesen, and Andrew Newman for communicating results prior to publication. We acknowledge Horst Domdey for the initial observation of the helix in the yeast actin gene. Finally, we thank Paul Siliciano, Robert Parker, Robert Jensen, and A.T.T. for help in the preparation of the manuscript. B. P. was supported by National Institutes of Health institutional training Grant GM 07810.

REFERENCES

Abovich, N., and Rosbash, M. (1984). *Mol. Cell. Biol.* **4,** 1871–1876.

Ares, M., Jr. (1986). *Cell (Cambridge, Mass.)* **47,** 49–59.

Bertrand, H., Bridge, P., Collins, R. A., Garriga, G., and Lambowitz, A. M. (1982). *Cell (Cambridge, Mass.)* **29,** 517–526.

Black, D. L., Chabot, B., and Steitz, J. A. (1985). *Cell (Cambridge, Mass.)* **42,** 737–750.

Boel, E., Hansen, M. T., Hjort, I., Hoegh, I., and Fiil, N. P. (1984). *EMBO J.* **3,**(7), 1581–1585.

Brown, J. W. S., Feix, G., and Frendeway, D. (1986). *EMBO J.* **5,** 2749–2758.

Cech, T. R. (1986). *Cell (Cambridge, Mass.)* **44,** 207–210.

Cellini, A., Felder, E., and Rossi, J. J. (1986). *EMBO J.* **5,** 1023–1030.

Chabot, B., Black, D. L., LeMaster, D. M., and Steitz, J. A. (1985). *Science* **230,** 1344–1349.

Dona, J. J. M., Mulder, G. H., Rouwendal, G. J. A., Springer, J., Bremer, W., and Wessels, J. G. H. (1984). *EMBO J.* **3**(9), 2201–2106.

Fouser, L. A., and Friesen, J. D. (1986). *Cell (Cambridge, Mass.)* **45,** 81–93.

Fouser, L. A., and Friesen, J. D. (1987). *Mol. Cell. Biol.* **7**(1), 225–230.

Frendeway, D., and Keller, W. (1985). *Cell (Cambridge, Mass.)* **42,** 355–367.

Fukamaki, Y., Ghosh, P. K., Benz, E. J., Reddy, V. B., Lebowitz P., Forget, B. G., and Sherman, M. W. (1982). *Cell (Cambridge, Mass.)* **28,** 585–593.

Gerke, V., and Steitz, J. A. (1986). *Cell (Cambridge, Mass.)* **47,** 973–984.

Green, M. (1986). *Annu. Rev. Genet.* **20,** 671–708.

Guthrie, C., Reidel, N., Parker, R., Swerdlow, H., and Patterson, B. (1986). *UCLA Symp. Mol. Cell. Biol.* **L33,** 301–321.

Hiraoka, Y., Toda, T., and Yanagida, M. (1984). *Cell (Cambridge, Mass.)* **39,** 349–358.

Jacquier, A., Rodriguez, J. R., and Rosbash, M. (1985). *Cell (Cambridge, Mass.)* **43,** 423–430.

Kaufer, N. F., Fried, H. M., Schwindinger, W. F., Jasin, M., and Warner, J. R. (1983). *Nucleic Acids Res.* **11**(10), 3123–3135.

Kinnaird, J. H., and Fincham, J. R. S. (1983). *Gene* **26,** 253–260.

Konarska, M. M., Padgett, R. A., and Sharp, P. A. (1985). *Cell (Cambridge, Mass.)* **42,** 165–171.

Langford, C. J., and Gallwitz, D. (1983). *Cell (Cambridge, Mass.)* **33,** 519–527.

Leer, R. J., van Raamsdonk-Duin, M. M. C., Molenaar, C. M. T., Cohen, L. H., Mager, W. H., and Planta, R. J. (1982). *Nucleic Acids Res.* **10**(19), 5869–5878.

Leer, R. J., van Raamsdonk-Duin, M. M. C., Hagendoorn, M. J. M., Mager, W. H., and Planta, R. J. (1984). *Nucleic Acids Res.* **12**(17), 6685–6700.
Leer, R. J., van Raamsdonk-Duin, M. C., Molenaar, C. M. T., Witsenboer, H. M. A., Mager, W. H., and Planta, R. J. (1985). *Nucleic Acids Res.* **13,** 5027–5039.
Lustig, A. J., Lin, R.-J., and Abelson, J. (1986). *Cell (Cambridge, Mass.)* **47,** 953–963.
McGraw, P., and Tzagoloff, A. (1983). *J. Biol. Chem.* **258,** 9459–9468.
Maniatis, T., and Reed, R. (1987). *Nature (London)* **325,** 673–678.
Mattaj, I. W., Habets, W. J., and Venrooij, W. J. (1986). *EMBO J.* **5**(5), 997–1002.
Miller, A. M. (1984). *EMBO J.* **3,** 1061–1065.
Mitra, G., and Warner, J. A. (1984). *J. Biol. Chem.* **259**(14), 9218–9224.
Molenaar, C. M. T., Woudt, L. P., Jansen, A. E. M., Mager, W. H., and Planta, R. J. (1984). *Nucleic Acids Res.* **12**(19), 7345–7358.
Mount, S. (1982). *Nucleic Acids Res.* **10**(2), 459–472.
Mount, S. M., Pettersson, I., Hinterberger, M., Kormas, M., and Steitz, J. (1983). *Cell (Cambridge, Mass.)* **33,** 509–518.
Ng, R., and Abelson, J. (1980). *Proc. Natl. Acad. Sci. U.S.A.* **77,** 3912–3916.
Parker, R. (1985). Ph.D. Thesis, University of California at San Francisco.
Parker, R., and Guthrie, C. (1985). *Cell (Cambridge, Mass.)* **41,** 107–118.
Patterson, R. B., and Guthrie, C. (1987). *Cell (Cambridge, Mass.)* **49,** 613–624.
Peebles, C. L., Perlman, P. S., Kecklenburg, K. L., Petrillo, M. L., Tabor, J. H., Jarrell, K. A., and Cheng, H.-L. (1986). *Cell (Cambridge, Mass.)* **44,** 213–233.
Pikielny, C. W., and Rosbash, M. (1985). *Cell (Cambridge, Mass.)* **41,** 119–126.
Pikielny, C. W., Rymond, B. C., and Rosbash, M. (1986). *Nature (London)* **324,** 341–345.
Piller, T., Lang, B. F., Steinburger, I., Vogt, B., and Kaudewitz, F. (1983). *J. Biol. Chem.* **258,** 7954–7959.
Reed, R., and Maniatis, T. (1985). *Cell (Cambridge, Mass.)* **41,** 95–105.
Ruskin, R., and Green, M. (1985). *Cell (Cambridge, Mass.)* **43,** 131–142.
Ruskin, R., Greene, J. M., and Green, M. R. (1985). *Cell (Cambridge, Mass.)* **41,** 833–844.
Rymond, B. C., and Rosbash, M. (1985). *Nature (London)* **317,** 735–737.
Schaap, P. J., Molenaar, M. T., Mager, W. H., and Planta, R. J. (1984). *Curr. Genet.* **9,** 47–54.
Schatz, P. Z., Pillus, L., Grisafi, P., Solomon, F., and Botstein, D. (1986). *Mol. Cell. Biol.* **6,** 3711–3721.
Shoemaker, S., Schweickart, V., Ladner, M., Gelfand, D., Kwok, S., Myambo, K., and Innis, M. (1983). *Bio/Technology* **1,** 691–696.
Simon, M., and Faye, G. (1984). *Proc. Natl. Acad. Sci. U.S.A.* **81,** 8–12.
Simon, M., Seraphin, B., and Faye, G. (1986). *EMBO J.* **5,** 2697–2704.
Solnick, D. (1985). *Cell (Cambridge, Mass.)* **42,** 157–164.
Tazi, J., Ailbert, C., Temsamani, J., Reveillaud, I., Cathala, G., Brunel, C., and Jeanteur, P. (1986). *Cell (Cambridge, Mass.)* **47,** 755–766.
Teem, J. L., and Rosbash, M. (1983). *Proc. Natl. Acad. Sci. U.S.A.* **80,** 4403–4407.
Teem, J. L., Abovich, N., Kaufer, N. F., Schwindinger, W. F., Warner, J. R., Levy, A., Woolford, J., Leer, R. J., van Raamsdonk-Duin, M. M., Mager, W. H., Planta, R. J., Schultz, L., Friessen, J. D., and Rosbash, M. (1984). *Nucleic Acids Res.* **12**(22), 8295–8312.
Toda, T., Adachi, Y., Hiraoka, Y., and Yanagida, M. (1984). *Cell (Cambridge, Mass.)* **37,** 233–242.
Vijayraghavan, U., Parker, R., Tamm, J., Iimura, Y., Rossi, J., Abelson, J., Guthrie, C. (1986). *EMBO J.* **5**(7), 1683–1695.
Woudt, L., Pastink, A., Kempers-Veebstra, A. E., Jansen, A. E. M., Mager, W., and Planta, R. J. (1983). *Nucleic Acids Res.* **11**(16), 5347–5360.

10

RNA Joining and Trypanosome Gene Expression

NINA AGABIAN,[1] KAREN L. PERRY,[1] AND WILLIAM J. MURPHY[2]

Department of Biomedical and Environmental Health Science
School of Public Health
University of California
Berkeley, California 94720
and
Naval Bioscience Laboratory
Berkeley, California 94720

I. INTRODUCTION

The molecular biology of protozoan parasites has many novel features that affect our current understanding of eukaryotic gene expression, parasite virulence and host immunity. Among the forerunners of parasite systems to be analyzed using molecular methods were the African trypanosomes, particularly *Trypanosoma brucei*. The African trypanosomes cause disease in vast proportions on the African continent to man as well as to his livestock and, in combination with the related American trypanosomes and leishmanias, are a major detrimental factor in the social and economic well-being of the Third World.

[1] Present address: Intercampus Program, Molecular Parasitology, School of Pharmacy, University of California, Laurel Heights, San Francisco, California 94143-1204.

[2] Present address: Department of Microbiology, University of Kansas, Kansas City, Kansas 66103.

African trypanosomes are most widely recognized for their ability to undergo antigenic variation—a process by which *T. brucei* and related organisms evade clearance by the host immune system (for recent reviews, see Boothroyd, 1985; Donelson and Rice-Ficht, 1985; Borst, 1986). The phenomenon of antigenic variation is characterized by the sequential expression of antigenically distinct variable-surface glycoproteins (VSGs); a single VSG forms a monolayer of molecules that coat the entire surface of the parasite. Each VSG is encoded by a separate gene and there are some 1000 VSG genes per trypanosome genome (Van der Ploeg *et al.*, 1982b). Sequence analysis of a number of mRNAs and cDNAs encoding antigenically distinct VSGs led to the observation that each began with the same 35-nucleotide sequence (Boothroyd and Cross, 1982; Van der Ploeg *et al.*, 1982a). Furthermore, this 35-nucleotide sequence was not encoded at the corresponding position in genomic copies of these VSG genes, suggesting that this leader sequence is joined to the body of the mRNA posttranscriptionally. Subsequent experiments have demonstrated that other, non-VSG genes also have this spliced leader (SL) at their 5′ ends (Parsons *et al.*, 1984; De Lange *et al.*, 1984a; Sather and Agabian, 1985; Tschudi *et al.*, 1985). In fact, the presence of the SL appears to be a universal feature not only of mRNAs in *T. brucei,* but all other trypanosome species as well (Nelson *et al.*, 1984; De Lange *et al.*, 1984b; Agabian *et al.*, 1985).

In this report we present a characterization of the events that lead to the formation of mature mRNAs in *T. brucei.* We find that there are approximately 200 copies of the SL in the trypanosome genome, each of which is encoded within a 1.4-kb unit and organized into one or a few tandem arrays. A 135- to 140-nucleotide small RNA is transcribed from the 1.4-kb unit and contains the SL at its 5′ end; this molecule has been termed the SL RNA. Our characterization of the SL RNA indicates that it has the same 5′ cap structure as found on mature mRNAs, suggesting that donation of a cap during mRNA synthesis may be one role for the SL RNA in the cell. The studies we describe demonstrate that the SL RNA and structural gene transcripts are joined by a process which, in many respects, appears analogous to RNA splicing in other eukaryotes, the primary difference appearing to be that the intervening sequences in trypanosomes are discontinuous. Because the substrates remain unlinked we have termed the process of RNA joining between the SL RNA and structural gene transcripts "trans-RNA splicing." In light of these observations, it is intriguing that the SL RNA is found in a small ribonucleoprotein particle with properties similar to those of the small nuclear ribonucleoprotein particles (snRNPs) of other eukaryotes.

II. MATERIALS AND METHODS

A. Trypanosome Isolation, DNA and RNA Preparation

IsTat 1.1 trypanosomes were grown in male rats until a density of 2–3 $\times 10^9$ cells/ml was reached. The trypanosome-infected blood was collected by cardiac puncture and immediately mixed with 10 volumes of ice-cold 6 *M* urea, 3 *M* LiCl (Auffray and Rougeon, 1980). The RNA was allowed to precipitate overnight at 0°C and then pelleted at 18,000 *g* at 0°C for 1 h. The precipitated RNA was resuspended in buffer containing 25 m*M* Tris-HCl (pH 8.0), 100 m*M* NaCl, 5 m*M* EDTA, 1% SDS, and 200 μl/ml proteinase K. After digestion for 1 h the RNA was phenol-$CHCl_3$ extracted, ether extracted, and ethanol precipitated. Poly$(A)^+$ RNA was prepared from a portion of this RNA as described (Aviv and Leder, 1972). RNA for the species Northern blot shown in Fig. 2 and genomic DNA were isolated as previously described (Milhausen *et al.*, 1984). *T. cruzi* RNA was kindly provided by Drs. P. Lizardi and N. Nogueira.

B. S1 Nuclease Analysis

Sl nuclease digestions were carried out as described (Maniatis *et al.*, 1982). RNA [20 μg of total or 6 μg of poly$(A)^+$] was hybridized with an M13 clone which contained the complementary strand of the entire SL reiteration unit (Murphy *et al.*, 1986) at 56°C for 3 hr before addition of nuclease Sl. Digestion was carried out at 30°C for 30 min. Reactions were stopped by the addition of 0.25 volumes of 2.5 *M* NH_4OAc, 50 m*M* EDTA (pH 8.0), 10 μg of carrier tRNA and an equal volume of isopropyl alcohol.

C. Northern Blot Analysis

Precipitated RNA or Sl-protected DNA fragments were centrifuged, washed with 70% ethanol, dried, and resuspended in 90% formamide, 1 m*M* EDTA, 0.5% xylene cyanol, 0.5% bromphenol blue. After boiling, the samples were electrophoresed on 5% polyacrylamide (acrylamide/bis, 20 : 1), 7 *M* urea gels (10% polyacrylamide, 7 *M* urea for species blot shown in Fig. 2). The gels were electro-transferred to Nytran membranes (Schleicher & Schuell) as per the manufacturer's protocol. The filters were hybridized and washed as described in Maniatis *et al.* (1982). RNA-containing filters were probed either with oligonucleotides complementary to positions 10–31 of the SL (Fig. 2) or complementary to positions 61–95 of the SL RNA (Fig. 5A and 6). The oligonucleotides were labeled with $[\gamma\text{-}^{32}P]$ATP and polynucleotide kinase (Pharmacia-PL) as described (Maniatis *et al.*, 1982). Filters containing Sl-protected DNA (Fig. 5B) were probed with uniformly labeled single-strand DNA generated from an M13 clone containing the sequence complementary to the SL RNA between the *Rsa*I restriction site (position 27 in Fig. 3) and a *Sau*3A site (not shown) located approximately 40 nucleotides downstream from the 3′ end of the SL RNA (clone M13RS−). The DNA was labeled by primer extension of the −20 sequencing primer (New England Biolabs) using all four ^{32}P-labeled nucleoside triphosphates as described by Hu and Messing (1982). The probe was then size-selected by elution from a 5% acrylamide 7 *M* urea gel.

D. Southern Blot Analysis

Genomic DNA was digested with various restriction enzymes and fractionated by agarose gel electrophoresis. After transfer to a nitrocellulose membrane (Southern, 1975) the blot was probed with the ^{32}P-labeled, 22-base oligonucleotide complementary to the SL (see above) as described by Nelson *et al.* (1983).

E. Debranching Enzyme Analysis

The debranching enzyme preparation used was purified 700-fold from HeLa cell cytosolic fraction and generously provided by Drs. Jaime Arenas and Jerard Hurwitz. The enzyme preparation was free from contaminating endo- or exonucleases, phosphatases, and nonbranched, 2′-5′ phosphodiesterase activity. RNA samples [20 μg of total and 6 μg of poly(A)$^+$] were digested in 20 μl of 20 m*M* HEPES (pH 7.0), 3 m*M* $MgCl_2$, 0.25 mg/ml bovine serum albumin for 30 min at 30°C. Samples were then phenol-$CHCl_3$ extracted, ether extracted, and ethanol precipitated. The RNA was electrophoresed, blotted to Nytran, and probed with the oligonucleotide complementary to the 100-mer portion of the SL RNA as described above.

F. 5′ Cap Analysis

For the *in vivo* labeling of procyclic-form *T. brucei,* cells were incubated in phosphate-free BSM medium (Bienen *et al.*, 1981) to which ^{32}P-orthophosphate (New England Nuclear) was added at a final concentration of 0.1 mCi/ml. Incubations were overnight at 26°C. Total RNA was purified from the labeled cells by the guanidinium/hot phenol extraction as described by Maniatis *et al.* (1982). Poly(A)$^+$ RNA was purified from the total RNA as described (Aviv and Leder, 1972). SL RNA was hybrid selected from total RNA using the M13RS− clone (see above) as described (Maniatis *et al.*, 1982). The labeled RNAs were digested to completion with RNase T2 in 50 m*M* NaOAc (pH 5.2), 2 m*M* EDTA at 37°C for several hours. Digestion products were spotted on PEI cellulose plates and chromatographed for 18 h in 2 *M* pyridinium formate (pH 3.4). Cap spots were located by autoradiography and eluted with 2 *M* triethylammonium bicarbonate (Silberklang *et al.,* 1979). For the analysis of the T2-resistant cap structures, the isolated oligonucleotides were digested with nuclease P1 (in 50 m*M* NaOAc, pH 5.2), nucleotide pyrophosphatase (in 10 m*M* Tris-HCl, pH 7.6, 10 m*M* $MgCl_2$), or a combination of the two enzymes. The digestion products were chromatographed on cellulose thin-layer plates as described (Silberklang *et al.,* 1979).

G. Ribonucleoprotein Particle Analysis

IsTat 1.1 trypanosomes were isolated from blood by centrifugation and DE-52 chromatography as described by Lanham (1968). Trypanosome nuclei were prepared by dounce homogenizing cells in hypotonic buffer with 0.5% NP40. Nuclei were resuspended in 60 m*M* KCl, 2.5 m*M* $MgCl_2$, 35 m*M* HEPES (pH 7.5), 1 m*M* DTT (buffer B), and sonicated for 15 sec at setting 175 with a Braun 200 sonifier. The extract was clarified by centrifugation at 8000 *g* for 20 min. The supernatant was then immediately loaded onto 10–30% sucrose gradients in buffer B. The gradients were spun at 35,000 rpm in an SW-41 rotor for 12 hr at 4°C and then fractionated (typically 36 fractions per gradient tube). The fractions were deproteinized by proteinase K digestion and phenol/$CHCl_3$ extraction as above and ethanol precipitated. Duplicate dot blots of each gradient fraction were prepared and hybridized with either a 35-base oligonucleotide complementary to the entire SL or a uniformly labeled probe complementary to the SL RNA gene (see above). S values were determined by comparing the sedimentation of ^{3}H-labeled ribosomal RNA markers (28 S, 18 S, and 4 S) run in parallel gradients.

III. RESULTS

A. Organization of the SL genes in *Trypanosoma brucei*

Because the SL sequence appeared to be unlinked to any known structural genes, our first concern was to determine where the SL sequence was encoded in the trypanosome genome. Figure 1A shows complete and partial *Apa*I restriction endonuclease digestions of trypanosome DNA hybridized to an oligonucleotide probe which is complementary to the SL sequence. In the complete digest the 35-nucleotide SL sequence is located within a 1.4-kb fragment. Partial restriction digestion of genomic DNA reveals a ladder of hybridizing fragments, each increasing in size by 1.4

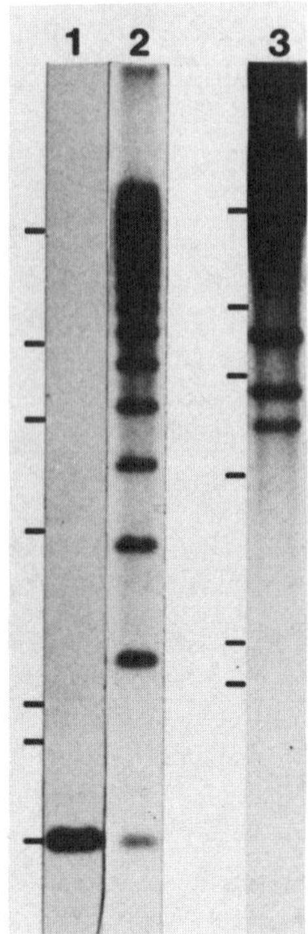

Fig. 1. Identification and genomic organization of SL sequence in *T. brucei*. Genomic Southern analysis using the complementary SL oligonucleotide was performed as in Materials and Methods. DNA was completely (lane 1) and partially (lane 2) digested with *Apa*I and completely digested with *Ava*I (lane 3). The hash marks indicate DNA markers (from top to bottom) of 23.3, 9.4, 6.6, 4.4, 2.3, 2.0, and 1.4 kb. (From Agabian *et al.*, 1985.)

kb. These results demonstrate that the SL gene is organized within tandemly linked repeat units. Digestion of genomic DNA with the enzyme *Ava*I (Fig. 1B), which does not cleave within the repeat unit, reveals not only these large tandem arrays, but also three or four orphons which contain one or a few copies of the SL sequence (Nelson *et al.*, 1983). Quantitative hybridization analysis has shown that the genome of *T. brucei* contains approximately 200 such 1.4-kb reiterated sequences (De Lange *et al.*, 1983; Michiels *et al.*, 1983).

The organization of the SL described above indicates clearly that the SL and each of the structural gene sequences to which it is joined must interact via an inter- rather than intramolecular process and predicts that a discrete SL-containing RNA is transcribed from the 1.4-kb repeat units. This prediction is borne out by the Northern blot analysis in Fig. 2 which shows that in addition to the high-molecular-weight mRNA, a small RNA

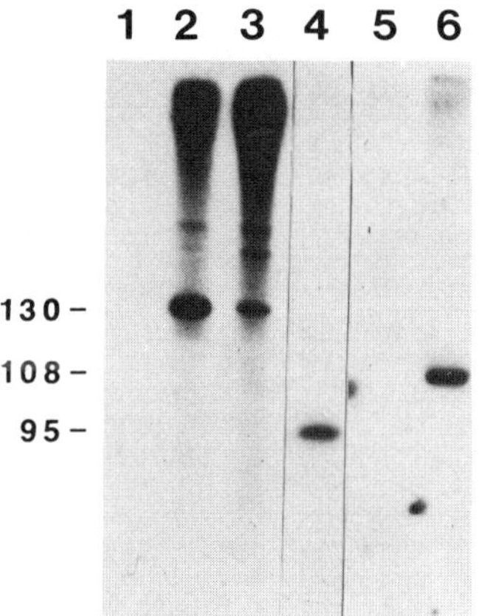

Fig. 2. Analysis of trypanosomatid small RNAs by gel electrophoresis and hybridization. Total RNA samples were electrophoresed, blotted, and probed with the complementary SL oligonucleotide (22-mer) described in Materials and Methods. Each lane contains 10 μg of RNA from *E. coli* (lane 1), *T. brucei* (bloodsteam form) (lane 2), *T. brucei* (procyclic form) (lane 3), *L. collosoma* (lane 4), *C. fasciculata* (lane 5), and *T. cruzi* (lane 6). The markers indicate the sizes in nucleotides of relevant *T. brucei* small ribosomal RNAs. Although RNA from *C. fasciculata* does not hybridize with the 22-mer employed here, this organism does contain repeated sequences which hybridize with oligonucleotides generated from another region of the SL (De Lange *et al.*, 1984b). (From Milhausen *et al.*, 1984. Copyright held by Cell Press.)

species of ~135 nucleotides is detected by the SL oligonucleotide probe. Nuclease S1 analysis demonstrates that this small RNA (the SL RNA) is transcribed from a continuous sequence within the reiteration unit (Milhausen *et al.*, 1984; Campbell *et al.*, 1984; Kooter *et al.*, 1984). Thus, only a very small portion of the SL gene is transcribed and gives rise to the SL RNA. Figure 2 further shows that two other species of trypanosomes, *Leptomonas collosoma* and *Trypanosoma cruzi* also possess small RNAs (95 and 105 nucleotides, respectively) which are homologous to the *T. brucei* SL probe. This observation has been extended to a number of trypanosome species and seems to be a universal feature of these organisms (Nelson *et al.*, 1984; De Lange *et al.*, 1984b).

A comparison of the SL genes from four trypanosome species (two from *T. cruzi*), is shown in Fig. 3. Sequence analysis of the SL RNAs from the four different species reveals that each has the SL at its 5′ end and each terminates 5′ to a run of thymidine residues in the DNA sequence (Milhausen *et al.*, 1984; De Lange *et al.*, 1984b). As predicted in Fig. 1 the SL sequence of each is highly conserved. In addition, the sequences are homologous for ~15 nucleotides downstream from the SL but diverge beyond this point. This region of homology includes the sequence imme-

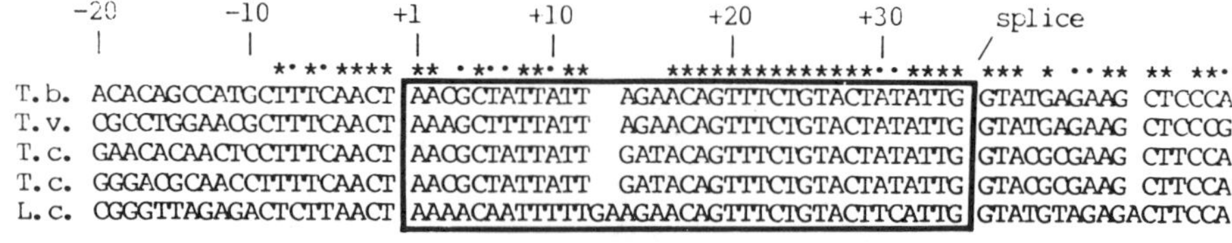

T.b. GTAGCAGCTGGGCCAACAGACGCATTGTGCTGTTGGTTCCTGCCGCATACTGCGGGAATCTGGAAGGTGGGGTCGG
T.v. GTCGCAAGACCGTGGTAATTTTTGGACACGGCCCTCGTGGCCGCGGCCCCGCGGCCAAACAACACAAAACAAATTT
T.c. ACCCGCCTCTGGTCGTATGTTTGGTCAATTTCTTTTGACCGGGGTCTACGGACCCCTTTTTTTTTTTTTTTTTTTT
T.c. AATCCGCGAGTCGCGGCTATGTTTGGTCAATTTCTTTTGACCGGGGTCCACGAACCCTTTTTTTTTTTTTTTTTTT
L.c. GAACCTAGTTCTGAAATTTTGGACCGAGCTTTCGGGCTTTTTTTTTACTTTTTCGGGTAC

T.b. ATGACCTCCACTCTTTTTATTTTTTTTTATTTTTTTTTCATTTTTT
T.v. TTGTTTTGTATATTTTTTTTTTTATTTTTTTATTATTATTATTATTT

Fig. 3. Sequence comparison of various trypanosomatid SL genes. From top to bottom: T.b., *T. brucei* (Milhausen *et al.*, 1984); T.v., *T. vivax* (De Lange *et al.*, 1984b); T.c., *T. cruzi* (Milhausen *et al.*, 1984); T.c., *T. cruzi* (De Lange *et al.*, 1984b); L.c., *L. collosoma* (Milhausen *et al.*, 1984). The SL sequence is boxed in. The reported 5′ end of the various SL RNAs is indicated by +1. The splice junction is also indicated. Other symbols: *, same nucleotide in all species; ., same nucleotide in all but one of the species.

diately adjacent to the SL, which conforms to the consensus sequence proposed for eukaryotic 5′ splice sites (Mount, 1982). Analysis of genomic clones of several *T. brucei* structural genes indicates that the site of SL joining also contains a 3′ consensus splice junction sequence. The conservation of the junction sequences in trypanosomes suggests that an RNA splicing mechanism, similar to that found in other eukaryotes, may be responsible for the production of mature mRNAs in these organisms.

B. The SL is Acquired via Trans-Splicing

Given the background information described above, one can propose several models for the generation of mature mRNAs in trypanosomes. As illustrated in Fig. 4, these fall into two general categories: primed transcription, whereby the SL RNA, or a portion thereof, acts as a primer for the transcription of the target gene; and independent transcription, whereby the SL RNA and target gene are transcribed separately and subsequently resolved by joining the SL to the target mRNA. Each of these models predict the formation of specific intermediates and end-products illustrated in Fig. 4. These predictions assume that the mechanism of SL joining in trypanosomes is, at least in some respects, analogous to the splicing of mammalian and yeast pre-mRNAs. The experiments described below were designed to detect specific intermediates and end-products that would allow us to differentiate between these potential mechanisms of SL joining in trypanosomes.

Review of the models outlined in Fig. 4 indicates that the structure and fate of the 100-mer sequence (the 100-nucleotide portion of the SL RNA immediately downstream from the SL) could distinguish which of the proposed pathways occur *in vivo*. Therefore we assayed for both the covalent association of the 100-mer with high-molecular-weight RNAs and the presence of either lariat or Y branch structures that would indicate cis- or trans-splicing, respectively. The lariat is formed during splicing of pre-mRNAs in mammalian cells and yeast by the cleavage of the pre-mRNA at the 5′ splice junction and concurrent formation of a 2′-5′ phosphodiester bond between the G residue located at the 5′ end of the intron and a nucleotide (usually an A) located a short distance upstream from the 3′ splice junction (for review, see Padgett *et al.*, 1986). Joining of the 5′ exon to the 3′ exon then occurs concurrently with the release of the intron as a lariat. The 2′-5′ phosphodiester branch is specifically cleaved by debranching enzyme to generate the linear intron species (Ruskin and Green, 1985). The trans-splicing of pre-mRNA would proceed similarly except that the intron sequence would be discontinuous, forming the Y structure.

A simple test of the covalent attachment of the 100-mer portion of the SL RNA to pre-mRNA can be made by comparing Northern blots of *T. brucei* RNA before and after S1 nuclease treatment. In this manner 100-mer, which is covalently attached to high-molecular-weight RNA would be detected as a discrete 100-nucleotide species only after its enzymatic release by nuclease S1. Without such treatment the 100-mer, attached to many different pre-mRNAs, would be dispersed throughout the entire pre-mRNA population and not resolved as a distinct species. In contrast, if there was no interaction with high-molecular-weight RNA, the 100-mer would remain free and be detected as a discrete 100-nucleotide species both before and after S1 nuclease treatment. As shown in Fig. 5, only the 135-nucleotide SL RNA is detected by directly blotting total *T. brucei* RNA; the SL RNA is absent from preparations of poly(A)$^+$ RNA. In contrast, a very strong 100-nucleotide species is detected if the RNA samples are hybridized with the complementary SL RNA sequence and then digested with nuclease S1. Primer extension of the 100-mer species detected in poly(A)$^+$ RNA reveals that the 5′ end of this molecule maps precisely to the G residue at the splice junction of the SL RNA (Murphy *et al.*, 1986). Thus an authentic 100-mer, derived from cleavage of the SL RNA immediately adjacent to the SL, is covalently attached to a population of poly(A)$^+$-containing RNA.

As noted above, if the SL RNA and target mRNA remain unlinked during splicing one would predict an intermediate Y rather than a lariat branch structure. A Y structure could be distinguished from a lariat by purified debranching enzyme: Enzymatic cleavage would release a 100-

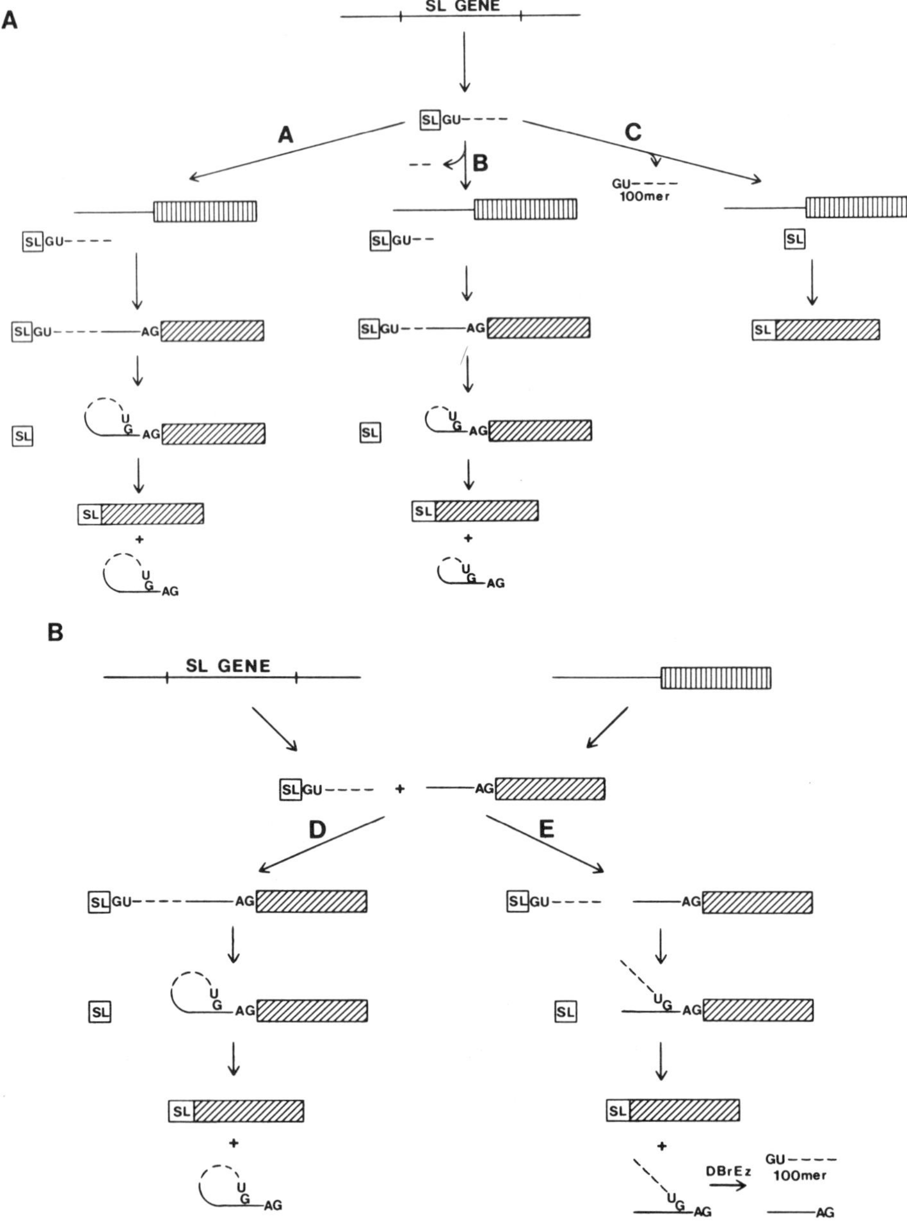

Fig. 4. Models of SL addition to target mRNAs. Two alternative models of SL RNA and structural gene transcription (primed versus independent) are suggested. In each case, possible pathways for the formation of mature mRNAs are schematized. The symbols are as follows: —|——|— , SLRNA gene (i.e., the reiteration unit); [SL]GU- - - - , the SL RNA (GU represents consensus 5′ splice site sequence); ——[||||||||], structural gene; —AG[////], putative pre-mRNA (AG represents consensus 3′ splice site sequence). (Top) Primed transcrip-

mer from a Y-branched pre-mRNA as a free 100-nucleotide species, which could be detected by Northern blot analysis. In contrast, if a lariat were formed, then cleavage of this structure with debranching enzyme would linearize the lariat but the 100-mer would remain covalently attached to the heterogeneous population of pre-mRNAs and a discrete 100-mer would not be resolved by Northern hybridization. The results of a debranching experiment, using either total or poly(A)$^+$ RNA, are shown in Fig. 6. Hybridization of an oligonucleotide probe complementary to the 100-mer portion of the SL RNA reveals the SL RNA in total RNA (lane 1) and, in addition, a small amount of ~95-nucleotide RNA. In contrast, hybridization of poly(A)$^+$ RNA reveals neither of these species (lane 3). Upon treatment of total RNA samples with debranching enzyme, however (lane 2), a stronger hybridization signal from a species slightly larger than 95 nucleotides is revealed. This same RNA species is detected in poly(A)$^+$ RNA upon debranching (lane 4). These results indicate that the 100-mer species is attached to poly(A)$^+$ RNA through a 2′-5′ phosphodiester branch which is readily cleaved by debranching enzyme; this interpretation is consistent with the idea that a Y-branch structure, rather than a lariat, is formed during SL addition. Furthermore, the detection of a splicing intermediate which contains the 100-mer portion of the SL RNA provides direct evidence that the SL RNA acts as donor of the SL sequence to mRNAs in trypanosomes.

C. Cap Structures of the SL RNA and Poly(A)$^+$ RNA

Recent evidence has indicated that the SL RNA contains a 5′-terminal modification with properties similar to those of a cap structure (Lenardo *et al.*, 1985; Laird *et al.*, 1985). If the SL RNA is indeed the donor of the SL sequence, then one would predict that the SL RNA and target mRNAs would have similarly capped 5′ ends. To specifically test this prediction,

---

tion. (A) The entire SL RNA transcript is used as a primer for transcription of structural genes resulting in a continuous pre-mRNA which is spliced in cis. (B) The SL RNA is first cleaved, releasing the 3′ end of the SL RNA (– –), and a smaller fragment of the SL RNA is used as a primer. The resulting pre-mRNA is spliced as in A. (C) The SL RNA is cleaved at the splice junction, releasing a free 100-mer, and the SL itself primes transcription immediately forming a mature mRNA. (Bottom) Independent transcription. Both the SL RNA and the structural genes are transcribed separately and the two are ligated head-to-tail to form a continuous pre-mRNA, which is spliced in cis (D), or the two remain as separate RNA molecules, but are spliced in trans (E). A, B, and D result in the formation of a lariat intron structural intermediate. In E a discontinuous, Y intron structure is formed. DBrEz, Debranching enzyme. (From Murphy *et al.*, 1986. Copyright held by Cell Press.)

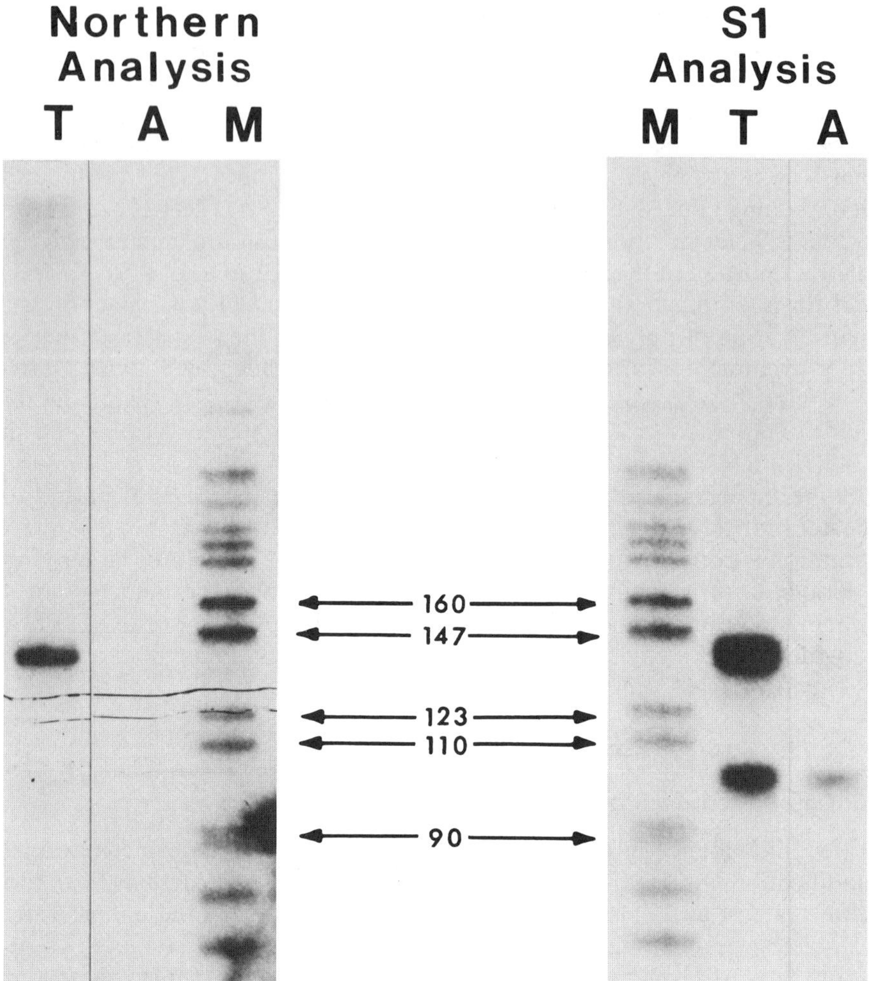

Fig. 5. Northern hybridization and S1 nuclease analysis of *T. brucei* RNA. Northern analysis. RNA samples [T, 20 μg total RNA; A, 6 μg poly(A)$^+$ RNA] were electrophoresed, blotted, and probed as described in Materials and Methods. S1 analysis. The same RNA preparations used for the Northern analysis were first hybridized with the complementary strand of the 1.4-kb reiteration unit and digested with nuclease S1. The S1-resistant DNA fragments were electrophoresed, blotted, and probed as described in Materials and Methods. The probe in this case contained only 148 nucleotides of the 1.4-kb reiteration unit immediately surrounding the 100-mer (see Materials and Methods). Thus, the protected DNA fragments did not arise from regions flanking the SL RNA sequence in the repeat unit. M, Markers prepared from *Hpa*II-digested pBR322.

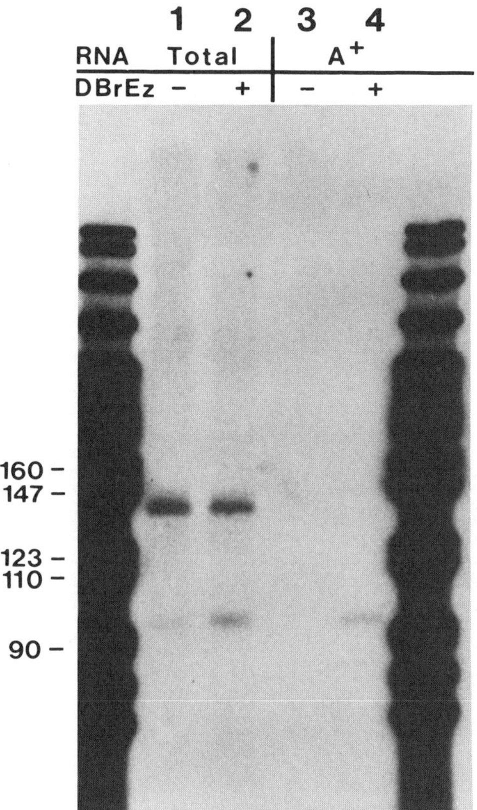

Fig. 6. Debranching analysis of *T. brucei* RNA. Total RNA (20 μg, lanes 1 and 2) or poly(A)$^+$ RNA (6 μg, lanes 3 and 4) was incubated in the presence (lanes 2 and 4) or absence (lanes 1 and 3) of debranching enzyme. The resultant products were electrophoresed, blotted, and probed as described in Materials and Methods. (From Murphy *et al.*, 1986. Copyright held by Cell Press.)

we have analyzed and compared the cap structures of purified SL RNA and mRNA labeled uniformly *in vivo* with ^{32}P. Digestion of the poly(A)$^+$ RNA with RNase T2 releases 3′ mononucleotides and a single T2-resistant fragment which are then separated by PEI cellulose thin-layer chromatography. Because RNase T2 cannot hydrolyze pyrophosphate bonds or 5′-3′ bonds adjacent to a 2′-*O* modification, this T2-resistant fragment is presumably an oligonucleotide with internal ribose modifications and/or a cap. The species was shown to be a capped oligonucleotide by digestion with nucleotide pyrophosphatase, which released a nucleotide that migrates with an unlabeled pm^{7}G standard. This result indicates that the

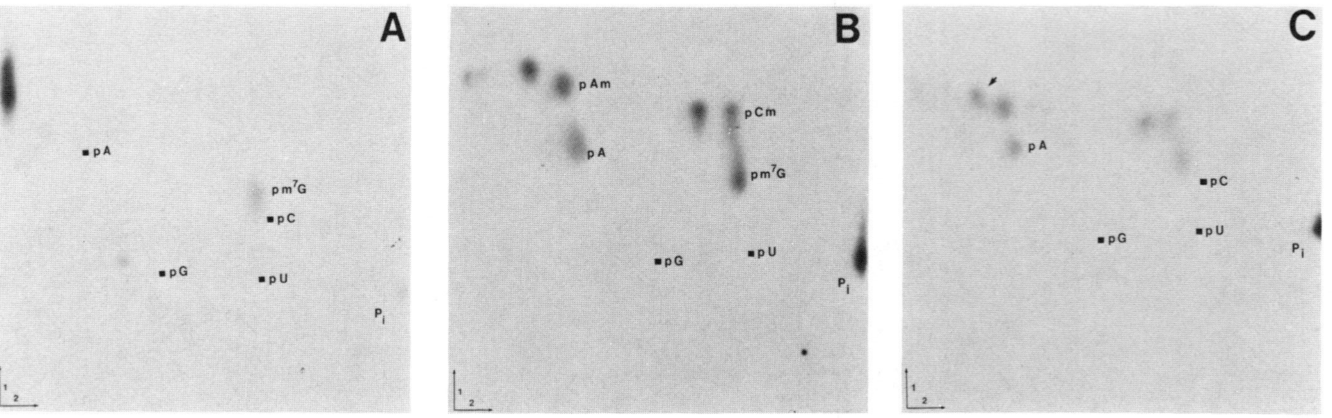

Fig. 7. Analysis of cap structures of *T. brucei* poly(A)$^+$ RNA and SL RNA. The T2-resistant cap structures from *T. brucei* poly(A)$^+$ RNA, labeled uniformly *in vivo,* were eluted from PEI cellulose and digested with nucleotide pyrophosphatase (A) or nucleotide pyrophosphatase and nuclease P1 (B). The T2-resistant cap structures from *T. brucei* SL RNA, labeled uniformly *in vivo,* were eluted from PEI cellulose and digested with nulceotide pyrophosphatase and nuclease P1 (C). The products were chromatographed on cellulose thin-layer plates in two dimensions using the solvent system developed by Silberklang *et al.* (1979). The positions of nonradioactive makers (visualized by UV irradiation) are indicated.

mRNA capping nucleotide in *T. brucei* is the same as is found in higher eukaryotes (Fig. 7A). Complete digestion of the T2-resistant species with nuclease P1 and pyrophosphatase revealed that there are modifications on the first four transcribed nucleotides of the SL (Fig. 7B). One of the modified nucleotides migrates with a pCm standard, while another migrates with pAm. A third nucleotide migrates with unmodified AMP, implying that the first unmodified nucleotide at the 5′ end of the SL (the 3′ end of the T2 fragment) is an A residue.

When *in vivo* labeled SL RNA is digested with RNase T2, the capped oligonucleotides released migrate as a diffuse spot or as two spots on PEI cellulose. The presence of these multiple species may be due to heterogeneity in the extent to which the SL RNAs are modified. When the major T2-resistant species is eluted and digested to completion with pyrophosphatase and nuclease P1, the same nucleotides as are present in the poly(A)$^+$ cap are released (Fig. 7C). This result is compatible with the hypothesis that the SL RNA acts as a donor of the mature mRNA cap structure.

D. A Ribonucleoprotein Particle, the SL RNP, Contains the SL RNA

In higher eukaryotes and yeast, the splicing of pre-mRNAs takes place within a large RNA–protein complex termed the spliceosome. This complex includes the pre-mRNA to be spliced, the well characterized heterogeneous nuclear RNP (hnRNP) proteins, several snRNPs which contain the U class of small RNAs, and, most likely, several as yet uncharacterized components (Padgett *et al.*, 1986). As splicing of the SL to target mRNAs in trypanosomes is related mechanistically to splicing of pre-mRNAs in higher eukaryotes, analogous complexes should be found in these organisms. In addition, the physical discontinuity between the SL and structural gene primary transcripts suggests that a specific protein complex may be formed with the SL RNA.

To test for the presence of an SL RNA-containing RNP (SL RNP) in trypanosomes, extracts prepared from *T. brucei* nuclei were sedimented through sucrose gradients and fractions from the gradient assayed for the presence of SL RNA sequences. Figure 8 shows the results of dot hybridization analysis of the gradient fractionated RNAs with sequences homologous either to the SL (Fig. 8A) or the 100-mer portion of the SL RNA (Fig. 8B). Sequences homologous to the SL are present in two regions of the gradient, one with a sedimentation coefficient of approximately 10S, the other with a sedimentation coefficient greater than 18 S. In contrast, the 100-mer sequences are confined predominantly to the 10 S region. As

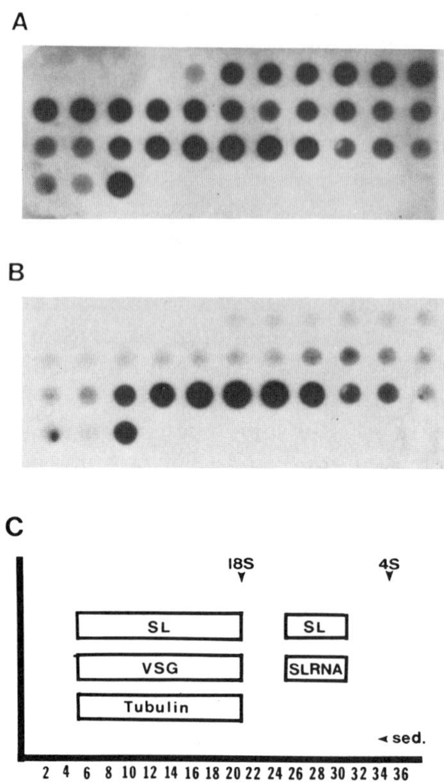

Fig. 8. Distribution of SL and SL RNA-containing particles in sucrose gradients. Extracts of *T. brucei* nuclei were prepared and sucrose gradient separation of the RNPs were performed as described in Section II. The 36 gradient fractions were dot blotted onto nitrocellulose (in duplicate) and hybridized either with the labeled oligonucleotide complementary to the SL (A) or a uniformly labeled probe complementary to the SL RNA (B). (C) Schematic representation of distribution of SL, SL RNA, VSG, and tubulin sequences in sucrose gradient. The data for VSG and tubulin hybridization are not shown.

a control, probes complementary to the tubulin and IsTat 1.1 VSG genes hybridize only to RNA with a particle sedimentation coefficient of greater than 18 S (data not shown; schematized in Fig. 8C). These results demonstrate that at least two types of RNA particles are present in nuclear extracts of trypanosomes, one which contains RNA with both the SL and 100-mer sequences (i.e., the SL RNA) and sediments at approximately 10 S and those which contain RNA with only SL sequences (i.e., mRNAs) and sediment at high S values. That these particles are RNA–protein complexes is suggested by the fact that proteinase K treatment of the nuclear extract prior to centrifugation results in a shift of SL RNA hybridization to densities of approximately 4 S (not shown). Thus, these organisms appear to possess a specific RNA–protein particle containing the SL RNA, the SL RNP, and may contain higher-order particles analogous to the hnRNPs of yeast and mammalian cells.

IV. DISCUSSION

Members of the family Trypanosomatidae have developed a unique method of mRNA production. Every trypanosome mRNA examined has a 35-nucleotide leader sequence, termed the spliced leader or SL. The SL sequence is not closely linked with any of the structural genes thus far characterized in trypanosomes. In two reported cases, the genes encoding the SL and two different expressed VSG genes are found on different chromosomes (Van der Ploeg *et al.,* 1984; Guyaux *et al.,* 1985). The SL sequence is encoded within a 1.4-kb unit which is repeated approximately 200 times in the trypanosome genome as part of one or a few tandem arrays. A portion of each reiteration unit is transcribed, yielding a 135-nucleotide transcript (the SL RNA) which bears the SL at its 5′ end.

Our data indicate that the SL is joined to the body of structural gene transcripts through a process of trans-splicing. We suggest that trans-addition of the SL differs mechanistically from the cis-splicing reaction described for other eukaryotes primarily in that the introns are discontinuous. We thus view the process as proceeding via a precise cleavage of the SL RNA which separates the SL from the 3′ 100-mer intron. The 5′ G residue of the 100-mer is then esterified, in a concerted fashion, to a 2′ hydroxyl group a short distance upstream from the 3′ splice junction of the target gene pre-mRNA. This intermediate Y structure is then resolved by joining the free SL to the target pre-mRNA at the 3′ splice junction with the concomitant release of the discontinuous intron as a Y structure. Coincidentally, the capped 5′ end of the SL RNA becomes the 5′ end of the newly formed mRNA.

As yet the precise location of the branch points is not known. We presume that, as in other eukaryotes, the branches will be located a short distance upstream from the 3′ splice junctions of trypanosome pre-mRNAs. The comparison of DNA sequences for numerous trypanosome genes has revealed no highly conserved sequence, such as the yeast TACTAAC box (Langford and Gallwitz, 1983), which might be involved in branch point formation. However, one can find sequences that resemble the branch point consensus of higher eukaryotes (unpublished observations). Whether or not these sequences have a function in the splicing of trypanosome pre-mRNAs remains to be determined.

Since all mRNAs in trypanosomes possess the SL, presumably added by trans-splicing to each pre-mRNA, one might expect that these organisms have developed a highly efficient mechanism for trans-splicing. In contrast, the *in vitro* trans-splicing of mRNA precursors in higher eukaryotes is a very inefficient process when compared to cis-splicing of the same sequence (Scolnick, 1985; Konarska *et al.*, 1985). Thus, trypanosomes must have developed one or more novel components of the splicing machinery that allow for the efficient trans-splicing of mRNA precursors. One such component, perhaps, is the SL RNP, the small, SL RNA-containing protein particle. This particle is distinct from the hnRNP particles of *T. brucei* and shares many physiochemical properties with the U class of RNPs in higher eukaryotes (S. Michaeli, personal communication). It is possible that the function of this particle is to shuttle the SL RNA directly to the target pre-mRNA branch point/splice junction. If this is the case then components that interact with the 5′ splice junction (such as the U1-snRNP in mammalian cells), or those which might serve to align 5′ and 3′ splice junction sequences in higher eukaryotes may be unnecessary in trypanosomes. In this manner, a requirement for fewer interacting components could in itself serve to increase the efficiency of trans-splicing in trypanosomes.

Why have trypanosomes developed this method of mRNA processing? While the SL ultimately provides the capped 5′ end of mature mRNAs, there is no direct evidence that allows one to assign a specific function to the SL sequence. However, the conservation of the SL sequence and of trans-splicing throughout the Trypanosomatidae suggest a fundamental role of the SL in trypanosome gene expression. RNA splicing, whether in cis or in trans, provides many sites for the regulatory interplay between RNA synthesis, processing, and nuclear export.

In this regard, it may be insightful to consider the organization of structural genes in trypanosomes. The majority of protein-encoding genes thus far examined are arranged as two or more tandem copies. These include the α and β-tubulins (Thomashow *et al.*, 1983), calmodulin (Tschudi *et al.*, 1985), aldolase (Clayton, 1985), glyceraldehyde-3-phosphate dehydrogenase (Swinkels *et al.*, 1986), phosphoglycerate kinase (Osinga *et al.*, 1985), hsp70 (L. H. T. Van der Ploeg, personal communication), hsp83 (J.C. Mottram, personal communication), and a highly expressed gene of unknown function from *T. cruzi* (Gonzalez *et al.*, 1985). There is evidence that this latter gene family is transcribed as a polycistronic pre-mRNA which is subsequently resolved into single capped mRNAs. If this phenomenon is also true of the other tandemly repeated gene families, then the function of trans-SL addition may be to provide a means for generating discrete mRNAs from the polycistronic transcripts, and to supply the processed mRNAs with a 5′ cap. The relative abundance of different mRNAs derived from the same transcription unit may then be regulated at the level of SL addition or, perhaps, degradation. SL addition, therefore, confers an increased flexibility on the transcriptional organization of the trypanosome genome.

V. SUMMARY

Each mRNA of trypanosomes has the same 35-nucleotide sequence, the spliced leader or SL, at its 5′ end. This sequence is acquired through an intermolecular RNA joining mechanism, trans-splicing. The SL sequence is transcribed from one of approximately 200 tandemly arranged SL genes as a 135-nucleotide species (the SL RNA) which has the SL at its 5′ end. Our data suggests that the SL is joined to target mRNAs by a process which is analogous to splicing in other eukaryotes. The SL RNA is cleaved in a manner that precisely separates the SL from the adjacent 100-mer sequence (i.e., the 3′, non-SL portion of the SL RNA). The 5′ end of the 100-mer is esterified with a 2′ hydroxyl of the target gene transcript, resulting in a Y-branched intermediate. This intermediate is presumably resolved by joining of the SL to the structural gene transcript thus displacing the intron Y structure. In support of the role of the SL RNA as donor of the SL sequence to target mRNAs is the fact that both the SL RNA and mRNAs of *T. brucei* share identical 5′ cap structures. The SL RNA exists in the cell as an RNA–protein complex (the SL RNP) with a sedimentation coefficient of approximately 10 S. The possible role of the SL RNP in trans-RNA splicing is discussed.

ACKNOWLEDGMENTS

The authors wish to thank Dr. Shulamit Michaeli for the kind use of her data on the SL RNP particle (Fig. 8) prior to publication and Dr. Rick Nelson for providing Figs. 1 and 2 for this chapter. We are also indebted to Dr. George Newport and Kenny Watkins for photographic assistance and Martha Evans and Ester Witte for technical assistance throughout most of these studies. This work was supported by National Institutes of Health Grants AI-21975 to N.A. and AI07242-03 to K.L.P., and the John D. and Catherine T. MacArthur Foundation. N.A. is a Burroughs Wellcome Scholar of Molecular Parasitology.

REFERENCES

Agabian, N., Nelson, R. G., Parsons, M., Milhausen, M., Sather, S., Selkirk, M., Watkins, K. P., Myler, P., Rothwell, V., and Stuart, K. (1985). *In* "Genome Rearrangement" (M. Simon and I. Herskowitz, eds.), Vol. 20, pp. 153–172. Alan R. Liss, New York.

Auffray, C., and Rougeon, R. (1980). *Eur. J. Biochem.* **107,** 303–314.
Aviv, H., and Leder, P. (1972). *Proc. Natl. Acad. Sci. U.S.A.* **69,** 1408–1412.
Bienen, E. J., Hammadi, E., and Hill, G. C. (1981). *Exp. Parasitol.* **51,** 408–417.
Boothroyd, J. C. (1985). *Annu. Rev. Microbiol.* **39,** 475–502.
Boothroyd, J. C., and Cross, G. A. M. (1982). *Gene* **20,** 281–289.
Borst, P. (1986). *Annu. Rev. Biochem.* **55,** 701–732.
Campbell, D. A., Thornton, D. A., and Boothroyd, J. C. (1984). *Nature* (*London*) **311,** 350–355.
Clayton, C. E. (1985). *EMBO J.* **4,** 2997–3003.
De Lange, T., Liu, A. Y. C., Van der Ploeg, L. H. T., Borst, P., Tromp, M. C., and Van Boom, J. H. (1983). *Cell* (*Cambridge, Mass.*) **34,** 891–900.
De Lange, T., Michels, P. A. M., Veerman, H. J. G., Cornelissen, A. W. C. A., and Borst, P. (1984a). *Nucleic Acids Res.* **12,** 3777–3790.
De Lange, T., Berkvens, T. M., Veerman, H. J. G., Frasch, A. C. A., Barry, J. D., and Borst, P. (1984b). *Nucleic Acids Res.* **12,** 4431–4443.
Donelson, J. E., and Rice-Ficht, A. C. (1985). *Microbiol. Rev.* **49,** 107–125.
Gonzalez, A., Lerner, T. J., Huecas, M., Sosa-Pineda, B., Nogueira, N., and Lizardi, P. M. (1985). *Nucleic Acids Res.* **13,** 5789–5804.
Guyaux, M., Cornelissen, A. W. C. A., Pays, E., Steinert, M., and Borst, P. (1985). *EMBO J.* **4,** 995–998.
Hu, N.-T., and Messing, J. (1982). *Gene* **17,** 271–277.
Konarska, M. M., Padgett, R. A., and Sharp, P. A. (1985). *Cell* (*Cambridge, Mass.*) **42,** 165–171.
Kooter, J., De Lange, T., and Borst, P. (1984). *EMBO J.* **3,** 2387–2392.
Laird, P. W., Kooter, J. M., Loosbroek, N., and Borst, P. (1985). *Nucleic Acids Res.* **13,** 4253–4266.
Langford, C. J., and Gallwitz, D. (1983). *Cell* (*Cambridge, Mass.*) **33,** 519–527.
Lanham, S. M. (1986). *Nature* (*London*) **218,** 1273–1274.
Lenardo, M. J., Dorfman, D. M., and Donelson, J. E. (1985). *Mol. Cell. Biol.* **9,** 2487–2490.
Maniatis, T., Fritsch, E. F., and Sambrook, J. (1982). "Molecular Cloning: A Laboratory Manual." Cold Spring Harbor Lab. Cold Spring Harbor, New York.
Michiels, F., Matthyssens, G., Kronenberger, P., Pays, E., Dero, B., Van Assel, S., Darvill, M., Cravador, A., Steinert, M., and Hamers, R. (1983). *EMBO J.* **2,** 1185–1192.
Milhausen, M., Nelson, R. G., Sather, S., Selkirk, M., and Agabian, N. (1984). *Cell* (*Cambridge, Mass.*) **38,** 721–729.
Mount, S. M. (1982). *Nucleic Acids Res.* **10,** 459–472.
Murphy, W. J., Watkins, K. P., and Agabian, N. (1986). *Cell* (*Cambridge, Mass.* **47,** 517–525.
Nelson, R. G., Parsons, M., Barr, P. J., Stuart, K., Selkirk, M., and Agabian, N. (1983). *Cell* (*Cambridge, Mass.*) **34,** 901–909.
Nelson, R. G., Parsons, M., Selkirk, M., Newport, G., Barr, P. J., and Agabian, M. (1984). *Nature* (*London*) **308,** 665–667.
Osinga, K. A., Swinkels, B. W., Gibson, W. C., Borst, P., Veeneman, G. H., Van Boom, J. H., Michels, P. A. M., and Opperdoes, F. R. (1985). *EMBO J.* **4,** 3811–3817.
Padgett, R. A., Grabowski, P. J., Konarska, M. M., Seiler, S., and Sharp, P. A. (1986). *Annu. Rev. Biochem.* **55,** 1119–1150.
Parsons, M., Nelson, R. G., Watkins, K. P., and Agabian, N. (1984). *Cell* (*Cambridge, Mass.*) **38,** 309–316.
Ruskin, B., and Green, M. R. (1985). *Science* **229,** 135–140.
Sather, S., and Agabian, N. (1985). *Proc. Natl. Acad. Sci. U.S.A.* **82,** 5695–5699.

Scolnick, D. (1985). *Cell (Cambridge, Mass.)* **42,** 157–164.

Silberklang, M., Gillum, A. M., and RajBhandary, U. L. (1979) *In* "Methods in Enzymology" (K. Moldave and L. Grossman, eds.), Vol. 59, pp. 58–109. Academic Press, New York.

Southern, E. M. (1975). *J. Mol. Biol.* **98,** 503–517.

Swinkels, B. W., Gibson, W. C., Osinga, K. A., Kramer, R., Veeneman, G. H., Van Boom, J. H., and Borst, P. (1986). *EMBO J.* **5,** 1291–1298.

Thomashow, L. S., Milhausen, M., Rutter, W. J., and Agabian, N. (1983). *Cell (Cambridge, Mass.)* **32,** 35–43.

Tschudi, C., Young, A. S., Ruben, L., Patton, C. L., and Richards, F. F. (1985). *Proc. Natl. Acad. Sci. U.S.A.* **82,** 3998–4002.

Van der Ploeg, L. H. T., Liu, A. Y. C., Michels, P. A. M., De Lange, T., Borst, P., Majumder, H. K., Weber, H., Veeneman, G. H., and Van Boom, J. (1982a). *Nucleic Acids Res.* **10,** 3591–3604.

Van der Ploeg, L. H. T., Valerio, D., De Lange, T., Bernards, A., Borst, P., and Grosveld, F. G. (1982b). *Nucleic Acids Res.* **10,** 5905–5923.

Van der Ploeg, L. H. T., Cornelissen, A. W. C. A., Michels, P. A. M., and Borst, P. (1984). *Cell (Cambridge, Mass.)* **39,** 213–221.

III

RNA Viruses

11

The Poliovirus Genome: A Unique RNA in Structure, Gene Organization, and Replication

STEVEN E. PINCUS,*[,1] RICHARD J. KUHN,* CHEN-FU YANG,* HARUKA TOYODA,* NAOKAZU TAKEDA,† AND ECKARD WIMMER*

** Department of Microbiology*
School of Medicine
State University of New York at Stony Brook
Stony Brook, New York 11794
† Central Virus Diagnostic Laboratory
National Institute of Health
Musashimurayama, Tokyo 190-12, Japan

I. INTRODUCTION

The *Picornaviridae* comprise one of the largest and most widely characterized families of RNA-containing animal viruses. This family is divided into four genera: the Cardioviruses (e.g., encephalomyocarditis virus, mengovirus), the Aphthoviruses (foot-and-mouth disease virus), the Rhinoviruses (whose members cause the common cold and number at least 115 distinct serotypes), and the Enteroviruses (e.g., poliovirus, coxsackievirus, echovirus, hepatitis A virus). These viruses cause a multitude of illnesses such as upper respiratory infections, poliomyelitis, aseptic meningitis, hepatitis, and pericarditis. One member of this family, enterovirus 71, has had a number of clinical syndromes ascribed to infections including hand-foot-and-mouth disease, aseptic meningitis, enceph-

[1] Present address: Wadsworth Center for Laboratories and Research, New York State Department of Health, Albany, New York 12201.

Molecular Biology of RNA
New Perspectives

alitis polio-type paralysis, and upper respiratory infections (Melnick, 1985).

The chemical structure of poliovirus was solved in 1981 (Kitamura *et al.*, 1981). More recently, the three-dimensional structures of two picornaviruses, human rhinovirus 14 and poliovirus, have been described in atomic detail (Rossman *et al.*, 1985; Hogle *et al.*, 1985). The picornavirion is a nonenveloped icosahedral particle consisting of 60 copies each of four virus-specific coat polypeptides (VP1, VP2, VP3, and VP4) which encapsidates a single-stranded messenger-sense genomic RNA with an average chain length of 7500 nucleotides. The primary structure, gene organization, and mechanism of replication of the picornavirus genome is unusual when compared with other RNA viruses. In the following discussions we will consider these unusual features, using poliovirus as a model system (for a recent review of picornavirus replication, see Kuhn and Wimmer, 1986).

Infection of a cell by poliovirus is initiated by attachment to a specific receptor. Although no information concerning the precise molecular structure of the poliovirus receptor has been obtained, this field of research is actively growing. The isolation of monoclonal antibodies that are specific for the poliovirus receptor has been reported (Minor *et al.*, 1984; Nobis *et al.*, 1985), and recently it was demonstrated that cells which normally lack the poliovirus receptor can acquire the ability to be infected by poliovirus if transfected with the DNA from a cell that expresses the receptor (Mendelsohn *et al.*, 1986).

The mechanisms through which poliovirus enters the cell and is uncoated remain obscure, although it has been proposed that uptake occurs via the endosomal pathway (Madshus *et al.*, 1984). After uncoating, the single, large open reading frame of RNA is translated into a polyprotein which is cleaved by two virus-encoded proteinases. Once the replication proteins have become available, the infecting RNA is replicated. At this stage, translation and replication occur in parallel. The final steps in the infectious cycle involve the packaging of viral RNA into a capsid structure. The precise pathway of morphogenesis is poorly understood (Putnak and Phillips, 1981).

The genome of poliovirus is a single-stranded RNA with a heteropolymeric region of 7441 nucleotides and a homopolymeric 3′ end consisting of approximately 60 adenylate residues (Kitamura *et al.*, 1981; Yogo and Wimmer, 1975). The 5′ terminus of the RNA is covalently linked to a small basic protein of 22 amino acids termed VPg (genome-linked *V*iral *P*rotein), by a phosphodiester bond between the O^4 hydroxyl group of tyrosine and the 5′-terminal uridylate residue (Lee *et al.*, 1977; Rothberg

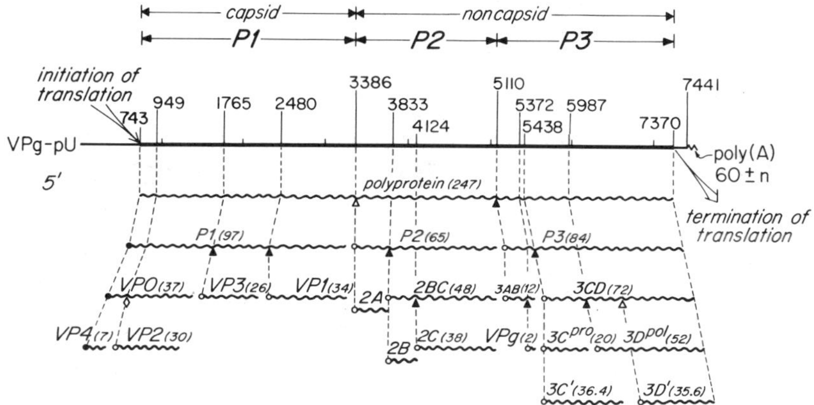

Fig. 1. Gene organization of poliovirus and map of proteolytic processing. Virion RNA, terminated at the 5′ end with the genome-linked protein VPg and at the 3′ end with poly(A), is shown as a solid line, the translated region being more pronounced than the noncoding regions. The numbers above the virion RNA refer to the first nucleotide of the codon specifying the amino-terminal amino acid for the viral-specific proteins. The coding region has been divided into three regions (P1, P2, P3), corresponding to rapid cleavages of the polyprotein. The newly adopted nomenclature of polypeptides is according to Rueckert and Wimmer (1984). Numbers in parentheses are calculated molecular weights. Open circles indicate that the exact position of the terminal amino acid is known. The amino termini are glycine in all cases except for VP2 where it is serine. The carboxy-terminal amino acid of 3CD is phenylalanine. Closed circles indicate that the amino termini are known to be blocked. Closed triangles, Gln-Gly pairs that are cleaved during proteolytic processing of a polypeptide by the virus-coded proteinase 3C; open triangles, Tyr-Gly pairs cleaved by viral proteinase 2A; open diamond, Asn-Ser pair cleaved only during morphogenesis. Polypeptides 3C′ and 3D′ are products of an alternative cleavage, the biological significance of which is unknown.

et al., 1978). Poliovirus type 1 (Mahoney) was the first picornaviral genome to be completely sequenced (Kitamura and Wimmer, 1980; Kitamura *et al.*, 1981; Racaniello and Baltimore, 1981a). Sequencing revealed that the primary structure and gene organization is unusual when compared to the genomes of other RNA viruses (Fig. 1). An example is the occurrence of a 743-nucleotide noncoding region at the 5′ end of the poliovirus RNA. Attempts to modify or shorten this region have usually led to nonviable or conditionally lethal mutants and observations suggesting that this region has an important role in the growth of poliovirus (Stanway *et al.*, 1986; Semler *et al.*, 1986). The function of VPg and the role of the poly(A) tail are unknown, although several hypotheses implicate these features in replication, as will be discussed below.

II. TRANSLATION AND PROCESSING OF THE POLYPROTEIN

Although the early events in the virus replication cycle are unclear, it is safe to assume that after penetration and uncoating of the virion the incoming viral RNA is first translated and subsequently utilized as a template by viral-specific proteins to produce additional copies of the genome. Poliovirus mRNA differs from vRNA by the absence of VPg (Nomoto *et al.,* 1977b). The removal VPg from the 5′ end of mRNA is most likely the result of its post-transcriptional cleavage from newly made RNA. A cellular protein termed the unlinking enzyme has been identified which removes VPg leaving a pU at the 5′ end of the RNA (Ambros *et al.,* 1978). VPg released from RNA is rapidly degraded by cellular enzymes (Dorner *et al.,* 1981; Sangar *et al.,* 1981). It is not known whether VPg is cleaved from the infecting vRNA prior to or during the initial virus-specific protein synthesis, however, VPg-linked RNA can form an initiation complex of translation *in vitro* (Golini *et al.,* 1980).

Why is VPg cleaved from those plus strands destined to become mRNA? One hypothesis suggests that VPg might serve as a recognition signal for encapsidation of RNA into virions (Nomoto *et al.,* 1977b). The nucleotide sequences of mRNA and vRNA are identical, thus an encapsidation signal consisting solely of a unique nucleotide sequence might lead to encapsidation of mRNA and hence would cause an inhibition of translation. If VPg is recognized as an additional signal in morphogenesis, the difference in the 5′ ends of newly synthesized RNA would allow the capsid proteins to distinguish between virion RNA and mRNA and insure the separation between encapsidation and translation. In this regard it is worth noting that the genome linked protein in bacteriophage ϕ29 of *Bacillus subtilis* has an essential role in DNA packaging (Guo *et al.,* 1986).

Poliovirus mRNA has a pU 5′ terminus instead of the ^{m7}G(5′)ppp(5′)N capping group seen on cellular mRNA (Nomoto *et al.,* 1976, 1977b; Pettersson *et al.,* 1977). Shortly after infection by poliovirus, there is a rapid decline in host protein synthesis, concurrent with a steady increase in virus-specific protein synthesis. Available evidence suggests that the inhibition of host protein synthesis occurs at the level of cap recognition (Sonenberg *et al.,* 1981; Hansen and Ehrenfeld, 1981). Analysis of the cellular translational machinery in poliovirus-infected cells revealed a defect in the cap-binding protein complex, which normally consists of three polypeptides of 24 kDa (eIF-4E), 46 kDa (eIF-4A), and 220 kDa (p220) (Tahara *et al.,* 1981; Grito *et al.,* 1983; Edery *et al.,* 1983; Etchison *et al.,* 1982). In poliovirus-infected HeLa cells p220 is degraded (Etchison and Foyt, 1985). Neither viral proteinase 3C nor 2A is directly involved in the degradation of p220 (Lloyd *et al.,* 1985; Lee *et al.,* 1985). However, studies with a small-plaque mutant of poliovirus that fails to elicit the cleavage of p220 has suggested that the virus may activate a normal cellular mechanism responsible for the turnover of p220 (Bernstein *et al.,* 1985).

Earlier suggestions that all viral proteins are derived from a single precursor molecule that is cleaved by proteinase(s) to generate the viral gene products (Jacobson and Baltimore 1968a; Summers and Maizel, 1968) were confirmed by sequencing of the poliovirus genome and the mapping of virus-specific polypeptides. This work identified an open reading frame of 2207 consecutive triplets spanning most (89%) of the viral genome (Kitamura *et al.,* 1981; Semler *et al.,* 1981a; Racaniello and Baltimore, 1981a). Translation of the poliovirus type 1 polyprotein commences 743 nucleotides downstream from the 5′ end of the RNA at the ninth AUG codon (Dorner *et al.,* 1982). At the present time there is no evidence that the AUG codons upstream from N743 serve to initate short viral polypeptides, the largest having the potential of being a protein of 7000 MW. Neither these AUGs nor the coding sequences following them are conserved in the RNAs of the three poliovirus serotypes (Toyoda *et al.,* 1984).

In vitro, poliovirus RNA can initiate protein synthesis several thousand nucleotides downstream from the 5′ end. This phenomenon is observed only when translation occurs in the reticulocyte lysate and is abolished when unknown HeLa cell factors are added or translation is carried out in HeLa or ascites cell extracts (Dorner *et al.,* 1984; Svitkin *et al.,* 1985; Phillips and Emmert, 1986). It is not clear at the present time whether initiation at the AUG at N743 follows the "scanning" of the small ribosomal subunit along the noncodong region (Kozak, 1983) or is the result of "internal" initiation. Direct binding of a ribosome to the proper initiation codon of the polyprotein cannot be excluded at present.

The poliovirus polyprotein can be divided into three regions according to function and processing: P1 corresponds to the capsid proteins and their precursors, P2 corresponds to the central portion of the genome encoding nonstructural proteins, and P3 to a group of nonstructural proteins that are involved in replication (Fig. 1). Processing of the polyprotein into functional proteins is accomplished through proteolytic cleavages (see the reviews by Nicklin *et al.,* 1986; Toyoda *et al.,* 1986a). Three types of cleavages were found to occur: (1) between glutamine-glycine (Q-G) amino acid pairs at nine processing sites within the polyprotein, (2) between tyrosine-glycine (Y-G) at two sites, and (3) between asparagine-serine (N-S) at one site (Kitamura *et al.,* 1980; 1981; Semler *et al.,* 1981a,b; Emini *et al.,* 1982; Larsen *et al.,* 1982; Pallansch *et al.,* 1984).

Cleavages at Q-G sites are carried out by the virus-encoded protease

3C^{pro}, a polypeptide of 20 kDa (Hanecak *et al.*, 1982). Since 3C^{pro} itself is a product of Q-G cleavage, and deproteinized vRNA is infectious (Nomoto *et al.*, 1977b), it appears likely that 3C^{pro} is generated in the initial stages of translation by an intramolecular cleavage of the polyprotein. Hanecak *et al.* (1984) have constructed an expression plasmid that contains a cloned segment of the poliovirus genome corresponding to a segment slightly larger than the 3C^{pro} coding region. Upon induction, three virus-specific polypeptides are produced that can be immunoprecipitated by anti-3C^{pro}. The two smaller polypeptides are the result of proteolytic processing at Q-G amino acid pairs, the smallest polypeptide is authentic 3C^{pro}. The kinetics of the cleavage reaction suggest intromolecular cleavage. Insertion of a DNA linker into the 3C^{pro} coding region abolished the cleavage activity.

Toyoda *et al.* (1986a,b) have used a similar approach to that of Hanecak et al. to find the proteinase responsible for cleavage at Y-G sites. An expression plasmid was constructed containing the P1 region plus part of the P2 region of poliovirus type 3. Expression of the poliovirus-specific sequence yields a protein that is rapidly processed at exactly the Y-G site which severs P1 from P2 (see Fig. 1; Toyoda *et al.*, 1986a,b). Insertion of 4 amino acids or deletion of 9 amino acids in the 2A coding region eliminates this activity, an observation suggesting that a Y-G proteinase activity maps in polio protein 2A. In experiments using *in vitro* translations, antibodies prepared against purified 2A^{pro} inhibited cleavage at a Y-G within polypeptide 3CD but did not prevent the Y-G cleavage immediately aminoterminal of 2A^{pro} (see Fig. 1). Failure of anti-2A^{pro} antibodies to inhibit the Y-G cut between P1 and P2 in the *in vitro* translation has been interpreted to mean that this cleavage is too rapid and possibly intramolecular. Cleavage with polypeptide 3CD to yield 3C′ and 3D′ (Fig. 1) occurs intermolecularly and can therefore be prevented with specific antisera.

Cleavage of VPO at the N-S amino acid pair occurs only during morphogensis (Rueckert, 1976). It has been proposed that this cleavage is a triggered, autocatalytic process (Rossman *et al.*, 1985).

III. RNA REPLICATION

A. *In Vivo* RNA Synthesis

1. Time Course of Events

Early in infection the infecting RNA genome must function as RNA in the synthesis of virus-specific proteins and then as template for the replication of viral RNA. Since translation and transcription proceed in opposite directions, a mechanism must exist for clearing the mRNA of ribosomes during the switch from translation to transcription in the initial phase of replication. Later in infection, translation and transcription do not compete for template strands as enough RNA is produced to serve either function. Moreover, RNA replication appears to occur in a complex unaccessible to ribosomes (see below).

Under special circumstances viral RNA synthesis can be measured directly by 75 min postinfection (p.i.), and its onset can be extrapolated to about 30 min p.i. (Baltimore *et al.,* 1966). Viral RNA synthesis occurs in three phases (Baltimore *et al.,* 1964). Initially one observes an exponential phase that begins 30–40 min p.i. and lasts up to 3 hr p.i. until approximately 20% of the final RNA yield has been synthesized. This is followed by a phase in which the kinetics show a linear, rather than an exponential, increase of synthesized RNA. During this linear phase, which lasts approximately 1–1.5 hr, the bulk of RNA synthesis occurs. Finally, by 4–5 hr p.i. the rate of synthesis rapidly declines. The events leading to the change in the observed kinetics of RNA accumulation remain obscure.

2. *RNA Structures Involved in Replication*

The RNA species induced by poliovirus infection have been the most accessible and well-defined element in the viral replication cycle. Three virus-specific RNAs, shown schematically in Fig. 2, can be found in the infected cell: (1) single-stranded RNA (ssRNA) that is either virion or mRNA (Zimmerman *et al.,* 1963), (2) a hydrogen-bonded, double-stranded, or replicative form RNA (RF; Montagnier and Sanders, 1963), and (3) replicative intermediate (RI), a multistranded RNA that is partially double-stranded, partially single-strand RNA and heterogeneous in size (Baltimore and Girard, 1966). The properties of these structures will be briefly discussed below.

Single-stranded RNA is the most abundant RNA formed after infection. This RNA sediments at 35 S as does purified poliovirus RNA (Zimmerman *et al.,* 1963). All ssRNA isolated from infected HeLa cells or virions has plus-strand polarity, that is it has the same polarity as RNA. This conclusion was based on translation experiments in *Escherichia coli* cell-free extracts in which polypeptides were produced that had properties similar to those of virus-specific polypeptides isolated from infected HeLa cells (Rekosh *et al.,* 1969). When mRNA was isolated from polyribosomes of infected cells and compared to viral RNA (vRNA), it was found that they have identical size and base composition (Penman *et al.,* 1964; Summers and Levintow, 1965; Summers *et al.,* 1967).

As was discussed above mRNA differs from vRNA in that it lacks VPg at the 5′ end (Nomoto *et al.,* 1977a,b). The 3′ end of the genomic RNA of poliovirus contains a poly(A) tract that is heterogeneous in size (Yogo and

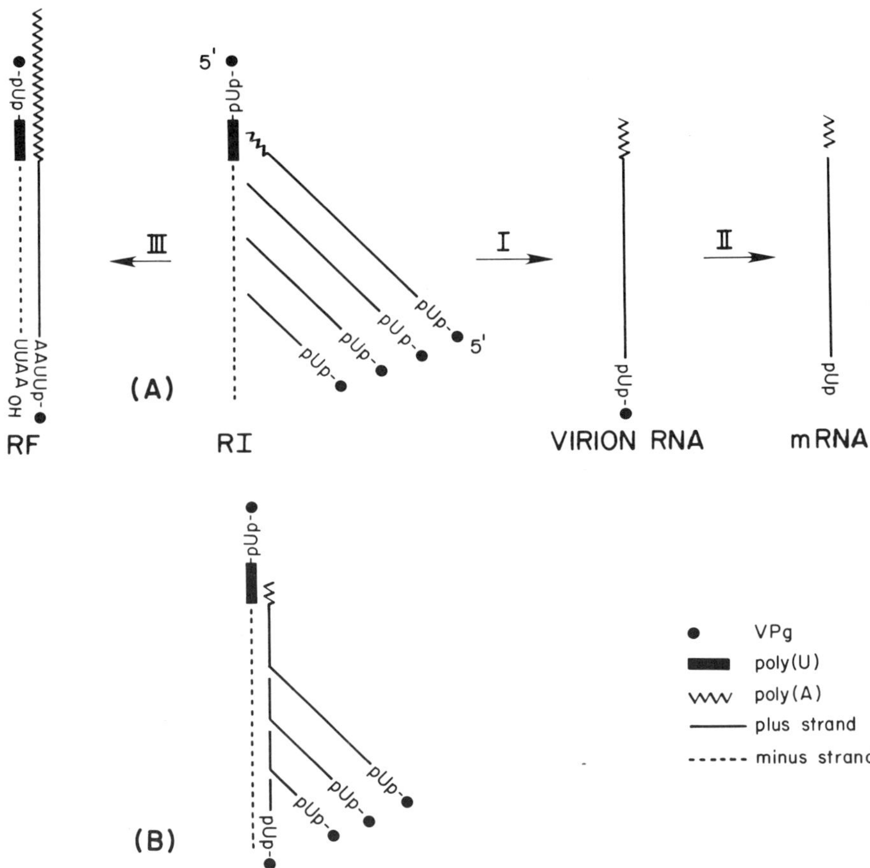

Fig. 2. Intracellular structures of poliovirus RNA. The RNAs are represented with the known sequence structures at their termini. The replicative intermediate (RI) is shown in two possible forms: (A) the "open" form (double-stranded regions are supposedly very short) and (B) the "closed" form. Form B requires that strand replacement occurs during replication, a mechanism strongly favored by the data of Yogo and Wimmer (1975) and Lundquist and Maizel (1978), although several investigations also support form A (see text). Arrows indicate the possible relationship between the structures (for example, RI can give rise to RF and VPg containing ssRNA). Various forms of the replicative form (RF) that occur in small quantitites in the infected cell (see text) have been omitted. (Reproduced, with permission, from Kitamura *et al.*, 1980.)

Wimmer, 1972; Miller and Plagemann, 1972; Armstrong *et al.*, 1972) with an average chain length of 60 nucleotides (Alquist and Kaesberg, 1979). Unlike cellular mRNA which is polyadenylated postranscriptionally by nuclear adenylate transferase, the homopolymeric tract in poliovirus RNA is genetically encoded, that is, it is transcribed from a poly(U) tract at the 5′ end of minus RNA strands (Dorsch-Hasler *et al.*, 1975). The poly(A) tract of polyribosomal mRNA isolated late from poliovirus-infected cells may be five to eight times longer than that of virion RNA (Spector and Baltimore, 1975a). The function of the poly(A) tract is not known. However, a reduction of the poly(A) size through the use of oligo(dT) and RNase H has been reported to abolish the infectivity of the virion RNA (Spector and Baltimore, 1974). A role of poly(A) in the initiation of minus-strand synthesis has been proposed and will be discussed below.

Montagnier and Sanders (1963) identified a 20 S double-stranded RNA (dsRNA) in encephalomyocarditis virus (EMCV)-infected cells that contains a complete copy of the genomic plus-stranded RNA that is hydrogen bonded to a full-length complementary strand (Baltimore, 1966). This molecule termed replicative form (RF) proved more complex in structure than originally anticipated: one end is a poly(A)/poly(U) homopolymeric duplex, the other end a heteropolymeric duplex (Yogo and Wimmer, 1973, 1975; Yogo *et al.*, 1974). It was originally reported that both 5′ ends are linked to VPg (Wu *et al.*, 1978); in the case of minus strands the VPg is linked to poly(U) (Nomoto *et al.*, 1977a). Richards *et al.* (1979) found a population of poliovirus RF that seemed to lack VPg at one end; their RF preparation, however, originated from cells late in infection (7 hr). The 3′-terminal poly(A) tract in RF is always longer than the 5′-terminal poly(U) of minus strands, and thus some 100 adenylate residues protrude as a single-stranded tail (Larsen *et al.*, 1980; see Fig. 2). Larsen *et al.* (1980) reported that the heteropolymeric end of RF is a perfect blunt-ended duplex; Richards and Ehrenfeld (1980), on the other hand, observed that the 3′ end of the minus strand in a small fraction of RF molecules, carries several extra A residues. Young *et al.* (1985) reported that a fraction of RF is apparently covalently linked at one end through a hairpin, thereby migrating in denaturing gels as a single-stranded RNA molecule twice the length of genomic RNA. The unusual structures of RF clearly indicate that the existence of these molecules is not the result of an artifact of the rapid hybridization between plus and minus strands occurring during isolation. Some of these structures have been seen as intermediates in an *in vitro* model of RNA replication (see below).

Poliovirus RF has a 30-fold higher specific infectivity than ssRNA (Bishop and Koch, 1967). This may reflect its decreased sensitivity to ribonuclease or an increased ability of the double-stranded nucleic acid to penetrate into the cell. Treatment of cells with actinomycin D decreases the infectivity of RF while having the opposite effect on ssRNA (Koch *et al.*, 1967). Pretreatment of cells with DEAE-dextran prior to transfection results in equal specific infectivity of RF and ssRNA. The infectivity of the dsRNA requires a nuclear function, since enucleated HeLa cells cannot be transfected with polio RF (Detjen *et al.*, 1978). How RF molecules initiate an infectious cycle is not known, but it appears that the plus strand must be separated from the minus strand. The following observations support this idea: (1) dsRNA is not an active message in an *in vitro* translation system (Baltimore, 1968); (2) minus strands isolated from dsRNA molecules are not infectious (Roy and Bishop, 1970); (3) when a gentic hybrid RF molecule was used to infect cells, the resulting virus had the phenotype that corresponded to the genotype of the positive strand (Best *et al.*, 1972), and (4) RF molecules in which one of the strands has been inactivated are infectious only if the plus strand is still intact (Chumakov and Agol, 1976). Proteins that unwind nucleic acid are known to be required for DNA replication and may be required for the replication of poliovirus RNA. At some time in the replication cycle continued synthesis of RNA may require the displacement of a strand in a hydrogen-bonded, double-stranded region. Whether this is accomplished by the viral polymerase $3D^{pol}$, another viral protein (for example, 2C), or a cellular protein remains to be determined.

The third species of RNA observed in poliovirus-infected cells is the RI. Although it occurs only in small amounts, it can be easily identified after a pulse with a suitable radiolabeled precursor. RI sediments heterogeneously in a sucrose gradient between 20 S and 70 S (Baltimore and Girard, 1966). When isolated from infected cells, RI molecules are found to be partially ribonuclease resistant and consist of a full-length minus-sense template strand with six to eight nascent plus-sense daughter strands (Baltimore, 1969; see Fig. 2). RI molecules containing full-length plus-strand template and nascent minus strands have never been found (Bishop *et al.*, 1969). Two types of structures of RI have been suggested: a fully double-stranded backbone or a predominantly single-stranded molecule (see Fig. 1), and both have received experimental support (Nilsen *et al.*, 1981; Richards *et al.*, 1984, and references therein).

All nascent plus strands of RI molecules are linked to VPg (Nomoto *et al.*, 1977a; Flanegan *et al.*, 1977; Pettersson *et al.*, 1978). The minus strands in RI also contain a VPg molecule bound to the 5′-terminal poly(U) (Yogo and Wimmer, 1975; Nomoto *et al.*, 1977a). Since nascent RNA strands with a 5′-terminal pppUp have never been found (Nomoto *et al.*, 1977a), it has been suggested that poliovirus RNA synthesis may not be initiated *de novo* (which would yield 5′-terminal pppNp but may be a primer dependent event (see Wimmer, 1982).

Throughout both the exponential and linear phases of RNA synthesis, single-stranded viral RNA is at least 10-fold more abundant than RF and RI. During the exponential phase, RI is found in greater amount than RF. After 3 hr p.i., the ratio of RF to RI is reversed, and the proportion of RF increases continually to the end of infection (Baltimore *et al.*, 1966; Oberg and Philipson, 1969).

3. Cellular Components Implicated in RNA Replication

Apart from small molecules provided by the host cell, such as the nucleotide precursors, very few cellular components have been implicated to function in poliovirus RNA replication.

One of the most striking chemical changes in poliovirus-infected cells is the proliferation of newly formed smooth membranes which show an 8-fold increase in protein content and more than a 13-fold increase in lipid phosphorous between 2.5 and 6 hr after infection (Mosser *et al.*, 1972). If a cytoplasmic extract of poliovirus-infected HeLa cells is subjected to centrifugation, all activity synthesizing poliovirus-specific RNA is associated with the membrane fraction. This fact, which has been know for many years (Girard *et al.*, 1967; Caliguiri and Tamm, 1970; Dorsch-Hasler *et al.*, 1975; Takegami *et al.*, 1983a), indicates that RNA replication complexes are associated with membranous structures. Fractionation of the membranous material on a discontinuous sucrose gradient has shown that the highest RNA-synthesizing activity bands in the fraction containing smooth membranes (Caliguiri and Tamm, 1970; Dorsch-Hasler *et al.*, 1975; Takegami *et al.*, 1983a). These membraneous complexes can synthesize authentic (VPg-linked and polyadenylated) plus-strand RNA, and this capacity is destroyed if nonionic detergent such as NP-40 is added to as little as 0.5% (Takeda *et al.*, 1986). It is therefore believed that poliovirus RNA replication requires a membraneous environment.

Whether cellular proteins are involved in poliovirus RNA replication remains an unanswered question. Two cellular enzymes have been purified and shown to allow the $3D^{pol}$ to initiate RNA synthesis *in vitro*. The first protein termed "host factor" was recently described to be a protein kinase (Morrow *et al.*, 1985). Andrews *et al.* (1985) have described a terminal uridylyl transferase (TUT) that allows transcription to occur *in vitro*. However, recent results suggest that this reaction has little relevance to the mechanism of poliovirus RNA replication (see below).

4. Viral Proteins Involved in Replication

Cellular RNA synthesis proceeds on a DNA template employing DNA-dependent RNA polymerases. Therefore it was assumed that the RNA polymerase(s) responsible for replication of poliovirus RNA is virus induced. Baltimore and co-workers (1963) identified a poliovirus-induced RNA polymerase in infected cells; its appearance correlated with the time course of infection, an observation suggesting that the enzyme was virus-encoded. The virus-induced polymerase sediments in a membrane-associated complex of 60–250 S (Baltimore, 1964; Girard *et al.*, 1967; Ehrenfeld *et al.*, 1970). Since neither puromycin nor EDTA changed the sedimentation rate of the complex, and because radiolabeled ribosomes did not associate with it, it was concluded that the replication complex was free of ribosomes. Initial attempts to isolate polymerase in an active form were unsuccessful due to the lack of an assay dependent upon the addition of exogenous RNA. Partial purification of the replication complex by precipitation in 2 *M* LiCl and gel filtration resulted in the identification of a protein of 58,000 MW, which was shown to be the virus-specific polypeptide 3D (Lundquist *et al.*, 1974).

Direct identification of a poliovirus-encoded RNA polymerase was accomplished upon the realization that the 3′-terminal poly(A) is genetically coded (Yogo and Wimmer, 1973, 1975; Dorsch-Hasler *et al.*, 1975; Spector and Baltimore, 1975b). Thus, the first step of replication would be the transcription of the 3′-terminal poly(A) tract of vRNA into the 5′-terminal poly(U) of minus strands. Initial attempts to isolate a soluble poly(A)-dependent "poly(U)polymerase" failed. However, addition of oligo(U) to the poly(A)-containing assay mixture resulted in the strong stimulation of the expected polymerase activity (Flanegan and Baltimore, 1977). This led to the discovery that the poliovirus-encoded RNA polymerase is not only template but also primer dependent. The purified activity was associated exclusively with viral polypeptide $3D^{pol}$ (52,500 MW); polypeptide 3CD (72,000 MW), on the other hand, was found to be inactive. $3D^{pol}$ can transcribe vRNA as long as an oligo(U) primer is added (Dasgupta *et al.*, 1979). Although the soluble membrane-free polymerase can be purified in an active form, attempts to purify a membrane-bound form of the polymerase with poly(U) polymerase activity have been unsuccessful (N. Takeda, T. Takegami, and E. Wimmer, unpublished results).

A second virus-encoded protein thought to be involved in RNA replication is the genome-linked protein VPg. This protein is found at the 5′ end of all newly synthesized viral RNAs. A high-energy (−9.6 Kcal/mole) phosphodiester bond links the phenolic hydroxyl group of the tyrosine at position three of the VPg to the 5′ phosphoric group of the terminal uridylic acid residue (Ambros and Baltimore, 1978; Rothberg *et al.*, 1978; Vartapetian *et al.*, 1980).

Although during the course of a poliovirus infection more than 6×10^6 VPg molecules should be synthesized, VPg is very difficult to detect in extracts of poliovirus-infected HeLa cells. VPg escapes detection because it is rapidly degraded in cell-free extracts (Dorner *et al.*, 1981), and also its small size and acid solubility cause it to elute easily from polyacrylamide gels (Lee *et al.*, 1976; Crawford and Baltimore, 1983). Membrane complexes of poliovirus-infected HeLa cells that contain an active RNA-synthesizing complex are rich in precursor polypeptides to VPg; the most prominent is 3AB (Takegami *et al.*, 1983a, Semler *et al.*, 1982). Two other membrane-associated viral polypeptides, 2ABC-3AB and 2C-3AB, that share primary sequence with 3AB have also been found in these replication complexes. These membrane-anchored proteins may serve to direct other viral polypeptides to the site of RNA synthesis or deliver VPg to the ends of nascent RNA strands in a hydrophobic environment (Semler *et al.*, 1982; Takegami *et al.*, 1983a,b).

An apparent high affinity between viral polypeptides 2C and 3AB has frustrated attempts to purify 3AB from infected HeLa cells. These polypeptides copurify through ion-exchange chromatography, gel-filtration, and antibody affinity chromatography, even in the presence of NP-40 and urea (R. J. Kuhn, T. Takegami, and E. Wimmer, unpublished results). Although 2C copurifies with the membrane bound replication complex (Butterworth *et al.*, 1976; Traub *et al.*, 1976), it is rapidly degraded with trypsin in the absence of detergents (Takegami *et al.*, 1983a). The tight association of 2C and 3AB to membranes cannot be reversed by treatment with high-ionic strength buffer containing EDTA or 4 *M* urea (Tershak, 1984). Since 2C lacks a hydrophobic region, it may be associated with membranes through its affinity to 3AB. In addition to its association with the membrane-bound replication complex, 2C has been directly implicated to play a role in RNA replication (see below).

The viral proteinase $3C^{pro}$ has also been identified as a component of the membrane-bound replication complex (Caliguiri and Mosser, 1971; Takegami *et al.*, 1983a). Since VPg is the exclusive polypeptide linked to viral RNA (3AB, for example, is never found attached even to nascent strands), cleavage of a VPg precursor may directly affect the synthesis of RNA molecules. $3C^{pro}$ may regulate the release of VPg from its precursor and, in so doing, limit the rate of initiation of RNA synthesis.

In addition to these nonstructural proteins, the capsid proteins VP0, VP1, and VP3 are also associated with the replication complex (Wright and Cooper, 1976; Loesch and Arlinghaus, 1975; Caliguiri and Mosser, 1971; Takegami *et al.*, 1983a). What role, if any, these proteins play in the replication complex is unknown. Virus maturation is considered an event occurring on membranes, and only newly synthesized viral RNA is incorporated into viral particles (Caliguiri and Tamm, 1968b).

5. Inhibition by Guanidine

Guanidine hydrochloride acts as a protein denaturant at concentrations of 5.0 *M* or higher, but at concentrations (0.1–2.0 m*M*) that have no detectable effects on cell growth, it selectively blocks the growth of several togaviruses (Friedman, 1970), several plant viruses (Dawson, 1975-1976; Varma, 1968), and many picornaviruses including poliovirus and foot-and-mouth disease virus (FMDV) (Crowther and Melnick, 1961; Loddo *et al.,* 1962; Pringle, 1964; Rightsel *et al.,* 1961). The period of sensitivity to guanidine in the growth cycle of poliovirus has been determined by infecting HeLa cells with poliovirus in the presence of guanidine and removing the drug by washing the cells various times after inoculation. Removal of guanidine 1 hr after inoculation results in a normal growth cycle. However, removal of guanidine at 2 hr is associated with a significant delay in the onset of multiplication of poliovirus. When the drug is added after 4 hr, the extent of inhibition of virus production rapidly diminishes (Caliguiri and Tamm, 1968a, and references therein). Although guanidine has been shown to interfere with maturation of poliovirus (Jacobson and Baltimore, 1968b), prevent capsid proteins from associating with smooth membranes (Yin, 1977), and cause a reduction of choline incorporation into membranes (Penman and Summers, 1965), the major effect appears to be blockage of synthesis of viral RNA, particularly the production of ssRNA and replicative intermediate (Caliguiri and Tamm, 1968a, 1970; Nobel and Levintow, 1970). If the replicative intermediate is the precursor form of viral RNA, a net reduction in this species of RNA indicates that guanidine blocks initiation of new RNA chains and allows previously initiated chains to be completed (Caliguiri and Tamm, 1968a; Tershak, 1982). However, the release of completed RNA chains might also be blocked under certain conditions (Huang and Baltimore, 1970).

Evidence that guanidine inhibits the growth of poliovirus through an interaction with virus-specified component(s) was obtained when the existence of several classes of guanidine-resistant mutants displaying different degrees of resistance was reported (Ledinko, 1963). Three classes of guanidine-resistant mutants were found and designated $gua^{r/20}$ (low resistance, 0.21 m*M*), $gua^{r/40}$ (medium resistance, 0.42 m*M*), and $gua^{r/100}$ (high resistance, 1.05 m*M*) (Ledinko, 1963; Sergiescu *et al.,* 1972). Strains of poliovirus were also obtained that required guanidine (guanidine dependent) for growth (Ledinko, 1963; Loddo *et al.,* 1963). The guanidine-dependent strains required the drug during the period of the growth cycle when wild-type poliovirus is sensitive to guanidine, suggesting that the drug-requiring process in the multiplication of dependent virus is analagous to the drug-sensitive process in the mutiplication of sensitive virus (Eggers *et al.,* 1965).

Three questions have dominated research on the inhibition of poliovirus growth by guanidine: (1) What is the viral-specific component effected by guanidine? (2) What role does this component play in viral replication? and (3) How does guanidine interfere with the function of this component? In the following discussions we will present evidence that polypeptide 2C is the viral component responsible for guanidine sensitivity and will speculate on the role of this protein in RNA replication and the mechanism of action of guanidine.

The viral component responsible for poliovirus sensitivity to guanidine has been the subject of considerable controversy. Lwoff (1965) proposed that guanidine may interfere with morphogenesis of the active viral RNA polymerase or may interact with fully formed viral RNA polymerase and cause allosteric changes that inhibit enzyme activity. However, in these experiments RNA polymerase activity was measured on an aggregate enzyme that consisted of template RNA in association with enzyme. This enzyme preparation neither responds to nor requires added primer RNA, and therefore the synthesis represents most likely the completion of previously initiated RNA chains. Inhibition of initiation by guanidine would deplete RNA ploymerase of RNA template and appear as a reduced level of polymerase in this assay.

Genetic and biochemical studies on guanidine resistant (g^r) mutants of poliovirus implicated the capsid proteins as the site of action of guanidine. Cooper (1969) found that g^r mutants were linked genetically to temperature-sensitive mutations that were believed to be carried by capsid proteins. In some cases the capsid protein mutants themselves exhibited increased sensitivity to guanidine, whereas strains with temperature-sensitive mutations in nonstructural genes failed to display such a change (Cooper *et al.,* 1970a). Korant (1977) observed capsid protein modifications in viruses made g^r by multiple passage in the presence of guanidine. The properties of deletion mutants (''defective interfering particles'') of poliovirus, however, seem to rule out the direct involvement of capsid proteins. Due to deletions in the genome region, coding for the capsid proteins (Nomoto *et al.,* 1979), these mutants make no detectable capsid protein, yet still retain the ability to synthesize RNA by a mechanism that is fully sensitive to guanidine (Cole and Baltimore, 1973).

Guanidine-resistant mutants of FMDV and poliovirus have been isolated after a minimal number of passages in the presence of drug. Analysis of virus-induced polypeptides in guanidine-resistant FMDV mutants showed that polypeptide 2C was altered in 5 out of 10 mutants, an observation suggesting that 2C may be the target of the antiviral action of guanidine (Saunders and King, 1982). Similar results were described for poliovirus; 75% of guanidine-resistant mutants contained modifications in protein 2C (Anderson-Sillman *et al.*, 1984). Recombinants between appropriate poliovirus strains and sequence analyses of mutants unambiguously mapped the g^r mutation(s) to the region of the viral genome specifying polypeptide 2C (Emini *et al.*, 1985; Pincus *et al.*, 1986a).

What kinds of mutations give rise to g^r, do they indeed map to 2C and, if so, are they clustered in one region of 2C or scattered throughout its sequence? We have sequenced poliovirus type 1 genomic RNA isolated from (1) seven mutants resistant to 2 m*M* guanidine, (2) one mutant resistant to 0.5 m*M* guanidine, and (3) three mutants whose growth is dependent upon the addition of guanidine (Pincus *et al.*, 1986a,b).

The mutants resistant to 2 m*M* guanidine all contain an amino acid substitution in 2C at the same position (amino acid 179, see Fig. 3). Interestingly, resistance to this "high" concentration required two nucleotide substitutions in the same codon that resulted in an exchange of either N > A (mutant GH-41, Fig. 3) or N > G (Pincus *et al.*, 1986a). This mutation is contained within an amino acid sequence MDDLN*QNP (where the N carrying an asterisk is subject to mutation to g^R) that is completely conserved between poliovirus and rhinovirus type 14 and differs in encephalomyocarditis virus, mengovirus, and foot-and-mouth disease viruses only in that the N* is substituted for a glycine (see Fig. 4). A strong correlation exists between the amino acid at this position and the sensitivity of the picornavirus to 2.0 m*M* guanidine. Both poliovirus and

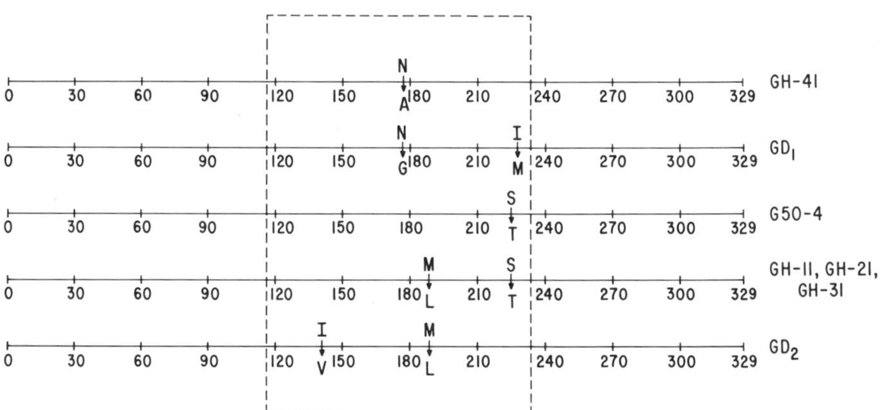

Fig. 3. Summary of the mutations seen in different classes of mutants. Numbers indicate the amino acid position from the amino terminus of p2C. The asparagine-to-alanine mutation at amino acid 179 in GH-41 has been observed in several mutants resistant to 2.0 m*M* guanidine, although in some cases this amino acid is changed to glycine (Pincus *et al.*, 1986a). The dashed box indicates a region in poliovirus 2C (amino acids 118–231) that is highly conserved among picornaviruses (see Fig. 4).

rhinovirus type 14 are strongly inhibited by 2.0 m*M* guanidine, whereas mengovirus, encephalomyocarditis virus, and foot-and-mouth-disease virus, as predicted from the analysis of polio g^r mutants, are resistant to this concentration of drug (Pringle, 1964; Tamm and Eggers, 1963). Hepatitis A (in which this sequence is IDDIGQNT) is also strongly inhibited by 2.0 m*M* guanidine (Siegel and Eggers, 1982). Both rhinovirus type 2 (MDDIMQNP) and coxsackie B1 (MDDLCQNP) are inhibited by 2.0 m*M* guanidine (Caliguiri and Tamm, 1973), an observation suggesting that methionine and cysteine at this position may function in the same manner as asparagine.

To eliminate the possibility of multiple mutations in virus isolates, a cDNA segment containing the N > G mutation was cloned into the wild-type background of an infectious poliovirus cDNA clone, and the virus isolated from transfected cells was found to express the g^r phenotype (Pincus and Wimmer, 1986). This proved that the N > G mutation alone can be responsible for the g^r phenotype.

The N > G mutation was also observed in a mutant that requires 2.0 m*M* guanidine for growth. This dependent mutant (GD_1, Fig. 3), which has a 3 $\log_{10}$ higher titer in the presence of guanidine than in its absence (Pincus *et al.*, 1986a), contains a second mutation (I > M) at amino acid 227. A cDNA segment containing the N > G and I > M mutations was cloned into an infectious poliovirus cDNA clone, and the virus isolated from transfected cells was found to express the guanidine-dependent phenotype (Pincus and Wimmer, 1986). This proved that these two mutations were responsible for the dependent phenotype and suggests that guanidine dependence can arise from high resistance. To examine the effect of the I > M mutation, we have constructed an infectious cDNA clone containing just this mutation. The virus isolated from transfected cells is highly sensitive to inhibition by low concentrations of guanidine (0.2 m*M*) and more so than is the parental poliovirus type 1 Mahoney (S. E. Pincus and E. Wimmer, unpublished data).

Genomic RNA encoding 2C of a mutant (G50-4, Fig. 3) resistant to 0.5 m*M* guanidine was sequenced, and a mutation (S > T) was found at amino acid 225. This mutant was found to be partially resistant to growth in 2.0 m*M* guanidine (Pincus *et al.*, 1986a). Passage of this mutant in the presence of 2.0 m*M* guanidine resulted in the appearance of a mutant (GH-11, GH-21, GH-31, Fig. 3) that is dependent on guanidine for growth (Pincus *et al.*, 1986b). This mutant contained the S > T change at amino acid 225 and a second change (M > L) at amino acid 187 (Fig. 3). The titer of this mutant was 1 $\log_{10}$ higher in the presence of 0.5 m*M* or 2.0 m*M* guanidine than in the absence of drug. This result suggests that guanidine dependence can arise from a low level of resistance. The results also demon-

```
             121                 131                 141                      151                 161                 171
Polio     R I E P V C L L V H G S P G T G K S V A T N L I A R A I A E - - - R E N - T S T Y S L P P D P S H F D G Y K Q Q G V V I
CB1       R I E P V C L L L H G S P G A G K S V A T N L I G R S L A E - - - K L N - S S V Y S L P P D P D H F D G Y K Q Q A V V I
Rhino 14  R T E P V C V L I H G T P G S G K S L T T S I V G R A I A E - - - H F N - S A V Y S L P P D P K H F D G Y Q Q Q E V V I
Rhino 2   R C E P V A I V I H G P P G A G K S I T T N F L A K M I T N - - - D - - - S D I Y S L P P D P K Y F D G Y D Q Q S V V I
HepA      R C E P V V C Y L Y G K R G G G K S L T S I A L A T K I C K H Y G V E P E K N I Y T K P V A S D Y W D G Y S G Q L V C I
EMC       R C E P V V I V L R G D A G Q G K S L S S Q V I A Q A V S K T I - F G R - Q S V Y S L P P D S D F F D G Y E N Q F A A I
FMD       R P E P V V V C L R G K S G Q G K S F L A N V L A Q A I S T H F - T G R I D S V W Y C P P D P D H F D G Y N Q Q T V V V
CPMV      R K M P F T I F F Q G K S R T G K S L L M S Q V T K D F Q D H Y G L G G - E T V Y S R N P C D Q Y W S G Y R R Q P F V L

                         181                 191                 201                 211                 221                 231
Polio     M D D L N Q - - N P D G A D M K L F C Q M V S T V E F I P P M A S L E E K G I L F T S N Y V L A S T N S - S R I S P P T
CB1       M D D L C Q - - N P D G K D V S L F C Q M V S S V D F V P P M A A L E E K G I L F T S P F V L A S T N A - G S I N A P T
Rhino 14  M D D L N Q - - N P D G Q D I S M F C Q M V S S V D F L P P M A S L D N K G M L F T S N F V L A S T N S - N T L S P P T
Rhino 2   M D D I M Q - - N P A G D D M T L F C Q M V S S V T F I P P M A D L P D K G K A F D S R F V L C S T N H - S L L T P P T
HepA      I D D I G Q - - N T T D E D W S D F C Q L V S G C P M R L N M A S L E E K G R H F S S P F I I A T S N W - S N P S P K T
EMC       M D D L G Q - - N P D G S D F T T F C Q M V S T T N F L P N M A S L E R K G T P F T S Q L V V A T T N L - P E F R P V T
FMD       M D D L G Q - - N P D G K D F K Y F A Q M V S T T G F I P P M A S L E D K G K P F N S K V I I A T T N L Y S G F T P R T
CPMV      M D D F A A V V T E P S A E A Q - M I N L I S S A P Y P L N M A G L E E K G I C F D S Q F V F V S T N F - L E V S P E A
```

strated that guanidine-dependent mutants could belong to more than one class. Thus, GD_1, which displays a 3 $\log_{10}$ higher titer in the presence of guanidine than in its absence, requires 2.0 m*M* guanidine, while GH-11, which only displays a 1 $\log_{10}$ higher titer in the presence of guanidine than in its absence, grows equally well in the presence of 0.5 m*M* or 2.0 m*M* guanidine. We have studied a second mutant that displays a 3 $\log_{10}$ higher titer in the presence of guanidine than in its absence (GD_2, Fig. 3). This mutant was found to display a third class of dependency. It prefers 0.5 m*M* guanidine to 2.0 m*M* guanidine (Pincus *et al.*, 1986b). Sequence analysis revealed that this mutant contains the M > L change at amino acid 187 seen in GH-11, but it lacks the S > T change at amino acid 225. Instead it contains a I > V change at amino acid 142 (Fig. 3).

All of these mutations are clustered in a region encompassing approximately 25% of the amino acid sequence of 2C (amino acids 142–227, see Fig. 3). A strong amino acid homology was previously observed between the 58-kDa polypeptide of cowpea mosaic virus and the 2C proteins of encephalomyocarditis virus, foot-and-mouth disease virus, poliovirus, and rhinovirus types 2 and 14 over regions totaling 115 residues (amino acids 119–231 in the poliovirus 2C sequence) (Argos *et al.*, 1984; Franssen *et al.*, 1984; Pincus *et al.*, 1986a). In Fig. 4 we have extended the comparison to include mengo virus (A. Palmenberg, personal communication), hepatitis A virus (Najarian *et al.*, 1985; Cohen *et al.*, 1986; Paul *et al.*, 1986), and coxsackie B1 (Iizuka *et al.*, 1986). The strong amino acid homology of this region suggests that it may be the active site of 2C. The affinity of 2C for 3AB suggests one possible function for 2C. In this model

Fig. 4. Comparison of 2C sequences. Alignment of the amino acid sequences of the 2C regions of encephalomyocarditis virus (EMCV), poliovirus, foot-and-mouth disease virus (FMDV), rhinovirus (Rhino) types 14 and 2, hepatitis A virus (HepA), coxsackie B1 (CB1), and the 58K polypeptide encoded by cowpea mosaic virus B-RNA. Numbers in the poliovirus sequence represent the amino acid positions from the amino terminus of poliovirus polypeptide 2C. Thus, the third amino acid in the poliovirus sequence (glutamic acid) represents amino acid 121. Boxed sequences identify those amino acids conserved among the viruses indicated. The bar above the poliovirus sequence (amino acids 122–142) shows the region containing the "ATP consensus sequence" found in the nucleotide-binding pockets of many ATP-utlilizing enzymes. The poliovirus sequence represents the consensus sequence of poliovirus type 1 (Mahoney) and Sabin 1, 2, and 3 strains. The underlined amino acids are different in Sabin 2 and Sabin 3. Arginine (amino acid 149) is a lysine in these two strains, and serine (amino acid 228) is a threonine. The FMD sequence is the consensus sequence of strains O1K, $A_{10}61$, and $A_{12}119$. The underlined isoleucine is a threonine in $A_{12}119$. The sequence of mengovirus is the same as EMC except that the underlined phenylalanine in EMC is a leucine in mengovirus. A third HepA isolate sequenced by our laboratory (A. Paul and E. Wimmer, unpublished) contains a leucine at the underlined methionine. (Modified from Pincus *et al.*, 1986a.)

3AB/2C acts as an anchor for the viral replication complex to membranes. Guanidine, although it does not release 2C from membranes (Tershak, 1984), could cause a conformational change that alters the interaction of viral and/or cellular proteins with 2C (see Fig. 6).

Gorbalenya *et al.* (1985) observed a sequence within 2C (amino acids 122–142, see Fig. 4) similar to the "ATP consensus," a universal sequence found in nucleotide binding pockets of several ATP-utilizing enzymes. This sequence could be involved in binding to nucleotides or allow 2C to bind to RNA. We have explored both of these possibilities. Extracts of poliovirus-infected HeLa cells were applied to columns containing GTP, ATP, or UTP covalent attached to agarose and eluted under various conditions. Polypeptide 2C did not bind to any of these columns under our experimental conditions. However, we observed binding of $3D^{pol}$ to ATP and GTP agarose but not to UTP agarose in these experiments (S. E. Pincus and E. Wimmer, unpublished data). This result suggests that $3D^{pol}$ may have a site that interacts specifically with purine nucleotides. To explore the possibility that 2C binds to ssRNA, we have incubated minus-strand RNA in extracts of infected HeLa cells in the presence of an excess of a nonviral RNA to act as a competitor. The sedimentation of the minus-strand RNA was altered after incubation in the infected cell extract, suggesting that there were proteins binding to this molecule. When the profile of bound protein was observed by SDS gel electrophoresis, however, there was no enrichment for any viral-specific protein including 2C (S. Obert, H. Toyoda, and E. Wimmer, unpublished data).

The potential nucleotide binding site in poliovirus 2C displays homology to a site in the coliphage T7 gene *4* protein (Husain *et al.*, 1986). The T7 gene *4* protein has a dual role at the replication fork. First, it enables T7 DNA polymerase to use duplex DNA as templates. In this reaction the gene *4* protein uses the energy derived from hydrolysis of NTPs to unwind the DNA helix ahead of the replication fork. Gene *4* protein initiates lagging-strand DNA synthesis by catalyzing the synthesis of RNA primers which are utilized by the T7 DNA polymerase (Hillenbrand *et al.*, 1978, and references therein). Continued initiation of poliovirus RNA synthesis may require the action of an "RNA helicase" to displace the nascent strand in a hydrogen-bonded, double-stranded region (see Fig. 2). If polypeptide 2C acts in this manner, the inhibition of its function by guanidine could block continued initiations while allowing the $3D^{pol}$ to complete the synthesis of previously initiated chains. This function could also be required for release of nascent strands under certain conditions. It is entirely possible that 2C could have a second (or alternative) function in the attachment of VPg to the 5′-end of nascent RNA (see below). Our understanding of the function of 2C awaits the isolation and characterization of temperature-sensitive mutants in this protein. The infectious cDNA clones of poliovirus should allow us to direct mutations to 2C, a goal that our laboratory is actively exploring.

The mechanism of inhibition of the function of 2C by guanidine remains unknown. Guanidine could bind directly to 2C and alter its conformation. This could prevent 2C from performing a catalytic function, or it could prevent specific protein–protein interactions between 2C and other viral or cellular proteins. Both possibilities require that the active conformation of 2C is sensitive and may be relatively difficult to maintain. The occurrence of dependent mutants suggests that guanidine can effect the conformation of 2C such that a normally inactive conformation of this protein can be converted to an active conformation by the binding of guanidine. The various degrees of guanidine resistance and dependence would reflect either differences in the ability of guanidine to bind and alter the conformation or differences in the degree of conformational change upon guanidine binding. Of interest in this regard is the observation that guanidine has been found to decrease the optimal growth temperature of various isolates of FMDV (Nick and Ahl, 1976) and that guanidine-resistant mutants of poliovirus display a range of sensitivities to growth at intraoptimal or supraoptimal temperatures (Tershak, 1985).

B. *In Vitro* Studies of RNA Replication

A genetic analysis of poliovirus RNA replication based on suitable mutants has not been possible. Temperature-sensitive mutants have been identified which are defective in the synthesis of dsRNA or ssRNA at the restrictive temperature (Cooper *et al.*, 1970b). The identification of the defective viral protein(s) in these mutants has been difficult since poliovirus mutants do not form complementation groups (Cooper, 1977). Cooper and his colleagues were able to obtain a genetic recombination map of poliovirus and correctly suggested that a region coding for RNA replication function(s) was located at the opposite end of the genome from that coding for the capsid proteins. The development of infectious DNA clones of poliovirus (Semler *et al.*, 1984; Racaniello and Baltimore, 1981b) will allow the identification of the altered proteins in previously characterized mutants (as we have shown for guanidine resistance) and will allow the isolation of new mutants through site-directed mutagenesis (Sarnow *et al.*, 1986; Bernstein *et al.*, 1985).

In the absence of genetic information numerous biochemical experiments have been performed over a period of 25 years whose objective was to separate and purify components of the infected cell that are involved in viral RNA synthesis. Most investigators have tried to reproduce the classical experiments by S. Spiegelman and his colleagues: the incubation of a "purified viral protein" with genome RNA, in this case with bacteriophage Qβ RNA and nucleotide precursors which yielded authentic phage RNA synthesized *de novo* (Spiegelman and Hayashi, 1963). It was subsequently discovered that the "purified viral protein" used in this replication reaction was a complex of the phage-encoded RNA polymerase and three bacterial proteins, the latter normally being involved in the translation of cellular mRNA (Kamen, 1975). These results suggested that any polypeptide of the host cell has the potential to be involved in poliovirus RNA replication either in a complex with $3D^{pol}$ or as a separate entity. This kind of reasoning led to the study of two cellular proteins mentioned above and which will be discussed in more detail below.

In spite of a great deal of effort, no *in vitro* replication system of poliovirus RNA has been developed. We define *in vitro* replication of RNA as the synthesis of progeny RNA that has the same polarity as the input template RNA and is linked to VPg at the 5′ terminus. Starting with virion RNA, that implies the pathway: plus strands → minus strands → plus strands. We consider that development of an *in vitro* replication system for poliovirus will be more difficult than for the RNA phage since (1) poliovirus replication is not solely catalytic since progeny RNA is VPg linked and (2) it is likely that membranes are an intrinsic component of the poliovirus replication machinery.

Two separate strategies have been followed in attempts to understand the mechanism of poliovirus RNA replication. The first involves the purification of polypeptides from virus-infected cells or uninfected cells. Those various polypeptides and factors are added to viron template RNA, and the resulting products are analyzed. The second strategy is a study of a membraneous replication complex isolated from infected cells that is capable of synthesizing authentic virion RNA. Both strategies will be discussed separately.

1. Soluble Systems of RNA Replication

The first strategy employed to understand the mechanism of poliovirus RNA replication has been based on attempts to reconstitute a replication system *in vitro* from purified protein and RNA components. This approach was dependent upon the purification of the viral polymerase, $3D^{pol}$, in a template-dependent form using an assay in which poly(A) was supplied as template along with oligo(U) as a potential primer for RNA synthesis (Flanegan and Baltimore, 1977, 1979). These experiments led to the important realization that $3D^{pol}$ is not only a template-dependent but also a primer-dependent RNA polymerase which catalyzed only RNA chain elongation; for example, polyadenylated polio RNA template was transcribed by this enzyme only if a synthetic oligo(U) primer was provided (Flanegan and Baltimore, 1977, 1979; Flanegan and Van Dyke, 1979; Tuschall *et al.,* 1982; Van Dyke and Flanegan, 1980).

Subsequent work was directed at identifying factors that would allow $3D^{pol}$ to initiate as well as elongate RNA chains. The strategy employed was to assay fractions of cell extracts for their ability to stimulate RNA synthesis by $3D^{pol}$ in the absence of oligo(U). Dasgupta *et al.* (1980) described the discovery of a host factor (HF, 67,000 MW) from uninfected HeLa cells that allowed initiation of RNA synthesis on a poliovirus RNA template with $3D^{pol}$ in the absence of oligo(U). The RNA products of this reaction were complementary to the plus-strand polio RNA template and were heterogeneous in length, up to the full size of the template (Dasgupta, 1983). In spite of the fact that RNA synthesis was carried out with only virion RNA, purified $3D^{pol}$, and purified HF, some of the RNA products were linked to a 45-kDa, VPg-related protein that cannot originate from the virion RNA template (Morrow and Dasgupta, 1983; Morrow *et al.,* 1984a,b). Quite similar results were reported by Baron and Baltimore (1982a,b). The mechanism of action of this host factor remained obscure. Morrow *et al.* (1985) described this host factor as a protein kinase that, apart from autophosphorylation, is capable of specifically phosphorylating an initiation factor of translation (eIF-2). Phosphorylation of some protein component in the reconstituted reaction mixture appeared to be a requirement in the priming reaction.

In contrast, Young *et al.* (1985) and Hey *et al.* (1986a) reported product RNA from their HF-mediated reactions that in a very small fraction was up to twice the length of the template RNA, containing both nascent plus- and minus-strand sequences (Hey *et al.,* 1986a). The larger product RNA resulted from addition of nucleotides directly onto the template to produce what appeared to be full-length RNA sequences covalently linked to the template via a hairpin structure. In this system the appearance of newly synthesized, VPg-linked RNA strands is the result of end-labeling of fragmented VPg-linked template RNA. If VPg is removed from the template prior to transcription by proteolysis, no product RNA could be immunoprecipitated with anti-VPg antibody (Andrews and Baltimore, 1986a; Young *et al.,* 1986).

J. B. Flanegan and colleagues have proposed a model for the priming of replication by $3D^{pol}$. This hypothesis was based upon the observation that in Flanegan's transcription system up to one-half of the reaction product synthesized in the presence of virion RNA, $3D^{pol}$, and HF is covalently linked at one end, presumably by a hairpin-like structure to the template RNA as was mentioned before. Figure 5 (kindly provided by Dr. J. B.

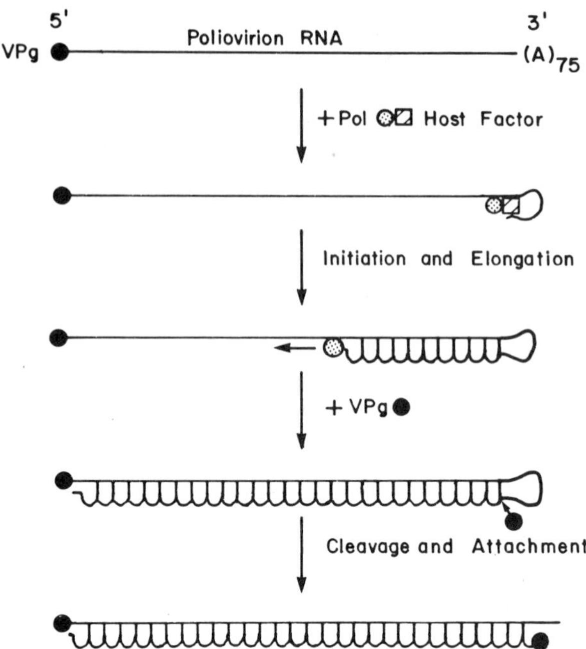

Fig. 5. Self-priming by template RNA. A host factor (either a kinase or a terminal uridylate transferase) produces a hairpin at the 3′ end of the plus-strand template RNA that serves as a primer for $3D^{pol}$. After some elongation has taken place, VPg (or its precursor) will cleave the hairpin, thereby attaching itself to the 5′ end of the nascent RNA strand. If nicking does not occur, an end-linked RF molecule is formed; as shown here, the homopolymeric segments of the RF would be covalently bound. We have named those structures "homo-linked" RF. Hairpin-mediated initiation and complete transcription of minus-stranded template would yield "hetero-linked" RF molecules. (This drawing was kindly provided by Dr. J. B. Flanegan.)

Flanegan) describes this model. It was speculated that HF might be able to allow the 3′-terminal poly(A) to bend such that it can form a hairpin which can be used by $3D^{pol}$ for the elongation reaction. Andrews *et al.* (1985) and Andrews and Baltimore (1986b) presented evidence that supported the hairpin priming model. They described the purification of a 68-kDa TUT that has HF activity. These authors proposed that virion template RNA is uridylylated by TUT at the 3′-terminal poly(A); the resulting oligo(U) tail then forms a snap-back structure whose 3′ end serves as primer for $3D^{pol}$.

The self-priming model does not provide a clue as to how VPg is eventually linked to the nascent RNA strand. Young *et al.* (1985) and Andrews *et al.* (1985) propose that VPg or a precursor polypeptide to VPg may nick the hairpin with the concomitant formation of a phosphodiester bond to the newly synthesized RNA strand (see Fig. 5). Such a mechanism of nucleic acid modification has been observed previously in interactions between supercoiled DNA strands and topoisomerases and the replication of parvoviruses (Berns *et al.,* 1985).

There are several problems that this model fails to address. If TUT is involved in replication, one would expect to find extra uridylate residues at the ends of the minus strands in RF. Instead, the heteropolymeric end of RF is mostly blunt-ended, and only a small fraction contains extra adenylate residues (Larsen *et al.,* 1980; Richards and Ehrenfeld, 1980). The addition of U residues to 3′ termini of RNA by TUT was found to be most efficient on poly(A) tails. This fits well with the 3′-terminal poly(A) of poliovirus plus strands but makes it difficult to explain initiation of plus strands at the 3′ end of minus strands. It has been suggested that HF could induce or stabilize a self-priming hairpin on minus strands (Young *et al.,* 1985; Hey *et al.,* 1986a). However, the nicking activity required in this model must recognize two very different hairpins.

Most recently, Hey *et al.* (1986b) presented avidence suggesting that the HF activity that has previously been described may have little relevance to polio RNA replication *in vivo*. The authors utilized a host factor preparation from uninfected HeLa cells purified according to Dasgupta and his colleagues, that allowed $3D^{pol}$ to transcribe virion RNA and produced some products larger than the template, apparently initiated by a template hairpin self-priming mechanism. This study demonstrated that preincubation of polio RNA with HF followed by phenol extraction yielded modified RNA that could now be transcribed by $3D^{pol}$ in the absence of HF or oligo(U). Analysis of HF-modified RNA showed that those moledules that were activated for subsequent elongation no longer contained their original 3′ poly(A) sequences. Instead, the 3′ termini of elongating templates displayed an approximately equal distribution of each of the four bases, an observation suggesting that initiation occurred randomly at different sites. RNA molecules that were chemically blocked at their original 3′ ends prior to their use as template showed no reduction in priming activity after preincubation with HF or when used in a reaction with $3D^{pol}$ plus HF. These observations argued against a TUT reaction being involved in template elongation. Indeed, no detectable TUT was observed in the HF preparation used by Hey *et al.* (1986b).

Hey *at al.* (1986b) proposed a model that explains these observations. They postulate that, during either preincubation with HF or in the HF-mediated reaction, random nucleolytic cleavages of the template occur. If some of the resulting templates contain 3′ terminal hairpins, these structures could be elongated by $3D^{pol}$. The generation of products that appear dimer size in length would arise from elongation of hairpins created near the original 3′ end of the template.

Several observations by this group and others support this model. Transcription of polio RNA in the presence of HF is inefficient. If one considers the number of A residues in the template and the number of U residues incorporated, it appears that only a small fraction of input RNA serves as template. Hey *et al.* (1986b) observed that after incubation with HF the vast majority of the template RNA remains intact. A very low level of nuclease in the system could generate active templates yet leave the majority of the RNA intact. Initiation on nicked RNA molecules would also explain the very heterogeneous nature of the transcription products. Lubinski *et al.* (1986) have shown that $3D^{pol}$ can extend RNA templates from an internal site, an observation that is consistent with the model presented by Hey *et al.* (1986b). Andrews and Baltimore (1986a) suggested that anti-VPg immunoprecipitable [α-^{32}P]UTP-labeled products can be formed if TUT uridylylates 5′-terminal fragments that contaminate virion RNA preparations. However, a small amount of nuclease in the TUT preparation could generate these fragments, capable of hairpin formation, which can be utilized by $3D^{pol}$ for transcription.

Thus, it appears that the HF activities that have been studied so far may have little relevance to polio RNA replication *in vivo*. An involvement of cellular protein in polio RNA replication is still quite possible. Elucidation of such factors requires a careful analysis of the products generated in the reconstituted system. Ideally an authentic *in vitro* system would be capable of synthesizing plus-sense, full-length RNA linked to VPg at its 5′ end and polyadenylated at its 3′ end.

2. *Membraneous Replication System*

The second strategy employed to understand the mechanism of poliovirus RNA replication involves the isolation of membraneous fractions, termed crude replication complexes, from infected host cells that are capable of synthesizing viral RNA (Caliguiri and Compans, 1973; McDonnel and Levintow, 1970; Takegami *et al.*, 1983a,b). This strategy is based on the observation that poliovirus RNA replication occurs on membranes (Girard *et al.*, 1967; Caliguiri and Tamm, 1970; Dorsch-Hasler *et al.*, 1975). Isopycnic centrifugation of the crude replication complex in a stepwise sucrose gradient results in the fractionation of the membranous material (Caliguiri and Tamm, 1970; Dorsch-Hasler *et al.*, 1975). The membranous material in each of the fractions contains virtually all the known poliovirus-specific polypeptides (Takegami *et al.*, 1983a). However, when assayed for the ability to synthesize viral RNAs, the fraction containing smooth membranes was found to have the highest activity

(Caliguiri and Tamm, 1970; Dorsch-Hasler *et al.,* 1975; Takegami *et al.,* 1983a; Takeda *et al.,* 1986).

The "crude replication complex" can synthesize all forms of poliovirus-specific RNA structures: RI, RF, and ssRNA (McDonnel and Levintow, 1970; Etchinson and Ehrenfeld, 1981; Takeda *et al.,* 1986). Addition of detergent (for example, the nonionic detergent NP-40 to 0.5%) alters the synthetic capability in that now only RF is synthesized (Etchison and Ehrenfeld, 1981; Takeda *et al.,* 1986). The entire RNA synthesized *in vitro* (in the absence of detergent) was analyzed by fingerprinting. Most of the viral RNA synthesized is of plus-strand polarity, although several spots were identified that are diagnostic for minus-stranded RNA (Takeda *et al.,* 1986). Takeda *et al.* (1986) were able to identify in two-dimensional fingerprints the VPg-linked, 5′-terminal RNase T1-resistant oligonucleotide from *in vivo*-labeled virion RNA (VPg-pUUAAAACAGp). This proved difficult as the peptidyl-oligonucleotide migrated aberrantly in the first dimension of the gel. When *in vitro*-synthesized RNA was isolated and analyzed in the same manner, the 5′-terminal oligonucleotide was also observed. Takeda *et al.* (1986) concluded that the crude membrane system can initiate RNA synthesis *de novo*.

What is the mechanism of initiation of RNA synthesis in the crude membrane system? A hypothesis, first proposed in 1977 (Lee *et al.,* 1977; Nomoto *et al.,* 1977a; Flanegan *et al.,* 1977), predicted that a uridylylated protein may function as a primer in RNA replication (Wimmer, 1982). Subsequent work on RNA synthesis in complexes of the infected cell led to the formulation of a model shown in Fig. 6. In this model VPg-pU(pU) is formed in the membranous replication complex and then serves as a primer for $3D^{pol}$ (Takegami *et al.,* 1983b). This model resembles the events that lead to the initiation of DNA synthesis in adenovirus and certain bacteriophages, that is, certain viruses with linear, double-stranded genomes whose 5′ termini are also protein linked (Lichy *et al.,* 1981; Pincus *et al.,* 1981; Hermoso and Salas, 1980; Penalva and Salas, 1982; Desiderio and Kelly, 1981; also see discussion by Wimmer, 1982, and Challberg and Kelly, 1982). In the case of these viruses, a terminal protein is first linked to dAMP or dCMP, a reaction catalyzed by the viral DNA polymerase. The resulting deoxynucleotidylyl protein then serves as a primer for DNA synthesis.

The best evidence for the priming mechanism shown in Fig. 6 came from the observation that VPg-pU and VPg-pUpU can be synthesized in the membranous replication complex (Takegami *et al.,* 1983b). It was supported by the discovery that VPg-pUpU can be found in infected HeLa cells (Crawford and Baltimore, 1983). Takegami *et al.* (1983b)

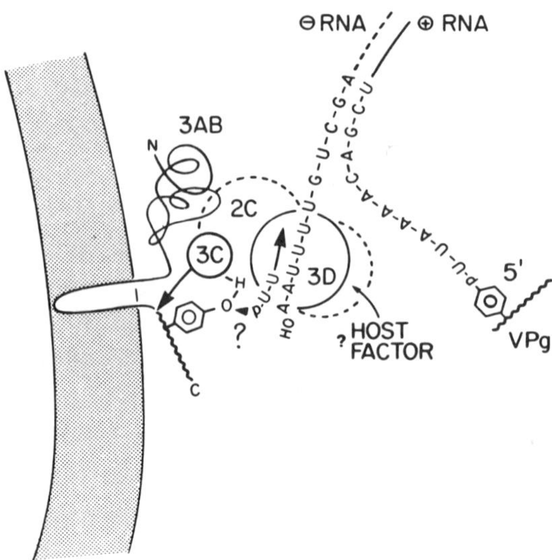

Fig. 6. Model of VPg-pUpU-primed initiation of RNA synthesis. 3AB is the membrane-bound poliovirus protein whose carboxyl terminus (wavy line) is VPg, 3C is the virus-encoded proteinase (responsible for the cleavage between 3A and 3B), 3D is the primer-dependent RNA polymerase, and 2C is an auxiliary viral protein of unknown function, mutations of which (g^r) lead to an altered phenotype of RNA synthesis *in vivo*. The possibility of the involvement of a "host factor" is indicated, although there is no evidence to support such a factor in the mebrane-bound replication complex. This model can account for the initiation of both plus- and minus-strand RNA. (Modified from Takegami *et al.*, 1983b.)

found that each fraction of the sucrose gradient-separated crude replication complex is capable of synthesizing VPg-pUpU, but that the smooth-membrane fraction (fraction I) had the highest activity. The synthesis of VPg-pUpU was completely inhibited when small amounts of NP-40 (0.5%) was added. Moreover, the formation of VPg-pUpU was found to be most efficient under conditions less favorable to the RNA elongation reaction (Takegami *et al.*, 1983b; Takeda *et al.*, 1986). These data suggested that synthesis of the uridylyl-oligopeptide and the $3D^{pol}$-catalyzed elongation of RNA strands are distinct processes.

Attempts to "chase" preformed VPg-pUpU into longer RNA strands met with difficulties due to the low yield of labeled nucleotidyl protein. The efficiency of the system was greatly stimulated when the membrane fractions (either crude replication complex or smooth-membrane fraction) were treated with DEAE-cellulose to remove endogenous nucleotide triphosphates and an ATP regenerating system was included into the reaction mixture (Takeda *et al.*, 1986). Under those conditions individual steps in the synthesis VPg-pU → VPg-pUpU → VPg-pUUAAAACAGp can be analyzed. An important result of this study was that preformed [α-^{32}P]VPg-pUpU, when incubated with nucleoside triphosphates (and UTP in large excess), can be "chased" into the elongation product VPg-pUUAAAACAGp (isolated after RNase T1 digestion of the elongation products). This reaction, but not the formation of VPg-pUpU, is stimulated by a cell extract of uninfected HeLa cells (Takeda *et al.*, 1986).

These observations strongly support the priming mechanism shown in Fig. 6. However, the data do not explain how VPg is uridylylated. Andrews *et al.* (1985) have cautioned that the occurrence of VPg-pUpU *in vitro* and *in vivo* may simply be the result of an aborted nicking/elongation reaction. We consider this to be less likely since the assumed "cleavage" of nascent strands by VPg would have to occur in the absence of a hairpin, an assumption that would hardly fit into the model shown in Fig. 5. At the present time it is not known whether the synthesis of VPg-pUpU is template dependent because the RNAs contained in the crude replication complex or smooth-membrane fraction are very resistant to digestion with micrococcal nuclease (or other RNases) (Takeda *et al.*, 1986). A solution to this problem will come if minus strands can be used in a reconstitution experiment in which the 3′ termini of exogenously added RNA can be analyzed after the reaction.

Two recent observations in our laboratory may shed light on the mechanism of uridylylation of VPg. First, we have begun to study a mutant of poliovirus that contains a mutation in 2C (I → M amino acid 227) which renders this virus more sensitive to inhibition by 0.2 m*M* guanidine hydrochloride than is our parental Mahoney virus. We have used this mutant to prepare crude replication complex and are examining the effect of guanidine on *in vitro* replication. Preliminary results (S. E. Pincus, C. F. Yang and E. Wimmer, unpublished) indicate that the synthesis of VPg-pUpU can be inhibited by 1.0–2.0 m*M* guanidine. If further experiments confirm these observations, then the results would suggest that 2C plays a role in uridylylation of VPg. Site-directed mutagenesis of 2C will then be performed to determine its function.

A second recent observation in our laboratory concerns the ability of crude replication complex from Sabin 1 to catalyze the uridylylation of VPg. Crude replication complexes from Mahoney- or Sabin-infected cells were compared for their ability to synthesize VPg-pUpU at temperatures ranging from 30 to 39°C. The Mahoney-derived complex displayed an increase in the amount of VPg-pUpU synthesized as the reaction temperature was increased from 30 to 37°C (which was the optimal condition) and a decreased synthesis at 38°C or 39°C. The Sabin 1-derived complex displayed optimal synthesis of VPg-pUpU at 36°C and a stronger inhibition at 38° or 39°C than observed with the Mahoney complex (H. Toyoda and E. Wimmer, unpublished). Analysis of the activity of crude replication complexes from recombinant viruses between Mahoney and Sabin 1 is being pursued in an attempt to determine the region(s) responsible for this difference. A recombinant containing the P1 and P2 regions of Mahoney and the P3 region of Sabin 1 displays the temperature sensitivity of Sabin 1. There are a number of nucleotide differences in this region between Mahoney and Sabin 1, however, only four lead to amino acid changes, and these map exclusively in $3D^{pol}$ (Nomoto *et al.,* 1982). By changing nucleotides in the P3 region of this recombinant back to the nucleotides seen in Mahoney, we hope to determine if the difference in VPg-pUpU formation between these two viruses is due to the amino acid changes in $3D^{pol}$ or the result of nucleotide changes in the template RNA. Results from such studies may implicate $3D^{pol}$ in uridylylation of VPg.

IV. CONCLUSIONS

Basic research on poliovirus has not ceased to be a challenge for the molecular biologist. Many fascinating and unique problems remain, including the mechanism of polyprotein processing, the mechanism of RNA replication, and the mechanism of morphogenesis. The answers to these questions will be important for those concerned with public health since picornaviruses are the cause of very frequent serious diseases in man. Some of the tools developed for the study of poliovirus, for example, cDNA clones and antiserum against various viral polypeptides, may prove useful in the development of diagnostic tests for picornaviral infections. As the mechanism of poliovirus replication is understood, it may be possible to identify or design drugs that can interfere with steps that are specific for picornaviruses and thus lead to chemotherapeutic treatments.

ACKNOWLEDGMENTS

We thank our colleagues, particularly Martin Nicklin, Aniko Paul, Michael Murray, and Hans Krausslich, for many stimulating discussions. We are grateful to Lynn Zawacki for the preparation of the manuscript. Work described here was supported in part by National Institutes of Health Grants AI 15122 and CA 28146.

REFERENCES

Alquist, P., and Kaesberg, P. (1979). *Nucleic Acids Res.* **1,** 1195–1204.
Ambros, V., and Baltimore, D. (1978). *J. Biol. Chem.* **253,** 5263–5266.

Ambros, V., Pettersson, R. F., and Baltimore, D. (1978). *Cell (Cambridge, Mass.)* **15,** 1439–1446.
Anderson-Sillman, K., Bartal, S., and Tershak, D. R. (1984). *J. Virol.* **50,** 922–928.
Andrews, N. C., and Baltimore, D. (1986a). *J. Virol.* **58,** 212–215.
Andrews, N. C., and Baltimore, D. (1986b). *Proc. Natl. Acad. Sci. U.S.A.* **83,** 221–225.
Andrews, N. C., Levin, D., and Baltimore, D. (1985). *J. Biol. Chem.* **260,** 7628–7635.
Argos, P. G., Kamer, G., Nicklin, M. J. H., and Wimmer, E. W. (1984). *Nucleic Acids Res.* **12,** 7251–7267.
Armstrong, J. A., Edmonds, M., Nagazato, H., Phillips, B. A., and Vaughan, M. H. (1972). *Science* **176,** 526–528.
Baltimore, D. (1964). *Proc. Natl. Acad. Sci. U.S.A.* **51,** 450–456.
Baltimore, D. (1966). *J. Mol. Biol.* **18,** 421–428.
Baltimore, D. (1968). *J. Mol. Biol.* **32,** 359–368.
Baltimore, D. (1969). *In* "The Biochemistry of Viruses" (H. Levi, ed.), pp. 101–176. Dekker, New York.
Baltimore, D., and Girard, M. (1966). *Proc. Natl. Acad. Sci. U.S.A.* **56,** 741–748.
Baltimore, D., Franklin, R. M., Eggers, H. J., and Tamm, I. (1963). *Proc. Natl. Acad. Sci. U.S.A.* **49,** 843–849.
Baltimore, D., Girard, M., and Darnell, J. E. (1984). *Science* **143,** 1034.
Baltimore, D., Girard, M., and Darnell, J. E. (1966). *Virology* **29,** 179–189.
Baron, M. H., and Baltimore, D. (1982a). *Cell (Cambridge, Mass.)* **30,** 745–752.
Baron, M. H., and Baltimore, D. (1982b). *J. Biol. Chem.* **257,** 12351–12358.
Berns, K. I., Muzyczka, N., and Hauswirth, W. W. (1985). *In* "Virology" (B. N. Fields, ed.), pp. 415–432. Raven Press, New York.
Bernstein, H. D., Sonenberg, N., and Baltimore, D. (1985). *Mol. Cell. Biol.* **5,** 2913–2923.
Best, M., Evan, B., and Bishop, J. M. (1972). *Virology* **47,** 592–603.
Bishop, J. M., and Koch, G. (1967). *J. Biol. Chem.* **242,** 1736–1743.
Bishop, J. M., Koch, G., Evans, B., and Merriman, M. (1969). *J. Mol. Biol.* **46,** 235–249.
Butterworth, B. E., Shinshick, E. J., and Yin, F. H. (1976). *J. Virol.* **19,** 457–466.
Caliguiri, L. A., and Compans, R. W. (1973). *J. Gen. Virol.* **21,** 99–108.
Caliguiri, L. A., and Mosser, A. G. (1971). *Virology* **46,** 375–386.
Caliguiri, L. A., and Tamm, I (1968a). *Virology* **35,** 408–417.
Caliguiri, L. A., and Tamm, I. (1968b). *Virology* **36,** 223–231.
Caliguiri, L. A., and Tamm, I (1970). *Virology* **42,** 100–111.
Caliguiri, L. A., and Tamm, I. (1973). *In* "Selective Inhibitors of Viral Functions" (W. Carter, ed.), pp. 287–294. CRC Press, Boca Raton, Florida.
Challberg, M. D., and Kelly, T. J., Jr. (1982). *Annu. Rev. Biochem.* **51,** 901–934.
Chumakov, K. M., and Agol, V. I. (1976). *Virology* **73,** 528–531.
Cohen, J. I., Ticehurst, J. R., Purcell, R. H., Buckler-White, A., and Baroudy, B. M. (1986). *J. Virol.* **61,** 50–59.
Cole, C. N., and Baltimore, D. (1973). *J. Mol. Biol.* **76,** 325–343.
Cooper, P. D. (1969). *In* "The Biochemistry of Viruses" (H. B. Levy, ed.), pp. 177–218. Dekker, New York.
Cooper, P. D. (1977). *In* "Comprehensive Virology" (H. Fraenkel-Conrat and R. R. Wagner, eds.), Vol. 9, pp. 133–207. Plenum, New York.
Cooper, P. D., Wentworth, B. B., and McCahon, D. (1970a). *Virology* **40,** 486–493.
Cooper, P. D., Stancek, D., and Summers, D. F. (1970b). *Virology* **40,** 971–977.
Crawford, N. M., and Baltimore, D. (1983). *Proc. Natl. Acad. Sci. U.S.A.* **80,** 7452–7455.
Crowther, D., and Melnick, J. L. (1961). *Virology* **15,** 65–74.
Dasgupta, A. (1983). *Virology* **128,** 245–251.

Dasgupta, A., Baron, M. H., and Baltimore, D. (1979). *Proc. Natl. Acad. Sci. U.S.A.* **76,** 2679–2683.

Dasgupta, A., Zabel, P., and Baltimore, D. (1980). *Cell (Cambridge, Mass.)* **19,** 423–429.

Dawson, W. O. (1975-1976). *Intervirology* **6,** 83–89.

Disiderio, S. V., and Kelly, T. J., Jr. (1981). *J. Mol. Biol.* **145,** 319–337.

Detjen, B. M., Lucas, J. J., and Wimmer, E. (1978). *J. Virol.* **27,** 582–586.

Dorner, A. J., Rothberg, P. G., and Wimmer, E. (1981). *FEBS Lett.* **132,** 219–223.

Dorner, A. J., Dorner, L. F., Larsen, G. R., Wimmer, E., and Anderson, C. W. (1982). *J. Virol.* **42,** 1017–1028.

Dorner, A. J., Semler, B. L., Jackson, R. J., Hanecak, R., Duprey, E., and Wimmer, E. (1984). *J. Virol.* **50,** 507–514.

Dorsch-Hasler, K., Yogo, Y., and Wimmer, E. (1975). *J. Virol.* **16,** 1512–1527.

Edery, I., Humbelin, M., Darveau, A., Lee, K., Milburn, S., Hershey, J., Trachsel, H., and Sonenberg, N. (1983). *J. Biol. Chem.* **258,** 11398–11403.

Eggers, H. J., Ikegami, N., and Tamm, I (1965). *Ann. N.Y. Acad. Sci.* **130,** 267–281.

Ehrenfeld, E., Maizel, J. V., and Summers, D. F. (1970). *Virology* **40,** 840–846.

Emini, E. A., Elzinga, M., and Wimmer, E. (1982). *J. Virol.* **42,** 194–199.

Emini, E. A., Leibowitz, J., Diamond, D. C., Bonin, J., and Wimmer, E. (1985). *Virology* **137,** 74–85.

Etchison, D., and Ehrenfeld, E. (1981). *Virology* **111,** 33–46.

Etchison, D., and Fout, S. (1985). *J. Virol.* **54,** 634–638.

Etchison, D., Milburn, S. C., Edery, I., Sonenberg, N., and Hershey, J. W. B. (1982). *J. Biol. Chem.* **257,** 14806–14810.

Flanegan, J. B., and Baltimore, D. (1977). *Proc. Natl. Acad. Sci. U.S.A.* **74,** 3677–3680.

Flanegan, J. B., and Baltimore, D. (1979). *J. Virol.* **29,** 352–360.

Flanegan, J. B., and Van Dyke, T. A. (1979). *J. Virol.* **32,** 155–161.

Flanegan, J. B., Pettersson, R. F., Ambros, V., Hewlett, M. J., and Baltimore, D. (1977). *Proc. Natl. Acad. Sci. U.S.A.* **74,** 961–965.

Franssen, H., Leunissen, J., Goldbach, R., Lomonassof, G., and Zinmern, D. (1984). *EMBO J.* **3,** 855–861.

Friedman, R. M. (1970). *J. Virol.* **6,** 628–636.

Girard, M., Baltimore, D., and Darnell, J. E. (1967). *J. Mol. Biol.* **24,** 59–74.

Golini, F., Semler, B. L., Dorner, A. J., and Wimmer, E. (1980). *Nature (London)* **287,** 600–603.

Gorbaleya, A. E., Blinov, V. M., and Koonin, E. V. (1985). *Mol. Genet.* **11,** 30–36.

Grito, J. A., Tahara, S. M., Morgan, M. A., Shatkin, A. J., and Merrick, W. C. (1983). *J. Biol. Chem.* **258,** 5804–5810.

Guo, P., Grimes, S., and Anderson, D. (1986). *Proc. Natl. Acad. Sci. U.S.A.* **83,** 3505–3509.

Hanecak, R., Semler, B. L., Anderson, C. W., and Wimmer, E. (1982). *Proc. Natl. Acad. Sci. U.S.A.* **79,** 3973–3977.

Hanecak, R., Semler, B. L., Ariga, H., Anderson, C. W., and Wimmer, E. (1984). *Cell (Cambridge, Mass.)* **37,** 1063–1073.

Hansen, J., and Ehrenfeld, E. (1981). *J. Virol.* **38,** 438–445.

Hermoso, J. M., and Salas, M. (1980). *Proc. Natl. Acad. Sci. U.S.A.* **77,** 6425–6428.

Hey, T. D., Richards, O. C., and Ehrenfeld, E. (1986a). *J. Virol.* **58,** 790–796.

Hey, T. D., Richards, O. C., and Ehrenfeld, E. (1986b). *J. Virol.* **61,** 802–811.

Hillenbrand, G., Morelli, G., Lanka, E., and Scherzinger, E. (1978). *Cold Spring Harbor Symp. Quant. Biol.* **43,** 449–459.

Hogle, J. M., Chow, M., and Filman, D. J. (1985). *Science* **229,** 1358–1365.

Huang, A. S., and Baltimore, D. (1970). *J. Mol. Biol.* **47,** 275–291.

Husain, I., Van Houten, B., Thomas, D. C., and Sancar, A. (1986). *J. Biol. Chem.* **261,** 4895–4901.

Iizuka, N., Kuge, S., and Nomoto, A. (1986). *Virology* **156,** 64–73.

Jacobson, M. F., and Baltimore, D. (1968a). *Proc. Natl. Acad. Sci. U.S.A.* **61,** 77–84.

Jacobson, M. F., and Baltimore, D. (1968b). *J. Mol. Biol.* **33,** 369–378.

Kamen, R. I. (1975). *In* "RNA Phages" (N. D. Zinder, ed.), pp. 203–234. Cold Spring Harbor Lab., Cold Spring Harbor, New York.

Kitamura, N., and Wimmer, E. (1980). *Proc. Natl. Acad. Sci. U.S.A.* **77,** 3196–3200.

Kitamura, N., Adler, C., and Wimmer, E. (1980). *Ann. N.Y. Acad. Sci.* **354,** 183–201.

Kitamura, N., Semler, B. L., Rothberg, P. G., Larsen, G. R., Adler, C. J., Dorner, A. J., Emini, E. A., Hanecak, R., Lee, J. J., van der Werf, S., Anderson, C. W., and Wimmer, E. (1981). *Nature (London)* **291,** 547–553.

Koch, G., Quentrell, N., and Bishop, J. M. (1967). *Virology* **31,** 388–390.

Korant, B. D. (1977). *Virology* **81,** 17–28.

Kozak, M. (1983). *Microbiol. Rev.* **47,** 1–45.

Kuhn, R. J., and Wimmer, E. (1986). *In* "The Molecular Biology of Positive Strand RNA Viruses" (D. J. Rowlands, B. W. J. Mahy, and M. Mayo, eds.). Academic Press, New York.

Larsen, G. R., Dorner, J. A., Harris, T. J. R., and Wimmer, E. (1980). *Nucleic Acids. Res.* **8,** 1217–1229.

Larsen, G. R., Anderson, C. W., Dorner, A. J., Semler, B. L., and Wimmer, A. (1982). *J. Virol.* **41,** 340–344.

Ledinko, N. (1963). *Virology* **20,** 107–119.

Lee, K. A. W., Edery, I., and Sonenberg, N. (1985). *J. Virol.* **54,** 515–524.

Lee, Y. F., Nomoto, A., and Wimmer, E. (1976). *Prog. Nucleic Acid Res. Mol. Biol.* **19,** 89–96.

Lee, Y. F., Nomoto, A., Detjen, B. M., and Wimmer, E. (1977). *Proc. Natl. Acad. Sci. U.S.A.* **74,** 59–63.

Lichy, J. H., Horwitz, M. S., and Hurwitz, J. (1981). *Proc. Natl. Acad. Sci. U.S.A.* **78,** 2678–2682.

Lloyd, R. E., Etchison, D., and Ehrenfeld, E. (1985). *Proc. Natl. Acad. Sci. U.S.A.* **82,** 2723–2727.

Loddo, B., Ferrari, W., Brotzu, G., and Spanedda, A. (1962). *Nature (London)* **193,** 97–98.

Loddo, B., Muntoni, S., Spanedda, A., Brotzu, G., and Ferrari, W. (1963). *Nature (London)* **197,** 315.

Loesch, W. T., and Arlinghaus, R. B. (1975). *Arch. Virol.* **47,** 201–215.

Lubinski, J. M., Kaplan, G., Racaniello, V. R., and Dasgupta, A. (1986). *J. Virol.* **58,** 459–467.

Lundquist, R. E., and Maizel, J. V. (1978). *Virology* **85,** 434–444.

Lundquist, R. E., Ehrenfeld, E., and Maizel, J. V. (1974). *Proc. Natl. Acad. Sci. U.S.A.* **71,** 4773–4777.

Lwoff, A. (1965). *Biochem J.* **96,** 289–301.

McDonnel, J. P., and Levintow, L. (1970). *Virology* **42,** 999–1006.

Madshus, I. H., Olsnes, S., and Sandvig, K. (1984). *J. Cell Biol.* **98,** 1194–1200.

Melnick, J. L. (1985). *In* "Virology" (B. N. Fields, ed.), pp. 739–794. Raven Press, New York.

Mendelsohn, C., Johnson, B., Lionetti, K. A., Nobis, P., Wimmer, E., and Racaniello, V. R. (1986). *Proc. Natl. Acad. Sci. U.S.A.* **83,** 7845–7849.

Miller, R. L., and Plagemann, P. G. W. (1972). *J. Gen. Virol.* **17,** 349–353.

Minor, P. D., Pipkin, P. A., Hockley, D., Schild, G. C., and Almond, J. W. (1984). *Virus Res.* **1,** 203–212.
Montagnier, L., and Sanders, F. K. (1963). *Nature (London)* **199,** 664.
Morrow, C. D., and Dasgupta, A. (1983) *J. Virol.* **48,** 429–439.
Morrow, C. D., Hocko, J., Navab, M., and Dasgupta, A. (1984a). *J. Virol.* **50,** 515–523.
Morrow, C. D., Navab, M., Peterson, C., Hocko, J., and Dasgupta, A. (1984b). *Virus Res.* **1,** 89–100.
Morrow, C. D., Gibbons, G. F., and Dasgupta, A. (1985). *Cell (Cambridge, Mass.)* **40,** 913–921.
Mosser, A. G., Caliguiri, L. A., Scheid, A. S., and Tamm, I. (1972). *Virology* **47,** 30–38.
Najarian, R., Caput, D., Gee, W., Potter, S. J., Renard, A., Merryweather, J., Van Nest, G., and Dina, D. (1985). *Proc. Natl. Acad. Sci. U.S.A.* **82,** 2627–2631.
Nick, H., and Ahl, R. (1976). *Arch. Virol.* **52,** 71–83.
Nicklin, M. J. H., Toyoda, H., Murray, M. G., and Wimmer, E. (1986). *Bio Technology* **4,** 33–42.
Nilsen, T. W., Wood, D. L., and Baglioni, C. (1981). *Virology* **109,** 82–93.
Nobel, J., and Levintow, L. (1970). *Virology* **42,** 634–642.
Nobis, P., Zibirre, R., Meyer, G., Kuhne, J., Warnecke, G., and Koch, G. (1985). *J. Gen. Virol.* **66,** 2563–2569.
Nomoto, A., Lee, Y. J., and Wimmer, E. (1976). *Proc. Natl. Acad. Sci. U.S.A.* **73,** 375–380.
Nomoto, A., Detjen, B., Pozzatti, R., and Wimmer, E. (1977a). *Nature (London)* **268,** 208–213.
Nomoto, A., Kitamura, N., Golini, F., and Wimmer, E. (1977b). *Proc. Natl. Acad. Sci. U.S.A.* **74,** 5345–5349.
Nomoto, A., Jacobson, A., Lee, Y. F., Dunn, J. J., and Wimmer, E. (1979). *J. Mol. Biol.* **128,** 179–196.
Nomoto, A., Omata, T., Toyoda, H., Kuge, S., Horie, H., Kataoka, Y., Genba, Y., Nakano, Y., and Imura, N. (1982). *Proc. Natl. Acad. Sci. U.S.A.* **79,** 5793–5797.
Oberg, B., and Philipson, L. (1969). *Eur. J. Biochem.* **11,** 305.
Pallansch, M., Kew, O. M., Semler, B. L., Omilianowski, D. R., Anderson, C. W., Wimmer, E., and Rueckert, R. R. (1984). *J. Virol.* **49,** 873–880.
Paul, A., Tada, H., von der Helm, K., Lezius, A., Wimmer, E., and Deinhardt, F. (1986). Submitted for publication.
Penalva, M. A., and Salas, M. (1982). *Proc. Natl. Acad. Sci. U.S.A.* **79,** 5522–5526.
Penman, S., and Summers, D. (1965). *Virology* **27,** 614–620.
Penman, S., Becker, Y., and Darnell, J. E. (1964). *J. Mol. Biol.* **8,** 541–555.
Pettersson, R. F., Flanegan, J. B., Rose, J. K., and Baltimore, D. (1977). *Nature (London)* **268,** 270–272.
Pettersson, R. F., Ambros, V., and Baltimore, D. (1978). *J. Virol.* **27,** 357–365.
Phillips, B. A., and Emmert, A. (1986). *Virology* **148,** 255–267.
Pincus, S. E., and Wimmer, E. (1986). *J. Virol.* **60,** 793–796.
Pincus, S. E., Robertson, W., and Rekosh, D. (1981). *Nucleic Acids Res.* **9,** 4919–4938.
Pincus, S. E., Diamond, D. C., Emini, E. A., and Wimmer, E. (1986a). *J. Virol.* **57,** 638–646.
Pincus, S. E., Rohl, H., and Wimmer, E. (1986b). *Virology* **157,** 83–88.
Pringle, C. R. (1964). *Nature (London)* **204,** 1012–1013.
Putnak, J. R., and Phillips, B. A. (1981). *Microbiol. Rev.* **45,** 287–315.
Racaniello, V. R., and Baltimore, D. (1981a). *Proc. Natl. Acad. Sci. U.S.A.* **78,** 4887–4891.
Racaniello, V. R., and Baltimore, D. (1981b). *Science* **214,** 916–919.

Rekosh, D., Lodish, H. F., and Baltimore, D. (1969). *Cold Spring Harbor Symp. Quant. Biol.* **34,** 747–753.
Richards, O. C., and Ehrenfeld, E. (1980). *J. Virol.* **36,** 387–394.
Richards, O. C., Ehrenfeld, E., and Manning, J. (1979). *Proc. Natl. Acad. Sci. U.S.A.* **76,** 676–680.
Richards, O. C., Martin, S. C., Jense, H. G., and Ehrenfeld, E. (1984). *J. Mol. Biol.* **173,** 235–240.
Rightsel, W. A., Dice, J. R., McAlpine, R. J., Timm, E. A., McLean, I. W., Dixon, G. J., and Schabel, F. M. (1961). *Science* **134,** 558–559.
Rossmann, M. G., Arnold, E., Erickson, J. W., Frankenberger, E. A., Griffith, J. P. Hecht, H. J., Johnson, J. E., Kamer, G., Luo, M., Mosser, A. G., Rueckert, R. R., Sherry, B., and Vriend, G. (1985). *Nature (London)* **317,** 145–153.
Rothberg, P. G., Harris, T. J. R., Nomoto, A., and Wimmer, E. (1978). *Proc. Natl. Acad. Sci. U.S.A.* **75,** 4868–4872.
Roy, P., and Bishop, D. H. L. (1970). *J. Virol.* **6,** 604–609.
Rueckert, R. R. (1976). *In* "Comprehensive Virology" (H. Fraenkel-Conrat and R. R. Wagner, eds.), Vol. 6, pp. 131–213. Plenum, New York.
Rueckert, R. R., and Wimmer, E. (1984). *J. Virol.* **50,** 957–959.
Sangar, D. V., Bryant, J., Harris, T. J. R., Brown, F., and Rowlands, D. J. (1981). *J. Virol.* **39,** 67–74.
Sarnow, P., Bernstein, D., and Baltimore, D. (1986). *Proc. Natl. Acad. Sci. U.S.A.* **83,** 571–575.
Saunders, K., and King, A. M. Q. (1982). *J. Virol.* **42,** 389–394.
Semler, B. L., Hanecak, R., Anderson, C. W., and Wimmer, E. (1981a). *Virology* **114,** 589–594.
Semler, B. L., Anderson, C. W., Kitamura, N., Rothberg, P. G., Wishart, W. L., and Wimmer, E. (1981b). *Proc. Natl. Acad. Sci. U.S.A.* **78,** 3464–3468.
Semler, B. L., Anderson, C. W., Hanecak, R., Dorner, L. F., and Wimmer, E. (1982). *Cell (Cambridge, Mass.)* **28,** 405–412.
Semler, B. L., Dorner, A. J., and Wimmer, E. (1984). *Nucleic Acids Res.* **12,** 5123–5141.
Semler, B. L., Johnson, V. H., and Tracy, S. (1986). *Proc. Natl. Acad. Sci. U.S.A.* **83,** 1777–1781.
Sergiescu, D., Horodniceanu, F., and Aubert-Combiescu, A. (1972). *Prog. Med. Virol.* **14,** 123–199.
Siegel, G., and Eggers, H. J. (1982). *J. Gen. Virol.* **61,** 111–114.
Sonenberg, N., Skup, D., Trachsel, H., and Millward, S. (1981). *J. Biol. Chem.* **256,** 4138–4141.
Spector, D. H., and Baltimore, D. (1974). *Proc. Natl. Acad. Sci. U.S.A.* **71,** 2985–2987.
Spector, D. H., and Baltimore, D. (1975a). *J. Virol.* **15,** 1418–1431.
Spector, D. H., and Baltimore, D. (1975b). *Virology* **67,** 498–505.
Spiegelman, S., and Hayashi, M. (1963). *Cold Spring Harbor Symp. Quant. Biol.* **28,** 161.
Stanway, G. Hughes, P. J., Westrup, G. P., Evans, D. M. A., Dunn, G., Minor, P. D., Schild, G. C., and Almond, J. W. (1986). *J. Virol.* **57,** 1187–1190.
Summers, D. F., and Levintow, L. (1965). *Virology* **27,** 44–53.
Summers, D. F., Maizel, J. V., Jr. (1968). *Proc. Natl. Acad. Sci. U.S.A.* **59,** 966–971.
Summers, D. F., Maizel, J. V., Jr., and Darnell, J. E. (1967). *Virology* **31,** 427.
Svitkin, Y. V., Maslova, S. V., and Agol, V. I. (1985). *Virology* **147,** 243–252.
Tahara, S. M., Morgan, M. A., and Shatkin, A. J., (1981). *J. Biol. Chem.* **256,** 7691–7694.

Takeda, N., Kuhn, R. J., Yang, C. F., Takegami, T., and Wimmer, E. (1986). *J. Virol.* **60,** 43–53.
Takegami, T., Semler, B. L., Anderson, C. W., and Wimmer, E. (1983a). *Virology* **128,** 33–47.
Takegami, T., Kuhn, R. J., Anderson, C. W., and Wimmer, E. (1983b). *Proc. Natl. Acad. Sci. U.S.A.* **80,** 7447–7451.
Tamm, I., and Eggers, H. J. (1963). *Science* **142,** 24–33.
Tershak, D. R. (1982). *J. Virol.* **41,** 313–318.
Tershak, D. R. (1984). *J. Virol.* **52,** 777–783.
Tershak, D. R. (1985). *Can. J. Microbiol.* **31,** 1166–1168.
Toyoda, H., Kohara, M., Kataoka, Y., Suganuma, T., Omata, T., Imura, N., and Nomoto, A. (1984). *J. Mol. Biol.* **174,** 561–585.
Toyoda, H., Nicklin, M. J. H., Murray, M. G., and Wimmer, E. (1986a). *In* "Protein Engineering: Applications in Science, Medicine and Industry" (M. Inouye and R. Sarma, eds.), pp. 319–337. Academic Press, New York.
Toyoda, H., Nicklin, M. J. H., Murray, M. G., Anderson, C. W., Dunn, J. J., Studier, F. W., and Wimmer, E. (1986b). *Cell (Cambridge, Mass.)* **45,** 761–770.
Traub, A., Diskin, B., Rosenberg, H., and Kalmar, E. (1976). *J. Virol.* **18,** 375–382.
Tuschall, D. M., Heibert, E., and Flanegan, J. B. (1982). *J. Virol.* **44,** 209–216.
Van Dyke, T. A., and Flanegan, J. B. (1980). *J. Virol.* **35,** 732–740.
Varma, J. P. (1968). *Virology* **36,** 305–308.
Vartapetian, A. B., Drygin, Y. F., Chumakov, K. M., and Bogdanov, A. A. (1980). *Nucleic Acids Res.* **8,** 3729–3741.
Wimmer, E. (1982). *Cell (Cambridge, Mass.)* **28,** 199–201.
Wright, P. J., and Cooper, P. D. (1976). *J. Gen. Virol.* **30,** 63–71.
Wu, M., Davidson, N., and Wimmer, E. (1978). *Nucleic Acids Res.* **5,** 4711–4723.
Yin, F. H. (1977). *Virology* **82,** 299–307.
Yogo, Y., and Wimmer, E. (1972). *Proc. Natl. Acad. Sci. U.S.A.* **69,** 1877–1882.
Yogo, Y., and Wimmer, E. (1973). *Nature (London) New Biol.* **242,** 171–174.
Yogo, Y., and Wimmer, E. (1975). *J. Mol. Biol.* **92,** 467–477.
Yogo, Y., Teng, M. H., and Wimmer, E. (1974). *Biochem. Biophys. Res. Commun.* **61,** 1101–1109.
Young, D. C., Tuschall, D. M., and Flanegan, J. B. (1985). *J. Virol.* **54,** 256–264.
Young, D. C., Dunn, B. M., Tobin, J., and Flanegan, J. B. (1986). *J. Virol.* **58,** 715–723.
Zimmerman, E. F., Heeter, M., and Darnell, J. E. (1963). *Virology* **19,** 400.

12

Permanent Expression of Influenza Virus Genes Coding for Transcriptase Complexes: Complementation of Viral Mutants

MARK KRYSTAL AND PETER PALESE

Department of Microbiology
Mount Sinai School of Medicine
New York, New York 10029

Influenza virus is the only eukaryotic RNA virus besides RNA tumor viruses that replicates through a nuclear phase. In addition, influenza virus encodes a unique "cap snatching" system whereby viral mRNA is primed through cleavage and usage of the first 10–13 nucleotides of host RNA (1). Another interesting feature of influenza viral transcription is that at least two genomic segments of RNA contain overlapping genes. One of the genes is coded by an mRNA which is co-linear with the RNA segment, whereas the other gene product derives from an mRNA that is regulated through RNA splicing (2). The transcriptase complex as it exists in viral cores consists of not one but three distinct "polymerase" proteins (designated PB1, PB2, PA) present in low but equimolar amounts along with large quantities of the basic nucleoprotein (NP). These proteins are complexed with the eight single-stranded RNAs of negative polarity which make up the influenza A and B virus genomes. Previous studies have suggested roles for PB2 and PB1 in primary transcription (cap binding and endonuclease activity and elongation, respectively) (reviewed in ref. 1), and more recently the NP protein has been implicated in the

Molecular Biology of RNA
New Perspectives

readthrough of the poly(A) addition site, which is a prerequisite for replication (3). It is not known whether any additional viral-encoded proteins are used for the transcription/replication of viral RNAs, although it is tempting to speculate that one or more of the three nonstructural proteins (NS1, NS2, and M2) transiently interact with the transcription complex. The NS1 and NS2 proteins seem the most likely candidates because both migrate soon after transcription to the cell nucleus which is the site of viral RNA transcription/replication (4,5).

I. ESTABLISHMENT OF A FUNCTIONAL EXPRESSION SYSTEM

Our laboratory has attempted to study the transcription complex of influenza virus through the constitutive expression of functional viral proteins in tissue culture cells (6). Cloned DNA copies coding for the influenza viral polymerase genes were inserted into a bovine papilloma virus (BPV) vector downstream of a mouse metallothionein (MT) promoter and used to transform C127 cells. Cells expressing the PB2 polymerase protein were isolated and tested for their ability to correct the defect in a number of previously defined temperature-sensitive (ts) mutants defective in PB2 (7–9). The PB2 cell line (PB2-5) expressed functional enzyme which could complement enzyme activity in mutant-infected cells. As a matter of fact, the level of protein synthesis seen in PB2 mutant-infected PB2-5 cells (at nonpermissive temperature) approaches that seen in wild-type infected cells (6). Although the PB2-5 cell line was able to rescue enzyme activity, it was not able to complement significantly the growth of the PB2-mapped ts mutant (Table I). However, C127 cell lines that were established through cotransfection of BPV vectors containing the three different polymerase genes were able to complement the growth of certain mutants. The two cell lines studied most intensely were designated 3P-133 and 3P-38. The 3P-133 cell line generated titers after infection with the PB2 ts mutant that were two to three logs higher than control cells and almost one log greater than titers produced in 3P-38-infected cells (Table I). On the other hand, the 3P-38 cells infected with the PA mutant generated consistently higher titers than the 3P-133-infected cells, and one log greater titers than control-infected cells (not shown) (6). These phenotypic differences in growth complementation are presumably the result of the differential expression of gene-specific RNA in each cell line, since cell line 3P-133 expresses about nine-fold more PB2-specific RNA than 3P-38 cells and about one-ninth as much PA RNA as cell line 3P-38 (6). Although significant levels of PB1 RNA were expressed in these cell lines,

TABLE I

Complementation of PB2 Mutants[a]

Cell line	Expressed protein	Virus	
		Wild-type	PB2 mutant (ts1)
C127	—	1[b]	1
PB2-5	PB2	0.14	5.6
3P-133	PB2, PB1, PA	0.79	326
3P-38	PB2, PB1, PA	0.79	42

[a] Data taken from Ref. (6).
[b] Data expressed as titer in transformed cells/titer in C127 cells.

no growth complementation was observed with PB1 mutant-infected cells (6).

II. ADDITION OF NUCLEAR PROTEIN TO THE TRANSCRIPTION COMPLEX

It is evident that the presence of the three polymerase proteins greatly enhances the growth complementation of certain groups of ts mutants. Presumably this is due to the formation of active complexes prior to virus infection that are used for both transcription and packaging. Therefore, it was of interest to determine if the presence of another protein from the transcription complex, namely the NP, would increase the growth complementation effect. A BPV vector containing the cloned NP gene of A/PR/8/34 virus downstream of the mouse MT promoter was cotransfected into 3P-133 cells with a plasmid containing the bacterial neomycin resistance gene cloned downstream of the mouse MT promoter. G418-resistant cells were selected, cloned out, and analyzed for the presence of the influenza virus NP through immunoprecipitation and immunofluorescence analysis (data not shown). One cell line, designated 3PNP-4, was selected for further study on the basis of its high level expression of NP. The level of NP produced in 3PNP-4 was compared with that produced during viral infection. Figure 1 is an immunoblot analysis comparing the steady-state amount of NP in the 3PNP-4 cell line with that of NP produced in wild-type virus-infected cells at various times after infection. It should be noted that the virus-infected samples contain both infected cells and supernatants (which contain released virus) so that the total amount of NP produced can be analyzed. The level of NP in infected samples

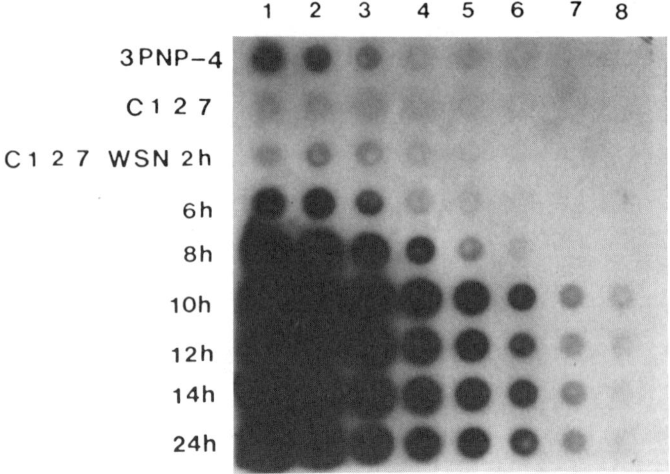

Fig. 1. Quantitation of NP produced in 3PNP-4 cells. Cell extract from 3PNP-4 cells was immunoblotted and compared to equivalent amounts of extract from A/WSN/33 virus-infected C127 cells. For virus-infected C127 cells both cell extract and supernatant were used. The filter was reacted with a mixture of monoclonal antibodies against the NP protein and developed with an ^{125}I-labeled antimouse κ-chain monoclonal antibody (gifts of J. L. Schulman and C. Bona).

peaks at about 10 hr postinfection. At later time points the level of NP in the supernatants increases (data not shown) but the total amount of NP remains relatively constant. Figure 1 therefore shows that the steady-state level of NP constitutively expressed in 3PNP-4 cells corresponds to about 5% of the maximum amount of NP produced during viral infection.

The 3PNP-4 cell line was then tested for its ability to complement the growth of ts mutants. Since these cells are derivatives of 3P-133, it was of interest to determine whether the addition of the NP would effect the complementation of PB2, PB1, or PA mutant-infected cells. We also tested whether these cells would now complement the growth of NP mutants. Results of these experiments are shown in Table II. Addition of NP to 3P-133 cells slightly increases the complementation efficiency for the PB2 mutant, but had no effect on the complementation efficiency of PB1- or PA-infected cells (data not shown). Also, the high level of NP expression present in 3PNP-4 cells allows for significant growth complementation of the NP ts mutant. Infectious yield was up 40-fold as compared to control C127 cells. Studies are now under way to determine if the

TABLE II

Addition of NP to 3P-133 Cells and Effect on Complementation[a]

Cell line	Expressed protein	Virus		
		Wild-type	PB2 mutant (ts1)	NP mutant (ts56)
C127	—	1[b]	1	1
3P-133	PB2, PB1, PA	0.51	456	1.1
3PNP4	PB2, PB1, PA, NP	0.29	1125	40

[a] Data averaged from two to five experiments.
[b] Data expressed as titer in transformed cells/titer in C127 cells.

NP expressed by 3PNP-4 cells is specifically packaged into the complemented virus.

III. CONCLUSIONS AND OUTLOOK

We have developed a viral complementation system based on the *in vivo* expression of influenza virus genes required for functional transcription complexes and we have been able to express four different viral proteins (PB1, PB2, PA, and NP) in the same cell line. It will now be of interest to determine the effect on complementation when other viral proteins such as the NS1, NS2, or the M (matrix) proteins are permanently expressed in these cells. For instance, M protein has been shown to inhibit viral ribonucleoprotein (RNP) transcriptase activity *in vitro* (10–12) so expression of M in 3PNP-4 cells may be expected to abolish all complementation. The establishment of a tissue culture growth complementation system for influenza virus should enable us to study influenza virus in new ways. These "COS-like" cells (13) may allow for the isolation of "tight" host-range mutants of influenza virus for use in genetic analysis. The mutants presently available are too leaky for precise analysis of function. These cell lines constitutively expressing influenza viral proteins may also prove useful for the "rescue" of virus-specific RNA into infectious viral particles. If these permanently expressed transcription complexes can be made to recognize influenza viral RNA introduced into these cells, then superinfection with virus may allow for reassortment to take place so that the RNA gets introduced into virus. If successful, this technique would allow for precise genetic manipulation of the influenza viral genome.

ACKNOWLEDGMENTS

This work was supported in part by Grants AI-11823 and AI-18998 from the National Institutes of Health. M. K. received support from the Alexandrine and Alexander Sinsheimer Fund and from the Charles H. Revson Foundation.

REFERENCES

1. Krug, R. M., *in* "Genetics of Influenza Viruses" (P. Palese and D. W. Kingsbury, eds.), pp. 70–98. Springer-Verlag, Berlin and New York, 1983.
2. Lamb, R. A., *in* "Genetics of Influenza Viruses" (P. Palese and D. W. Kingsbury, eds.), pp. 21–69. Springer-Verlag, Berlin and New York, 1983.
3. Plotch, S. J., and Krug, R. M., *Proc. Natl. Acad. Sci. U.S.A.* **83,** 5444–5448 (1986).
4. Young, J. F., Desselberger, U., Palese, P., Ferguson, B., Shatzman, A. R., and Rosenberg, M., *Proc. Natl. Acad. Sci. U.S.A.* **80,** 6105–6109 (1983).
5. Greenspan, D., Krystal, M., Nakada, S., Arnheiter, H., Lyles, D. S., and Palese, P., *J. Virol.* **54,** 833–843 (1985).
6. Krystal, M., Ruan, L., Lyles, D., Pavlakis, G., and Palese, P., *Proc. Natl. Acad. Sci. U.S.A.* **83,** 2709–2713 (1986).
7. Sugiura, A., Tobita, K., and Kilbourne, E. D., *J. Virol.* **10,** 639–647 (1972).
8. Sugiura, A., Ueda, M., Tobita, K., and Enomoto, C., *Virology* **65,** 363–373 (1975).
9. Palese, P., Ritchey, M. B., and Schulman, J. L., *J. Virol.* **21,** 1187–1195 (1977).
10. Zvonaryov, A. Y., and Ghendon, Y. Z., *J. Virol.* **33,** 583–586 (1980).
11. Milheeva, A. V., and Ghendon, Y. Z., *J. Gen. Virol.* **64,** 305–312 (1983).
12. Melnikov, S. Y., Mikleeva, A. V., Leneva, I. A., and Ghendon, Y. Z., *Virus Res.* **3,** 353–365 (1985).
13. Gluzman, Y., *Cell (Cambridge, Mass.)* **23,** 175–182 (1981).

13

Molecular Mechanisms of Pathogenesis by HTLV-III

FLOSSIE WONG-STAAL AND ROBERT C. GALLO

Laboratory of Tumor Cell Biology
National Cancer Institute
Bethesda, Maryland 20892

Human T-lymphotropic virus, type III (HTLV-III/LVA/ARV/HIV) the etiological agent of the acquired immune deficiency syndrome (AIDS), is a member of the retrovirus family. HTLV-III shares with HTLV-I and HTLV-II its target cell tropism for the T4 helper cell as well as a number of other characteristics, including impairment of T-cell function and modes of transmission (1). The underlying defect in HTLV-III-infected patients is the depletion of T4 cells, leading to severe immunosuppression. This phenomenon is mimicked in peripheral blood lymphocytes infected *in vitro* in which the T4 cell is also preferentially depleted. It has been observed that cell death is linked to an explosion of virus expression which is triggered by immune activation (2). There is also evidence that T4 expression is required for the cell to respond to the viral cytopathic effect (CPE) (3). The purpose of this chapter is to address the relationship between virus replication and cytopathicity and to define the viral genes necessary for both processes.

The genetic structure of HTLV-III is by far the most complex among retroviruses (Fig. 1). In addition to the three structural genes (*gag,* pol, and *env*) necessary for replication of all retroviruses, the genome of HTLV-III contains at least four additional genes (*sor, tat-3, trs/art,* and *3'orf*) (4,5). Two of these, *sor* and *3'orf,* were first identified as open reading frames by nucleotide sequence analyses (6–8) and subsequently verified to encode proteins that could be detected serologically (9,10). However, their functions remain unclear but are probably regulatory in

Molecular Biology of RNA
New Perspectives

Fig. 1. Genetic structure of HTLV-III.

nature. Deletions introduced into the *sor* and *3′orf* genes of biologically active genomes of HTLV-III did not abrogate virus production but significantly compromised or enhanced the level of expression, respectively (11,12). The *tat-3* gene was defined functionally in its capacity to activate in trans expression of genes linked to the viral long terminal repeat (LTR) (13). It is transcribed from three discontiguous gene segments into a 2.0-kb mRNA (14). The coding sequences reside in the second and third exons and encode a protein of 86 amino acids. Deletion mutants designed to be low and nonexpressors of *tat-3* produced accordingly low and non-detectable amounts of viral proteins and viral particles (15,16). Recent evidence suggests that *tat-3* regulates at both transcriptional and post-transcriptional levels (17,18). An alternate reading frame encoding a protein of 116 amino acids is contained in the same spliced *tat-3* mRNA and has been shown to be a functional gene requisite for expression of viral *gag* and *env* proteins (4,5). This gene originally named *art,* for antirepression of translation (4), was believed to reverse an intrinsic block on the translation of *gag* and *env* mRNA. Subsequent analysis of mutants lacking this gene function revealed an aberrant pattern of spliced mRNA that consisted of only 2.0 kb of mRNA and none of the higher-molecular-weight species, thus explaining the failure of the mutants to synthesize *gag* and *env* proteins (5). Since this novel gene product may have a role in the regulation of levels of spliced viral mRNA, we renamed it *trs* (trans regulator of splicing).

In addition to *tat-3* and *trs,* we have also demonstrated the presence of transcriptional activators in HTLV-III-infected cells (18). Such factors specifically promote transcription directed by the HTLV-III LTR, and are probably involved in the formation of the preinitiation complex. It is not known whether this factor is virus coded or is a cellular transcriptional factor modified by a viral product. In any event, it appears that HTLV-III expression is regulated at multiple levels. The interplay of these regulatory genes may modulate the lytic and latent phases of infection and assure efficient virus expression when cells are immune activated. One practical implication of this finding is that the investigator has multiple targets for inhibiting viral expression. For example, antagonists of *tat-3* and *trs* should also be effective antiviral agents.

What viral genes are critical for the cytopathicity of the virus? In this regard, much light has also been shed by studies utilizing deletion mutants of HTLV-III. In a series of experiments originally designed to determine the role of *3′orf* in virus replication and cytopathicity, deletions of various sizes in this gene were constructed and transfected into normal T lymphocytes or different T-cell lines (12). We found that all mutants with deletions exclusively in *3′orf* were highly replicative and cytopathic. However, two mutants with deletions extending into the carboxy-terminus of the neighboring *env* gene did not exert overall CPE while maintaining their replicative capacity (12). As summarized in Table I these infected cells produced copious amounts of virus, retained full trans-activating function, contained high levels of unintegrated DNA, and formed syncytia, indicating that while each of these parameters may contribute to CPE, none is sufficient in itself. Furthermore, these results point to the envelope gene as a critical determinant for CPE.

The *env* gene of HTLV-III encodes a precursor protein of 160 kDa which is then cleaved into the major extracellular envelope protein (gp120) and the small envelope protein (gp41). The sites responsible for binding to the cellular receptor as well as the major epitopes recognized by neutralizing antibodies reside in gp120. Significant polymorphism among isolates is also seen in this protein, and both hypervariable and conserved regions can be discerned (19–21). The gp41 protein is predicted to go through the membrane three times because it contains three hydrophobic domains. The part that is extracellular contains potential glycosylation sites and has been shown to contain the major immunodominant epitopes for gp41. The long, presumed cytoplasmic tail of gp41 is unique to HTLV-III and related viruses. Our results with the noncytopathic variants indicate that modifications of the carboxyl terminus of gp41 can render the virus noncytopathic. How may this fact fit with the actual events that lead to cell death? D. Zagury showed that virus infection per se, which involves binding of gp120 to T4, does not result in cell death even at high multiplicities of infection (personal communication). Within a week after infection, syncytial formation was usually observed in a few percent of the cells which eventually die, but overall CPE is not observed because the cells continued to proliferate and the rate of proliferation far exceeded cell death. In the presence of immune activation, virus expression occurred and a dramatic decline in cell viability set in at the peak of virus production usually 2–3 weeks after infection. Furthermore, although very often only 10–20% of cells were expressing virus, the whole

TABLE I

Properties of HTLV-III Clones Deleted in the *3′orf* Region of the Viral Genome

Plasmid constructs	Number of base pairs deleted at *Xho*I in *3′orf*	Genes deleted	Trans-activation	Syncytia	Virus expression in transfected H9 cells		Unintegrated viral DNA	Cytopathic effects in lymphocytes
					p17/p24/p41 expression	HTLV-III virions		
Δ HXB-2gpt	0	—	+	+	+	+	+	+
Δ X-A	55	*3′orf*	+	+	+	+	+	+
Δ X-B	109	*3′orf*	+	+	+	+	+	+
Δ X-C	85	*3′orf*	+	+	+	+	+	+
Δ X-D	100	*3′orf*	+	+	+	+	+	+
Δ X9-3	177	*3′orf* + *env*	+	+	+	+	+	–
Δ X10-1	200	*3′orf* + *env*	+	+	+	+	+	–

population of T4 cells are depleted. These observations are consistent with the infected cells producing a suppressor molecule that is specific for T4 cells. The depletion of T4 cells may be a combined result of syncytial formation, accelerated terminal differentiation, and more importantly, suppression of proliferation. How modification of the carboxyl terminus of gp41 relates to any one of these phenomena and results in loss of cytopathic capacity of the virus remains to be determined.

REFERENCES

1. Wong-Staal, F., and Gallo, R. C., Human T-Lymphotropic retroviruses. *Nature* (*London*) **317,** 395–403 (1985).
2. Zagury, D., Bernard, J., Leonard, R., Chepier, R., Feldman, M., Sarin, P. S., and Gallo, R. C., Longterm cultures of HTLV-III infected T-cells: A model of cytopathology of T-cell depletion in AIDS. *Science* **231,** 850–853 (1986).
3. Derossi, A., Franchini, G., Aldovini, A., Del Mistro, A., Chieco-Bianchi, L., Gallo, R. C., and Wong-Staal, F., Differential response to the cytopathic effects of HTLV-III super-infection in OKT4 and OKT8 cell clones transformed by HTLV-I. *Proc. Natl. Acad. Sci. U.S.A.* **83,** 4297–4301 (1986).
4. Sodroski, J. G., Goh, W. C., Rosen, C., Dayton, A., Terwilliger, E., and Haseltine, W., A second post-transcriptional trans-activator gene required for HTLV-III replication. *Nature* (*London*) **321,** 412–417 (1986).
5. Feinberg, M. B., Jarrett, R. F., Aldovini, A., Gallo, R. C., and Wong-Staal, F., HTLV-III expression and production involve complex regulation at the levels of splicing and translation of viral RNA. *Cell* (*Cambridge, Mass.*) **46,** 807–817 (1986).
6. Ratner, L., Haseltine, W., Patarda, R., Livak, K. J., Starcich, B., Josephs, S. F., Doran, E. R., Rafalski, J. A., Whitwhoen, E. A., Baumeister, K., Ivanoff, L., Petteway, S. R., Pearson, M. L., Lautenberger, J. A., Papas, T. S., Ghrayeb, J., Chang, N. T., Gallo, R. C., and Wong-Staal, F., Complete nucleotide sequence of the AIDS virus HTLV-III. *Nature* (*London*) **313,** 277–284 (1985).
7. Wain-Hobson, S., Sonigo, P., Danos, O., Cole, S., and Alizon, M., Nucleotide sequence of the AIDS virus. LAV. *Cell* (*Cambridge, Mass.*) **40,** 9–17 (1985).
8. Sanchez-Pescador, R., Power, M. D., Barr, P. J., Steimer, K. S., Stempten, M. M., Brown-Shimer, S. L., Gee, W. W., Renard, A., Randolph, A., Levey, J. A., Dian, D., and Luciw, P. A., Nucleotide sequence and expression of an AIDS-associated retrovirus (ARV-2). *Science* **227,** 484–492 (1985).
9. Franchini, G., Robert-Guroff, M., Wong-Staal, F., Ghrayeb, N., Kato, N., and Chang, N. Expression of the 3′ open reading frame of the HTLV-III in bacteria: Demonstration of its immunoreactivity with human sera. *Proc. Natl. Acad. Sci. U.S.A.* **83,** 5282–5285 (1986).
10. Kan, N. C., Franchini, G., Wong-Staal, F., Dubois, G. C., Robey, W. G., Lautenberger, J. A., and Papas, T. S., Identification of HTLV-III/LAV sor gene product and detection of antibodies in human sera. *Science* **231,** 1553–1555 (1986).
11. Sodroski, J., Goh, W. C., Rosen, C., Tartar, A., Portetelle, D., Burny, A., and Haseltine, W., Replicative and cytopathic potential of HTLV-III/LAV with sor gene deletions. *Science* **231,** 1549–1552 (1986).
12. Fisher, A. G., Ratner, L., Mitsuya, H., Marshelle, L. M., Harper, M. E., Broder, S.,

Gallo, R. C., and Wong-Staal, F., Infectious mutants of HTLV-III with changes in the 3′ region and markedly reduced cytopathic effects. *Science* **233,** 655–659 (1986).
13. Sodroski, J., Rosen, C., Wong-Staal, F., Salahuddin, S. Z., Popovic, M., Arya, S., Gallo, R. C., and Haseltine, W. A., *Trans*-acting transcriptional regulation of human T-cell leukemia virus type III long terminal repeat. *Science* **227,** 171–177 (1985).
14. Arya, S. K., Guo, C., Josephs, S. F., and Wong-Staal, F., *Trans*-activator gene of human T-lymphotropic virus type III (HTLV-III). *Science* **229,** 69–73 (1985).
15. Fisher, A. G., Feinberg, M. B., Josephs, S. F., Harper, M. E., Marselli, L. M., Reyes, G., Gonda, M. A., Aldovini, A., Debouck, C., Gallo, R. C., and Wong-Staal., F., The trans-activator gene of HTLV-III is essential for virus replication. *Nature* (*London*) **320,** 367–371 (1986).
16. Dayton, A. I., Sodroski, J. G., Rosen, C. A., Goh, W. C., and Haseltine, W. A., The trans-activator gene of the human T-cell lymphotropic virus type III is required for replication. *Cell* (*Cambridge, Mass.*) **44,** 941–947 (1986).
17. Rosen, C. A., Sodroski, J. G., Goh, W. C., Dayton, A. I., Lippke, J., and Haseltine, W. A., Post-transcriptional regulation accounts for the trans-activation of the human T-lymphotropic virus type III. *Nature* (*London*) **319,** 555–559 (1986).
18. Okamoto, T., and Wong-Staal, F., Demonstration of virus-specific transcriptional activator(s) in cells infected with human T-cell lymphotropic virus type III by *in vitro* cell-free system. *Cell* (*Cambridge, Mass.*) **47,** 29–35 (1986).
19. Starcich, B. R., Hahn, B. H., Shaw, G. M., McNeely, P. D., Modrow, S., Wolf, H., Parks, E. S., Parks, W. P., Josephs, S. F., Gallo, R. C., and Wong-Staal, F., Identification and characterization of conserved and variable region in the envelope gene of HTLV-III/LAV, The retrovirus of AIDS. *Cell* (*Cambridge, Mass.*) **45,** 637–648 (1986).
20. Hahn, B. H., Shaw, G. M., Taylor, M. E., Redfield, R. R., Markhan, P. D., Salahuddin, S. Z., Wong-Staal, F., Gallo, R. C., and Parks, W. P., Genetic variation in HTLV-III/LAV over time in patients with AIDS or at risk for AIDS. *Science* **232,** 1548–1553 (1986).
21. Modrow, S., Hahn, B. H., Shaw, G. M., Gallo, R. C., Wong-Staal, F., and Wolf, H., Computer assisted analysis of the envelope proteins sequences of seven HTLV-III/LAV isolates: Prediction of antigenic epitopes in conserved and variable regions. *J. Virol.* **61,** 570–578 (1987).

IV

RNA in DNA Replication

14

Changes in RNA Secondary Structure May Mediate the Regulation of IncFII Plasmid Gene Expression and DNA Replication

DAVID D. WOMBLE, XINNIAN DONG, AND ROBERT H. ROWND

Department of Molecular Biology
The Medical and Dental Schools
Northwestern University
Chicago, Illinois 60611

I. INTRODUCTION

Plasmid incompatibility, which is the inability of closely related plasmids to be maintained together in the descendants of a single bacterial cell, is a result of the mechanisms that regulate plasmid DNA replication and segregation at cell division (Scott, 1984). Plasmid NR1 (Rownd *et al.,* 1966) belongs to the FII incompatibility group, which also includes plasmids R1 and R6 (Datta, 1975). These multiple-antibiotic resistance plasmids have sizes near 90 kilobase pairs (kb) and are self-transmissible by conjugation. Plasmid NR1 has a low copy number of one to two per chromosome (Rownd and Womble, 1978; Rownd *et al.,* 1985; Womble *et al.,* 1977), yet is maintained stably even in the absence of selection (Miki *et al.,* 1980). The processes in the regulation of replication of plasmid NR1 and its IncFII relatives have been studied extensively both *in vivo* and *in vitro* (for reviews, see Nordstrom *et al.,* 1984; Scott, 1984; Timmis *et al.,*

Molecular Biology of RNA
New Perspectives

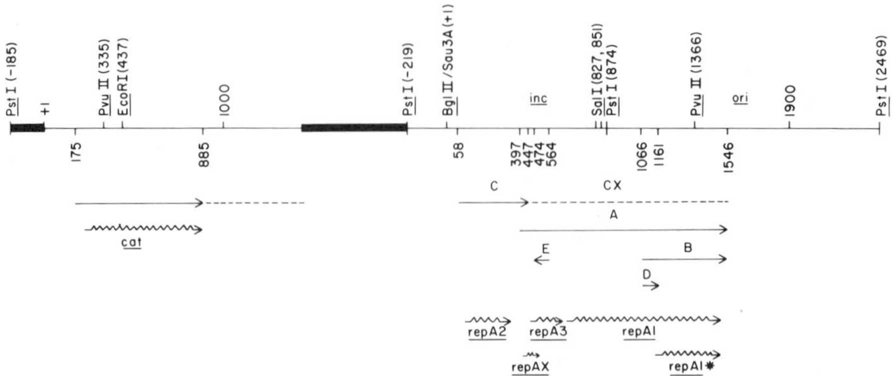

Fig. 1. Map of miniplasmid pRR949. pRR949 was constructed *in vitro* by ligation of three *Pst*I restriction fragments from derivatives of NR1 (Miki *et al.*, 1980). The 2.2-kb fragment coding for chloramphenicol acetyltransferase (Cat) is derived from the r-determinants portion of NR1 (Miki *et al.*, 1978, 1980). The *cat* gene is bounded by partial IS1 insertion elements (heavy lines) and is essentially homologous to a portion of the transposable element Tn*9* (Miki *et al.*, 1980; Shaw, 1983). Representative restriction sites along with base numbers from the nucleotide sequences of *cat* (Fig. 2) and the replication control region (Fig. 3) are shown above the line. The sites coding for replication control and incompatibility (*inc*) and the origin of replication (*ori*) are also shown. The numbers below the line refer to the estimated start and stop sites for the six transcripts produced *in vitro* (Easton *et al.*, 1981; Machida *et al.*, 1983; Rosen *et al.*, 1981; Womble *et al.*, 1985a) which are represented by the arrows. The five transcripts from the replication control region are designated A–E (Easton *et al.*, 1981). The protein coding regions of the transcripts are designated by the positions of the sawtooth arrows. The coding regions for *cat* (Alton and Vapnek, 1979), *repA1, repA2, repA3* (Rosen *et al.*, 1980), *repAX,* and *repA1** (Womble *et al.*, 1985a) were previously identified from the DNA sequences.

1981). The NR1 replication control system is particularly interesting because it involves a diversity of macromolecular interactions, including protein–protein, protein–nucleic acid, and nucleic acid–nucleic acid interactions. This information on the NR1 regulatory processes recently has been incorporated into a quantitative model for control of its replication in the bacterial cell division cycle (Womble and Rownd, 1986).

A map of an NR1-derived miniplasmid is illustrated in Fig. 1. This miniplasmid, pRR949, is composed of a 2.7-kb segment from the NR1 replication control region joined to a 2.2-kb segment that contains the gene for resistance to chloramphenicol, *cat*. The 2.7-kb segment from the NR1 replication control region contains all of the genes and DNA sites normally involved in the specific control of NR1 replication (Miki *et al.*, 1980; Rosen *et al.*, 1980; Taylor and Cohen, 1979). DNA synthesis begins in the origin region (*ori*) and proceeds unidirectionally around the plasmid

(Masai *et al.*, 1983; Ohtusubo *et al.*, 1977). Initiation of *in vivo* DNA replication at the NR1 *ori* requires the *de novo* synthesis of the cis-acting 33-kDa protein product of the *repA1* gene (Miki *et al.*, 1980; Morris *et al.*, 1974; Rosen *et al.*, 1980; Taylor and Cohen, 1979, Womble and Rownd, 1979). Replication of plasmid NR1 is controlled by regulating the amount of synthesis of the RepA1 initiator protein (Easton *et al.*, 1981; Easton and Rownd, 1982; Light and Molin, 1983).

In vivo transcription of the *repA1* initiator gene of NR1 begins at two different promoter sites, whose locations were determined from *in vitro* transcription experiments (Easton *et al.*, 1981; Rosen *et al.*, 1981; Womble *et al.*, 1985a). The *repA1* mRNAs are referred to as RNA-CX and RNA-A (Fig. 1). RNA-CX includes the coding sequences of both the *repA2* and *repA1* genes (Rownd *et al.*, 1985; Womble *et al.*, 1985a,b). We have sometimes referred to the *repA2* portion of this transcript as RNA-C and to the portion that extends into *repA1* as RNA-CX (Womble and Rownd, 1986). Transcription *in vivo* from the RNA-CX promoter is weak and constitutive whereas transcription from the RNA-A promoter is of intermediate strength but is regulated (Dong *et al.*, 1985; Rownd *et al.*, 1985; Womble *et al.*, 1985a). As a result of interference from convergent transcription in the *inc* region (see below), not all of the transcripts initiated at the RNA-CX promoter continue through the *repA1* gene. This interference is eliminated when convergent transcription is prevented by deletion or mutation of the opposing promoter (Womble *et al.*, 1984, 1985a).

Transcription from the RNA-A promoter is repressed by the trans-acting 10-kDa protein product of the *repA2* gene (Dong *et al.*, 1985; Light and Molin, 1982b; Liu *et al.*, 1983). The active form of the RepA2 transcription repressor is likely to be a tetramer that is formed cooperatively from inactive monomers (Dong *et al.*, 1985). The tetramers presumably bind to an operator site in the RNA-A promoter. Owing to the constitutive transcription of the *repA2* gene, the total amount of RepA2 protein synthesized in a cell is gene dosage dependent (Light and Molin, 1982a). Therefore, the amount of transcription from the RNA-A promoter is inversely related to the plasmid copy number, and the total amount of mRNA for the *repA1* initiator gene is the sum of the constitutively synthesized RNA-CX and the regulated RNA-A (Fig. 1). Examination of the nucleotide sequence of the NR1 replication region (Rosen *et al.*, 1980) reveals three additional coding sequences (Fig. 1) that could be translated from RNA-CX and RNA-A: *repAX, repA3,* and *repA1** (Womble *et al.*, 1985a). The functions, if any, of these additional hypothetical proteins are not presently understood.

The incompatibility (*inc*) region also is transcribed constitutively in the direction opposite to that of RNA-CX and RNA-A from a strong third promoter site (Fig. 1) (Easton *et al.,* 1981; Rosen *et al.,* 1981; Rownd *et al.,* 1985; Womble *et al.,* 1985a). The 91-base untranslated transcription product, referred to as RNA-E, is a trans-acting inhibitor of translation of the RNA-CX and RNA-A *repA1* mRNAs (Light and Molin, 1983; Womble *et al.,* 1984, 1985b). The inhibition of translation probably results from formation of an RNA–RNA duplex between the complementary transcripts, which may cause the mRNA to form an intramolecular secondary structure that is unfavorable for initiation of RepA1 translation (Rownd *et al.,* 1985). Examination of mutants with base-pair changes in the *inc* region (Brady *et al.,* 1983; Givskov and Molin, 1984; Rosen *et al.,* 1981; Rownd *et al.,* 1985; Ryder *et al.,* 1981) has led to the hypothesis that the initial, rate-determining step in the formation of the RNA–RNA duplex is a pairing of six-base, complementary regions of each transcript (Rownd *et al.,* 1985). Base-pairing of these regions, predicted to be in single-stranded loops of stem-loop structures (Rosen *et al.,* 1980, 1981), would provide the free energy for this initial interaction. Base substitutions that would change the base-stacking free energy for this interaction have resulted in mutants with alterations in plasmid copy number control. Because *in vivo* transcription from the RNA-E promoter is constitutive, the amount of inhibitor RNA-E present in a cell is a function of the plasmid copy number (Rownd *et al.,* 1985; Womble *et al.,* 1985a). RNA-E can be titrated in trans by synthesis of the portion of *repA1* mRNA that contains the complementary target sequence for binding to RNA-E (Light and Molin, 1983; Womble *et al.,* 1984, 1985b).

The function of the RepA1 initiator protein has not yet been determined. However, it is primarily cis-acting both *in vivo* (Miki *et al.,* 1980) and *in vitro* (Masai *et al.,* 1983). It is possible that there is normally a one-to-one correspondence between the translation of a *repA1* mRNA transcript and the initiation of replication at the *ori* of the same plasmid that served as the template for that mRNA transcript. Therefore, our model for control of NR1 replication considers the act of translation of *repA1* mRNA to be the trigger for the initiation of replication at the NR1 *ori*. The frequency with which that occurs is a product of the rate of *repA1* mRNA transcription and the fraction of *repA1* mRNA transcripts that escape interaction with RNA-E and therefore are allowed to be translated. We have used the information on the rates of *in vivo* transcription and translation of *repA1* mRNA to formulate a quantitative model that is based on statistical thermodynamics for control of IncFII plasmid replication during the bacterial cell division cycle (Womble and Rownd, 1986). In the present communication, we examine the possible roles of RNA secondary structure in the regulation of expression of the genes in the NR1 replication control region from miniplasmids such as pRR949 (Fig. 1). The computer predictions of the most stable RNA structures suggest an interesting mechanism of regulated gene expression that results in control of plasmid DNA replication.

II. RNA SECONDARY STRUCTURE PREDICTIONS

A. Description of the Transcripts

In vitro transcription experiments with miniplasmids such as pRR949 (Fig. 1) have identified six RNA products (Easton *et al.*, 1981; Machida *et al.*, 1983; Rosen *et al.*, 1981; Womble *et al.*, 1985a). The locations of the sites of transcription initiation and the directions of RNA synthesis are shown in Fig. 1 (Rownd *et al.*, 1980; Easton *et al.*, 1981; Rosen *et al.*, 1981; Lurz *et al.*, 1981; Chan and Lebowitz, 1982; Machida *et al.*, 1983; Womble *et al.*, 1985a). Only RNA-E is synthesized in the leftward direction, and it is not translated (Easton *et al.*, 1981; Rosen *et al.*, 1981; Stougaard *et al.*, 1981a; Womble *et al.*, 1985a). The proteins that could be coded by the rightward transcripts were deduced from the DNA sequences (Alton and Vapnek, 1979; Rosen *et al.*, 1980; Womble *et al.*, 1985a) and are shown in Fig. 1. These include the five RepA proteins that may be involved in replication control, RepA1 (284 amino acids, 33 kDa), RepA1* (121 amino acids, 14 kDa), RepA2 (84 amino acids, 10 kDa), RepA3 (61 amino acids, 7 kDa), and RepAX (26 amino acids, 3 kDa) and the Cat protein (219 amino acids, 25 kDa) that inactivates chloramphenicol. All five RepA proteins could be translated from a single mRNA, RNA-CX, whose termination site is likely to be near the plasmid origin of replication (Easton *et al.*, 1981; Womble *et al.*, 1985a). RNA-A encodes all of the RepA proteins with the exception of RepA2, and may be terminated at the same site as RNA-CX. A majority of the *in vivo cat* transcripts continue past their *in vitro* termination site (Womble *et al.*, 1985a). However, in pRR949 there is a transcription terminator located to the left of the RNA-CX promoter that prevents the extended *cat* transcripts from reading through the replication control region (Womble *et al.*, 1985a). RNAs B and D are not synthesized detectably *in vivo* (Womble *et al.*, 1985a).

1. Sequence of the cat Gene

The sequence of 1000 nucleotides from the *cat* coding region of pRR949 is presented in Fig. 2 (Alton and Vapnek, 1979). The transcript begins with the G at position number 175, which is preceded by −35 and −10

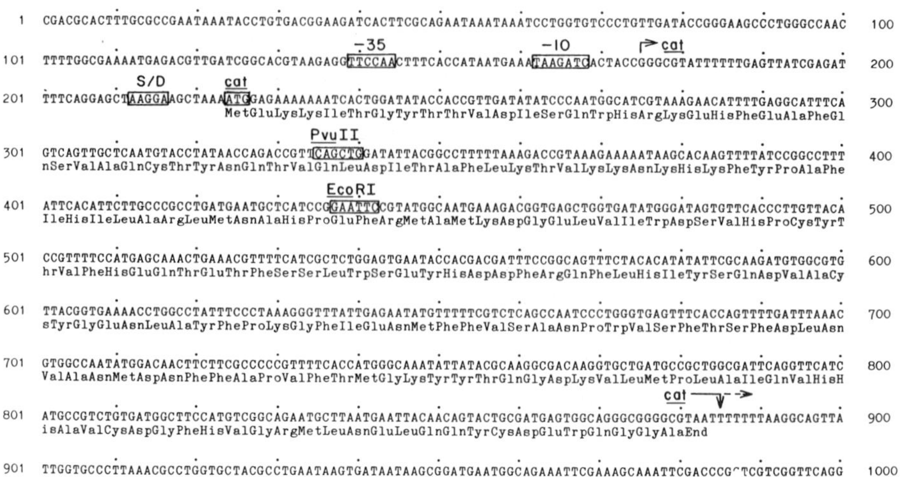

Fig. 2. The DNA sequence of the *cat* gene. The DNA sequence of the *cat* gene of Tn*9* (Alton and Vapnek, 1979), which is homologous to the *cat* gene of NR1 (Miki *et al.*, 1978, 1980; Shaw, 1983), is shown. The numbering corresponds to the left-hand portion of Fig. 1. Sequences for restriction enzyme sites, −35 and −10 transcription promoters (Reznikoff and Abelson, 1978; Rosenberg and Court, 1979), Shine–Dalgarno ribosome binding sites (S/D) (Shine and Dalgarno, 1974) and translation start codons are boxed. The estimated *in vitro* transcription start and stop sites are indicated by the solid arrows. The dashed arrow indicates transcription readthrough *in vivo*. The amino acid sequence is that predicted for Cat from the DNA sequence.

promoter sequences. This is a high-level constitutive transcription promoter both *in vitro* and *in vivo* (Easton *et al.*, 1981; Machida *et al.*, 1983; Womble *et al.*, 1985a). A Shine–Dalgarno ribosome binding sequence (S/D) precedes the *cat* AUG translation start codon. The translation termination codon, UAA, is located near the *in vitro* transcription termination site for RNA-*cat*. This site, near base 885, is preceded by a GC-rich sequence (13 of 14) and a U-rich sequence (8 of 10). There are no other transcription promoters active *in vitro* or *in vivo* to the right of the *Eco*RI site in the *cat* gene (Easton *et al.*, 1981; Machida *et al.*, 1983; Womble *et al.*, 1985a).

2. *Sequence of the NR1 Replication Control Region*

The sequence of 1900 bases from the NR1 replication control region is shown in Fig. 3 (Rosen *et al.*, 1980; Womble *et al.*, 1985a). Four promoters and transcription start sites are identified. The estimated start site for RNA-CX is the A at position number 58. The RNA-CX promoter is expressed constitutively at a low level (Easton *et al.*, 1981; Rownd *et al.*,

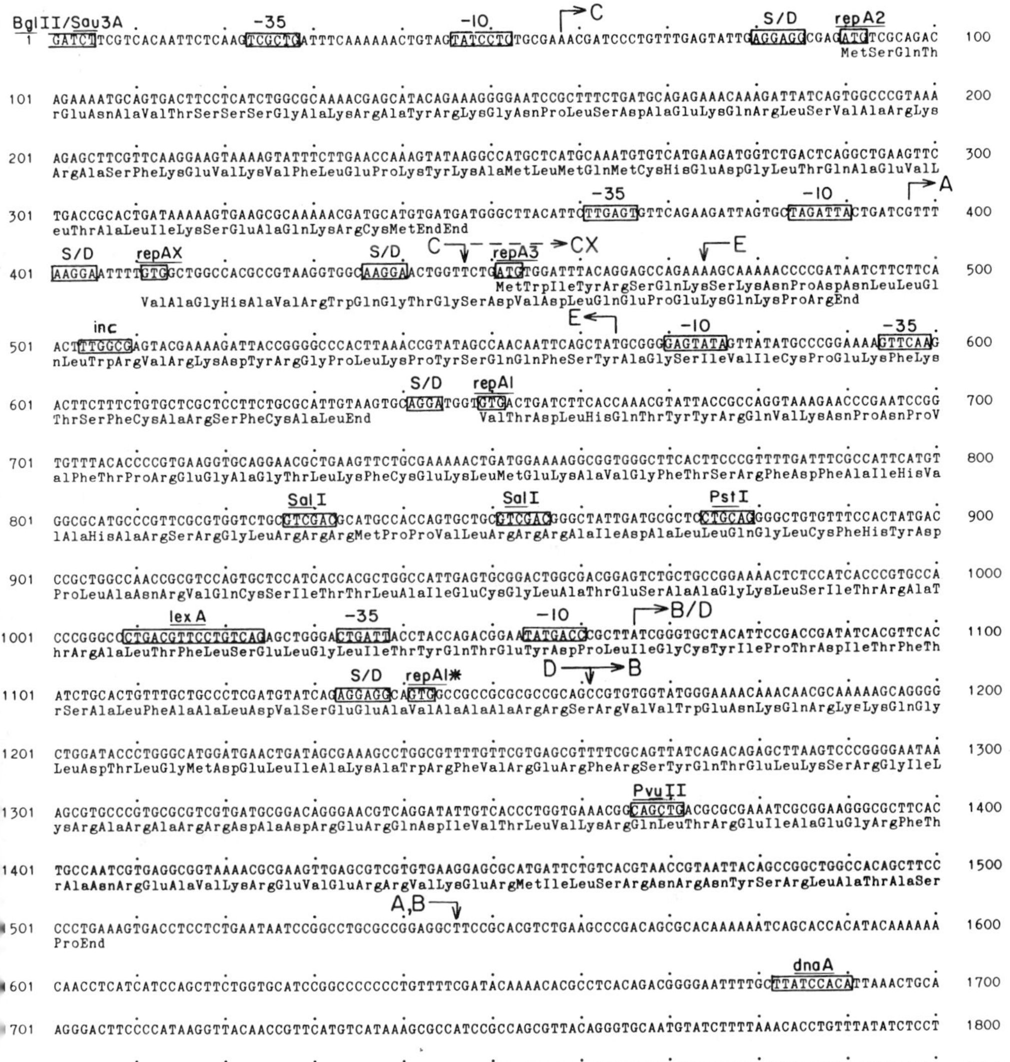

Fig. 3. The DNA sequence of the NR1 replication control region. The DNA sequence of Rosen *et al.* (1980) corrected to represent that of wild-type NR1 (Ryder *et al.*, 1981; Womble *et al.*, 1985a) is shown. The numbering corresponds to the right-hand portion of Fig. 1. Boxed sequences are as in Fig. 2, with the addition of *inc*, the incompatibility target sequence (see text), *lexA*, a possible binding site for the *E. coli* LexA repressor protein (Little and Mount, 1982), and *dnaA*, a possible binding site for the *E. coli* DnaA protein (Fuller and Kornberg, 1983) which is located near the NR1 replication origin (Masai *et al.*, 1983). The amino acid sequences are those predicted from the DNA sequence for the five RepA proteins (Fig. 1).

1985). A C → T transition at base 24 results in a 10-fold up promoter mutation (Rownd *et al.*, 1985; Womble *et al.*, 1985b). RNA-A begins at the G at position number 397. The RNA-A promoter is of intermediate strength, but is repressed *in vivo* by the RepA2 protein (Dong *et al.*, 1985; Easton *et al.*, 1981; Rownd *et al.*, 1985). RNA-E begins at the A at position number 564′ and is transcribed in the leftward direction. The RNA-E promoter is a high-level constitutive promoter. A G → A transition at base 597′ results in a 100-fold down promoter mutation (Rownd *et al.*, 1985; Womble *et al.*, 1984; 1985a). The estimated start site for RNA-B and RNA-D is the A at position number 1066, which is preceded by a −35 promoter sequence at base 1033 and a −10 sequence at base 1054. The sequence around the −35 region resembles the *lexA* repressor binding site in front of the *lexA* gene promoter of *Escherichia coli* (Miki *et al.*, 1981; Little and Mount, 1982). The consensus sequence for *lexA* binding sites, 5′-CTG$(X)_{10}$CAG-3′, is present at base 1009, and a second sequence match that is one base short is at base 1033, 5′-CTG$(X)_9$CAG-3′, overlapping the −35 promoter sequence.

The DNA sequence immediately upstream from the *Bgl*II site in the replication control region (Fig. 1) contains no translation start codons but has a total of 10 translation stop condons distributed among the three reading frames and a GC-rich region (13 of 14) starting at base −97 (not shown). This is a likely location for the transcription terminator mentioned above. Deletion of this region from pRR949 allows the extended *cat* mRNA to read through the NR1 replication region *in vivo,* causing an increase in plasmid copy number (Womble *et al.*, 1985a).

The origin of replication has been estimated, by electron microscopy and deletion mapping, to lie between bases 1693 and 1843 (Ohtsubo *et al.*, 1977, 1986; Masai *et al.*, 1983). Although NR1 replication does not require functional *dnaA* protein (Womble and Rownd, 1979), there is a perfect *dnaA* binding site (Fuller and Kornberg, 1983) in the DNA sequence at base 1682.

B. Strategies for RNA Secondary Structure Prediction

Using the nucleotide sequences from Fig. 2 and 3, the secondary structures of each of the transcripts from miniplasmid pRR949 were predicted by a computer program (Zucker and Stiegler, 1981) that uses the base-stacking and destabilizing rules of Salser (1977). The graphics plotting program of Luckow *et al.* (1984) was used to display the predicted structures. The program calculates the lowest energy "global fold" of any RNA sequence. Given the 5′ end of a transcript, predicted structures through various 3′ ends can be examined. Two strategies can be used for predicting the folded structures of large RNAs: sequential local folding or global folding. The first finds the most stable structure of each small domain, perhaps 50–100 bases at a time, while the second finds the most stable structure for the entire sequence. For many RNA sequences, both methods will predict essentially the same overall structure (see below). The folded structures near the translation start sites and transcription stop sites of pRR949 were examined. For predicting the interior folded structures of polycistronic mRNA, the upstream portions of the transcript that were expected to be translated were assumed to be withdrawn from other secondary structure interactions, and folding was continued from the stop codon onward. In Figs. 4–8, the folded structures of various portions of the transcripts from pRR949 are presented with the 5′ → 3′ direction as clockwise around the sequence. The nucleotide numbers above the folded structures refer to the positions in the DNA sequence of the *cat* gene (Fig. 2) or the replication control region (Fig. 3). Numbers from the DNA sequences complementary to those shown in Figs. 2 and 3 are denoted with a prime symbol, such as 565′.

1. The Structure of cat **mRNA**

Starting with the 5′ G of *cat* mRNA at position number 175 (Fig. 2), the base sequence was folded through various 3′ ends from base numbers 217–344. When the 3′ end reaches number 230, there is a stable secondary structure predicted with the S/D (5′-AAGGA-3′) for *cat* present in a seven-base single-stranded loop (Fig. 4A). The presence of this loop is predicted for all folds of *cat* mRNA through base 344. The AUG start codon is partially double-stranded in Fig. 4A, but for *cat* transcripts extending beyond number 280, the AUG start codon is also predicted to be single-stranded (not shown). Folds that begin with the A at position number 172 are not different from those starting at position 175. The ribosome recognition sequences of *cat* mRNA therefore are predicted to be readily available for ribosome binding, which would result in efficient translation. The *cat* gene is known to be expressed constitutively at a high level in *Escherichia coli* (Shaw, 1975; 1983).

Sequences around the 3′ end of the *in vitro cat* transcript were also examined. There are no GC-rich stem-loop structures often associated with transcription termination sites (Rosenberg and Court, 1979) between bases 780 and 980 (not shown). The structure predicted for the *in vitro* 3′ end is shown in Fig. 4B. There is a GC-rich loop followed by a U-rich stretch; the UAA translation stop codon for Cat separates these two regions. Translation of the *cat* message *in vivo* therefore would disrupt this structure, which may explain why RNA-*cat* is extended beyond this position *in vivo* (Womble *et al.*, 1985a).

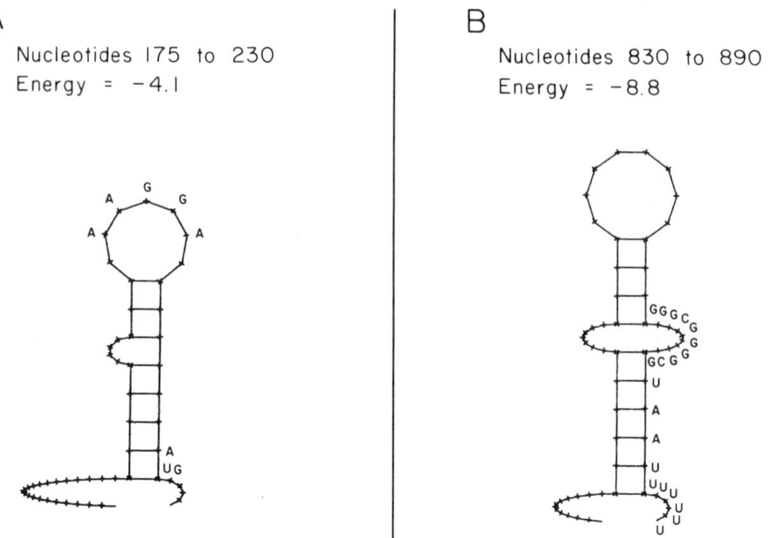

Fig. 4. The structure of RNA-*cat*. (A) The 5′ end of RNA-*cat*, nucleotides 175–230 (Fig. 2). The nucleotides in the S/D (5′-AAGGA-3′) and translation start codon (AUG) are labeled. (B) The 3′ end of the *in vitro cat* transcript. GC-rich and U-rich segments are labeled, including the UAA translation stop codon of *cat*.

2. The Structure of the 5′ Region of RNA-CX

The predicted structure of the 5′ end of RNA-CX is shown in Fig. 5A, for bases 58–112 (Fig. 3). The S/D (5′-AGGAGG-3′) and AUG start codon for the *repA2* coding sequence are both single-stranded. This would allow efficient ribosome recognition and translation, in agreement with the observed levels of RepA2 synthesis *in vivo* and *in vitro* (Brawner and Jaskunas, 1982; Light and Molin, 1982a). The six-base loop containing the AUG codon is predicted for all folds of RNA-CX through base 257.

A possible structure for RNA-CX near the *repAX* start site is shown in Fig. 5B. In the 5′ region of this structure are located a single-stranded

Fig. 5. The structures of RNA-CX and RNA-E. (A) The 5′ end of RNA-CX, nucleotides 58–112 (Fig. 3). The nucleotides in the S/D (5′-AGGAGG-3′) and translation start codon (AUG) for *repA2* are labeled. (B) Midsection of RNA-CX. The nucleotides in the S/D (5′-AAGGA-3′) and translation start codon (GUG) for *repAX* are labeled. (C) The structure of RNA-CX downstream from the *repAX* coding region (Figs. 1 and 3). The nucleotides in the incompatibility target loop (5′-UUGGCG-3′) are labeled. Reference points at nucleotides 524 and 564 are indicated. (D) The complete structure of RNA-E, nucleotides 564′–474′. The nucleotides in the inhibitor loop (5′-CGCCAA-3′) are labeled.

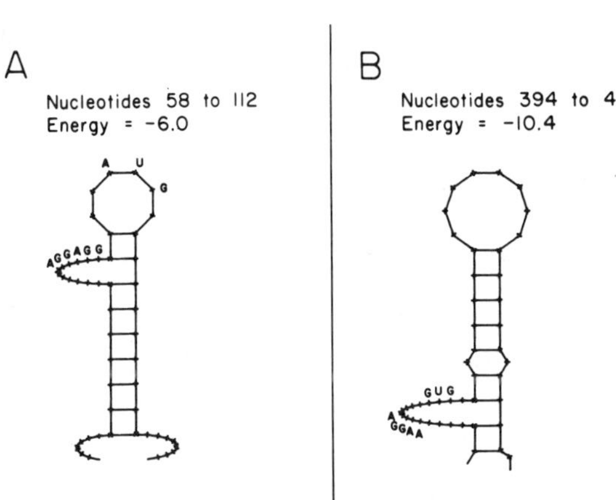
A
Nucleotides 58 to 112
Energy = -6.0
A U
G
AGGAGG
B
Nucleotides 394 to 442
Energy = -10.4
GUG
AGGAA

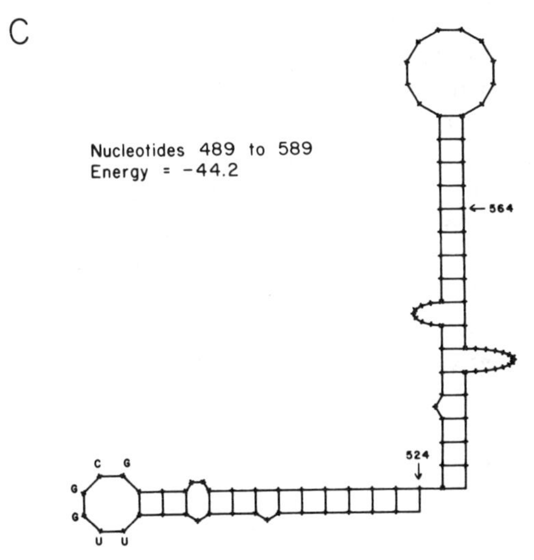
C
Nucleotides 489 to 589
Energy = -44.2
564
524
C G
G
G
U U

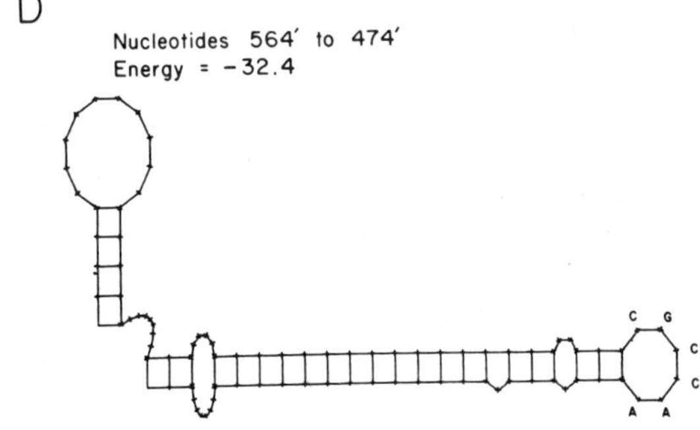
D
Nucleotides 564′ to 474′
Energy = -32.4
C G
C
C
A A

S/D (5′-AAGGA-3′) and GUG start codon for the short 26-amino-acid *repAX* reading frame (Figs. 1 and 3). Translation of *repAX* would remove bases of RNA-CX up to number 488 from further folding interactions. Starting at base 489 and continuing through 589, two folded domains are predicted by both sequential local and global schemes (Fig. 5C). Bases 489–524 form the first domain with a stability of −7.9 kcal, and bases 525–589 form the second domain with a stability of −36.3 kcal. The second, very stable domain is predicted for all folds of RNA-CX extending through base 717 (not shown). The first domain has a six-base loop, 5′-UUGGCG-3′, which is the incompatibility target sequence (Light and Molin, 1983; Womble *et al.*, 1984). That is, the single-stranded loop is the likely site of initial interaction between the *repA1* mRNA and the inhibitor, RNA-E. RNA-E has a complementary loop in its predicted structure, 5′-CGCCAA-3′ (Fig. 5D). The computer-predicted secondary structure of RNA-E has been confirmed experimentally by determination of the nuclease sensitivity of the various single- and double-stranded regions of the RNA molecule (Wagner and Nordstrom, 1986). Substitutions of the G or C residues in the single-stranded loop by A or U result in mutant replication phenotypes: higher copy numbers per cell and a change in incompatibility (Miki *et al.*, 1980; Stougaard *et al.*, 1981b; Easton and Rownd, 1982; Brady *et al.*, 1983; Givskov and Molin, 1984). The mutant plasmids are no longer incompatible with wild-type derivatives, but each is incompatible, although weakly, with its own derivatives (Miki *et al.*, 1980; Easton and Rownd, 1982; Brady *et al.*, 1983). These phenotypes are predicted precisely by the types of pairing allowed between the six-base single-stranded loops of RNA-CX and RNA-E. The complementary wild-type/wild-type pairing is strongest, the complementary mutant/mutant pairing is weaker, and either kind of mutant/wild-type pairing, with a sequence mismatch, is weaker yet. Pairing between sequences in the six-base loops is proposed to be the first step in the mechanism by which RNA-E inhibits translation of *repA1* mRNA (Rownd *et al.*, 1985; Womble *et al.*, 1984). This mechanism will be explored further below.

3. *The Structure of the 5′ Region of RNA-A*

Beginning at base number 397, RNA-A has the same nucleotide sequence as the corresponding part of RNA-CX (Fig. 3). However, lacking the sequences upstream from base 397, RNA-A might form secondary structures different from those of RNA-CX. The predicted structure for the 5′ end of RNA-A is shown in Fig. 6A. The S/D (5′-AAGGA-3′) and GUG start codon for the *repAX* peptide are in single-stranded regions, similar to the case for RNA-CX shown in Fig. 5B. In addition, the S/D (5′-AAGGA-3′) and AUG start codon for the *repA3* protein are also found in

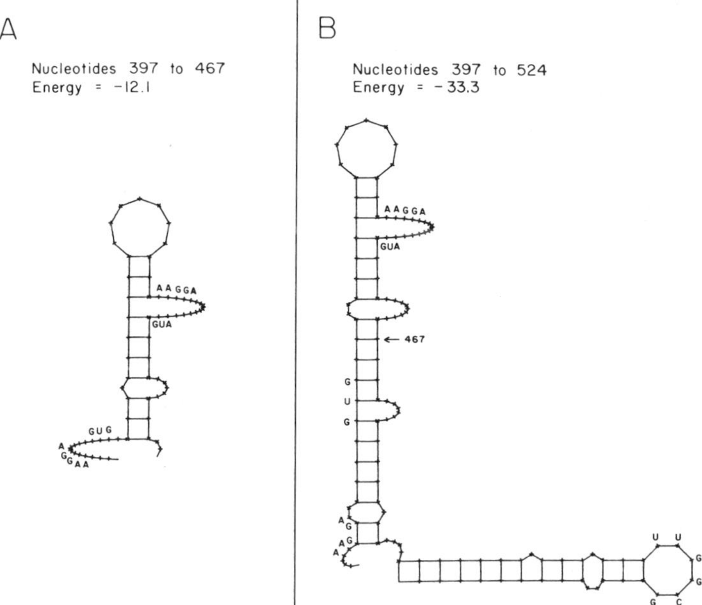

Fig. 6. The structure of RNA-A. (A) The 5′ end of RNA-A, nucleotides 397–467 (Fig. 3). The nucleotides in the S/D and translation start codons for *repAX* (5′-AAGGA-3′, GUG) and *repA3* (5′-AAGGA-3′, AUG) are labeled. (B) The structure of the 5′ end of RNA-A extended through nucleotide 524. The nucleotides in the S/D and translation start codons for *repAX* and *repA3* and in the incompatibility target loop (5′-UUGGCG-3′) are labeled.

a single-stranded region. Three possibilities exist for further folding of RNA-A. The first is the possibility for translation of *repAX,* in which case the structure for the rest of RNA-A would be identical to that of RNA-CX from base 489 on (Fig. 5C). The second possibility is for translation of *repA3*. The termination codon for *repA3* is immediately adjacent to the S/D for *repA1* (Figs. 1 and 3), providing for a possible intrastrand transfer of a 30 S ribosomal subunit from the *repA3* cistron to that for *repA1* (Schumperli *et al.,* 1982). The third possibility is for translation of neither *repAX* nor *repA3*. In that case, extension of RNA-A to base 524 predicts the formation of the incompatibility target domain (Fig. 6B), and further folding should be identical to that of RNA-CX. Formation of the incompatibility target domain would allow RNA-A to interact with the RNA-E inhibitor in a fashion similar to that described above for RNA-CX. However, the structure in Fig. 6B also predicts that translation of *repA3* should still be a possibility for RNA-A, because of the single-stranded S/D and AUG codon.

4. The Structure of repA1 mRNA

The predicted structure of *repA1* mRNA (RNA-CX and RNA-A) from bases 489 to 589 is to form initially the two domains shown in Fig. 5C. Further growth of the transcript to base 614 would not add any more secondary structure, with the 25 bases on the 3′ end remaining unpaired (Fig. 7A). With the addition of four more bases, to number 618, two alternate folding pathways are possible. The first is for the 29 bases on the 3′ end to fold independently, with a weak stability of −0.8 kcal (not shown). The second possibility is for the 3′ end to interact with the 5′ region, through the incompatibility target domain, to form a very stable structure with a total stability of −54.0 kcal (Fig. 7B). This second possibility represents a net gain of −9.0 kcal compared with the first, and therefore would be highly favored. It also would cause disassembly of the incompatibility target domain, decreasing the possibility of interaction with the RNA-E incompatibility inhibitor.

Growth of *repA1* mRNA to base 682 would allow formation of another domain, which includes the S/D (5′-AGGA-3′) and GUG start codon for *repA1* in single-stranded regions (Fig. 7C). This domain is predicted by both sequential local and global schemes, and has a stability of −19.9 kcal. This configuration should allow ribosome recognition at the single-stranded *repA1* S/D, which would result in translation of the *repA1* mRNA. Synthesis of RepA1 protein would then mediate replication initiation at the plasmid origin (Miki *et al.*, 1980; Ryder *et al.*, 1981; Masai *et al.*, 1983; Rownd *et al.*, 1985).

The incompatibility target domain of *repA1* mRNA (bases 489–524, Fig. 5C) is formed 94 bases prior to the possibility of the intramolecular 3′ → 5′ interaction described above (Fig. 7B). If the target loop of the mRNA had already paired with the complementary loop in RNA-E (Fig. 5D), then the 3′ end of RNA-CX would be prevented from interacting with this domain and would be forced to fold independently. All predicted structures from this pathway have both the S/D and GUG start codon for *repA1* paired in intramolecular double-stranded regions. One such possibility is for the *repA1* mRNA and RNA-E to have formed a complete 91-base-pair duplex through base 564. The predicted structure of the remain-

---

Fig. 7. The structure of *repA1* mRNA. (A) The structure of RNA-CX extended through nucleotide 614 (Fig. 3). The nucleotides in the incompatibility target loop (5′-UUGGCG-3′) are labeled. (B) The structure of RNA-CX extended through nucleotide 618. (C) The structure of RNA-CX extended through nucleotide 682. The nucleotides in the S/D (5′-AGGA-3′) and translation start codon (GUG) for *repA1* are labeled. (D) The structure of RNA-CX downstream from the region duplexed with RNA-E, nucleotides 565–682. The nucleotides in the S/D and translation start codon for *repA1* are labeled.

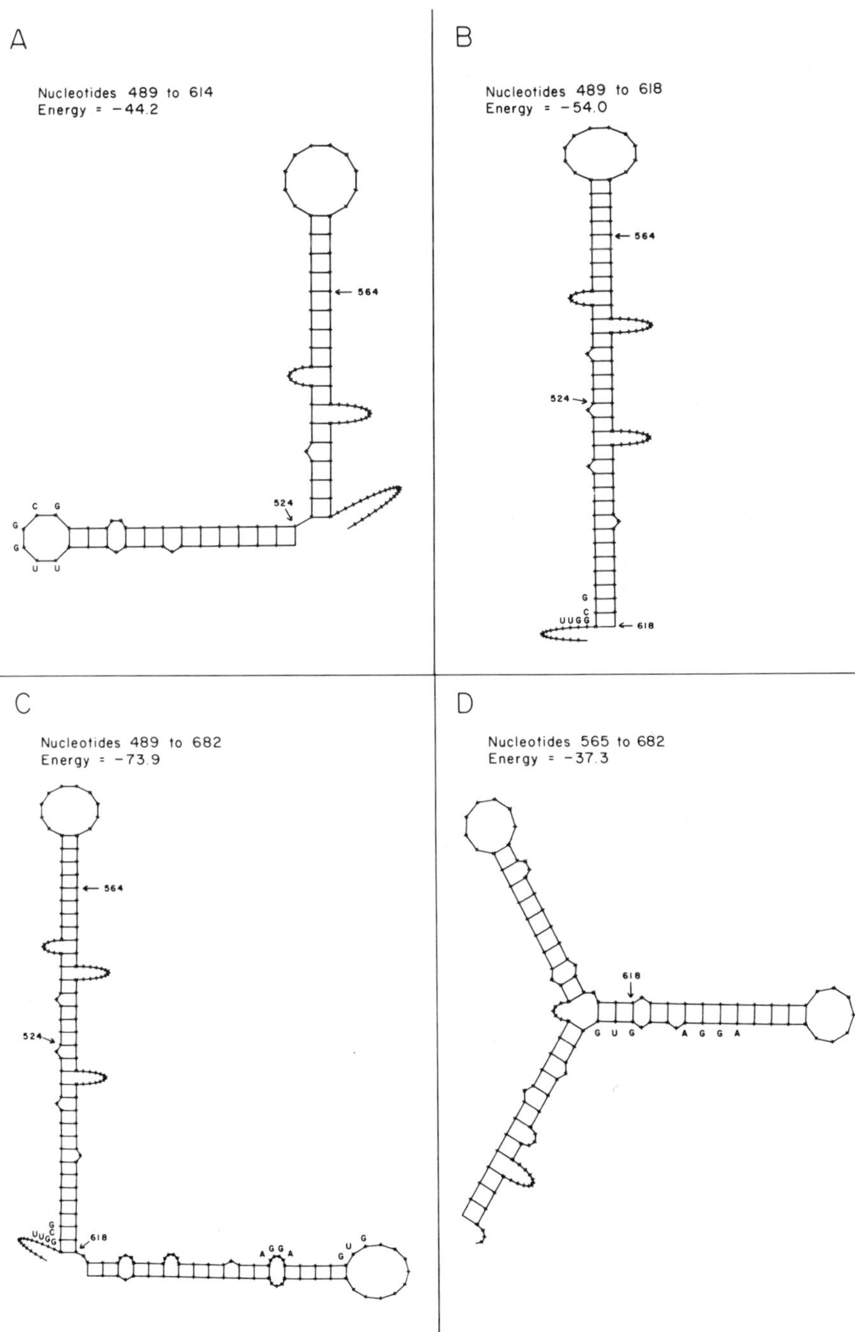
A
Nucleotides 489 to 614
Energy = −44.2
564
524
C G
G
G
U U
B
Nucleotides 489 to 618
Energy = −54.0
564
524
G
C
UUGG
618
C
Nucleotides 489 to 682
Energy = −73.9
564
524
G
C
UUGG
618
A G G A
G U G
D
Nucleotides 565 to 682
Energy = −37.3
618
G U G
A G G A

der of the mRNA for bases 565–682 is shown in Fig. 7D. This structure would preclude interaction with the ribosome, preventing *repA1* mRNA translation and, therefore, the initiation of plasmid replication. All other degrees of pairing between RNA-CX and RNA-E that prevent the intramolecular structure in Fig. 7B from forming also have predicted structures similar to that in Fig. 7D. Formation of a complete 91-base-pair duplex after initial pairing between RNA-E and RNA-CX would not be essential to prevent *repA1* mRNA translation, but formation of such a duplex structure would be highly favored energetically.

5. *The Structure of RNA-B and RNA-D*

The estimated start site for RNA-B and RNA-D is the A at position number 1066 (Fig. 3). The predicted structure for RNA-D and, therefore, the 5′ end of RNA-B is shown in Fig. 8A. Immediately preceding the GC-rich sequence near the RNA-D termination site, there are a S/D (5′-AGGAGG-3′) and GUG start codon in phase with the *repA1* coding sequence. Should translation begin here, it would produce a protein composed of the carboxy-terminal portion of the *repA1* protein, called *repA1** in Fig. 1. This translation would disrupt the structure shown in Fig. 8A and might be expected to cause RNA-D to be extended as RNA-B. The S/D and GUG start codon for *repA1** are predicted to be double-stranded, suggesting that translation would be unlikely or inefficient. However, because RNA-B/D has no incompatibility target similar to RNA-CX or RNA-A, translation of RNA-B/D would be independent of regulation by RNA-E.

The predicted structure for a possible termination site of RNA-A and RNA-B is shown in Fig. 8B. There is a stable GC-rich stem-loop structure. This occurs 40 bases downstream from the UGA stop codon for *repA1* protein and is approximately 130 bases upstream from the region identified as the replication origin (Masai *et al.*, 1983; Ohtsubo *et al.*, 1986). There are no open reading frames for translation extending through this region that might interfere with transcription termination *in vivo,* but we have not investigated the *in vivo* termination of these transcripts.

C. Pausing during Transcription of *repA1* mRNA

Pausing by RNA polymerase during transcription of *repA1* mRNA might prolong the existence of the transient single-stranded six-base target structure (Fig. 5C) formed during synthesis. This could be an important regulatory feature of the interaction of the mRNA with the inhibitor RNA-E for the translational control of *repA1* expression. Northern blot analysis of *in vivo* transcripts from the NR1 replication control region has

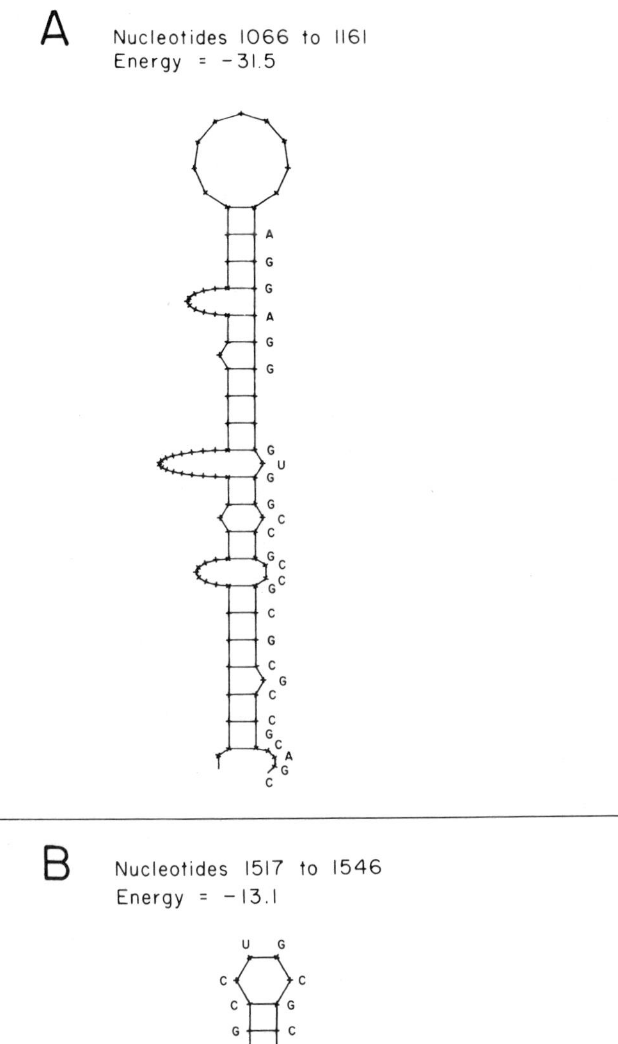

Fig. 8. The termination structures of RNA-A, RNA-B, and RNA-D. (A) The complete structure of *in vitro* transcript RNA-D, nucleotides 1066–1161 (Fig. 3). The nucleotides in the S/D (5′-AGGAGG-3′) and translation start codon (GUG) for *repA1** and the GC-rich segment are labeled. (B) The structure of the predicted 3′ end of transcripts RNA-A and RNA-B, nucleotides 1517–1546. The nucleotides in the GC-rich segment are labeled.

revealed *repA1* mRNA species that end at a position between the RNA-E target site in the *inc* gene and the ribosome binding site of the *repA1* gene (Dong *et al.*, 1987; Rownd *et al.*, 1986). *In vivo* transcripts from miniplasmid derivatives of NR1 were analyzed using labeled DNA probes that would hybridize to the transcripts of interest. In addition to the transcripts from the *cat* gene and RNA-E, which were observed for all miniplasmids, a transcript whose size was approximately 560 bases was also observed. The amount of the transcript was increased from a plasmid having a promoter up mutation for RNA-CX transcription, whereas the transcript was missing from a mutant deleted of that promoter. This suggests that the 560-base transcript is an intermediate of RNA-CX transcription. Another transcript of approximately 180 bases was observed only from the mutant deleted of the RNA-CX promoter. Because that mutant fails to produce the RepA2 repressor protein, the 180-base transcript may be an intermediate in the derepressed synthesis of RNA-A. Additional experiments have shown that the rate of transcription downstream from the 3′ position of these transcripts was not reduced, nor was the presence of these transcripts caused by convergent transcription in the *inc* gene or by interaction of the *repA1* mRNA with the inhibitor RNA-E (Dong *et al.*, 1987). This suggests that the observed transcript bands resulted from the pausing of RNA polymerase at this position rather than from termination. *In vitro* single-round transcription experiments have confirmed that RNA polymerase pauses during transcription of this region (Dong *et al.*, 1986). Taken together, these experiments strongly suggest that there is a pause in *repA1* mRNA synthesis just after the target site for the inhibitor RNA has been synthesized. This region, near base number 620, corresponds approximately to the transition point in the predicted refolding of *repA1* mRNA (Figs. 7A,B). The predicted secondary structure of *repA1* mRNA shown in Fig. 7A has a stable stem-loop structure, which is considered to be an important feature of transcription pause sites (Adhya and Gottesman, 1978; Kassavetis and Chamberlin, 1981; Rosenberg *et al.*, 1978).

III. DISCUSSION

There are six transcripts produced *in vitro* from a pRR949 miniplasmid template (Easton *et al.*, 1981; Womble *et al.*, 1985a). Analysis of the nucleotide sequences of the *cat* gene and replication control region of pRR949 (Alton and Vapnek, 1979; Rosen *et al.*, 1980) reveals six possible protein coding regions that could be translated from the *in vitro* transcripts. RepA1, RepA2, and chloramphenicol acetyltransferase have been identified as proteins synthesized *in vivo* (Rosen *et al.*, 1980; Shaw, 1975), whereas RepA3 is also synthesized by an *in vitro*–coupled transcription–translation system (Brawner and Jaskunas, 1982). We have examined the possible role of RNA secondary structure in regulating expression of the genes on pRR949. The secondary structures of the ribosome recognition sequences and translation start codons are expected to influence the efficiency of translation (Inserentant and Fiers, 1980; Gold *et al.*, 1981; Hall *et al.*, 1982). In addition, the act of translation of certain regions of a transcript might alter the predicted secondary structures for other portions of the molecule. This could result in antitermination of transcription (Yanofsky, 1981) or altered secondary structure of downstream ribosome recognition sequences (Horinouchi and Weisblum, 1982). Interaction of a mRNA with other molecules, either protein or RNA, could also affect its structure (Yates and Nomura, 1981; Winkler *et al.*, 1982), resulting in posttranscriptional regulation of gene expression.

The RepA1 protein of the IncFII plasmids is required for initiation of DNA replication at the plasmid origin (Ryder *et al.*, 1981; Masai *et al.*, 1983). The regulation of *repA1* expression is what controls NR1 plasmid replication. The *repA1* gene is regulated by control of both transcription and translation. Translation of the mRNA for *repA1*, either RNA-CX or RNA-A (Fig. 1), is inhibited by interaction with a small, RNA transcript made from the opposite DNA strand, RNA-E (Light and Molin, 1983; Rownd *et al.*, 1985; Womble *et al.*, 1984). The RNA secondary structure predictions suggest a possible mechanism by which this negative posttranscriptional regulation might occur.

Both the *cat* and *repA2* proteins are expressed constitutively and efficiently *in vivo* (Shaw, 1975; Light and Molin, 1982a). Examination of the 5′ ends of their mRNAs suggested that the ribosome recognition sequences of both mRNAs would be in single-stranded regions of the predicted secondary structures (Fig. 4A and 5A, respectively). Therefore, the predicted structures are consistent with known levels of expression of these proteins.

The *in vitro* transcription termination sites of pRR949 are preceded by GC-rich sequences. The *in vitro* termination site for RNA-*cat* approximately coincides with the stop codon UAA for *cat* translation (Fig. 1). Translation of RNA-*cat* would proceed through the GC-rich sequence and disrupt the possible transcription termination structure (Fig. 4B). This could explain the high level of extended RNA-*cat* synthesis observed *in vivo* (Womble *et al.*, 1985a).

The NR1 sequences to the left of the *Bgl*II site (Fig. 1) contain a transcription terminator. This could serve to insulate the replication region from outside transcription. The pRR949 miniplasmid has the same low copy number as its 90-kb parent, NR1 (Easton and Rownd, 1982; Miki *et al.*, 1980). However, deletion of the sequences left of the *Bgl*II site of pRR949 allows the extended *cat* transcript to proceed through the replication region, causing an increase in plasmid copy number (Womble *et al.*, 1985a). Insertion of foreign DNA at the *Bgl*II site can also cause an increase in copy number, presumably as a result of transcription read-through from the inserted DNA (Rownd *et al.*, 1980).

Normally, RNA-A transcription is almost completely repressed *in vivo* (Dong *et al.*, 1985; Rownd *et al.*, 1985). There is a low level of constitutive RNA-CX transcription that provides sufficient RepA2 repressor to block RNA-A transcription. The *repAX* open reading frame is located just downstream from the *repA2* sequence in RNA-CX and at the 5′ end of RNA-A. Immediately downstream from the *repAX* stop codon, the first folded domain of RNA-CX that is predicted includes the incompatibility target loop, or site of initial interaction with the incompatibility inhibitor, RNA-E (Fig. 5C,D). Formation of the target domain allows translation of the mRNA for the RepA1 replication initiation protein to be regulated by interaction with the complementary RNA-E.

RNA-E is synthesized in about a 20-fold excess compared with RNA-CX (Womble *et al.*, 1985a), and measurement of the intracellular concentrations of these transcripts suggested that nearly all RNA-CX molecules would be bound by RNA-E (Rownd *et al.*, 1985). The predicted structure of the bound RNA-CX molecules masks the *repA1* ribosome recognition sequence and GUG start codon in an intramolecular double-stranded structure (Fig. 7D). This would inhibit translation of the *repA1* mRNA. The nucleotide sequence of the *repA1* ribosome binding site is very similar to that of *lacI* mRNA, which is known to be inefficiently translated (Steege, 1977; Beyreuther, 1978). Masking the *repA1* site in a double-stranded secondary structure should effectively prevent synthesis of the RepA1 initiation protein.

For the rare RNA-CX transcript that escapes interaction with RNA-E, an alternate folding pathway is predicted. The free 3′ end of RNA-CX can interact with the incompatibility target to form a stable intramolecular pairing (Fig. 7B). The ribosome recognition sequence for *repA1* would then fold into a single-stranded structure, which would allow synthesis of the initiation protein (Fig. 7C). Synthesis of the RepA1 protein is the signal that mediates initiation of NR1 replication. This is predicted to happen, on average, once per plasmid per cell-cycle (Womble and Rownd, 1986). The competition for pairing with the incompatibility target of *repA1* mRNA, either intermolecularly by RNA-E or intramolecularly by the free 3′ end of the mRNA, is likely to be the switch that regulates initiation of

NR1 replication. We propose that the decision point in this molecular switch occurs after synthesis of the target stem-loop of *repA1* mRNA (Fig. 7A). Pausing of RNA polymerase at this position may help to synchronize and coordinate the alternate pathways that occur after this point, leading either to inactivation or translation of the *repA1* mRNA.

The other possible mRNA for *repA1*, RNA-A (Fig. 1), is only synthesized when the RepA2 repressor protein concentration is low (Dong *et al.*, 1985; Rownd *et al.*, 1985). This happens either when the copy number of the plasmid falls below normal, as when the plasmid is integrated into the chromosome or by chance fluctuations of plasmid segregation at cell division, or in mutants that fail to produce RepA2 repressor protein. The additional rightward transcription provided by derepression of RNA-A synthesis under these conditions increases the expression of *repA1* and, therefore, the frequency of replication initiation (Rownd *et al.*, 1985). Translation of RepA1 from RNA-A is also inhibited by interaction with RNA-E. The predicted secondary structures for RNA-A and RNA-CX are identical from base 489 on.

RNA-B possibly codes for RepA1*, composed of the carboxy-terminal 121 amino acids of RepA1 (Fig. 1). Transcription of RNA-B has not been detected *in vivo*, but the presence of a possible LexA repressor binding site in the RNA-B/D promoter (Fig. 3) suggests that synthesis of this transcript could be responsive to the host SOS system (Little and Mount, 1982). We are presently investigating this possibility. Translation of RepA1* protein from RNA-B would not be regulated by RNA-E and therefore would be independent of copy number. If the RepA1* protein has initiation protein activity, then the response of NR1 to host cell damage (SOS) may be to derepress transcription of RNA-B, which would allow synthesis of RepA1* protein and increase the rate of replication of the plasmid. We have observed that a fourfold increase in plasmid copy number is accompanied by a 100-fold increase in the frequency of conjugal transfer (unpublished data), which could provide a means for NR1 to escape from a damaged host cell.

ACKNOWLEDGMENTS

We wish to thank Verne A. Luckow for assistance with implementation of the computer programs. This work was supported in part by U.S. Public Health Services research grant GM30731 from the National Institutes of Health and by grants from the Otho S.A. Sprague Memorial Institute and Bane Biomedical Research Fund. The secondary structure analyses were performed on the VAX 11/780 computer with funds provided by the Vogelback Computing Center of Northwestern University.

REFERENCES

Adhya, S., and Gottesman, M. (1978). *Annu. Rev. Biochem.* **47,** 967–996.

Alton, N. K., and Vapnek, D. (1979). *Nature (London)* **282,** 864–869.

Beyreuther, K. (1978). *In* "The Operon" (J. H. Miller and W. S. Reznikoff, eds.), pp. 123–154. Cold Spring Harbor Lab., Cold Spring Harbor, New York.

Brady, G., Frey, J., Danbara, H., and Timmis, K. N. (1983). *J. Bacteriol.* **154,** 429–436.

Brawner, M. E., and Jaskunas, S. R. (1982). *J. Mol. Biol.* **159,** 35–55.

Chan, P. T., and Lebowitz, J. (1982). *Nucleic Acids Res.* **10,** 7295–7311.

Datta, N. (1975). *In* "Microbiology—1974" (D. Schlessinger, ed.), pp. 9–15. Am. Soc. Microbiol., Washington, D.C.

Dong, X., Womble, D. D., Luckow, V. A., and Rownd, R. H. (1985). *J. Bacteriol.* **161,** 544–551.

Dong, X., Womble, D. D., and Rownd, R. H. (1987). *J. Bacteriol.* (submitted for publication).

Easton, A. M., and Rownd, R. H. (1982). *J. Bacteriol.* **152,** 829–839.

Easton, A. M., Sampathkumar, P., and Rownd, R. H. (1981). *In* "The Initiation of DNA Replication" (D. S. Ray, ed.), pp. 125–141. Academic Press, New York.

Fuller, R. S., and Kornberg, A. (1983). *Proc. Natl. Acad. Sci. U.S.A.* **80,** 5817–5821.

Givskov, M., and Molin, S. (1984). *Mol. Gen. Genet.* **194,** 286–292.

Gold, L., Pribnow, D., Schneider, T., Shinedling, S., Singer, B. S., and Stormo, G. (1981). *Annu. Rev. Microbiol.* **35,** 365–403.

Hall, M. N., Gabay, J., Debarbouille, M., and Schwartz, M. (1982). *Nature (London)* **295,** 616–618.

Horinouchi, S., and Weisblum, B. (1982). *J. Bacteriol.* **150,** 815–825.

Iserentant, D., and Fiers, W. (1980). *Gene* **9,** 1–12.

Kassavetis, G. A., and Chamberlin, M. J. (1981). *J. Biol. Chem.* **256,** 2777–2786.

Light, J., and Molin, S. (1982a). *J. Bacteriol.* **151,** 1129–1135.

Light, J., and Molin, S. (1982b). *Mol. Gen. Genet.* **187,** 486–493.

Light, J., and Molin, S. (1983). *EMBO J.* **2,** 93–98.

Little, J. W., and Mount, D. W. (1982). *Cell (Cambridge, Mass.)* **29,** 11–22.

Liu, C.-P., Churchward, G., and Caro, L. (1983). *Plasmid* **10,** 148–155.

Luckow, V. A., Littlewood, R. K., and Rownd, R. H. (1984). *Nucleic Acids Res.* **12,** 665–673.

Lurz, R., Danbara, H., Ruckert, B., and Timmis, K. N. (1981). *Mol Gen. Genet.* **183,** 490–496.

Machida, C., Machida, Y., Wang, H.-C., Ishizaki, K., and Ohtsubo, E. (1983). *Cell (Cambridge, Mass.)* **34,** 135–142.

Masai, H., Kaziro, Y., and Arai, K. (1983). *Proc. Natl. Acad. Sci. U.S.A.* **80,** 6814–6818.

Miki, T., Easton, A. M., and Rownd, R. H. (1978). *Mol. Gen. Genet.* **158,** 217–224.

Miki, T., Easton, A. M., and Rownd, R. H. (1980). *J. Bacteriol.* **141,** 87–99.

Miki, T., Ebina, Y., Kishi, F., and Nakazawa, A. (1981). *Nucleic Acids Res.* **9,** 529–543.

Morris, C. F., Hoshimoto, H., Mickel, S., and Rownd, R. (1974). *J. Bacteriol.* **118,** 855–866.

Nordstrom, K., Molin, S., and Light, J. (1984). *Plasmid* **12,** 71–90.

Ohtsubo, E., Feingold, J., Ohtsubo, H., Mickel, S., and Bauer, W. (1977). *Plasmid* **1,** 8–18.

Ohtsubo, H., Ryder, T. B., Maeda, Y., Armstrong, K., and Ohtsubo, E. (1986). *Adv. Biophys.* **21,** 115–133.

Reznikoff, W. S., and Abelson, J. N. (1978). *In* "The Operon" (J. H. Miller and W. S. Reznikoff, eds.), pp. 221–243. Cold Spring Harbor Lab., Cold Spring Harbor, New York.

Rosen, J., Ryder, T., Inokuchi, H., Ohtsubo, H., and Ohtsubo, E. (1980). *Mol. Gen. Genet.* **179,** 527–537.
Rosen, J., Ryder, T., Ohtsubo, H., and Ohtsubo, E. (1981). *Nature (London)* **290,** 794–797.
Rosenberg, M., and Court, D. (1979). *Annu. Rev. Genet.* **13,** 319–353.
Rosenberg, M., Court, D., Shimatake, H., Brady, C., and Wuff, D. L. (1978). *Nature (London)* **272,** 414–423.
Rownd, R. H., and Womble, D. D. (1978). *In* "R Factor, Drug Resistance Plasmid" (S. Mitsuhasi, ed.), pp. 161–193. Univ. of Tokyo Press, Tokyo.
Rownd, R. H., Nakaya, R., and Nakamura, A. (1966). *J. Mol. Biol.* **17,** 376–393.
Rownd, R. H., Easton, A. M., Barton, C. R., Womble, D. D., McKell, J., Sampathkumar, P., and Luckow, V. (1980). *In* "Mechanistic Studies of DNA Replication and Genetic Recombination" (B. Alberts, ed.), pp. 311–334. Academic Press, New York.
Rownd, R. H., Womble, D. D., Dong, X., Luckow, V. A., and Wu, R.-P. (1985). *In* "Plasmids in Bacteria" (D. Helinski, S. N. Cohen, D. Clewell, D. Jackson, and A. Hollaender, eds.), pp. 335–354. Plenum, New York.
Rownd, R. H., Womble, D. D., and Dong, X. (1986). *Banbury Rep.* **24,** 179–194.
Ryder, T., Rosen, J., Armstrong, K., Davison, D., Ohtsubo, E., and Ohtsubo, H. (1981). *In* "The Initiation of DNA Replication" (D. S. Ray, ed.), pp. 91–111. Academic Press, New York.
Salser, W. (1977). *Cold Spring Harbor Symp. Quant. Biol.* **42,** 985–1002.
Schumperli, D., McKenney, K., Sobieski, D. A., and Rosenberg, M. (1982). *Cell (Cambridge, Mass.)* **30,** 865–871.
Scott, J. R. (1984). *Microbiol. Rev.* **48,** 1–23.
Shaw, W. V. (1975). *In* "Methods in Enzymology" (J. H. Hash, ed.), Vol. **43,** pp. 235–255. Academic Press, New York.
Shaw, W. V. (1983). *CRC Crit. Rev. Biochem.* **14,** 1–46.
Shine, J., and Dalgarno, L. (1974). *Proc. Natl. Acad. Sci. U.S.A.* **71,** 1342–1346.
Steege, D. A. (1977). *Proc. Natl. Acad. Sci. U.S.A.* **74,** 4163–4167.
Stougaard, P., Molin, S., and Nordstrom, K. (1981a). *Proc. Natl. Acad. Sci. U.S.A.* **78,** 6008–6012.
Stougaard, P., Molin, S., Nordstrom, K., and Hansen, F. G. (1981b). *Mol. Gen. Genet.* **181,** 116–122.
Stougaard, P., Light, J., and Molin, S. (1982). *EMBO J.* **1,** 323–328.
Taylor, D. P., and Cohen, S. N. (1979). *J. Bacteriol.* **137,** 92–104.
Timmis, K. N., Danbara, H., Brady, G., and Lurz, R. (1981). *Plasmid* **5,** 53–75.
Wagner, E. G. H., and Nordstrom, K. (1986). *Nucleic Acids Res.* **14,** 2523–2538.
Winkler, M. E., Mullis, K., Barnett, J., Stoynowski, I., and Yanofsky, C. (1982). *Proc. Natl. Acad. Sci. U.S.A.* **79,** 2181–2185.
Womble, D. D., and Rownd, R. H. (1986). *J. Mol. Biol.* **192,** 529–548.
Womble, D. D., and Rownd, R. H. (1979). *Plasmid* **2,** 95–108.
Womble, D. D., Taylor, D. P., and Rownd, R. H. (1977). *J. Bacteriol.* **130,** 148–153.
Womble, D. D., Dong, X., Wu, R.-P., Luckow, V. A., Martinez, A. F., and Rownd, R. H. (1984). *J. Bacteriol.* **160,** 28–35.
Womble, D. D., Sampathkumar, P., Easton, A. M., Luckow, V. A., and Rownd, R. H. (1985a). *J. Mol. Biol.* **181,** 395–410.
Womble, D. D., Dong, X., Luckow, V. A., Wu, R.-P., and Rownd, R. H. (1985b). *J. Bacteriol.* **161,** 534–543.
Yanofsky, C. (1981). *Nature (London)* **289,** 751–758.
Yates, J. L., and Nomura, M. (1981). *Cell (Cambridge, Mass.)* **24,** 243–249.
Zucker, M., and Stiegler, P. (1981). *Nucleic Acids Res.* **9,** 133–148.

15

Regulation of ColE1 DNA Replication by Antisense RNA

JUN-ICHI TOMIZAWA

National Institute of Diabetes and Digestive and Kidney Diseases
National Institutes of Health
Bethesda, Maryland 20892

I. INTRODUCTION

Regulation of gene expression by a specific cellular protein has been known for long. On the other hand, regulation by a specific RNA became known quite recently. Binding of an RNA transcript (antisense RNA) having a nucleotide sequence complementary to a sense RNA could affect expression of a particular gene function. Such regulation by an antisense RNA was first found in the system for replication of plasmid ColE1. Here I describe first the mechanism of formation of the RNA primer for ColE1 DNA replication and then the inhibition of primer formation by an antisense RNA.

II. RNA PRIMER FORMATION

ColE1 DNA can be synthesized in the absence of protein synthesis using a cell extract made from bacteria that do not carry the plasmid (Tomizawa *et al.*, 1975). This result indicates that plasmid-encoded proteins are not necessary for replication of the plasmid. A rifampicin-sensitive step precedes DNA synthesis (Sakakibara and Tomizawa, 1974a,b). In addition, some ribonucleotides are attached to the 5′ end of newly synthesized DNA strands. These results suggest primer formation by

Molecular Biology of RNA
New Perspectives

RNA polymerase (Tomizawa *et al.*, 1977; Bird and Tomizawa, 1978). Once this step has been carried out, further replication of ColE1 DNA does not require RNA polymerase (Tomizawa, 1975). *In vitro* replication of ColE1 DNA also requires functional DNA polymerase I, since extracts made from *pol*A1 bacteria cannot support replication of ColE1 DNA (Tomizawa, 1978; Itoh and Tomizawa, 1978; Staudenbauer, 1978). However, synthesis of ColE1 DNA by a mixture of RNA polymerase and DNA polymerase I is very inefficient. Later, a protein that was identified as RNase H was isolated from cell extracts by virture of its ability to stimulate DNA synthesis at the normal origin of replication (Itoh and Tomizawa, 1978, 1982). In the presence of RNase H, DNA synthesis initiates efficiently and almost exclusively at the normal origin.

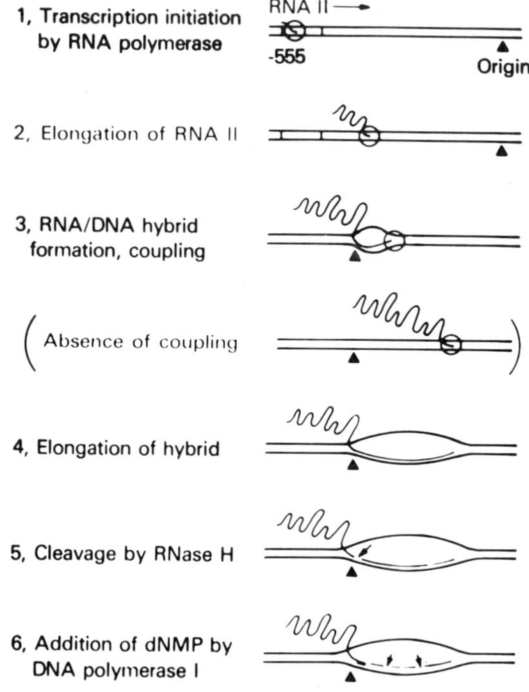

Fig. 1. Schematic illustration of primer formation and its inhibition by RNA I. An explanation of each process is presented in the figure. Step 3 consists of two alternative processes; the abortive process is presented in parentheses. Straight lines represent double strands of DNA; wavy lines, RNA transcripts; and small circles, RNA polymerase molecules. The DNA template forms an eye structure from the origin of DNA replication (indicated by ▲). Then the hybridized RNA elongates. Arrows in the eye structure in step 5 indicate cleavage of hybridized RNA by RNAase H. A thick line in the eye structure in step 6 is a newly synthesized DNA strand.

The analysis of products of *in vitro* transcription of ColE1 DNA revealed the roles played by RNase H and the polymerases in initiation of replication of ColE1 DNA (Itoh and Tomizawa, 1980). Transcripts (RNA II) that start 555 nucleotides upstream of the replication origin form hybrids spontaneously with the template DNA. Cleavage of these hybridized transcripts by RNase H generate 3′-OH termini at the origin region. The cleaved transcript is used as a primer for DNA synthesis by DNA polymerase I. Only nascent RNA II transcript is used for primer formation. A transcript that is separated from the template DNA can no longer be used (Itoh and Tomizawa, 1980). The process of initiation of replication of ColE1 DNA is schematically shown in Fig. 1.

Although ColE1 DNA initiates replication specifically from the normal origin in the reaction in which RNase H dictates the specificity, ColE1 DNA can be maintained in bacteria devoid of RNase H (Naito *et al.*, 1984), or even in bacteria lacking both RNase H and DNA polymerase I (Kogoma, 1984). These results suggest the presence of mechanisms for ColE1 DNA replication that are different from the mechanism that involves the RNase H action. Studies of ColE1 DNA replication with extracts made from these mutant bacteria showed that all these mechanisms share the same process, namely formation of a hybrid of RNA II with the template DNA. In the absence of RNase H, the hybridized RNA II activates synthesis of the lagging strands instead of providing the primer for synthesis of the leading strands. When RNase H is present, operation of this mechanism is suppressed (Dasgupta *et al.*, 1987).

III. REGULATION OF PRIMER FORMATION

Studies on the maintenance of various bacterial plasmids have defined two possibly related phenomena. One is that the number of molecules of a given plasmid found in the cell is relatively fixed. The other is that certain plasmids, said to be members of the same incompatibility group, are not stably maintained together in the same host cell. In principle, any incompatibility-specific mechanism that regulates replication or segregation of the plasmid molecules could be responsible for these two effects. The plasmid ColE1 displays both copy number control and incompatibility. A region of the plasmid 200–600 base pairs upstream of its origin of replication has been implicated in regulation of ColE1 DNA replication and incompatibility. Various mutations that alter the plasmid copy number had been found in this region of the genome.

A priori it seemed likely that the specificity of the incompatibility mechanism would be mediated by a protein encoded by one of the origin region transcripts. However, when we compared the nucleotide sequences of the region of ColE1 that is required for replication and for expression of specific incompatibility to the corresponding regions of other ColE1-like plasmids, we found no evidence for analogous open reading frames (Selzer *et al.,* 1983). This led us to consider the possibility that an RNA transcript is directly involved in regulation of replication.

A candidate for this role was the previously identified transcript called RNA I (Morita and Oka, 1979). Synthesis of RNA I starts 445 base pairs upstream of the replication origin, proceeds in the direction opposite to that of RNA II synthesis, and terminates near the initiation site of RNA II synthesis (Ito and Tomizawa, 1982). Thus, RNA I is an antisense transcript with respect to RNA II (Fig. 2).

A possible role for RNA I in regulation has been suggested based on the presence of mutations in the region that specify RNA I (Tomizawa and Itoh, 1981; Lacatena and Cesareni, 1981). However, because the region also specifies RNA II, these genetic results do not exclude a possibility of autoregulation of the RNA II function by RNA II. Therefore, we considered it crucial to demonstrate directly whether or not one of these RNAs is involved in regulation. Using the *in vitro* system of primer formation which we had developed, we found that RNA I inhibits primer formation by RNA II (Tomizawa *et al.,* 1981; Tomizawa and Itoh, 1981). Each RNA I of various ColE1-related plasmids can inhibit primer formation by the plasmid that specifies it, but has no effect on primer formation by heterologous plasmids. Mutations that affect the incompatibility property were found to alter the inhibitory activity of RNA I on *in vitro* primer formation. In other words, inhibition of primer formation by RNA I is incompatibility group specific. Thus, the regulation of primer formation for ColE1 DNA replication by an antisense RNA was established.

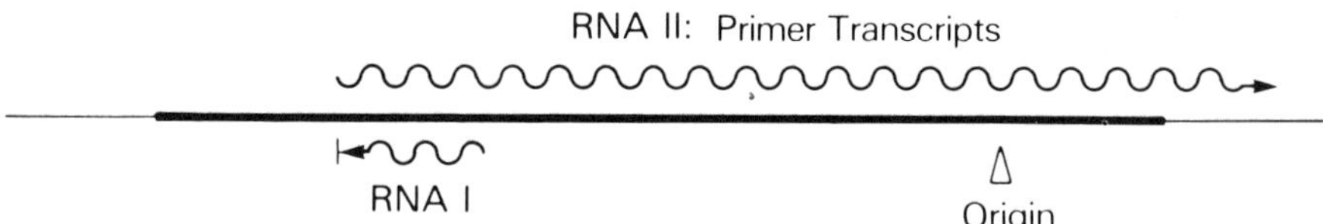

Fig. 2. Two kinds of transcripts made on the origin region of ColE1. RNA II (primer transcripts) starts 555 nucleotides upstream of the replication origin. RNA I starts 447 nucleotides upstream of the origin and terminates near the site where the primer transcripts start on the opposite strand. Thus, RNA I is an antisense RNA with respect to RNA II.

IV. BINDING OF RNA I TO RNA II

The first evidence that RNA I and RNA II bind each other was provided by the experiments that showed when RNA I and RNA II were mixed they both could be cleaved by RNase III that specifically cleaves double-stranded RNA (Tomizawa and Itoh, 1981). RNA I or RNA II alone was not cleaved by the enzyme. Later the product of binding of RNA I to RNA II was isolated and the process of binding was analyzed in detail (Tomizawa, 1984).

Both before and during formation of a bimolecular complex, RNA I and RNA II each contain a well-defined folded structure (Morita and Oka, 1979; Tomizawa, 1984; Masukata and Tomizawa, 1984, 1986; Wong and Polisky, 1985) (Fig. 3). Point mutations located at the center of any one of the three palindromes in these structures reduce the rate of binding of RNA I to RNA II without detectably altering the folded RNA structure (Tomizawa, 1984). This result suggests that these centers of the palin-

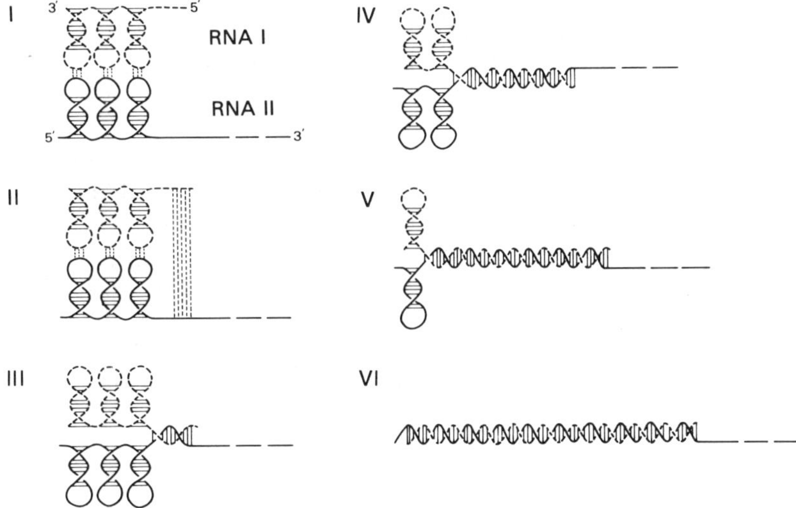

Fig. 3. Schematic illustration of the process of binding of the wild-type pair of RNA I and RNA II. RNA I and RNA II interact with each other between the loops of their folded structures (step I). This interaction facilitates pairing (step II) that starts at the 5′ end of RNA I (step III). At this stage, the loop-to-loop contacts may be broken. The pairing propagates progressively as the stem–loop structures are dissolved (steps IV and V). Finally, RNA I hybridizes along the entire length to RNA II (step VI). While RNA I will interact in the structure as illustrated, RNA II may interact in a similar alternative folded structure (Tomizawa, 1984).

dromes are involved in the interaction of RNA I and RNA II. However, the contact between the two RNAs in the first stable complex involves only a region very near the 5′ end of RNA I. The persistent pairing propagates toward the 3′ end of RNA I (Tomizawa, 1984). Based on these results, I proposed a stepwise model for the binding of RNA I to RNA II as illustrated schematically in Fig. 3 (Tomizawa, 1984). Initially, the two molecules interact at the single-stranded loops of the folded structures (step I). Although transient, this contact facilitates initiation of pairing at the 5′ end of RNA I, probably by bringing this end close to the complementary region of RNA II (step II). Pairing leading directly to formation of a stable complex starts at the 5′ end of RNA I (step III). Pairing propagates progressively into the palindromes as the stem–loop structures unfold (steps IV and V). Finally, RNA I is hybridized over its entire length to RNA II (step VI).

The most unique feature of the proposed mechanism of binding of RNA I to RNA II is that the initial contact of the RNAs is transient and does not itself provide a nucleus for complete pairing over the entire length of RNA I. The evidence that the initial interaction between the stem–loop structures is reversible was provided by the analysis of the effect of heterologous RNA I on binding of homologous RNA I and RNA II (Tomizawa, 1985). It was found that binding of RNA I to the homologous RNA II was inhibited by an RNA I specified by a plasmid of different compatibility. Kinetic analysis showed that inhibition to be due to competition between the homologous and heterologous RNA Is for reversible binding to the RNA II. This competition reduces the overall rate of stable binding of the two homologous RNAs. As predicted from these results, the copy numbers of both ColE1 and RSF1030 are increased when both plasmids are present in the same cell.

Replication of ColE1 DNA is affected by the presence of a region downstream of the replication origin (Twigg and Sherratt, 1980; Som and Tomizawa, 1983). It was initially assumed that this region expressed a product that inhibited primer transcription; accordingly the product was named Rop (*r*epressor *o*f *p*rimer) (Cesareni *et al.,* 1982). The region encodes a 63-amino-acid protein that enhances the inhibitory activity of RNA I on *in vitro* primer formation (Tomizawa and Som, 1984: Lacatena *et al.,* 1984). This protein has no effect on RNA II transcription (Tomizawa and Som, 1984; Lacatena *et al.,* 1984; Tomizawa, 1986). Instead, it generally increases the rate of binding of RNA I to RNA II, and increases the extent of inhibition of primer formation by RNA I (Tomizawa and Som, 1984; Tomizawa, 1986). Based on this property, the protein was named Rom (*R*NA *o*ne inhibition *m*odulator). The observation that led to the proposal of repressor function can be explained by the increased rate of binding of RNA I to RNA II in the presence of the protein (Tomizawa and Som, 1984; Tomizawa, 1986).

V. IMPORTANCE OF THE RATE OF BINDING OF RNA I TO RNA II

RNA I inhibits primer formation if added at the time of initiation of RNA II synthesis. However, when RNA I is added after RNA II has formed a hybrid with the template DNA (Step II in Fig. 1), it no longer affects primer formation. Pulse treatment of elongating RNA II with RNA I has shown that when nascent RNA II that is 100–360 nucleotides long is exposed to RNA I, subsequent primer formation is inhibited (Tomizawa, 1986). However, when the transcripts are shorter than 100 nucleotides or longer than 360 nucleotides, primer forms normally. Nascent RNA II molecules shorter than 100 nucleotides long are insensitive because RNA I cannot bind to them, while RNA II transcripts longer than 360 nucleotides are insensitive because binding of RNA I no longer affects primer formation. Thus, for inhibition by RNA I to be effective, it must bind to RNA II when elongating RNA II has a length between 100 and 360 nucleotides. This result indicates that the rate of binding of RNA I to RNA II has a critical importance to regulation of primer formation. The effect of mutational alterations on the rate of binding of RNA I to RNA II correlates well with their effect on plasmid replication *in vivo* (Tomizawa, 1984; Tomizawa and Som, 1984).

VI. SECONDARY STRUCTURE OF RNA II AND ITS ALTERATION BY BINDING OF RNA I

For a nascent RNA II to form a hybrid with the template DNA, it must form a specific folded structure (Masukata and Tomizawa, 1984, 1986). By combining the results obtained from partial digestion of the wild-type RNA II with various RNases, from mutational effects on primer formation and from computer analysis of RNA structure, the structure of functional RNA II presented in Fig. 4 was deduced. When RNA I binds to RNA II, it sequesters the region of RNA II that is complementary to RNA I from participation in formation of a folded structure (Fig. 4). As a result, RNA II is prevented from formation of the wild-type structure.

A region of RNA II of about 200 nucleotides that includes the sequence complementary to RNA I can be deleted without impairing the primer formation. Therefore, this region must exist exclusively for the regulatory purpose.

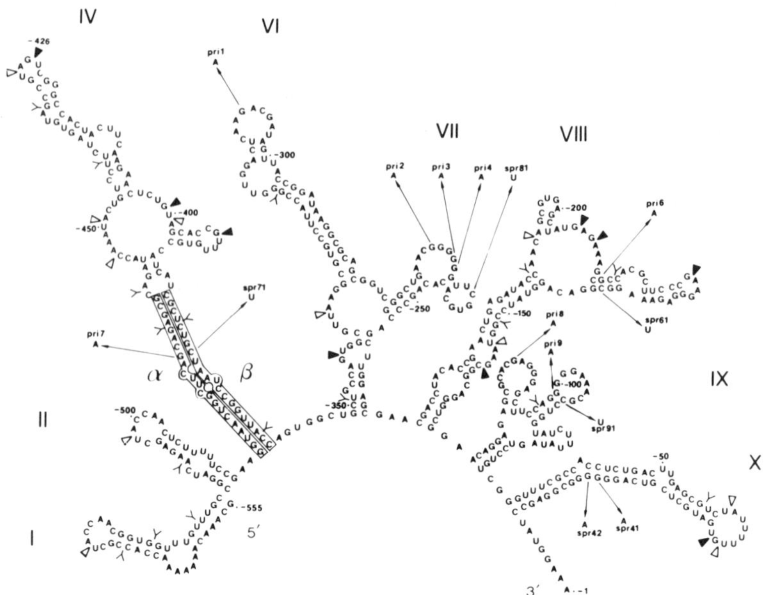

RNA I - bound RNA II

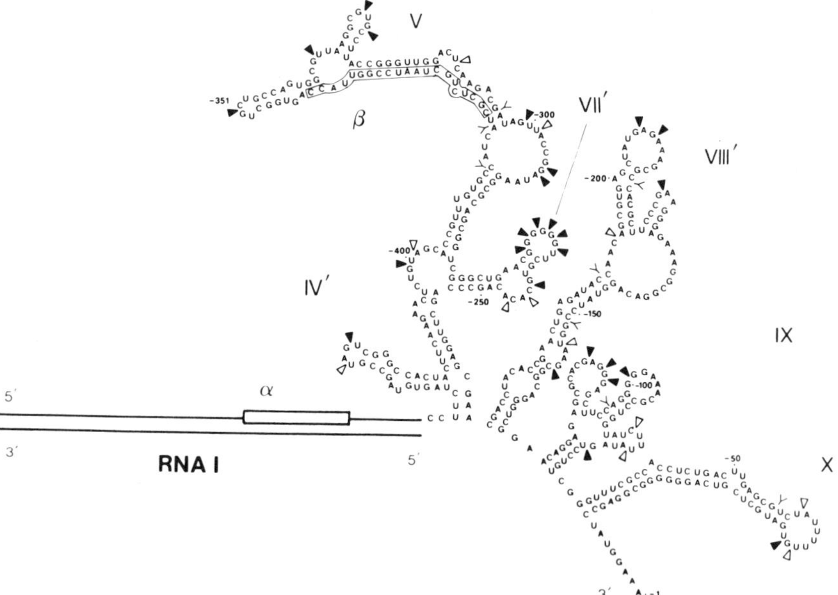

VII. CONCLUSIONS

In vitro studies on ColE1 primer formation show that some nascent transcripts (RNA II) that start 555 base pairs upstream of the replication origin form a hybrid with the template DNA. When cleaved by RNase H, these hybridized transcripts provide the primer for DNA synthesis by DNA polymerase I. To form a hybrid with the template DNA, RNA II must form a specific folded structure.

ColE1 primer formation is regulated by RNA I, an antisense RNA with respect to RNA II. RNA I binds to RNA II in the complementary region and inhibits formation of a structure that is necessary for RNA II to form the hybrid. RNA I does not inhibit primer formation when RNA II has elongated to more than 360 nucleotides prior to RNA I binding. Therefore, the rate of binding of RNA I and RNA II is important in regulation of primer formation. Rom protein affects the rate of binding of RNA I to RNA II. The copy number regulation and incompatibility of ColE1-related plasmids can be understood by operation of these mechanisms of regulation of primer formation as revealed by *in vitro* studies.

From the results on regulation on ColE1 DNA replication by RNA I, one can envisage certain general properties of regulation by an antisense RNA. Binding of antisense RNA could affect expression of a gene function by binding directly to a sequence recognized by a cellular component or alternatively, it could affect a gene function through alteration of the structure of the sense RNA. Generally, an RNA molecule can be folded into several structures that differ little from each other in their free energy of formation. Therefore, interaction of an RNA molecule with an antisense RNA could easily alter its structure. An effect of a structural change need not be confined within the region where the antisense RNA binds. Binding could affect the structure in a region far away from the binding sequence. Our analysis of the structure of RNA II of ColE1 provides an example of these basic properties of RNA folding and its alteration by binding to an antisense RNA.

Sense and antisense RNA molecules could interact with each other in extended structures where the specificity of interaction is solely determined by complementary nucleotide sequences. However, our studies

---

Fig. 4. Structures of free (top) and RNA I-bound (bottom) RNA II. The procedure of deduction of these structures is described in Masukata and Tomizawa (1986). Sites of preferential cleavage by RNase T1 (solid arrowhead), RNAse A (open arrowhead), and RNase VI (Y) are indicated. Base changes by *pri* and *spr* mutations are also shown. A *pri* mutation inactivates RNA II, and a *spr* mutation restores the activity lost by a specific *pri* mutation. In B the region where RNA I and RNA II pair each other is shown by two parallel lines.

with ColE1 show that they can also interact as folded structures. In this latter case, a minor change in the primary sequence can have a profound effect on the rate of interaction of RNAs and on the outcome of the regulation. This can provide extensive flexibility to the mechanism as for specificity and strength of regulation. In addition, modulation of the rate of binding of RNAs by a protein can further increase versatility in regulation by an antisense RNA. Mutational alterations of the rate of binding of RNA I to RNA II and modulation of the rate by the Rom protein provide clear examples of these mechanisms.

Increasing examples of regulation by antisense RNA (for review, see Green *et al.,* 1986) show the general importance of the mechanism in biological regulation (Tomizawa, 1983).

REFERENCES

Bird, R. E., and Tomizawa, J. (1978). *J. Mol. Biol.* **120,** 137–143.

Cesareni, G., Meusing, M. A., and Poliski, B. (1982). *Proc. Natl. Acad. Sci. U.S.A.* **79,** 6313–6317.

Dasgupta, S., Masukata, H., and Tomizawa, J. (1987). In preparation.

Green, P. J., Pines, O., and Inouye, M. (1986). *Annu. Rev. Biochem.* **55,** 569–597.

Itoh, T., and Tomizawa, J. (1978). *Cold Spring Harbor Symp. Quant. Biol.* **43,** 409–418.

Itoh, T., and Tomizawa, J. (1980). *Proc. Natl. Acad. Sci. U.S.A.* **77,** 2450–2454.

Itoh, T., and Tomizawa, J. (1982). *Nucleic Acids Res.* **10,** 5949–5965.

Kogoma, T. (1984). *Proc. Natl. Acad. Sci. U.S.A.* **81,** 7845–7849.

Lacatena, R. M., and Cesareni, G. C. (1981). *Nature (London)* **294,** 623–626.

Lacatena, R. M., Banner, D. W., Castagnori, L., and Casareni, G. (1984). *Cell (Cambridge, Mass.)* **37,** 1009–1014.

Masukata, H., and Tomizawa, J. (1984). *Cell (Cambridge, Mass.)* **36,** 513–522.

Masukata, H., and Tomizawa, J. (1986). *Cell (Cambridge, Mass.)* **44,** 125–136.

Morita, M., and Oka, A. (1979). *Eur. J. Biochem.* **97,** 435–443.

Naito, S., Kitani, T., Ogawa, T., Okazaki, T., and Uchida, H. (1984). *Proc. Natl. Acad. Sci. U.S.A.* **81,** 550–554.

Sakakibara, Y., and Tomizawa, J. (1974a). *Proc. Natl. Acad. Sci. U.S.A.* **71,** 802–806.

Sakakibara, Y., and Tomizawa, J. (1974b). *Proc. Natl. Acad. Sci. U.S.A.* **71,** 1403–1407.

Selzer, G., Som, T., Itoh, T., and Tomizawa, J. (1983). *Cell (Cambridge, Mass.)* **32,** 119–129.

Som, T., and Tomizawa, J. (1983). *Proc. Natl. Acad. Sci. U.S.A.* **80,** 3232–3236.

Staudenbauer, W. L. (1978). *Curr. Top. Microbiol. Imunol.* **83,** 94–156.

Tomizawa, J. (1975). *Nature (London)* **257,** 253–254.

Tomizawa, J. (1978). *In* "DNA Synthesis, Present and Future" (I. Molineaux and M. Kohiyama, eds.), pp. 797–826. Plenum, New York.

Tomizawa, J. (1983). *In* "Nucleic Acid Research: Future Development" (K. Mizobuchi, I. Watanabe, and J. Watson, eds.), pp. 475–485. Academic Press, New York.

Tomizawa, J. (1984). *Cell (Cambridge, Mass.)* **38,** 861–870.

Tomizawa, J. (1985). *Cell (Cambridge, Mass.)* **40,** 527–535.

Tomizawa, J. (1986). *Cell (Cambridge, Mass.)* **47,** 89–97.

Tomizawa, J., and Itoh, T. (1981). *Proc. Natl. Acad. Sci. U.S.A.* **78,** 6096–6100.
Tomizawa, J., and Itoh, T. (1982). *Cell (Cambridge, Mass.)* **31,** 575–583.
Tomizawa, J., and Som, T. (1984). *Cell (Cambridge, Mass.)* **38,** 871–878.
Tomizawa, J., Sakakibara, Y., and Kakefuda, T. (1975). *Proc. Natl. Acad. Sci. U.S.A.* **72,** 1050–1054.
Tomizawa, J., Ohmori, H., and Bird, R. E. (1977). *Proc. Natl. Acad. Sci. U.S.A.* **74,** 1865–1869.
Tomizawa, J., Itoh, T., Selzer, G., and Som, T. (1981). *Proc. Natl. Acad. Sci. U.S.A.* **78,** 1421–1425.
Twigg, A. J., and Sherratt, D. (1980). *Nature (London)* **283,** 216–218.
Wong, E. M., and Polisky, B. (1985). *Cell (Cambridge, Mass.)* **42,** 959–966.

16

A Transfer RNA Implicated in DNA Replication

JAMES R. WALKER

Department of Microbiology
The University of Texas at Austin
Austin, Texas 78712-1095

I. INTRODUCTION

The *Escherichia coli* $tRNA^{Arg}_{UCU}$, a minor species that corresponds to the rare arginine codon AGA, is the product of the *dnaY* gene (Mullin *et al.*, 1984; Garcia *et al.*, 1986), also called *argU* (Fournier and Ozeki, 1985). This gene is transcribed *in vitro,* in the presence of rho factor, from one promoter as primary transcripts of 180 and 190 nucleotides. The primary transcripts are thought to be processed *in vivo* at both 5′ and 3′ ends to generate a 77-nucleotide mature form which folds into the typical cloverleaf structure. Processing results in removal of 25 nucleotides from the 5′ end and 78 or 88 from the 3′ ends (Fig. 1). This tRNA is known to be encoded by the *dnaY* gene, because hybrid plasmids that carry *dnaY* direct the overproduction of an arginine-accepting tRNA (Garcia *et al.*, 1986). The *dnaY* gene is especially interesting because of the characteristics of its expression and because it has been implicated in DNA synthesis.

II. EXPRESSION OF THE *dnaY* GENE

The first unusual feature of *dnaY* gene expression is the fact that it is transcribed (in precursor form) as the only product of the gene (Fig. 1). Although this is not unique (Fournier and Ozeki, 1985), the overwhelming majority of tRNAs are encoded by operons that include several tRNAs,

Molecular Biology of RNA
New Perspectives

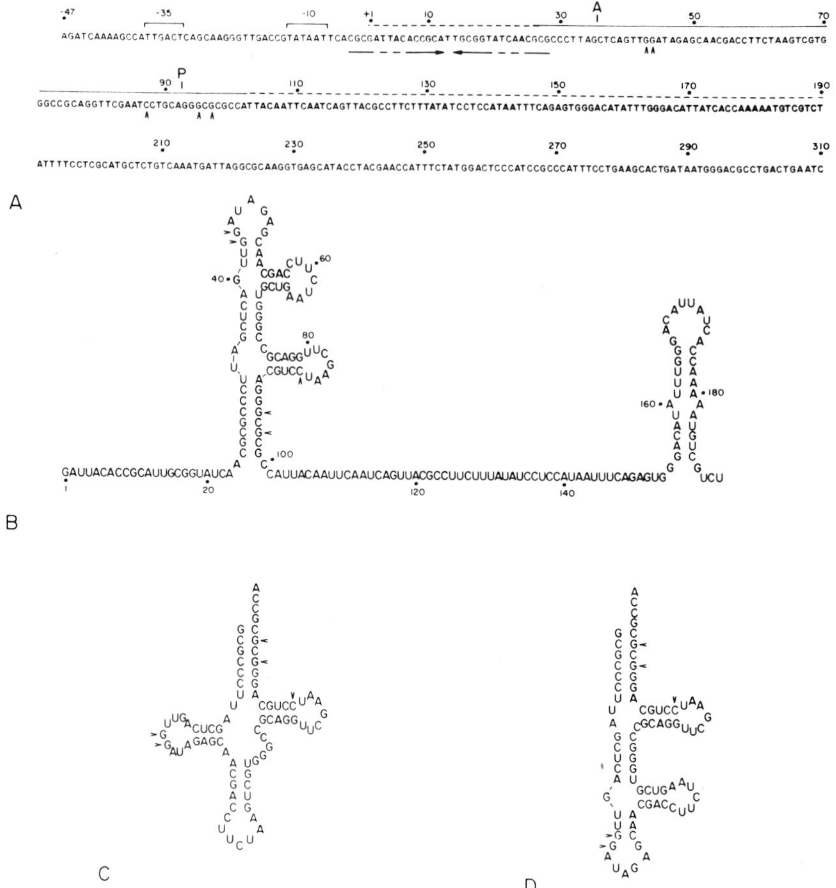

Fig. 1. Sequence of the *dnaY* gene and potential secondary structure of RNA products (Garcia *et al.*, 1986). (A) The sequence of the *dnaY* gene and flanking regions. The promoter is designated by brackets labeled −35 and −10; +1 indicates transcription start; the dashed line indicates transcript not present in the mature tRNA; the solid line indicates the mature tRNA sequence; the broken arrows indicate a 31-base-pair region of dyad symmetry. (B) The precursor transcript synthesized *in vitro* in the presence of rho factor. (C and D) Alternative potential secondary structures for the *dnaY* RNA mature form. Arrow points indicate G → A or C → U transitions which, in a cloned fragment, inactivate $dnaY^+$ complementing activity. (From Garcia *et al.*, 1986. Reprinted by permission of copyright holder, Cell Press.)

ribosomal RNAs, or proteins (e.g. Hsu *et al.*, 1984; Ikemura and Nomura, 1977).

Second, the stringency control region common to stable RNAs (Travers, 1980, 1984; Lamond and Travers, 1985) is weak or missing in *dnaY*. The consensus sequence CgCccC is located downstream and adjacent to the −10 promoter sequence in genes subject to stringent control (Lamond and Travers, 1985). The corresponding position in the *dnaY* promoter region is CACGCG, matching in only three of six positions and in only two of the three most highly conserved positions.

Third, the $tRNA^{Arg}_{UCU}$ is a minor species (Celis and Maas, 1971; Ikemura, 1981b; Garcia *et al.*, 1986), which suggests that the gene is transcribed inefficiently, but the promoter, based on similarity to the consensus sequence of *E. coli* promoters, should be strong (Fig. 1). The *dnaY* −35 sequence of TTGACT and −10 sequence of TATAAT match the consensus promoter sequences of TTGACA and TATAAT perfectly with the exception of one base—the 3′ position of the −35 region which is the most variable position of the promoter consensus sequence (Hawley and McClure, 1983). Moreover, the spacing between the *dnaY* −35 and −10 regions is 16 nucleotides, which also is within the usual range of 16–18 (Hawley and McClure, 1983). Why, then, is $tRNA^{Arg}_{UCU}$ rare? Perhaps the *dnaY* gene is subject to repression.

Situated downstream of the −10 region (from −3 to +28) of *dnaY* is a 31-base-pair region of dyad symmetry (Garcia *et al.*, 1986) characteristic of repressor-binding sites. This sequence is:

CGCGATTACACCGCATTGCGGTATCAACGCG

In organization, this 31-base-pair symmetrical region is similar to that of many operators (Gicquel-Sanzey and Cossart, 1982). The *lac* operator consists of 35 base pairs of dyad symmetry; 14 of 17 bases on either side of the axis are perfectly symmetrical (Gilbert and Maxam, 1973; Dickson *et al.*, 1975). The potential *dnaY* operator has 12 perfect matches in the 15-base-pair symmetrical arms. In sequence, the *dnaY* symmetrical region contains the sequence CACN$_9$GTA, which resembles the B-class conserved regulatory protein binding site proposed by Gicquel-Sanzey and Cossart (1982). The B consensus sequence CACN$_{5-10}$GTG was deduced from nine operator regions (Gicquel-Sanzey and Cossart, 1982). Thus, it is possible that a specific repressor limits expression of the *dnaY* gene.

The low concentration of active $tRNA^{Arg}_{UCU}$ might also be the result of incomplete modification of the processed form. This interpretation is suggested by the observation that the presence of a cloned *dnaY* gene on a hybrid plasmid directed 50- to 100-fold overproduction of a tRNA-size RNA, as measured by hybridization, but the amount of chargeable

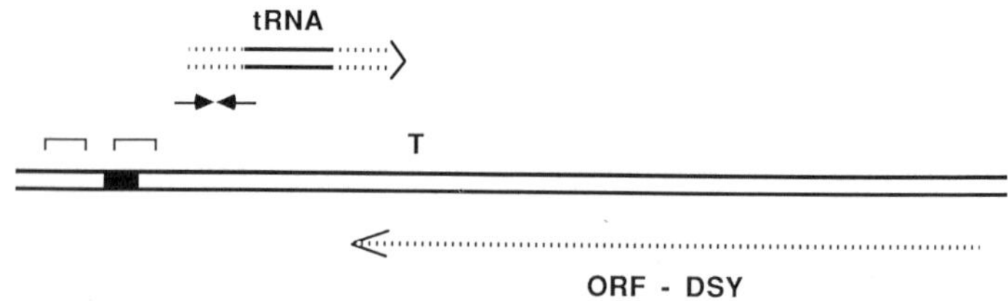

Fig. 2. Structure of *dnaY* and the downstream gene. The open wide bar represents chromosomal DNA, the darkened area is a decanucleotide repeat of a sequence in *oriC*, and the inverted solid arrows are a potential regulatory protein binding site. The open arrow from left to right is the primary transcript of *dnaY;* the dashed regions are removed by processing to generate the mature tRNA (solid lines). T is a rho-dependent terminator. ORF-DSY is an open reading frame of a gene of unknown function downstream of *dnaY*. The transcript of this gene overlaps at least a portion of the *dnaY* transcript.

$tRNA^{Arg}_{UCU}$ increased only three-to fourfold (Garcia *et al.*, 1986). Perhaps the majority of the overproduced RNA was not fully or normally modified.

The fourth unusual aspect of this tRNA gene expression is that immediately downstream there is an open reading frame that extends toward *dnaY*, overlaps the *dnaY* primary transcript, and terminates only 18 base pairs from the 3′ CCA of the mature tRNA (Fig. 2). This open reading frame consists of 387 codons which encode a 45,095-dalton protein of unknown function, but which has been detected in minicells (D. F. Lindsey and J. R. Walker, unpublished data). Thus, the transcript of the unknown function open reading frame must overlap the *dnaY* transcript, at least at their 3′ ends. The unknown function gene transcript might even extend into or through the tRNA part of the *dnaY* gene, in which case the 3′ end of this transcript is complementary to part or all of the tRNA. It is also surprising that the unknown function gene contains 8 AGA codons, the codons recognized by $tRNA^{Arg}_{UCU}$, and 5 AGG codons which presumably are also recognized by this tRNA.

In *E. coli,* preferred codons are recognized by the most abundant tRNA isoacceptor species (Post *et al.*, 1979; Ikemura, 1981a,b). Moreover, synonymous codons are nonrandomly distributed—abundantly expressed genes disciminate against rare codons (Guoy and Gautier, 1982) with the result that their translation depends on the abundant tRNAs and, in general, rare tRNAs are not required in the translation of abundantly expressed genes. On the other hand, regulatory protein genes and genes that are inefficiently expressed, such as replication genes (e.g., Smiley *et al.*, 1982), do not discriminate against rare codons (Guoy and Gautier, 1982; Sharp and Li, 1986) [although the presence of rare codons does not limit expressivity unless their frequency is higher than that occurring naturally (Holm, 1986)]. Inasmuch as the gene of unknown function which lies adjacent to, and is transcribed toward, *dnaY* contains 13 rare AGA/AGG arginine codons among a total of 387, it is probably that this gene encodes a rare product, such as a repressor or a replication factor. [In comparison, AGA is present only once and AGG zero times among 6240 codons of very highly expressed *E. coli* genes. Among 9223 codons in highly expressed genes, AGA and AGG are present a total of only five times (Sharp and Li, 1986).] Is coincidence the explanation for the observations that the gene for $tRNA^{Arg}_{UCU}$, one of the rarest tRNAs, lies adjacent to a gene that contains 13 rare AGA/AGG codons, and, furthermore, that the two transcripts overlap in a convergent manner?

III. IMPLICATION IN DNA REPLICATION

Another unexpected property of the $tRNA^{Arg}_{UCU}$ is its implication in DNA replication. The *dnaY* gene was defined by a mutation that created a temperature-sensitive defect preferentially in DNA synthesis (Henson *et al.*, 1979). The *dnaY*(ts) mutant stopped replication slowly at high temperature, but the defect was proposed to be in polymerization (rather than in initiation) because synthesis was inhibited at high temperature also in toluene-treated mutant cells (Henson *et al.*, 1979). Subcloning from a Clarke and Carbon (1976) plasmid demonstrated that a 118-base-pair fragment (nucleotides −17 to 101 of Fig. 1) provided wild-type complementing activity to a *recA*$^+$ or *recA dnaY*(ts) recipient. Moreover, single base mutations within the cloned *dnaY* gene abolished the complementing activity (Fig. 1; Mullin *et al.*, 1984). In addition, the *dnaY*(ts) defect is thought to be the result of a single mutation because the mutant reverted with a frequency of about 10^{-7} (i.e., a frequency characteristic of a single mutation) and the revertants regained wild-type phenotype (K.-S. Chen, L. Hays, and J. R. Walker, unpublished data). The *dnaY*(ts)10 allele (Henson *et al.*, 1979) has been cloned from the original *dnaY*(ts) mutant and shown to be a change of the 5′ G of the acceptor stem to an A (K.-S. Chen and J. R. Walker, unpublished data).

Is its role in replication related to an alternative secondary structure? This RNA has an alternative potential secondary structure of approximately the same or even greater stability than the tRNA cloverleaf (Fig. 1D). The standard cloverleaf is predicted to have a free energy of −46.2 kcal/mole, whereas the conformer shown in Fig. 1D should have a free energy of −47.6 kcal/mole (Tinoco *et al.*, 1973). Although there is no physical evidence that the structure of Fig. 1D in fact exists, it is possible that it forms and participates in a function other than translation. Moreover, sequestration of part of the *dnaY* transcript in a form different from the translation conformer (cloverleaf) might be part of the explanation for the failure of most of the overproduced *dnaY* RNA to be functional as an arginine acceptor.

Another interesting property of the *dnaY* gene that might be related to its role in replication is the perfect match of 10 nucleotides of the *dnaY* promoter region to 10 nucleotides in *oriC*, the chromosomal origin of replication. Nucleotides −18 to −9, GACCGTATAA (Fig. 1), correspond perfectly to nucleotides 154–163 of *oriC*. Inasmuch as transcription from the *oriC* promoter P*ori*L begins at position 164 (Lother and Messer, 1981), the second to eleventh nucleotides of the *oriC* transcript from P*ori*L will be (reading from the 11th to the 2nd position) 3′-CTGGCATATT-5′, which is homologous to the *dnaY* promoter region. Moreover, nucleotides 2–11 of the P*ori*R transcript (reading from 11 toward 2), 3′-CGGTCTTCTT-5′, are homologous in 6 of 10 positions to the same 10 nucleotides of *dnaY*. The significance of the perfect 10-base-pair repeat, if any, is unknown. By chance alone, that decanucleotide should occur six times in the *E. coli* chromosome.

IV. SUMMARY AND MODELS FOR REPLICATION ROLE

In summary, the *dnaY* gene encodes a precursor RNA which is processed to generate $tRNA^{Arg}_{UCU}$. This is the only known mature product of the gene, although an alternative conformer equally as stable as a tRNA can theoretically be formed. $tRNA^{Arg}_{UCU}$ is one of the rarest tRNAs, although the gene has a potentially strong promoter. The presence of a 31-base-pair region of dyad symmetry downstream of the promoter suggests that expression of the gene might be repressed by a regulatory factor which binds the symmetrical site. The *dnaY* transcript overlaps the transcript of a gene of unknown function which lies downstream of *dnaY* and which is transcribed toward *dnaY*.

Although there is no doubt that the gene for the $tRNA^{Arg}_{UCU}$ complemented the *dnaY*(ts) defect, that the gene does not encode a polypeptide, and that one product of the *dnaY* gene is a tRNA (Garcia *et al.*, 1986), it is unknown how this gene functions in replication. Does a gene product participate in replication or does the *dnaY* gene or a portion of the gene participate in replication by acting as a site. The wild-type gene (site) is known to be dominant, when present on a multicopy plasmid, over the mutant gene (site). Therefore, it is possible that the wild-type gene, acting as a site, binds a replication-inhibitor, thereby acting positively in DNA synthesis.

If a product of *dnaY* participates in replication, is the participation direct or indirect? A direct role would involve activity in the polymerization process, perhaps a property of the alternative conformer shown in Fig. 1D. An indirect role could be imagined if the *dnaY* product is required to effect positively the synthesis of a replication protein. Perhaps one function of the *dnaY* RNA is to regulate synthesis of the downstream gene product (Fig. 2). The regulation could be positive or negative, depending on whether the downstream gene product is required for, or acts to inhibit, replication. This gene of unknown function is thought to be rarely expressed, because of its content of rare codons, and it could be a replication gene. An even more indirect role could be played if the temperature-sensitive $tRNA^{Arg}_{UCU}$ limits DNA synthesis by limiting translation of replication proteins such as *dnaA, dnaN,* or *dnaG,* all of which contain AGA/AGG codons (Hansen *et al.,* 1982; Ohmori *et al.,* 1984; Smiley *et al.,* 1982). Presumably, such a protein would be especially labile or synthesized periodically.

ACKNOWLEDGMENTS

This work was supported by Public Health Service Grant GM34471, and, in part, by American Cancer Society Grant NP169 and Welch Foundation Grant F949.

REFERENCES

Celis, T. F. R., and Maas, W. K. (1971). *J. Mol. Biol.* **62,** 179–188.

Clarke, L., and Carbon, J. (1976). *Cell (Cambridge, Mass.)* **9,** 91–99.

Dickson, R. C., Abelson, J., Barnes, W. M., and Reznikof, W. S. (1975). *Science* **187,** 27–35.

Fournier, M. J., and Ozeki, H. (1985). *Microbiol. Rev.* **49,** 379–397.

Garcia, G. M., Mar, P. K., Mullin, D. A., Walker, J. R., and Prather, N. E. (1986). *Cell (Cambridge, Mass.)* **45,** 453–459.

Gicquel-Sanzey, B., and Cossart, P. (1982). *EMBO J.* **5,** 591–595.

Gilbert, W., and Maxam, A. (1973). *Proc. Natl. Acad. Sci. U.S.A.* **70,** 3581–3584.

Guoy, M., and Gautier, C. (1982). *Nucleic Acids Res.* **10,** 7055–7074.

Hansen, E. B., Hansen, F. G., and von Meyenburg, K. (1982). *Nucleic Acids Res.* **10,** 7373–7385.

Hawley, D. K., and McClure, W. R. (1983). *Nucleic Acids Res.* **11,** 2237–2255.

Henson, J. M., Chu, H., Irwin, C. A., and Walker, J. R. (1979). *Genetics* **92,** 1041–1059.

Holm, L. (1986). *Nucleic Acids Res.* **14,** 3075–3087.

Hsu, L. M., Klee, H. J., Zagorski, J., and Fournier, M. J. (1984). *J. Bacteriol.* **158,** 934–942.

Ikemura, T. (1981a). *J. Mol. Biol.* **146,** 1–21.
Ikemura, T. (1981b). *J. Mol. Biol.* **151,** 389–409.
Ikemura, T., and Nomura, M. (1977). *Cell (Cambridge, Mass.)* **11,** 779–793.
Lamond, A. I., and Travers, A. A. (1985). *Cell (Cambridge, Mass.)* **41,** 6–8.
Lother, H., and Messer, W. (1981). *Nature (London)* **294,** 376–378.
Mullin, D. A., Garcia, G. M., and Walker, J. R. (1984). *Cell (Cambridge, Mass.)* **37,** 669–674.
Ohmori, H., Kimura, M., Nagata, T., and Sakakibara, Y. (1984). *Gene* **28,** 159–170.
Post, L. E., Strycharz, G. D., Nomura, M., Lewis, H., and Dennis, P. O. (1979). *Proc. Natl. Acad. Sci. U.S.A.* **76,** 1697–1701.
Sharp, P. M., and Li, W.-H. (1986). *Nucleic Acids Res.* **14,** 7737–7749.
Smiley, B. L., Lupski, J. R., Svec, P. S., McMacken, R., and Godson, G. N. (1982). *Proc. Natl. Acad. Sci. U.S.A.* **79,** 4550–4554.
Tinoco, I., Jr., Borer, P. N., Dengler, B., Levine, M. D., Uhlenbeck, O. C., Crothers, D. M., and Gralla, J. (1973). *Nature New Biol.* **246,** 40–41.
Travers, A. A. (1980). *J. Bacteriol.* **141,** 973–976.
Travers, A. A. (1984). *Nucleic Acids Res.* **12,** 2605–2618.

V

RNA: Structure, Function, and Isolation

17

Stable Branched RNA Covalently Linked to the 5′ End of a Single-Stranded DNA of Myxobacteria

SUMIKO INOUYE, TEIICHI FURUICHI, ANIL DHUNDALE, AND MASAYORI INOUYE

Department of Biochemistry
Robert Wood Johnson Medical School at Rutgers
University of Medicine and Dentistry of New Jersey
Piscataway, New Jersey 08854

I. INTRODUCTION

Myxobacteria are gliding, gram-negative bacteria that undergo a unique developmental cycle. Upon starvation of nutrient on a solid surface, the cells aggregate to form fruiting bodies, as shown in Fig. 1 (see a review by Kaiser *et al.*, 1979, and a book edited by Rosenberg, 1984). Among myxobacteria, *Myxococcus xanthus* has been investigated extensively as a model system for differentiation. Its genome size has been determined to be approximately 60% larger than that of *Escherichia coli* (Yee and Inouye, 1981, 1982). *M. xanthus* contains a peculiar satellite DNA in total DNA preparations (Yee *et al.*, 1984). This DNA, consisting of a short single-stranded DNA, has been designated multicopy single-stranded DNA (msDNA), which exists at a level of 500–700 copies per genome (Yee *et al.*, 1984). Similar msDNAs have also been found in other myxobacteria (Yee *et al.*, 1984; Dhundale *et al.*, 1985).

In this chapter, we describe a most unusual structure at the 5′ end of msDNA—a branched RNA is covalently linked to the 5′ end. This struc-

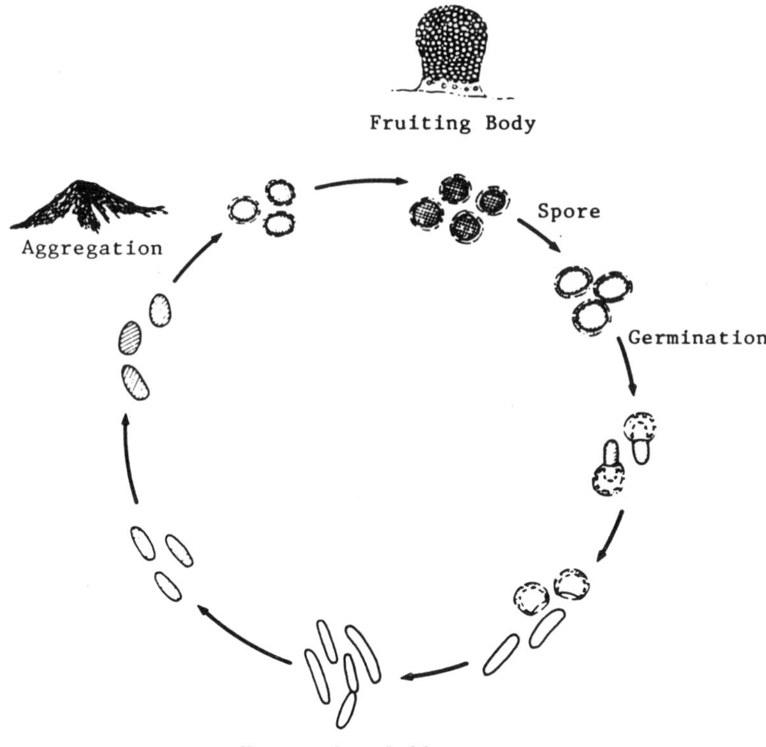

Fig. 1. Life cycle of Myxobacteria. When nutrients are depleted from a solid culture medium, vegetative cells aggregate to form a fruiting body. The rod-shaped vegetative cells convert to round or ovoid spores, which are resting cells resistant to many environmental factors such as desiccation and high temperature.

ture was found while we were determining the structure of msDNA from another myxobacterium, *Stigmatella aurantiaca* (Furuichi *et al.*, 1987a,b). The RNA (msdRNA) attached at the 5′ end of msDNA consists of 76 bases, and is linked at the 19th G residue of msdRNA to msDNA by a 2′-5′ phosphodiester linkage. The coding region for msdRNA was located downstream of the coding region of msDNA. These coding regions exist in opposite orientation with respect to each other and overlap by 8 bases at their 3′ ends. We also discuss possible mechanisms of biosynthesis of RNA-linked msDNA.

II. DNA STRUCTURE OF *STIGMATELLA AURANTIACA* msDNA

msDNA was prepared from *S. aurantiaca* by polyacrylamide gel electrophoresis after ribonuclease (A and T1) digestion of total DNA prepared according to Yee *et al.* (1984). To determine the DNA sequence, msDNA was labeled at the 5′ end with [γ-^{32}P]ATP using T4 polynucleotide kinase or at the 3′ end with [α-^{32}P]GTP using terminal deoxynucleotidyl transferase. These end-labeled msDNAs were found to be separated into two bands on a 6% polyacrylamide sequencing gel. Each of these bands was extracted from the gel and subjected to Maxam–Gilbert DNA sequencing reactions (Maxam and Gilbert, 1980). The DNA sequences of both bands were thus determined as shown in Fig. 5 (below). The two bands were found to be heterogeneous only at their 3′ ends with one consisting of 162 bases and the other 163 bases. When *S. aurantiaca* msDNA (Furuichi *et al.,* 1987a) was compared with *M. xanthus* msDNA (Yee *et al.,* 1984), it was found that they are approximately 81% homologous and there are 21 base substitutions, 5 insertions, and 4 deletions between them. Almost all these base changes are found in the stem region of the highly structured msDNA in such a way as the base changes simultaneously occurred on both strands of the stem structure to conserve the stem structure. In the course of these experiments, we found that for both bands, the majority of the radioactivity at the 5′ end was released by the sequencing reactions and migrated with the front on a 20% sequencing gel. This alkaline-labile material was identified as adenine ribonucleotide 5′-monophosphate (pA) after ribonuclease P1 treatment.

III. RNA SEQUENCE OF RNA-LINKED msDNA

The results obtained above indicated that an RNA molecule is attached at the 5′ end of msDNA. To estimate the size of the RNA, we prepared msDNA without ribonuclease treatment or with ribonuclease A or T1 treatment and applied these to a 5% polyacrylamide gel. As shown in Fig. 2, msDNA was found to migrate more slowly if it was not treated with ribonucleases before gel electrophoresis (lane 2) in comparison to msDNA treated with ribonuclease T1 (lane 3) and msDNA treated with ribonuclease A (lane 4). These results indicate that msDNA is attached to a long RNA and that ribonuclease A or T1 removes a part of the RNA molecule.

Next, to investigate how msDNA is synthesized with the RNA mole-

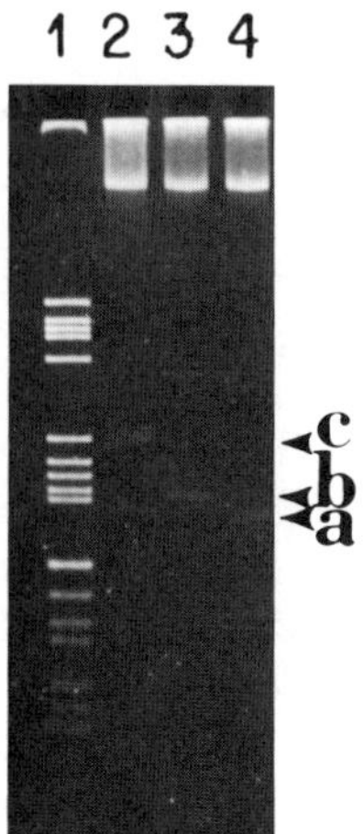

Fig. 2. Mobility of *S. aurantiaca* msDNA with or without ribonuclease treatment. RNA preparations were analyzed by 5% polyacrylamide gel electrophoresis. After electrophoresis, the gel was digested *in situ* with ribonuclease A. Lane 1, pBR322 *Hae*III digests as molecular-weight markers; lane 2, total RNA; lane 3, total RNA digested with ribonuclease T1; lane 4, total RNA digested with ribonuclease A. Arrows with a, b, and c indicate the positions of msDNA after different treatments, as described above. (From Furuichi *et al.*, 1987b. Copyright held by Cell Press.)

cule, cells were pulse-labeled with [^{3}H]thymidine for 30 min and chased with nonradioactive thymidine for 15, 35, 75, 150, and 270 min. Total RNA prepared for each time point was then analyzed by polyacrylamide gel electrophoresis with or without prior ribonuclease treatment. It was found that msDNA was first synthesized as a precursor form (P form) and chased to a mature form (M form) which migrated faster than P-form msDNA on a denaturing gel. The half-life of the precursor was approximately 20 min, and the mature form of msDNA was found to be extremely stable. It should be noted that both P and M forms gave rise to the same msDNA band if treated with ribonuclease (A + T1) prior to gel electrophoresis.

To determine the nucleotide sequences of RNA attached to precursor and mature forms of msDNA, they were purified by gel electrophoresis and were kinased with [γ-^{32}P]ATP at their 5′ end. Subsequently they were digested with *Rsa*I to shorten the DNA portion to 49 bases. They were then subjected individually to RNA sequencing enzyme reactions and applied to a 20% polyacrylamide sequencing gel (Fig. 3). To our surprise, there was a big gap within the sequencing ladders for both forms. As shown in Fig. 3, an 18-base sequence for the P form and 3-base sequence for M form were identified and then these sequences were interrupted by a gap. It should be noted that the 3-base sequence of M form is identical to the sequence from the 16th base to the 18th base (UCA) of the P-form RNA. This indicates that the M-form RNA is derived from the P-form RNA as a result of processing of the P-form RNA between A15 and U16.

The existence of the gap can be explained if the nucleotide next to residue A18 is linked to msDNA. When the same reaction mixtures were analyzed using an 8% sequencing gel, both P-form and M-form RNA gave the identical sequence. This result supports the notion that the gap was caused by msDNA which is branched out from the middle of the RNA molecule.

IV. DETERMINATION OF THE LINKAGE BETWEEN RNA AND msDNA

It was possible to label RNA linked msDNA at three different positions; (1) at the 3′ end of msDNA with terminal deoxynucleotidyl transferase using [α-^{32}P]GTP or [α-^{32}P]ddATP, (2) at the 5′ end of RNA with T4 polynucleotide kinase using [γ-^{32}P]ATP, and (3) at the 3′ end of RNA with T4 RNA ligase using [5′-^{32}P]pCp. These results are consistent with branched RNA linked to the 5′ end of msDNA. This structure would resemble the branch point of a lariat RNA which is an intermediate during RNA splicing (Ruskin *et al.*, 1984; Sharp, 1984). To test this possibility, an msDNA preparation treated with ribonucleases A and T1, was kinased at the 5′ end with [γ-^{32}P]ATP. The labeled msDNA was separated from unreacted [γ-^{32}P]ATP using a Sephadex G-50 column. The purified msDNA was then treated with a debranching enzyme (2′,5′-phosphodiesterase) from HeLa cells (a generous gift from Drs. J. Arenas and J. Hurwitz), which is capable of hydrolyzing the 2′-5′ phosphodiester linkage of lariat RNA to yield 2′-OH and 5′-phosphate. The reaction product was electrophoresed on an 8% polyacrylamide sequencing gel and the radioactive band that migrated much faster than the original msDNA was extracted from the gel.

The radioactive material now became sensitive to digestion by ribonuclease T1 and the product migrated one base shorter than the original material. This new product was found to correspond to a dinucleotide obtained by partial alkaline hydrolysis. The treatment of the original material with ribonuclease U2 yielded a mononucleotide, which was identi-

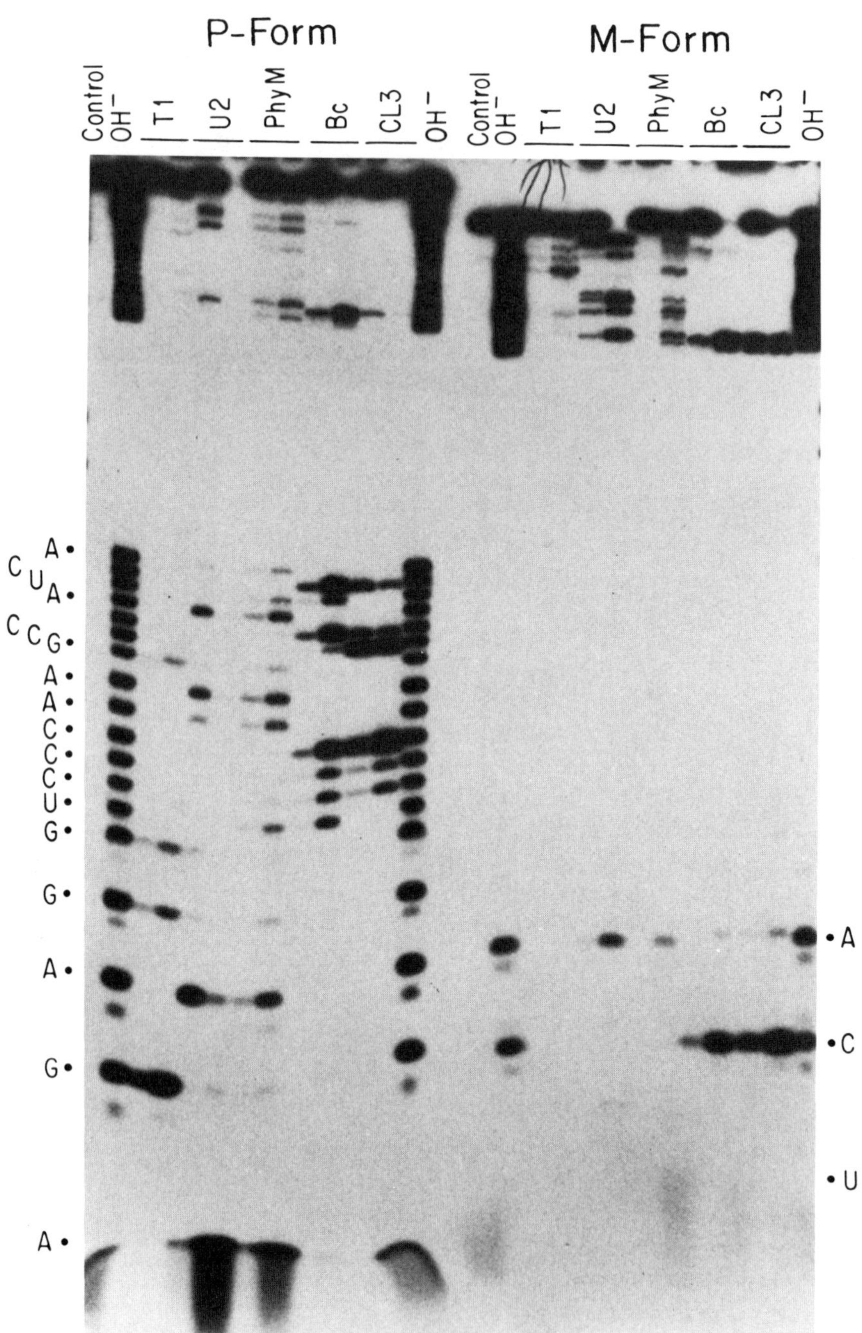
P-Form
M-Form
Control OH⁻
T1
U2
PhyM
Bc
CL3
OH⁻
Control OH⁻
T1
U2
PhyM
Bc
CL3
OH⁻
A
C U A
C C G
A
A
C
C
C
U
G
G
A
G
A
A
C
U

fied to be adenosine 5′,3′-bisphosphate. From these results, it was concluded that the nucleotide sequence of the triribonucleotide is AGY, where Y is assumed to be C from RNA sequence. Since the G residue was identified only after the treatment with the debranching enzyme, the 5′-OH position of the 5′-end residue of the DNA strand is concluded to be branched out from the 2′-OH position of the guanine ribonucleotide forming a 2′-5′ phosphodiester linkage.

To identify the 5′-end residue of the DNA strand, the reaction product, after the debranching enzyme treatment, was digested with bacterial alkaline phosphatase to remove the 5′-end monophosphate of the DNA strand. The DNA strand could only be kinased efficiently with [γ-^{32}P]ATP after the alkaline phosphatase treatment, since a phosphate remains attached at the 5′ end of the DNA strand after the debranching reaction. The 5′-end-labeled DNA was then digested with *Rsa*I (see Fig. 5, below). The resulting product was purified by 8% polyacrylamide gel electrophoresis. As shown in Fig. 4A, in addition to the original, undigested band (band 0), two new bands, bands 1 and 2, appeared. The distance between band 0 and band 1 is equivalent to three bands, which is consistent with the notion that a triribonucleotide was removed by the debranching enzyme treatment. This indicates that band 1 DNA contains the 5′-end deoxynucleotide which is directly linked to the triribonucleotide. Band 2 was probably derived from band 1 due to an exonuclease existing in the debranching enzyme preparation. Bands 1 and 2 were extracted from the gel and digested with nuclease P1. The resulting radioactive mononucleotides were identified by two-dimensional thin-layer chromotography (Fig. 4A). These were unambiguously identified to be deoxycytidine 5′-monophosphate for band 1 and thymidine 5′-monophosphate for band 2, respectively. Furthermore, the 5′-end sequence of msDNA was confirmed by sequencing the 3′-end-labeled *Rsa*I fragment of RNA-linked msDNA by the Maxam–Gilbert method. As shown in Fig. 4B, the 5′-end sequence of msDNA was clearly determined to be $^{5'}$CTTCTCACA-$^{3'}$. Thus, the structure of the RNA-linked msDNA was unambiguously determined as shown in Fig. 5.

Fig. 3. RNA sequence analysis of precursor (P form) and mature (M form) forms of msDNA labeled at the 5′ end of RNA. RNA-linked msDNA was labeled at the 5′ end of RNA with [γ-^{32}P]ATP using T4 polynucleotide kinase. The labeled msDNA was then digested with *Rsa*I, and labeled RNA–DNA fragments were gel-purified and subjected to RNA sequence analysis using various base-specific ribonucleases. The digests were electrophoresed on a 20% sequencing gel. Control, *Rsa*I-digested RNA-linked DNA; OH^-, partial alkaline hydrolysis; T1, U2, PhyM, Bc, and CL3 represent ribonucleases T1, U2, PhyM, *B. cereus,* and CL3, respectively. (From Furuichi *et al.,* 1987b. Copyright held by Cell Press.)

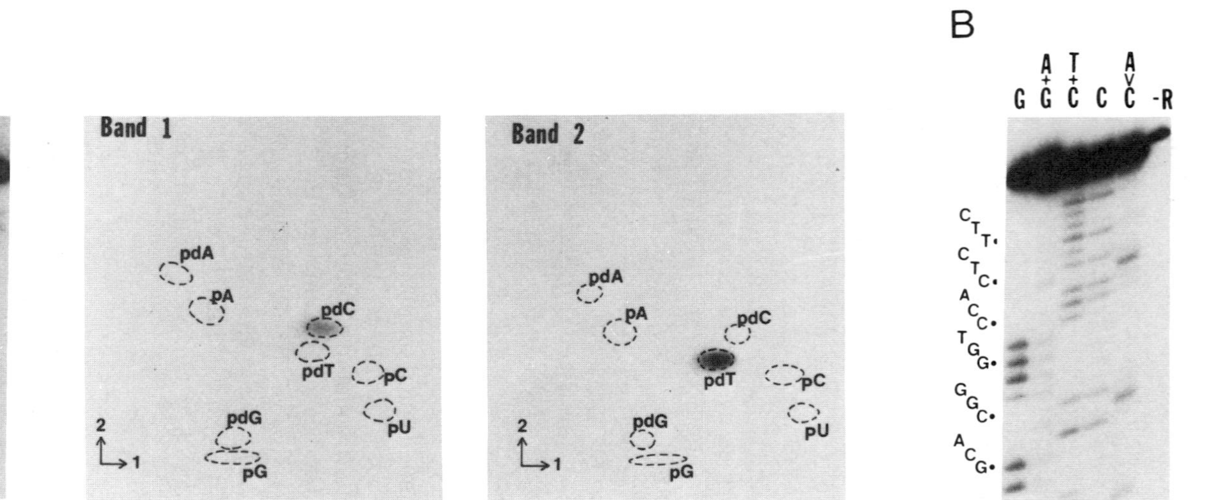

Fig. 4. Identification of the 5′-end deoxynucleotide of msDNA. (A) The msDNA preparation was treated with the debranching enzyme, followed by bacterial alkaline phosphatase treatment. The reaction products were then kinased with [γ-^{32}P]ATP, and digested with *Rsa*I. The resulting material was subjected to 8% polyacrylamide sequencing gel electrophoresis. Band 0 corresponds to the *Rsa*I fragment which was not digested by the debranching enzyme. Bands 1 and 2 represent two major products after the debranching reaction. Bands 1 and 2 were extracted from the gel and digested with nuclease P1. The resulting radioactive mononucleotides were subjected to two-dimensional thin-layer chromatography with nonradioactive standards. Cold standards were identified by UV shadowing as indicated by dotted lines. (B) An msDNA preparation was digested with *Rsa*I and labeled at the 3′ end with [α-^{32}P]ddATP. The reaction mixture was then subjected to 8% polyacrylamide sequencing gel electrophoresis to isolate the 5′-end fragment. The DNA sequence was determined according to the Maxam–Gilbert DNA sequencing method. Lane −R, which represents the sample without any sequencing reactions, migrated one base slower than the other lanes. This is because all the other samples lost the 5′-end rA as a result of piperidine treatment during the sequencing reaction. (From Furuichi *et al.*, 1987a. Copyright held by Cell Press.)

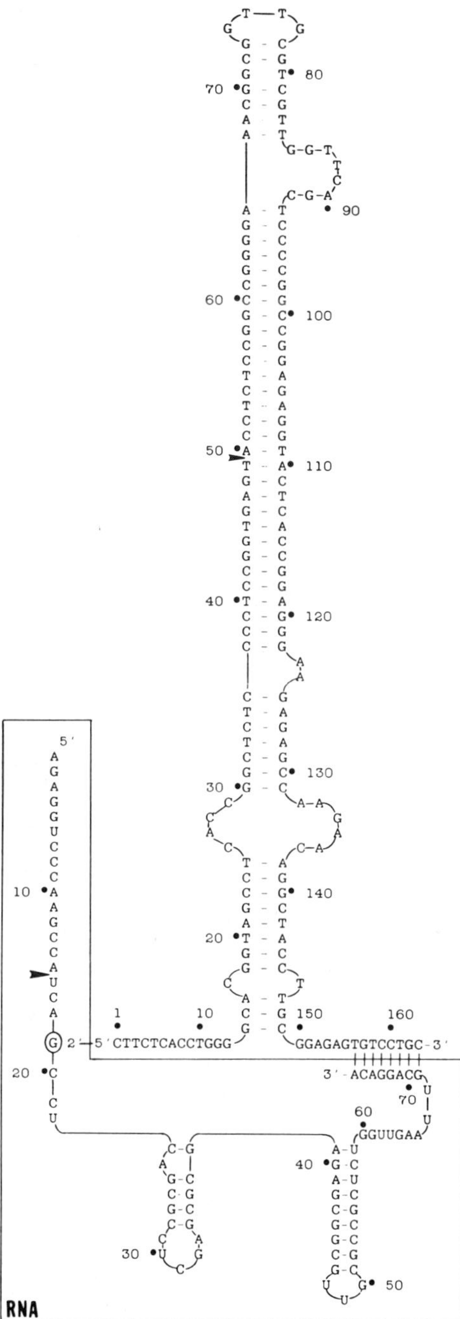

Fig. 5. Proposed secondary structure of branched RNA-linked msDNA. An arrow at the stem of msDNA indicates the *Rsa*I site. Complementary base pairings are shown by bars between two bases. A large arrow at the msdRNA indicates the processing site to produce the mature form. (From Furuichi *et al.*, 1987b. Copyright held by Cell Press.)

V. GENE ARRANGEMENT OF THE CODING REGIONS FOR msDNA (msd) AND msdRNA (msr) ON THE CHROMOSOME

A DNA fragment (1300 bp) encompassing the msDNA-coding region (*msd*) on the chromosome has been cloned and the DNA sequence of the *msd* region has been determined as shown in Fig. 6 (Furuichi *et al.*, 1987a). The completely identical sequence to msDNA was found between C_{122} at the 5′ end to G_{283} or C_{284} at the 3′ end. The RNA sequence of RNA (msdRNA) linked to msDNA did not match with the sequence immediately upstream of the msDNA coding region. Instead, this sequence was found downstream of the *msd* region in the opposite orientation and overlapping with the *msd* region at their 3′ ends. As shown in Fig. 6, the 18-

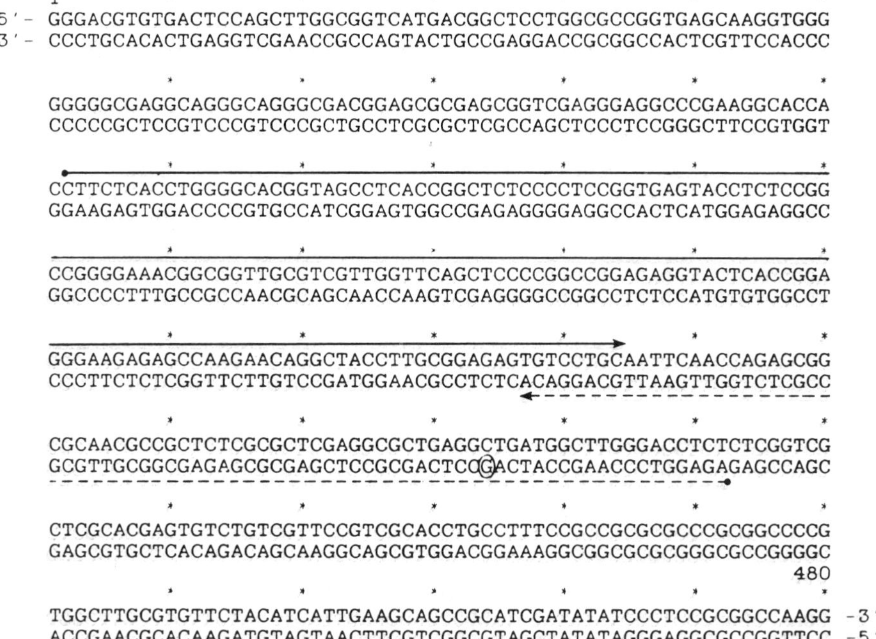

Fig. 6. Nucleotide sequence of the 480-base-pair DNA fragment encompassing the *msd* and *msr* regions from *S. aurantiaca* (Furuichi *et al.*, 1987b). A solid line and a broken line with arrows indicate the coding regions of DNA and RNA, respectively. The arrows indicate the direction from 5′ to 3′. The guanosine residue involved in forming the branching RNA is circled. (From Furuichi *et al.*, 1987b. Copyright held by Cell Press.)

base sequence before the gap of the P-form RNA (Fig. 3) is identical to the sequence from A_{352} to A_{335} on the lower strand (from right to left). The identical sequence determined after the gap was also found from C_{333} to A_{277} on the lower strand, although G_{334} in Fig. 6 could not be identified on the RNA sequencing gels. This G should correspond to the rG residue from which msDNA is branched out by a 2′-5′ phosphodiester linkage. Thus, it is reasonable to assume that the RNA molecule attached to msDNA is derived from the region downstream of the *msd* region from A_{352} to A_{277}.

Since the *msd* region and the *msd*RNA coding region (*msr*) overlap by 8 bases at their 3′ ends (see Fig. 6), complementary hybrid formation between msDNA and msdRNA at their 3′ ends is predicted. If so, the 3′ end portion of msdRNA should be susceptible to ribonuclease H, which digests RNA within RNA–DNA hybrids. Both precursor and mature forms were labeled with [α-^{32}P]ddATP at their 3′ ends of msDNA by using terminal deoxynucleotidyl transferase and subjected to ribonuclease H digestion. It was found that both of them were shortened by one to eight bases as a result of the ribonuclease H treatment. When msdRNAs were labeled at their 3′ ends by RNA ligase with [5′-^{32}P]pCp, the radioactivity was easily removed by ribonuclease H. On the basis of these results, it was concluded that the 3′ end position of msdRNA forms a duplex with msDNA and the secondary structure of the precursor msDNA is proposed in Fig. 5.

VI. BIOSYNTHESIS OF BRANCHED RNA-LINKED msDNA

To characterize the biosynthetic mechanism of msDNA, various antibiotics were tested on msDNA production. Total DNA was isolated from cells incubated with [^{3}H]thymidine in the presence and the absence of inhibitors and analyzed by polyacrylamide gel electrophoresis. The incorporation of [^{3}H]thymidine into the chromosomal DNA was severely inhibited by nalidixic acid (100 μg/ml) while the production of msDNA was little affected by the same drug.

In contrast to nalidixic acid, rifampicin (20 μg/ml) as well as chloramphenicol (100 μg/ml) showed a strong inhibitory effect on msDNA synthesis. These results indicate that msDNA synthesis requires a rifampicin-sensitive RNA polymerase and labile protein factor(s), but does not require DNA gyrase whose activity is blocked by nalidixic acid.

VII. HOW IS BRANCHED RNA-LINKED msDNA SYNTHESIZED?

At this stage it is not clear how msDNA is synthesized. It is possible that the DNA portion of msDNA is synthesized by a conventional mechanism with an RNA primer originating from the region immediately upstream of the msDNA coding region (*msd*) like the synthesis of Okazaki fragments (Kitani *et al.*, 1985). If this is the case, the RNA primer would be replaced with the msdRNA to form the branched structure. Another model is that the msdRNA molecule itself directly serves as the primer for msDNA synthesis. Since there is a significant homology between the 5′-end sequence of the msdRNA molecule (G_2 to A_{18}) and the DNA sequence immediately upstream of the *msd* region (Fig. 6), the msdRNA molecule could hybridize to chromosomal DNA. As a result, the G residue at position 19 of msdRNA is placed at a position, where it would be able to serve as a primer of msDNA synthesis (from residue C at position 122 on the chromosomal DNA shown in Fig. 5).

On the basis of this model, a possible mechanism of msDNA synthesis is illustrated in Fig. 7: The msdRNA region (RNA primer) is first transcribed by a rifampicin-sensitive RNA polymerase. The RNA primer then serves as a primer by hybridizing with the DNA sequence immediately upstream of the msDNA coding region. However, it is unknown how the 2′ position of G_{19} of msdRNA is activated for initiation of msDNA synthesis.

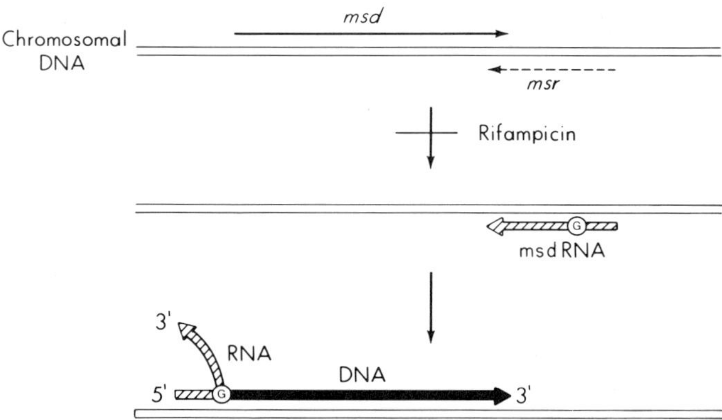

Fig. 7. Proposed mechanism for synthesis of RNA-linked msDNA. Solid and broken lines with arrows indicate the regions and the 5′ → 3′ orientations of the coding regions of msDNA and msdRNA, respectively. Solid and hatched bars indicate msDNA and msdRNA, respectively. The G residue involved in the formation of the branched RNA is circled.

Alternatively, the primary product of msdRNA may be much longer at either the 5′ and/or 3′ ends than the 76 nucleotides of msdRNA. Such an RNA may form a partial duplex at the region between G_2 and A_{18} msdRNA which may be able to initiate msDNA synthesis using RNA as a template by a reverse transcriptase type activity.

At present, we do not know the function of msDNA. However, in the view of the following facts that (1) msDNA has been found in various myxobacteria isolated from nature (Dhundale *et al.,* 1985) (2) msDNA of *M. xanthus* seems to have a similar branched RNA at the 5′ end (A. R. Dhundale, T. Furuichi, M. Inouye, and S. Inouye, unpublished results), and (3) *M. xanthus* has another smaller msDNA in addition to msDNA already identified (A. R. Dhundale, M. Inouye, and S. Inouye, unpublished results), it is likely that msDNA is playing an important role during the life cycle of myxobacteria. Genetic analysis of the *msd* and *msr* regions, and further biochemical characterization of the RNA-linked msDNA in the cell such as its biosynthesis and identification of protein components associated with msDNA, should provide insight into the function of msDNA.

REFERENCES

Dhundale, A. R., Furuichi, T., Inouye, S., and Inouye, M. (1985). Distribution of multicopy single-stranded DNA among myxobacteria and related species. *J. Bacteriol.* **164,** 914–917.

Furuichi, T., Dhundale, A. R., Inouye, M., and Inouye, S. (1987a). Branched RNA covalently linked to the 5′ end of a single-stranded DNA in *Stigmatella aurantiaca:* Structure of msDNA. *Cell (Cambridge, Mass.)* **48,** 47–53.

Furuichi, T., Inouye, S., and Inouye, M. (1987b). Biosynthesis and structure of stable branched RNA covalently linked to the 5′ end of multicopy single-stranded DNA (msDNA) of *Stigmatella aurantiaca. Cell* (*Cambridge, Mass.*) **48,** 55–62.

Kaiser, D., Manoil, C., and Dworkin, M. (1979). Myxobacteria: Cell interactions, genetics, and development. *Annu. Rev. Microbiol.* **33,** 595–639.

Kitani, T., Yoda, K., Ogawa, T., and Okazaki, T. (1985). Evidence that discontinuous DNA replication in *Escherichia coli* is primed by approximately 10 to 12 residues of RNA starting with a purine. *J. Mol Biol.* **184,** 45–52.

Maxam, A., and Gilbert, W. (1980). Sequencing end-labeled DNA with base-specific chemical cleavages. *In* "Methods in Enzymology" (L. Grossman and K. Moldave, eds.), Vol. 65, pp. 499–560.

Rosenberg, E., ed. (1984). "Myxobacteria, Development and Cell Interactions." Springer-Verlag, Berlin and New York.

Ruskin, B., Kainer, A. R., Maniatis, T., and Green, M. R. (1984). Excision of an intact intron as a novel lariat structure during pre-mRNA splicing *in vitro. Cell* (*Cambridge, Mass.*) **38,** 317–331.

Sharp, P. A. (1984). Lariat RNAs as intermediates and products in the splicing of messenger RNA precursors. *Science* **225,** 898–903.

Yee, T., and Inouye, M. (1981). Reexamination of the genome size of myxobacteria, including the use of a new method for genome size analysis. *J. Bacteriol.* **145,** 1257–1265.

Yee, T., and Inouye, M. (1982). Two-dimensional DNA electrophoresis applied to the study of DNA methylation and the analysis of genome size in *Myxococcus xanthus*. *J. Mol. Biol.* **154,** 181–196.

Yee, T., Furuichi, T., Inouye, S., and Inouye, M. (1984). Multicopy single-stranded DNA isolated from a gram-negative bacterium, *Myxococcus xanthus*. *Cell (Cambridge, Mass.)* **38,** 203–209.

18

Recognition of RNA by Proteins

OLKE C. UHLENBECK, HUEY-NAN WU, AND JEFFREY R. SAMPSON

Department of Chemistry and Biochemistry
University of Colorado
Boulder, Colorado 80309

The aim of the research in our laboratory is to understand the molecular basis for how RNA molecules interact specifically with proteins. The general approach has been to make changes in the sequence of an RNA molecule and measure the functional consequence with an appropriate biochemical assay. In this paper two examples of such "structure–function" experiments will be discussed and some recent results presented.

I. A SIMPLE RNA–PROTEIN INTERACTION

Relatively early after bacteriophage infection, the product of the R17 bacteriophage coat protein gene reaches a sufficiently high concentration to bind to a single hairpin loop on the phage genome that corresponds to the initiation site of the phage-encoded replicase subnunit. As a result of this interaction, ribosomes cannot bind to the replicase initiation site and translation of the replicase gene is repressed.

The solution properties of the interaction of R17 coat protein to a synthetic 21-nucleotide RNA hairpin loop has been studied extensively using a Millipore filter binding assay (Carey *et al.*, 1983a; Carey and Uhlenbeck, 1983). In addition, the specificity of the interaction was studied using a collection of about 50 different variants of the native sequence synthesized by laboriously joining short fragments with RNA ligase (Carey *et al.*, 1983b; Romaniuk *et al.*, 1987). More recently we have developed a new method for the synthesis of short RNA fragments using transcription

Molecular Biology of RNA
New Perspectives

of synthetic DNA with T7 RNA polymerase (Lowary *et al.*, 1985; Milligan *et al.*, 1987). This method uses a DNA template prepared by annealing an 18-nucleotide T7 promoter fragment to a 39-nucleotide template strand containing the T7 promoter followed by the complement of the sequence desired. Under optimal reaction conditions, more than 200 moles of the 21-nucleotide transcript per mole of template DNA can be obtained. This procedure allowed the facile synthesis of 20 additional variants that completed our understanding of how the specific interaction occurs.

Interpretation of coat protein binding to variant RNAs is complicated by the fact that the structure of the variant RNA may be quite different from the wild-type sequence, even when only a single nucleotide is changed. For example, Fig. 1 presents binding data for several variants where the bulged A residue present in the wild-type sequence is replaced by a G residue. It is clear that the A $\rightarrow$ G change results in a large decrease in the K_a to coat protein when the hairpin can refold into an alternate conformational form. When no alternate folding state can form, G can substitute for A in this position. The important conclusion of this example is that results from single base substitutions, as would be obtained from mutagenesis studies, can lead to an incorrect interpretation of molecular events.

The data from a large number of variant binding fragments are summarized in Fig. 2. Four of the single-stranded residues (circled) cannot be substituted without a reduced K_a to coat protein, whereas the remaining three must be there but can be substituted. Disrupting the upper five base pairs results in poor protein binding but, to a first approximation, any base pair can be tolerated at these positions. The simplest interpretation of these data is that the base pairs are needed to hold the essential single-stranded residues in a configuration so that they contact the protein. However, one cannot rule out the possibility that some of the single-stranded residues do not contact the protein directly, but are also required to ensure the correct overall conformation of the hairpin. For example, the bulged A (or G) residue may be intercalated into the helix, thus altering the distance of contacts on either side of the residue. C and U may not be able to intercalate.

Figure 3 gives a model for the R17 coat protein binding site based on the data in Fig. 2. It has been tested by preparing RNA fragments that have sequences very different from the wild-type sequence that fit the model, and in all cases they were shown to bind coat protein. Thus, we feel confident that in this sample case the structural requirements for protein binding are understood.

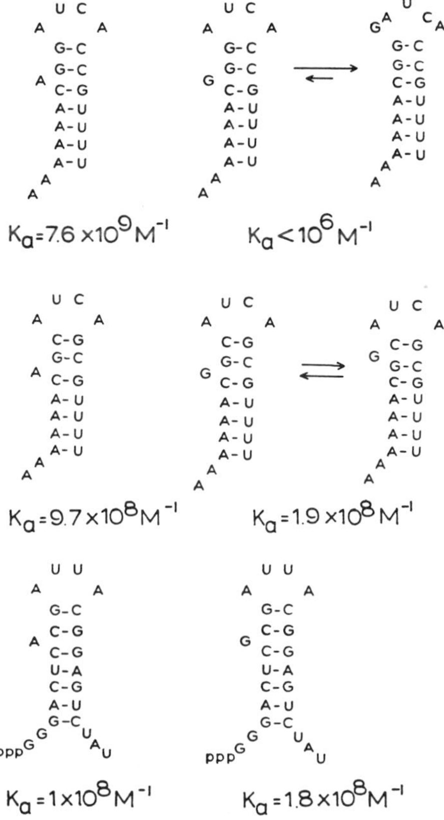

Fig. 1. Substitution of the bulged A residue with a G in three different binding fragments. In the top two cases, the G-containing fragments can take on two conformational forms, reducing K_a. [Data from Wu and Uhlenbeck (1987).]

In an attempt to identify which functional groups of the essential bulged A residue were needed for protein binding, a series of variants with modified purines at this position were constructed and assayed for coat protein binding (Wu and Uhlenbeck, 1987). As seen in Table I, in all cases where bulky substituents were introduced onto the purine ring, protein binding was reduced severely. Since these molecules would be expected to intercalate quite well, this result argues against the intercalation model discussed above and supports the idea that functional groups on the bulged A interact directly with the protein. However, the identification of essential hydrogen bond donor or acceptor sites by removal of the appropriate

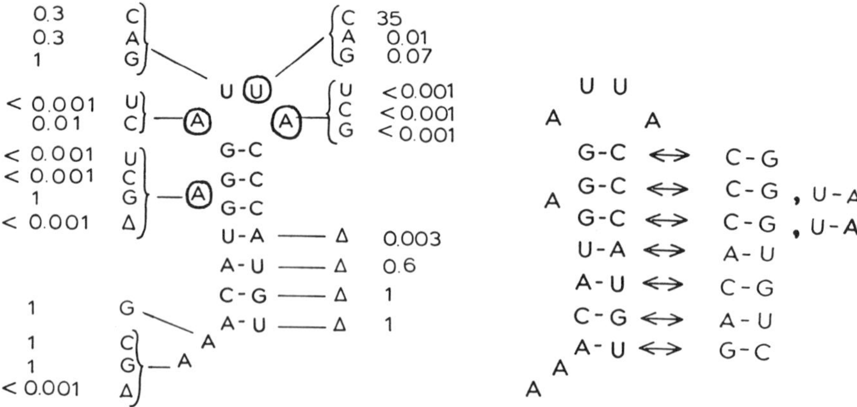

Fig. 2. The affinity of variants of the R17 replicase translational operator to R17 coat protein. Data on the left is normalized to the K_a of the wild-type sequence ($3 \times 10^8\ M^{-1}$). △ indicates that the nucleotide is deleted. On the right, the indicated base pairs were substituted without altering K_a substantially.

functional groups was not possible due to the very small effects on K_a (Table I).

The fact that R17 coat protein "recognizes" a combination of structure and sequence makes it difficult to identify sequences that would be capable of binding coat proteins unless one knows the requirements rather thoroughly. For example, Table II gives the sequences of eight fragments that bind coat protein followed by four that do not. Although one might be

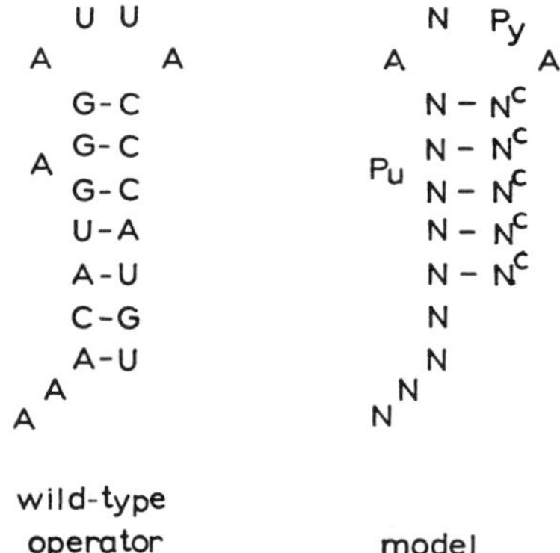

Fig. 3. A model for the R17 coat protein binding site. N^c indicates an arbitrary nucleotide complementary to N.

TABLE I

Coat Protein Binding to RNA Fragments with Substitutions of the Bulged A Residues[a]

Substitution	Relative K_a
A	1
G	1.8
U	<0.001
C	<0.001
m^6A	0.16
m^6_2A	<0.005
m^1A	<0.0005
7-deaza A	0.9
2′OmA	0.9
εA	<0.005
I	2.4
n^2Pu	1.8

[a] Data taken from Wu and Uhlenbeck (1987).

able to identify several of the essential nucleotides (circled in Fig. 2 and marked by arrows in Table II) by the "consensus" approach used for identifying DNA-binding proteins, it would require many more examples than the eight given. As an additional complexity, three of the four fragments that do not bind coat protein contain all the consensus nucleotides. The reason that these fragments do not bind relates to their structure in

TABLE II

Attempt to Develop a Consensus Binding Site

RNA fragments that bind R17 coat protein	
	↓ ↓ ↓↓
1	AAACAUGAGGAUUACCCAUGU
2	GGGACUCGCGAUUACGGAGUC
3	AAAAAACAGCAACAGCGUUUU
4	GAGACUGACGAUUACGCAGUC
5	GGGCAUGAGGAUUACCCAUGC
6	GGGACUGACGAUCACGCAGUC
7	GGGACUUAGGAUUACCAAGUC
8	GGGACUUAUGAUUACAAAGUC
RNA fragments that do not bind R17 coat protein	
9	AAACAUGAGGCUUACCCAUGU
10	AAAAAAUAUAAACAUAAUUUU
11	AAAAAACGGGAUUACCGUUUU
12	GGGACUGAGGAUUACCUCAGU

each case. Due to its high AU content, fragment 10 does not have a stable enough secondary structure to form the hairpin. Fragment 11 is one of those in Fig. 1 that takes on an alternate conformation. Fragment 12 has a U residue opposite the bulged A that can pair with it to form an uninterrupted helix. If the R17 coat protein binding site is characteristic of other protein-binding sites on RNA, it will be difficult to identify these sites by simple sequence comparisons.

II. A COMPLEX RNA–PROTEIN INTERACTION

Each of the 20 aminoacyl tRNA synthetases must be able to distinguish its cognate group of isoacceptor tRNAs from more than 100 different tRNAs in the cell. This synthetase–tRNA interaction must be highly specific to ensure the low error rate in protein synthesis. This example of protein–RNA recognition is likely to be much more complex than the R17 system. Not only are the tRNA and synthetase molecules much larger, but the reaction pathway undoubtedly includes multiple steps, including a proofreading step responsible for the high accuracy of the reaction. It is quite possible that different specific contacts between RNA and protein may occur at each step, with accompanying conformational changes of either macromolecule. Nevertheless, it appears likely that, like the R17 case, contacts between the protein and the RNA are spread out over the surface of the two macromolecules and the overall structure of the tRNA will be essential for specific interaction. The major available model proposes that the area of contact lies along the inside of the L-shaped tRNA tertiary structure (Rich and Schimmel, 1977).

A good deal of evidence suggests that one or more of the anticodon nucleotides interact specifically with the synthetase in several cases (Kisselev, 1985). Using tRNAs modified enzymatically in their anticodon loops, our lab has provided two types of experiments that indicate that yeast phenylalanine synthetase makes specific anticodon contacts with yeast $tRNA^{Phe}$. First, changing any one of the three anticodon nucleotides to another nucleotide reduces the K_m of the aminoacylation reaction (Bruce and Uhlenbeck, 1982). Second, when the ψ-35 of yeast $tRNA^{Tyr}$ is changed to an A, the resulting tRNA has the same anticodon as $tRNA^{Phe}$ and shows substantially increased misacylation with phenylalanine synthetase (Bare and Uhlenbeck, 1985). However, since the altered $tRNA^{Tyr}$ does not show a rate of aminoacylation similar to $tRNA^{Phe}$, it is clear that other nucleotides on $tRNA^{Phe}$, not present on $tRNA^{Tyr}$, are needed for optimal interaction with the synthetase. To change nucleotides elsewhere in $tRNA^{Phe}$ not accessible to the nuclease–ligase methodology, we have

TABLE III

Aminoacylation of tRNA Variants[a]

tRNA	K_m (n*M*)	V_{max}
$tRNA^{Phe}$	100	(1.0)
T7 RNA^{Phe}	300	0.94
$G_{20} \rightarrow U$	2000	0.46
$G_{19} \rightarrow C$	1360	0.76
$C_{56} \rightarrow G$	1250	0.86
$G_{19}C_{56} \rightarrow C_{19}G_{56}$	310	2.0

[a] Data taken from Sampson and Uhlenbeck (1987).

constructed a plasmid that contains the T7 promoter precisely adjacent to the $tRNA^{Phe}$ gene (Lowary *et al.*, 1985; Sampson and Uhlenbeck, 1987). Cleavage of the DNA and *in vitro* runoff transcription results in a transcript with a size and sequence identical to $tRNA^{Phe}$ but without modified nucleotides (Fig. 4). This transcript shows virtually identical aminoacylation kinetics as fully modified yeast $tRNA^{Phe}$ (Table III). No increase in the amount of misacylation of yeast tyrosine synthetase is observed. Thus, the modified nucleotides do not appear to affect the aminoacylation reaction.

The above observation permits a systematic approach to understanding the specificity of the aminoacylation reaction by the same approach used in the R17 project. Unlike the analysis of mutant suppressor tRNAs where substitutions at certain positions could not be examined due to incomplete RNA processing of poor transcription, every possible mutant $tRNA^{Phe}$ can, in principle, be prepared and assayed using the T7 transcription approach. Considering that nearly 80 variants of the R17 coat protein binding site were made before the principles for recognizing this 21-nucleotide site were elucidated, one might expect an even larger number of $tRNA^{Phe}$ variants will have to be prepared before the synthetase recognition site can be understood. However, the knowledge of the $tRNA^{Phe}$ tertiary structure and the availability of a good deal of comparative sequence data promise to simplify the problem.

Our approach to understanding the recognition site of yeast phenylalanine synthetase follows the same logic used in the R17 project and the anticodon loop substitution experiments. Specific substitutions are made into the $tRNA^{Phe}$ sequence with the goal of disrupting its interaction with the synthetase and reducing the rate of aminoacylation. It seems highly likely that substitutions that disrupt the tertiary folding of $tRNA^{Phe}$ will

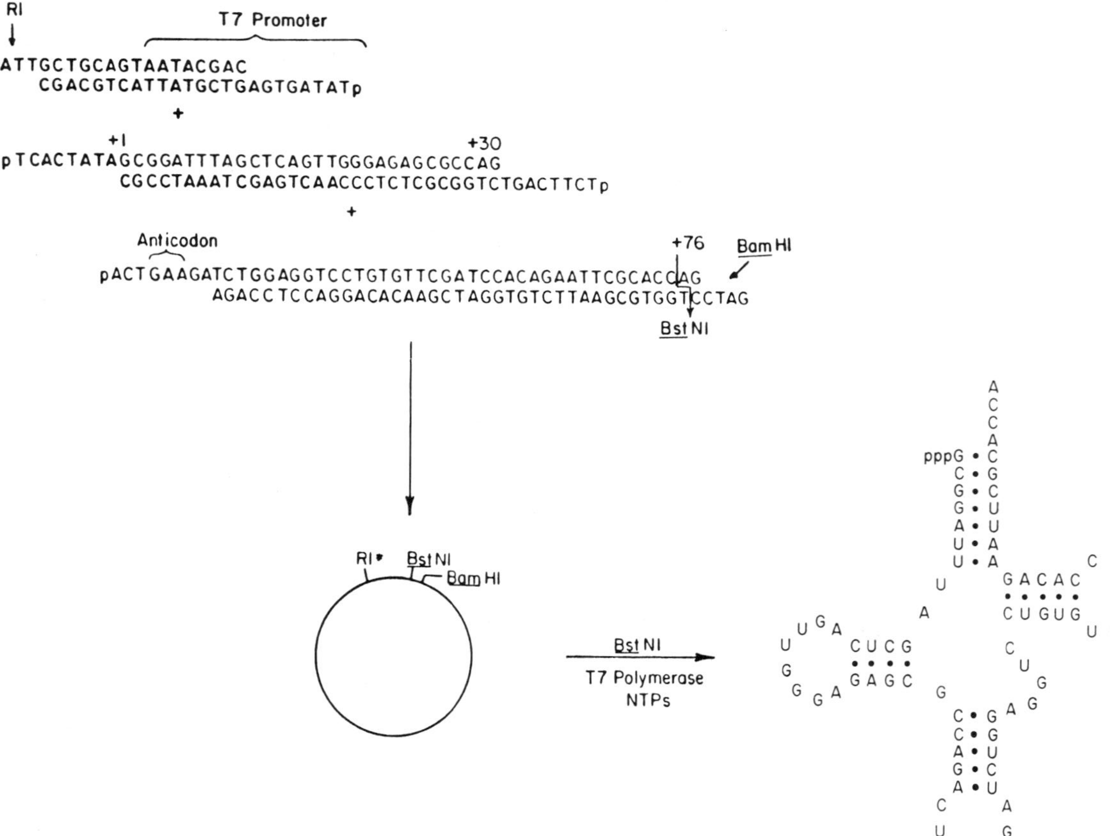

Fig. 4. Construction of the plasmid to produce the T7 transcript with the sequence of $tRNA^{Phe}$.

reduce the aminoacylation rates for the trivial reason that essential nucleotides will no longer be positioned correctly relative to one another. Therefore, we have concentrated on studying variants that are not expected to disrupt the overall structure of the molecule. This means that only positions where single base substitutions can be legitimately made are those 13 nucleotides in $tRNA^{Phe}$ that are not involved in hydrogen bond interactions with other nucleotides. These are D_{16}, D_{17}, G_{20}, G_{34}, A_{36}, Y_{37}, U_{47}, U_{59}, C_{60}, A_{73}, C_{74}, C_{75}, A_{76}. In the remaining positions, at least two nucleotides must be changed to maintain the tertiary structure. When the 18 "simple" Watson–Crick base pairs are involved, three different double substitutions are possible. However, when the more complex tertiary interactions are involved, very few substitutions can be made which do not disrupt the polynucleotide chain. Inspection of other tRNA sequences can suggest possible isomorphic substitutions, but in most cases the crystal structure is not known so these remain problematical.

Our progress thus far is limited to a few positions. We have substituted G_{20} by a U and found that this greatly reduces the rate of aminoacylation (Table III). Since G_{20} is not involved in any interaction with other nucleotides in $tRNA^{Phe}$, it is unlikely that the substitution disrupts the overall folding of the molecule and therefore suggests that the synthetase contacts G_{20} directly. This appears to contradict the Rich and Schimmel model, since G_{20} lies on the opposite side of $tRNA^{Phe}$. We have also prepared the single base variants $G_{19} \rightarrow C$ and $C_{56} \rightarrow G$ and found that both of them showed a substantially lower rate of aminoacylation. However, either single substitution would disrupt the G_{19}–C_{56} Watson–Crick tertiary base pair that connects the D loop and the TψCG loop. Thus, the poor aminoacylation may only be due to improper folding at $tRNA^{Phe}$ and not to direct interaction of the enzyme with either nucleotide. This view is supported by the fact that the C_{19}, G_{56} double mutant, which restores the base pair, was found to aminoacylate normally (Table III).

Although a large number of additional variants will have to be prepared before a recognition model similar to Fig. 3 can be proposed for yeast phenylalanine synthetase, we expect rapid progress due to the ease of preparing variants and assaying them.

REFERENCES

Bare, L. B., and Uhlenbeck, O. C. (1985). *Biochemistry* **24,** 2354–2360.
Bruce, A. G., and Uhlenbeck, O. C. (1982). *Biochemistry* **21,** 3921–3926.
Carey, J. C., and Uhlenbeck, O. C. (1983). *Biochemistry* **22,** 2610–2615.

Carey, J. C., Cameron, V., de Haseth, P. L., and Uhlenbeck, O. C. (1983a). *Biochemistry* **22,** 2601–2610.

Carey, J. C., Lowary, P. T., and Uhlenbeck, O. C. (1983b). *Biochemistry* **22,** 4723–4730.

Kisselev (1985). *Prog. Nucleic Acid Res. Mol. Biol.* **32,** 239–266.

Lowary, P., Sampson, J., Milligan, J., Groebe, D., and Uhlenbeck, O. C. (1985). *NATO ASI Ser., C* **110,** 69–76.

Milligan, J., Groebe, D., Witherall, G., and Uhlenbeck, O. C. (1987). Submitted for publication.

Rich, A., and Schimmel, P. R. (1977). *Nucleic Acids Res.* **4,** 1649–1665.

Romaniuk, P. J., Lowary, P., Wu, H. N., Stormo, G., and Uhlenbeck, O. C. (1987). *Biochemistry* **26,** 1563–1568.

Sampson, J., and Uhlenbeck, O. C. (1987) Submitted for publication.

Wu, H., and Uhlenbeck, O. C. (1987). *Biochemistry* (in press).

19

A New Role for Transfer RNA: A Chloroplast Transfer RNA Is a Cofactor in the Conversion of Glutamate to Delta-Aminolevulinic Acid

ASTRID SCHÖN,* GUIDO KRUPP,*
SIMON GOUGH,† C. GAMINI KANNANGARA,†
AND DIETER SÖLL*

**Department of Molecular Biophysics and Biochemistry*
Yale University
New Haven, Connecticut 06511
†Department of Physiology
Carlsberg Laboratory
2500 Copenhagen-Valby, Denmark

I. INTRODUCTION

Chlorophyll is the major pigment in photosynthesis. The mechanism of its formation has been studied extensively in many different organisms (Castelfranco and Beale, 1983). Its multistep biosynthesis is well characterized, and mutants affecting many of these steps have been identified (reviewed in von Wettstein and Oliver, 1985; Simpson and von Wettstein, 1980; Simpson *et al.*, 1985). One molecule of chlorophyll is synthesized from eight molecules of delta-aminolevulinic acid (ALA), the universal precursor of porphyrins. In animals, yeast, and eubacteria this compound is formed by the enzyme ALA-synthase via condensation of succinyl-

Molecular Biology of RNA
New Perspectives

coenzyme A and glycine. However, a different pathway operates in chlorophyll synthesis in higher plants, algae, cyanobacteria, and archaebacteria, where ALA is derived from glutamate (reviewed by Castelfranco and Beale, 1983; Friedman and Thauer, 1986).

Earlier biochemical studies on the enzymatic activities required for ALA synthesis in chloroplasts from barley (Wang *et al.*, 1981) and *Chlamydomonas* (Wang *et al.*, 1983) showed that the activities could be separated into three fractions (Fig. 1). Further examination of these materials led to the unexpected finding, that one of the "enzyme" fractions (Fraction H in Fig. 1) had a UV spectrum characteristic of nucleic acids. Subsequently it was shown by nuclease inactivation and reconstitution experiments that the active principle in this fraction was an RNA (Kannangara *et al.*, 1984; Huang *et al.*, 1984). This RNA, designated RNA^{DALA}, was characterized as a chloroplast $tRNA^{Glu}$ (Schön *et al.*, 1986). Our current knowledge on ALA formation is summarized in the reaction scheme outlined in Fig. 2. In the first step glutamate is esterified to RNA^{DALA} by a ligase in a linkage likely to resemble a normal aminoacyl-tRNA bond (Schön *et al.*, 1986). The nature of the ligase is not known at present. In the next step, the glutamyl-$tRNA^{DALA}$ is a substrate for the dehydrogenase. This enzyme, which has so far been refractory to purification, appears to be unique because there is no other enzyme known in interme-

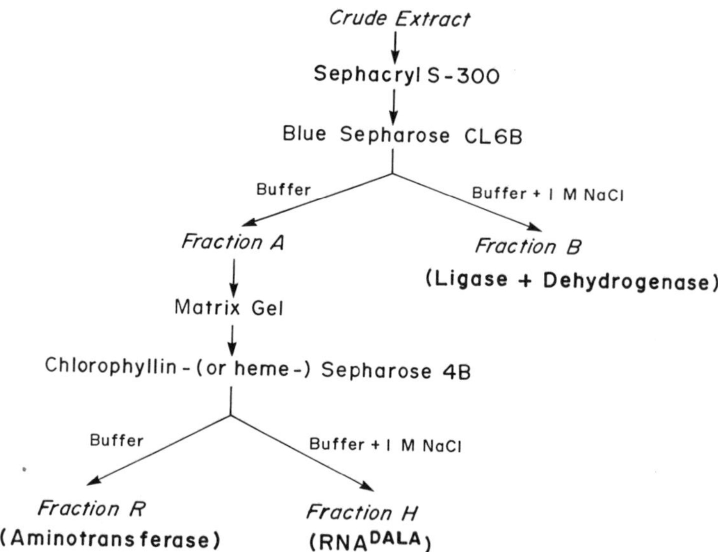

Fig. 1. Fractionation of ALA-synthesizing activities from barley chloroplast stroma extracts. For details see Kannangara *et al.* (1984).

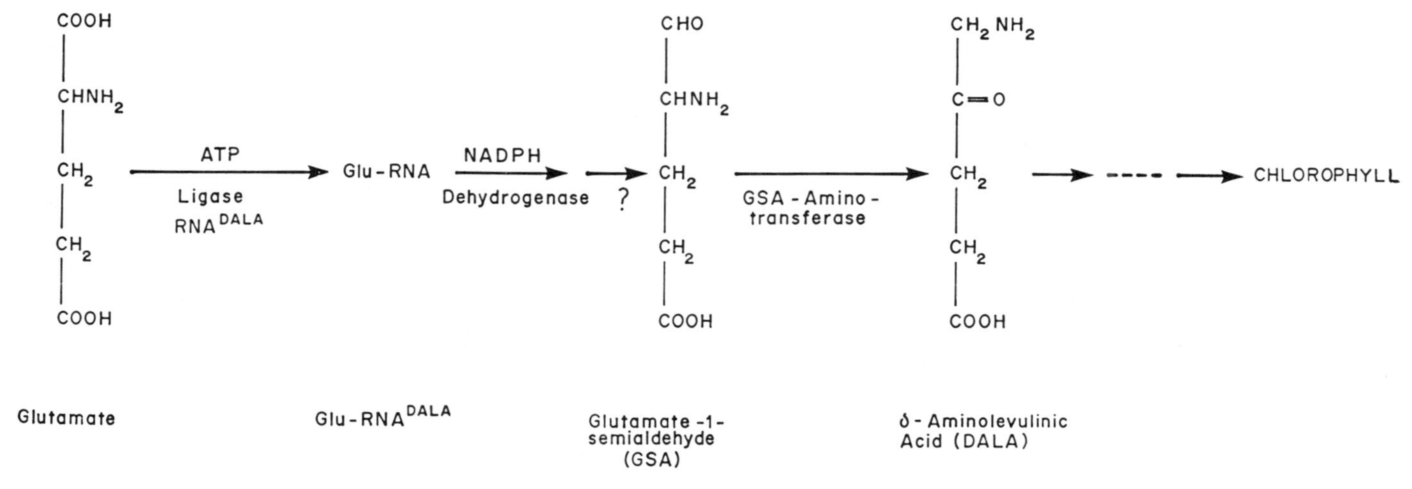

Fig. 2. Scheme of the first steps of chlorophyll synthesis. For details see text.

diary metabolism that produces glutamate-1-semialdehyde. RNA^{DALA} is a specific required cofactor of this reaction; two other glutamyl-tRNA species from barley chloroplasts cannot serve as substrates in this reaction (Schön *et al.*, 1986). However, heterologous RNA of tRNA size (presumably of chloroplast origin) from *Chlamydomonas, Chlorella,* and *Euglena* can fulfill the RNA requirement of *in vitro* ALA formation in RNA-depleted extracts from *Euglena, Chlorella,* and *Cyanidium* (Weinstein *et al.*, 1986). Extending earlier observations (Kanangara *et al.*, 1984) *E. coli*, yeast, or wheat germ tRNA could not reconstitute ALA synthesis (Weinstein *et al.*, 1986).

II. RNA^{DALA} IS A UNIQUE GLUTAMATE-ACCEPTING tRNA

Like tRNA, RNA^{DALA} can be aminoacylated enzymatically (Fraction B in Fig. 1) with glutamate in the presence of Mg^{2+} and ATP. Separation of the RNA contained in Fraction H by high-pressure liquid chromatography (HPLC) revealed three major glutamate accepting RNA fractions. Only one of them supported ALA synthesis in the reconstituted *in vitro* system. After further purification of the RNA by polyacrylamide gel electrophoresis, a pure RNA species was obtained. This RNA^{DALA} was identified as a tRNA^{Glu} species by RNA sequence analysis (Fig. 3). Hybridization of fragments of this RNA to nuclear and chloroplast DNA confirmed that it is of chloroplast origin. The essential nature of the 3′-terminal CCA sequence (a structural feature of all tRNAs) in tRNA^{DALA} for *in vitro* ALA synthesis was ascertained in the following experiment. Limited degradation with a 3′-exonuclease (removing the 3′-terminal CCA sequence) abolished ALA synthesis. The capacity to be active in ALA formation could be restored by treatment of the truncated tRNA^{DALA} with tRNA nucleotidyl transferase in the presence of CTP and ATP (i.e., restoration of the CCA sequence). In a parallel experiment [α-^{32}P]ATP was used; this allowed the unambiguous identification of the ^{32}P-labeled RNA as the active tRNA^{DALA} species.

At present tRNA^{DALA} is the only chloroplast tRNA^{Glu} species sequenced at the RNA level. Comparison with known tRNA^{Glu} gene sequences (Fig. 3) showed high homology to tRNA^{Glu} genes from chloroplasts of higher plants: the gene from wheat (*Triticum aestivum*) has a sequence identical to the barley RNA^{DALA} (Quigley and Weil, 1985). The sequence from tobacco (*Nicotiana tabacum*) differs in only three bases (Ohme *et al.*, 1985), the sequences from broad bean (*Vicia faba*), pea

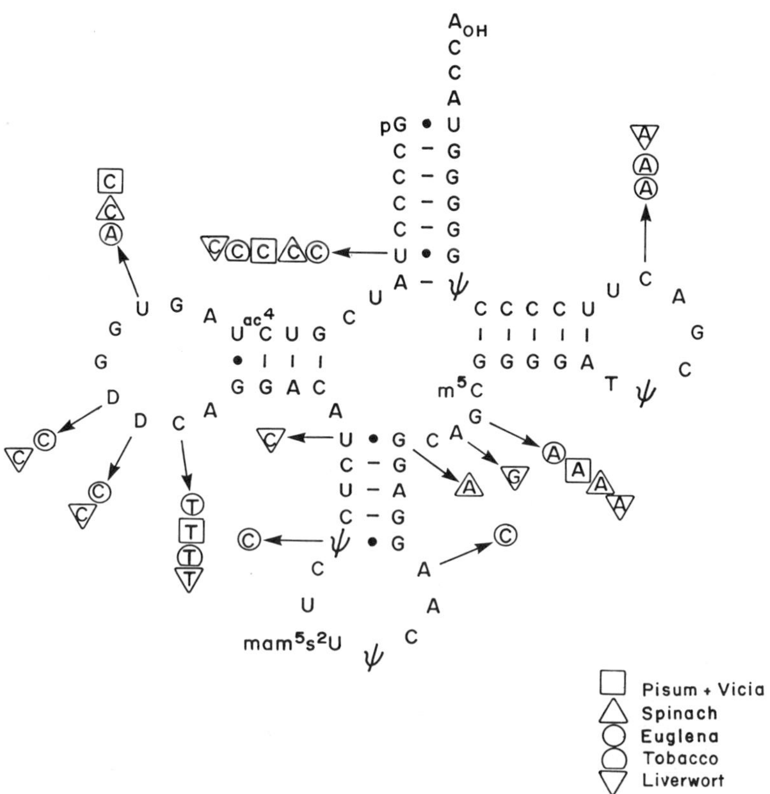

Fig. 3. Nucleotide sequence of RNA^{DALA}. Arrows designate differences in primary structure if compared with chloroplast tRNA^{Glu} genes of the following species: tobacco (Ohme *et al.*, 1985), broad bean (Kuntz *et al.*, 1984), pea (Rasmussen *et al.*, 1984), spinach (Holschuh *et al.*, 1984), *Euglena gracilis* (Hollingsworth and Hallick, 1982), and liverwort (Ohyama *et al.*, 1986). The sequence is identical to a wheat chloroplast tRNA^{Glu} gene (Quigley and Weil, 1985).

(*Pisum sativum*), and spinach (*Spinacea oleracea*) differ in four positions (Kuntz *et al.*, 1984; Rasmussen *et al.*, 1984; Holschuh *et al.*, 1984), that from liverwort (*Marchantia polymorpha*) in eight bases (Ohyama *et al.*, 1986), whereas the sequence from *Euglena gracilis* shows nine differences (Hollingsworth and Hallick, 1982). One distinctive feature of tRNA^{Glu} of chloroplast origin is also present in this tRNA: the otherwise universally present G53:C61 is replaced by a A53:U61 pair (Kuntz *et al.*, 1984; Rasmussen *et al.*, 1984).

III. RNA AS A COFACTOR: POSSIBLE REACTION MECHANISM

The activation of glutamate is clearly achieved by the formation of an ester bond with the 3′-terminal adenosine of RNA^{DALA}. The linkage between glutamate and RNA was identified as a glutamyl-adenosine ester bond after labeling the 3′ end of the purified RNA with [α-^{32}P]ATP by use of nucleotidyl transferase and charging with [^{14}C]glutamate using fraction B. The terminal [^{14}C]glutamyl-[^{32}P]adenosine 5′-phosphate was recovered after cleavage with nuclease P_1 and identified by thin-layer chromatography. However, it still needs to be established whether glutamate is linked through its 1- or 5-carboxyl group to the tRNA.

As pointed out by Haselkorn (1986) this glutamyl-$tRNA^{DALA}$ ester is a suitably activated form of the amino acid to allow the highly endergonic reduction to the aldehyde. It is interesting to note that the energy level of aminoacyl-tRNA is about 3 kcal/mole higher ($\Delta G_{hydrolysis} = -7$ kcal/mole) than in simple esters like ethyl acetate or sugar phosphates (Lehninger, 1975). Even though one can assume that the reduction of glutamyl-$tRNA^{DALA}$ to glutamate-1-semialdehyde is endergonic, a study of the detailed reaction mechanism may reveal coupling to another energy-generating step. It seems possible that the second carboxyl group is also activated, e.g., by adenylation, and the reaction is driven by the action of a hydrolase (short arrow in Fig. 2). Fortunately, there are experiments that will answer these questions.

The elucidation of the role of $tRNA^{DALA}$ in the dehydrogenase reaction is a very important question. There is no precedent as yet for RNA involvement in such a redox reaction. What determines the specificity of the RNA recognition by the dehydrogenase and what is the mechanism of the reduction? Is the glutamate group reduced while bound to the tRNA, or is the activated amino acid transferred to a site on the enzyme in an intermediate step?

IV. OUTLOOK

The major role of tRNA is in protein biosynthesis. Only a few examples are known for the involvement of aminoacyl-tRNA in other reactions: e.g., the synthesis of aminoacyl-phosphatidyl-glycerol (Nesbitt and Lennartz, 1968), the amino-terminal addition of amino acids to proteins (Soffer, 1974), and the formation of pentapeptide bridges in bacterial cell walls (Bumsted *et al.*, 1968). Analogous to the case of $tRNA^{DALA}$, the transamination of glutamyl-$tRNA^{Glu}$ to yield glutaminyl-$tRNA^{Gln}$ in *Bacillus megatherium* is an example of enzymatic conversion of an amino acid already linked to a tRNA (Wilcox and Nirenberg, 1968). However, in contrast to the utilization of the glutamate metabolite in ALA synthesis (i.e., the conversion of glutamate into another low-molecular-weight metabolite), glutamine formed in the transamination reaction is subsequently donated to peptide linkage during the process of protein biosynthesis. For the normal requirements of intermediary metabolism in *Bacillus,* glutamine is formed by glutamine synthase (Henner and Hoch, 1980).

Although the outline of the pathway from glutamate to ALA is now available, many questions about the detailed mechanism remain to be solved. Some of them will be discussed briefly. As there is little information on the enzymes, their purification is important and the elucidation of their properties will provide the key answers. For instance, is there a specific ligase for RNA^{DALA} or are all isoacceptors charged by the same enzyme?

The molecular nature of the differences between the three major glutamate accepting tRNAs is of particular interest. Sequence analysis showed a high degree of homology to known tobacco or liverwort chloroplast $tRNA^{Gln}$ genes, suggesting a mischarging with glutamate by a mechanism similar to the one observed in gram-positive bacteria (Wilcox and Nirenberg, 1968). In this case, a single aminoacyl-tRNA synthetase is responsible for the glutamylation of all isoaccepting Glu-tRNA species (Lapointe *et al.,* 1986).

The usage of aminoacyl-tRNA as an activated form of glutamate in ALA synthesis leads to competition and possible interference with protein synthesis. The different requirements of both systems may necessitate a specialized tRNA species for porphyrin synthesis. At present we do not know whether RNA^{DALA} serves only the special function of ALA formation or whether it is also used in protein biosynthesis, although a comparison of the total chloroplast tRNA with polysome-bound tRNA strongly suggests the involvement of all tRNA species in protein biosynthesis (Pfitzinger *et al.,* 1987).

How many $tRNA^{Glu}$ genes are encoded in the barley chloroplast? This is not known at present. However, very recently the total nucleotide sequence of the chloroplast genomes of liverwort and tobacco has been established. These genomes encode only one $tRNA^{Glu}$ gene (Ohyama *et al.,* 1986; Shinozaki *et al.,* 1986). Codon recognition rules require only one $tRNA^{Glu}$ isoacceptor (anticodon UUC) to satisfy the needs of protein biosynthesis, i.e., wobble allows recognition of both glutamate codons, GAA and GAG. Although s^2U derivatives in the first position of the tRNA usually restrict wobble to GAA, this is not always the case (Raba *et al.,* 1979). The question remains open whether any of the nucleotide modifications of $tRNA^{DALA}$ are required for its special function, i.e., the recognition by the enzymes of the ALA synthesis pathway.

Another question concerns the transcription mechanism of the gene encoding $tRNA^{DALA}$. Is it transcribed in a monomeric precursor RNA (i.e., with the possibility of independent transcriptional regulation) or is it part of a multimeric tRNA transcript? In tobacco it has been shown (Ohme *et al.,* 1985) that the chloroplast $tRNA^{Glu}$ gene is transcribed as part of a multimeric tRNA precursor. The other available data make it likely that a similar transcription unit may exist in spinach (Holschuh *et al.,* 1984), pea (Rasmussen *et al.,* 1984), broad bean (Kuntz *et al.,* 1984), and *Euglena* (Hollingsworth and Hallick, 1982) that in most cases includes a $tRNA^{Tyr}$.

There is considerable variation in the sequences of chloroplast $tRNA^{Glu}$ genes from different organisms (Fig. 3). This indicates a flexibility in the enzymes recognizing RNA^{DALA}: nucleotide modification enzymes, ligase, and dehydrogenase.

The regulation of RNA^{DALA} synthesis is another interesting problem. Chlorophyll synthesis can occur in the dark in a number of plant species, including gymnosperms, mosses, ferns, and most algae (Kirk and Tilney-Bassett, 1978). However, in flowering plants (angiosperms), chlorophyll synthesis in dark grown tissue is limited by the low level of available ALA. The enzymes required for the consecutive steps are already present in etiolated plants (Schneider 1976; Kannangara and Gough, 1979). Is this also the case of RNA^{DALA}, the plastid-encoded cofactor of ALA synthesis, or is its expression/nucleotide modification an intricate part of the light induction of ALA synthesis?

ACKNOWLEDGMENTS

This paper is dedicated to Professor H. J. Gross on the occasion of his fiftieth birthday. A. S. was supported by a fellowship of the Deutscher Akademischer Austauschdienst, and G. K. was a Postdoctoral Fellow of the Deutsche Forschungsgemeinschaft. This work was supported in part by a grant from the National Institutes of Health.

REFERENCES

Bumsted, R. M., Dahl, J. M., Söll, D., and Strominger, J. L. (1968). *J. Biol. Chem.* **243,** 779–782.

Castelfranco, P. A., and Beale, S. I. (1983). *Annu. Rev. Plant Physiol.* **34,** 241–278.

Friedman, H. C., and Thauer, R. K. (1986). *FEBS Lett.* **207,** 84–88.

Haselkorn, R. (1986). *Nature (London)* **322,** 208.

Henner, D. J., and Hoch, J. A. (1980). *Microbiol. Rev.* **44,** 57–82.
Hollingsworth, M. J., and Hallick, R. B. (1982). *J. Biol. Chem.* **257,** 12795–12799.
Holschuh, K., Bottomley, W., and Whitfield, P. R. (1984). *Plant Mol. Biol.* **3,** 313–317.
Huang, D.-D., Wang, W.-Y., Gough, S. P., and Kannangara, C. G. (1984). *Science* **225,** 1482–1484.
Kannangara, C. G., and Gough, S. P. (1979). *Carlsberg Res. Commun.* **44,** 11–20.
Kannangara, C. G., Gough, S. P., Oliver, R. P., and Rasmussen, S. K. (1984). *Carlsberg Res. Commun.* **49,** 417–437.
Kirk, J. T. O., and Tilney-Bassett, R. A. E. (1978). "The Plastids." Elsevier, Amsterdam.
Kuntz, M., Weil, J. H., and Steinmetz, A. (1984). *Nucleic Acids Res.* **12,** 5037–5047.
Lapointe, J., Duplain, L., and Proulx, M. (1986). *J. Bacteriol.* **165,** 88–93.
Lehninger, A. L. (1975). "Biochemistry." Worth, New York.
Nesbitt, J. A., and Lennartz, W. J. (1968). *J. Biol. Chem.* **243,** 3088–3095.
Ohme, M., Kamogashira, T., Shinozaki, K., and Sugiura, M. (1985). *Nucleic Acids Res.* **13,** 1045–1056.
Ohyama, K., Fukuzawa, H., Kohchi, T., Shirai, H., Sano, T., Umesono, K., Shiki, Y., Takeuichi, M., Chang, Z., Aota, S., Inokuchi, H., and Ozeki, H. (1986). *Nature (London)* **322,** 572–574.
Pfitzinger, H., Guillemaut, P., Weil, J. H., and Pillay, D. T. N. (1987). *Nucleic Acids Res.* **15,** 1377–1386.
Quigley, F., and Weil, J. H. (1985). *Curr. Genet.* **9,** 495–503.
Raba, M., Limburg, K., Burghagen, M., Katze, J. R., Simsek, M., Heckman, J. E., RajBhandary, U. L., and Gross, H. J. (1979). *Eur. J. Biochem.* **97,** 305–318.
Rasmussen, O. F., Stummann, B. M., and Henningsen, K. W. (1984). *Nucleic Acids Res.* **12,** 9143–9153.
Schneider, H. A. W. (1976). *Z. Naturforsch., C: Biosci.* **31C,** 55–63.
Schön, A., Krupp, G., Gough, S., Berry-Lowe, S., Kannangara, C. G., and Söll, D. (1986). *Nature (London)* **322,** 281–284.
Shinozaki, K., Ohme, M., Tanaka, M., Wakasugi, T., Hayashida, N., Matsubayashi, T., Zaita, N., Chunwongse, J., Obokata, J., Yamaguchi-Shinozaki, K., Ohto, C., Torazawa, K., Meng, B. Y., Sugita, M., Deno, H., Kamogashira, T., Yamada, K., Kusuda, J., Takaiwa, F., Kato, A., Tohdoh, N., Shimada, H., and Sugiura, M. (1986). *EMBO J* **5,** 2043–2049.
Simpson, D. J., and von Wettstein, D. (1980). *Carlsberg Res. Commun.* **45,** 283–314.
Simpson, D. J., Machold, A., and Hoyer-Hansen, G. (1985). *Carlsberg Res. Commun.* **50,** 223–238.
Soffer, R. L. (1974). *Adv. Enzymol.* **40,** 91–139.
von Wettstein, D., and Oliver, R. P. (1985). *Ber. Dtsch. Bot. Ges.* **98,** 261–287.
Wang, W.-Y., Gough, S. P., and Kannangara, C. G. (1981). *Carlsberg Res. Commun.* **46,** 243–257.
Wang, W.-Y., Huang, D.-D., Stachon, D., Gough, S. P., and Kannangara, C. G. (1983). *Plant Physiol.* **74,** 569–575.
Weinstein, J. D., Mayer, S. M., and Beale, S. I. (1986). *Plant Physiol.* **80,** Suppl. 52 (Abstr.).
Wilcox, M., and Nirenberg, M. (1968). *Proc. Natl. Acad. Sci. U.S.A.* **61,** 229–236.

20

Natural Suppressor Transfer RNA in Eukaryotes: Its Implication in the Evolution of the Genetic Code and Expression of Specific Genes

YOSHIYUKI KUCHINO*, HILDBURG BEIER†, NAOHIRO HANYU*, AND SUSUMU NISHIMURA*

**Biology Division*
National Cancer Center Research Institute
Tokyo, Japan
†Institute für Biochemie
Baverische Julius-Maximillans-Universität
D-8700 Würzburg, Federal Republic of Germany

I. INTRODUCTION

Transfer RNA (tRNA) is a macromolecule that can provide a considerable amount of information with respect to studies on molecular evolution. This information is derived not only from a simple comparison of nucleotide sequences but also from features that may be distinctive according to whether the RNAs are from prokaryotic, unicellular eukaryotic, or multicellular eukaryotic organisms. These distinctive features are observed both in the structure of the tRNAs and in the mechanisms by which their transcription is controlled.

Studies on structural aspects of tRNA might be expected to provide a rich harvest. Although tRNA consists of less than 100 nucleotide residues, it possesses unique structural features that determine species-spe-

Molecular Biology of RNA
New Perspectives

cific characteristics (1). tRNA contains numerous modified nucleosides in specific locations in the molecule. Therefore, not only the nucleotide sequence but also the structure and location of modified nucleosides can be regarded as good markers for the study of the molecular evolution of tRNA. In the case of some hypermodified nucleosides, their biosyntheses are multistep processes and many enzymes must be needed. Therefore, modified nucleosides in tRNA also reflect evolutionary diversity of proteins.

For some time we have been interested in the relationship between tRNA structure and its species specificity. For this study, the initiator methionine tRNA was thought to be the best choice for the nucleotide sequence comparison since, contrary to other amino acid-specific tRNAs which have isoaccepting species, there is only a single initiator methionine tRNA. Some years ago we showed that initiator methionine tRNAs from archaebacteria have a common unique structural feature that differs from both the true bacterial and eukaryotic initiator methionine tRNAs (2): that is, that initiator methionine tRNAs from archaebacteria contained a unique modified nucleoside, 1-methylpseudouridine, which replaces ribothymidine in the TψC loop (3).

In parallel with this work, we isolated and obtained the nucleotide sequence of the initiator methionine tRNA from *Tetrahymena thermophilia* and found that the nucleotide residue in the position next to the 5′ end of the anticodon is uridine instead of the normal cytidine found in cytoplasmic initiator methionine tRNAs from other multicellular eukaryotic organisms (4). Sequence comparison among initiator methionine tRNAs from other organisms revealed that the presence of uridine in this region is a common characteristic of unicellular eukaryotic initiator tRNAs. This suggests that the divergence of the protozoan *Tetrahymena* occurred before the animal–plant divergence.

As mentioned previously, in addition to there being distinct structural differences between tRNA methionine from eukaryotes and prokaryotes, the manner of their biosynthesis also differs. In prokaryotic cells, RNA polymerase recognizes a promoter sequence proximal to the 5′ side of the tRNA gene, whereas in the case of eukaryotes, the recognition sequence for RNA polymerase III usually resides within the coding sequence for the tRNA (5). It should be noted, however, that particular eukaryotic cells, such as in *Bombyx mori,* are known to have a recognition signal for RNA polymerase III in the 5′-flanking region, indicating that the biosynthesis of tRNA has diverged in a complex manner during evolution (6). It should also be noted that some eukaryotic tRNA genes as well as tRNA genes from archaebacteria contain an intron (7,8).

To obtain more information on the phylogeny of unicellular eukaryotes, especially *Tetrahymena,* we set out to isolate tRNA genes from *Tetrahymena* and to study their transcriptional control mechanism. These experiments led us to some unexpected findings.

II. ISOLATION FROM *TETRAHYMENA* OF THE tRNA GENE AND tRNA CORRESPONDING TO THE TERMINATION CODON UAA (9)

Tetrahymena is a ciliated protozoan, which possesses unique biological properties, such as the presence of two different gene organizations, i.e., micronuclear and macronuclear genes and the self-splicing system of pre-rRNA in the absence of proteins (10). For the isolation of tRNA genes macronuclear DNA isolated from *Tetrahymena* was partially digested with *Eco*RI and ligated into bacteriophage Charon 21A. The gene library thus obtained was screened with 5′-^{32}P-labeled *Tetrahymena* cytoplasmic tRNA partially fractionated by BD-cellulose column chromatography. From 10,000 recombinant phages examined, three positive clones were selected. One of the clones contained a DNA insert of 2.5 kbp, which hybridized to a tRNA fraction eluted by 1.5 *M* NaCl–6% ethanol and was used as a probe. Southern hybridization analysis using the same ^{32}P-labeled tRNA as a probe indicated that the DNA insert contained three tRNA genes (Fig. 1). Subsequent nucleotide sequence analysis by the dideoxy chain-termination method showed that the 2.5-kbp DNA fragment contained three identical tRNA genes in the same polarity. A most unusual feature of this tRNA gene is that its anticodon is TTA, corresponding to the UAA termination codon (Fig. 2).

To prove that a UAA-recognizing tRNA gene is actually transcribed and not merely a silent gene, an attempt was made to isolate the corresponding tRNA from total unfractionated cytoplasmic tRNA. Isolation of the corresponding tRNA and determination of its nucleotide sequence is

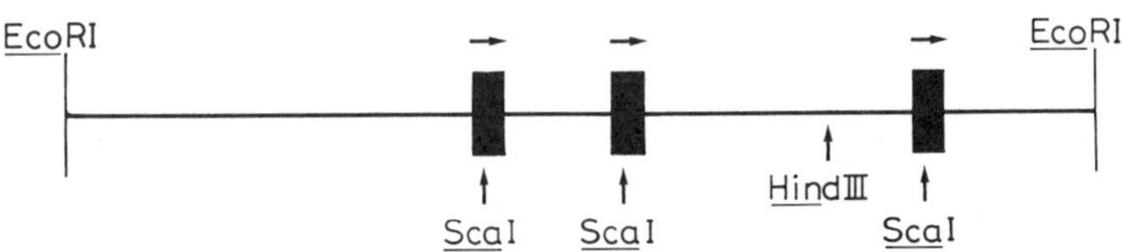

Fig. 1. Three *Tetrahymena* tRNA genes having TTA as the anticodon. The black box shows the region of the tRNA gene. The arrow shown in the upper portion of the box indicates the polarity of the tRNA gene. The cleavage sites of restriction endonucleases are indicated by vertical bars in the map.

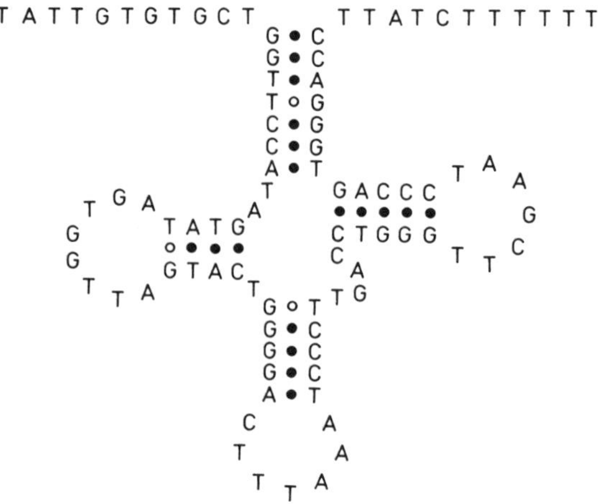

Fig. 2. Cloverleaf structure of $tDNA_{TTA}$ from *Tetrahymena*, as predicted from the DNA sequence.

also important for characterization of the actual codon recognition properties of the tRNA. For example, in the case of an *Escherichia coli* isoleucine minor tRNA species, the anticodon structure of its gene is CAT, corresponding to the methionine codon, AUG. However, the cytidine residue of the first position of the anticodon of the minor species of $tRNA^{Ile}$ is posttranscriptionally modified, thus permitting it to recognize the isoleucine codon AUA (11).

The fraction of tRNA contained in the 2 *M* NaCl–20% ethanol eluate from a BD-cellulose column was further purified by RPC-5 column chromatography, followed by two-dimensional polyacrylamide gel electrophoresis. A pure tRNA species that specifically hybridized to the cloned tRNA gene was finally obtained. The total nucleotide sequence of the purified tRNA was determined using the postlabeling procedure as described previously (12). It was found that the sequence of the tRNA coincided completely with the DNA sequence of the tRNA gene. The first position of the anticodon, residue 34, was modified to 2′-*O*-methyluridine (Um). Since 2′-*O*-methylation of the ribose residue should not make any significant change in base-pairing properties of uridine, the tRNA species corresponding to $tDNA_{TTA}$ should recognize the UAA nonsense codon. Subsequently, it was shown that this tRNA is exclusively charged by glutamine when a crude mixture of aminoacyl-tRNA synthetase isolated from *Tetrahymena* is used, indicating that it is glutamine tRNA ($tRNA^{Gln}_{UmUA}$).

III. DEVIATION OF THE GENETIC CODE OF *TETRAHYMENA* FROM THE UNIVERSAL GENETIC CODE

At the time that we found $tRNA^{Gln}_{UmUA}$ from *Tetrahymena,* several other groups reported independently that in ciliated protozoans the termination codons TAA and TAG are present internally in several structural genes. Furthermore they showed that these termination codons can be read as glutamine or glutamic acid, based on a comparison of the nucleotide sequences of those genes with the amino acid sequences or amino acid compositions of the proteins corresponding to or closely related to those genes (13–16). These results agreed well with our findings and imply that $tRNA^{Gln}_{UmUA}$ from *Tetrahymena* is actually used in protein synthesis *in vivo,* recognizing UAA as an internal codon.

Deviations in the genetic code have been found previously in the mitochondrial genomes of mammalian cells, yeast, *Neurospora,* and *Drosophila* (17–20). In the mitochondria of these organisms, the termination codon UGA is used as a tryptophan codon. In addition, codon usage for some amino acids differs from the universal genetic code. Another divergence from the universal code was recently found in *Mycoplasma,* in which UGA is used exclusively as a tryptophan codon (21). However, there had been no clear evidence to show divergence from the universal code in the nuclear genome of eukaryotes. Therefore the finding of UAA as a codon for glutamine is the first case of deviation from the universal code in eukaryotes in nuclear genomes and raises an interesting question as to the evolution of *Tetrahymena*.

IV. EVOLUTION OF GLUTAMINE tRNAs RECOGNIZING UAA AND UAG TERMINATION CODONS IN *TETRAHYMENA* (22)

According to the wobble hypothesis, U in the first anticodon position of tRNA can recognize G as well as A (23). The question, therefore, arises as to whether $tRNA^{Gln}_{UmUA}$ can also recognize the UAG codon, and if not, whether *Tetrahymena* contains another $tRNA^{Gln}$ with an unusual anticodon. Another point to be clarified is to what extent the sequence of $tRNA^{Gln}_{UmUA}$ differs from that of normal *Tetrahymena* glutamine tRNA. To answer these questions, we purified all the glutamine tRNA species from *Tetrahymena* and determined their nucleotide sequences. In addition, the codon recognition properties of these glutamine tRNAs were examined in an *in vitro* translation system using mRNAs with known termination codons, i.e., tobacco mosaic virus (TMV) RNA, which contains a UAG stop codon, and α-globin mRNA, which is terminated by a UAA stop codon.

Tetrahymena tRNA was first fractionated by BD-cellulose column chromatography. As shown in Fig. 3, glutamine acceptor activity was roughly separated into three fractions (fraction I, II, and III). The glutamine tRNAs present in fractions I, II, and III were further purified by RPC-5 column chromatography and polyacrylamide gel electrophoresis. It was found that each fraction contained one major glutamine tRNA species and also small amounts of isoacceptors corresponding to the major species of neighboring fractions. Several other minor glutamine tRNA species were also isolated during the fractionation, but their yield was very low and preliminary analyses of their nucleotide sequences indicated that they were undermodified variants of the major glutamine tRNAs. Thus, three major glutamine tRNAs were isolated from the *Tetrahymena* cytoplasmic fraction, and their nucleotide sequences were determined by the postlabeling techniques (Fig. 4). They were designated as $tRNA^{Gln}_{UmUG}$

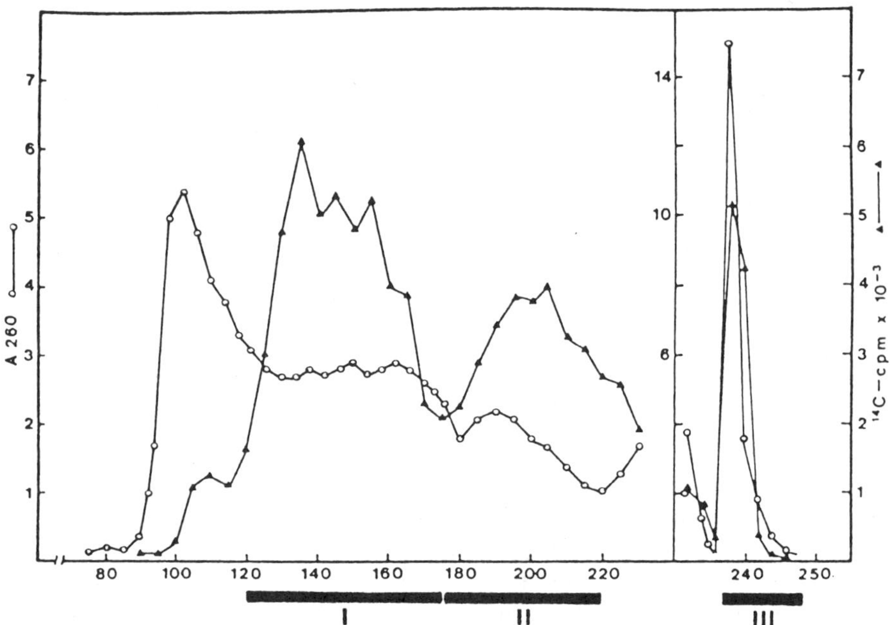

Fig. 3. Chromatographic profile of glutamine acceptor activity of *Tetrahymena* tRNA after fractionation by BD-cellulose column chromatography. The column (1.5 × 80 cm) was eluted with a linear gradient of 0.3–1.2 *M* NaCl in 0.02 *M* NaOAc, pH 6.0 (total volume, 2 liters). After 1.9 liters of effluent has been collected (up to fraction 230), elution was continued with a second linear gradient of 1.2 *M* NaCl–2 *M* NaCl/20% ethanol in 0.02 *M* NaOAc, pH 6.0 (total volume, 1 liter). The bars indicate the fractions.

(from fraction I) and $tRNA^{Gln}_{UmUA}$ (fraction III). Transfer RNA^{Gln}_{UmUA} isolated from fraction III turned out to be the one that was previously isolated by hybridization with the gene for the tRNA having TTA as the anticodon sequence.

The UAG and UAA readthrough activities of the three purified glutamine tRNA species were assayed in a rabbit reticulocyte lysate with TMV RNA or globin mRNA as messengers. Two high-molecular-weight proteins of 126K and 183K are synthesized *in vivo* when TMV RNA is translated. The 183K protein is the product of a UAG readthrough at the end of the 5′-proximal cistron coding for 126K protein (24,25). The globin mRNA preparation used is a mixture of α- and β-globin mRNAs. α-Globin mRNA is terminated by a UAA codon, and readthrough of this termination codon yields a protein that differs significantly in size from the major translation products (26,27). As shown in Fig. 5, $tRNA^{Gln}_{UmUA}$ was found to recognize both codons, UAA and UAG, whereas $tRNA^{Gln}_{CUA}$ recognized only UAG. Contrary to these glutamine tRNAs, $tRNA^{Gln}_{UmUG}$ had no readthrough activity at all, and thus it must be only used for the recognition of the two standard glutamine codons CAA and CAG.

The detection of a second *Tetrahymena* glutamine tRNA with unusual codon recognition properties, i.e., $tRNA^{Gln}_{CUA}$, was rather unexpected. The UAG readthrough activity of $tRNA^{Gln}_{CUA}$ is extremely high, showing almost complete readthrough of the UAG codon. It is also noteworthy to mention that the amount of $tRNA^{Gln}_{CUA}$ is quite large, almost comparable to that of the normal $tRNA^{Gln}_{UmUG}$. These results suggest that recognition of the UAG codon by glutamine in *Tetrahymena* is an efficient process.

The sequence homology between *Tetrahymena* $tRNA^{Gln}_{UmUA}$ and $tRNA^{Gln}_{CUA}$ is much higher than that between them and the normal $tRNA^{Gln}_{UmUG}$; namely, the sequence homology between $tRNA^{Gln}_{UmUG}$ and $tRNA^{Gln}_{UmUA}$ is 81%, whereas $tRNA^{Gln}_{UmUA}$ differs from $tRNA^{Gln}_{UmUG}$ in only four nucleotides, resulting in a 95% homology. Therefore, it is likely that one of the suppressor tRNAs (probably $tRNA^{Gln}_{UmUA}$) evolved from the normal glutamine tRNA ($tRNA^{Gln}_{UmUG}$) and then $tRNA^{Gln}_{CUA}$ was generated from $tRNA^{Gln}_{UmUA}$ to strengthen the recognition of the UAG codon in *Tetrahymena*. Speculation can be made as to why and how the use of UAA and UAG as glutamine codons evolved in *Tetrahymena*. Because of the many unique features of this organism, it is likely that it branched off in a very early stage of eukaryotic evolution. Since most of the eubacteria and eukaryotes use UAA and UAG as termination codons, it is more likely that *Tetrahymena* lost two of these codons as termination codons in its evolution rather than that these termination codons were acquired by the other organisms by convergent evolution. Perhaps the stop codons UAA and UAG were rarely used during prociliate evolution, and weak suppressor tRNAs existed for these codons at a very early stage. Spontaneous

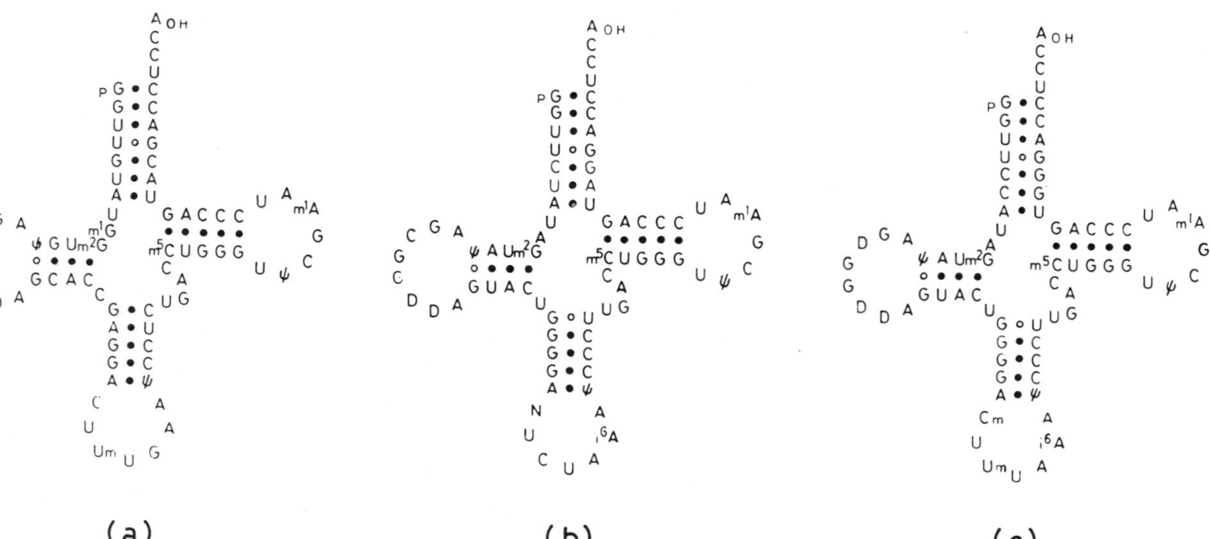

Fig. 4. Cloverleaf structures of the three glutamine tRNAs from *Tetrahymena*. (a) $tRNA^{Gln}_{UmUG}$; (b) $tRNA^{Gln}_{CUG}$; (c) $tRNA^{Gln}_{UmUA}$. N in position 32 of the sequence b is probably a derivative of C.

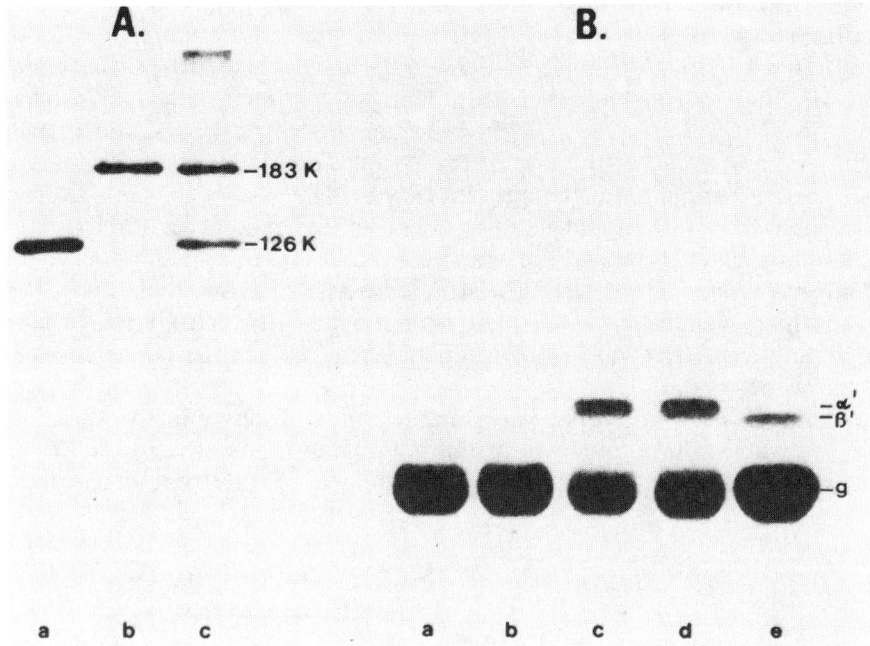

Fig. 5. Readthrough ability of UAG and UAA codons *in vitro* by purified *Tetrahymena* glutamine tRNAs. (A) Translation of TMV RNA in the presence of 25 μg/ml of $tRNA^{Gln}_{UmUG}$ (a), $tRNA^{Gln}_{CUA}$ (b), and $tRNA^{Gln}_{UmUA}$ (c). (B) Translation of globin mRNA in the presence of 25 μg/ml of $tRNA^{Gln}_{UmUA}$ (a), $tRNA^{Gln}_{CUA}$ (b), and $tRNA^{Gln}_{UmUA}$ (c), in parallel with 50 μg/ml of UAA yeast suppressor tRNA (d) and UGA yeast suppressor tRNA (e).

point mutations creating UAA and UAG codons within protein genes would then not have been lethal for those prociliates. Once these early suppressor tRNAs acquired mutated anticodons for perfect UAA and UAG recognition, the use of these codons as amino acid codons became fixed.

V. ISOLATION OF A NATURAL UAG SUPPRESSOR GLUTAMINE tRNA FROM MOUSE CELLS (28)

The finding of unique glutamine suppressor tRNAs in *Tetrahymena* prompted us to look for similar glutamine tRNAs in other eukaryotes. Recently, Pure *et al.* reported that yeast glutamine tRNA, produced *in vivo* by transfection of multiple copies of tRNA genes having a TTG anticodon sequence, can weakly suppress the UAA termination codon (29). In addition, Yoshinaka *et al.* have shown that a viral protease, coded by Moloney murine leukemia virus (Mo-MuLV), is a readthrough product of the UAG termination codon between the *gag* and *pol* genes, with glutamine inserted at the site of the termination codon (30). These data suggested that glutamine suppressor tRNA might also be present in organisms that have the normal genetic code, as a sign of some shared ancestral trait, although, in ciliates, the use of UAG and UAA as a glutamine codon may have been more strongly pronounced throughout their evolution. Therefore, we attempted to isolate all glutamine tRNAs from mouse cells, and determine their nucleotide sequences as well as their codon recognition properties.

Figure 6 shows the chromatographic profile of glutamine acceptor activity, when a large quantity of unfractionated mouse liver tRNA (8000 A_{260} units) was fractionated by BD-cellulose column chromatography.

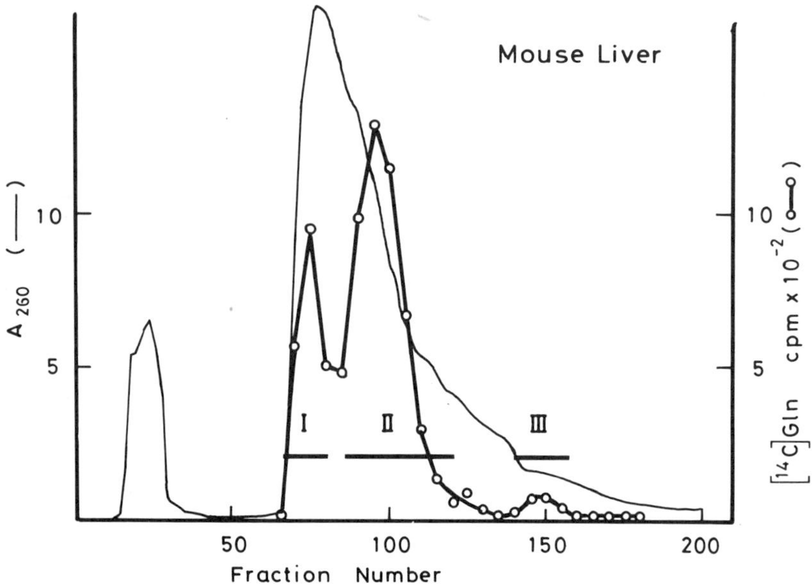

Fig. 6. Chromatographic profile of glutamine acceptor activity of mouse liver tRNA after being fractionated by BD-cellulose column chromatography. The column (2 × 100 cm) was eluted with a linear gradient produced by 1 liter of 0.02 *M* NaOAc buffer (pH 6.0)–1.2 *M* NaCl in the reservoir, and 1 liter of 0.02 *M* NaOAc buffer (pH 6.0)–0.3 *M* NaCl in the mixing chamber. Fractions of 10 ml of effluent were collected. Horizontal bars indicate fractions used for further purification.

Three peaks of glutamine acceptor activity were obtained, which were designated as fractions I, II, and III in the order of their elution from the column. Glutamine tRNAs in the three fractions were further purified by RPC-5 column chromatography, followed by successive steps of polyacrylamide gel electrophoresis. During the former purification step, glutamine-accepting tRNA in fraction II was separated into a major and a minor portion. The glutamine tRNA species that constituted the major portion of the glutamine acceptor activity turned out to be a hypomodified species of the major species of glutamine tRNA eluted in fraction I, based on its nucleotide sequence analysis. On the other hand, the minor glutamine tRNA species in fraction II seems to be the same as that eluted in fraction III. The nucleotide sequences of all the purified glutamine tRNAs were determined by postlabeling techniques. The nucleotide sequence data indicated that mouse liver essentially contains two species of glutamine tRNA (major $tRNA^{Gln}_{CUG}$ and minor $tRNA^{Gln}_{UmUG}$), designated according to their anticodon sequence (Fig. 7). The amount of $tRNA^{Gln}_{UmUG}$ was approximately one-fiftieth to one-hundredth of that of the major species, $tRNA^{Gln}_{CUG}$, of which nucleotide sequence is the same as that of rat $tRNA^{Gln}_{CUG}$ reported by Yang *et al.* except for some differences in posttranscriptional modification (31). Both $tRNA^{Gln}_{UmUG}$ species from fractions

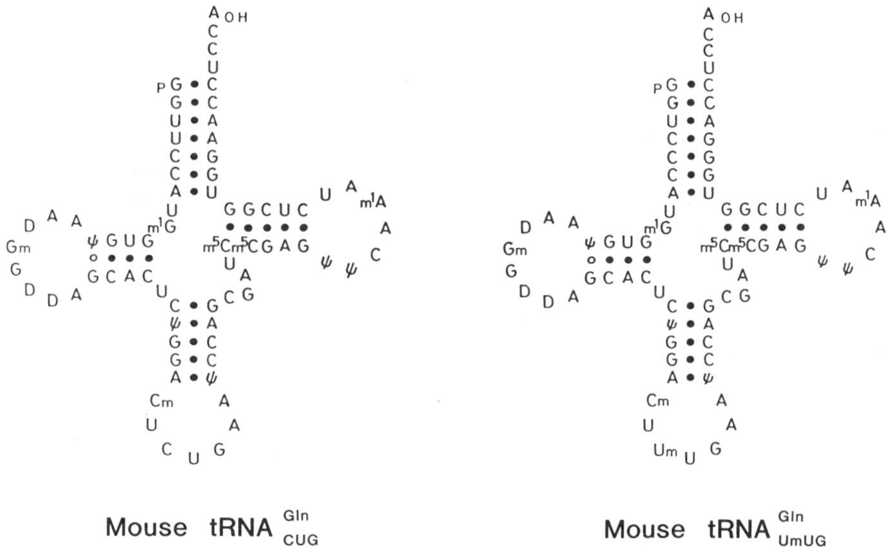

Fig. 7. Cloverleaf structures of two glutamine tRNAs from mouse liver. (a) $tRNA^{Gln}_{CUG}$, (b) $tRNA^{Gln}_{UmUG}$.

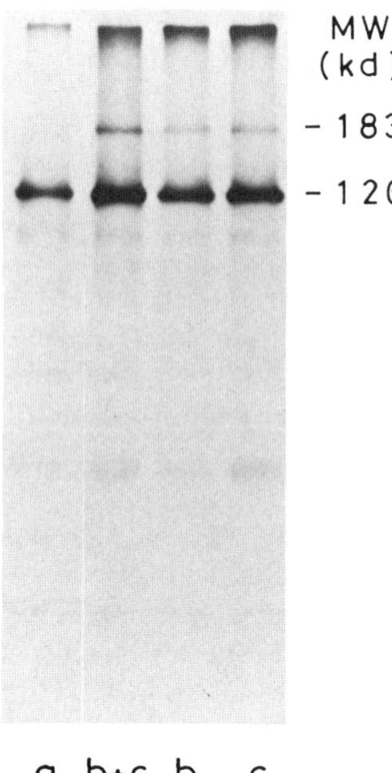

Fig. 8. Readthrough ability of UAG codon *in vitro* by purified mouse liver glutamine tRNAs. The translation of TMV RNA was carried out in the presence of 25 μg/ml of each glutamine tRNA; (a) normal $tRNA^{Gln}_{CUG}$; (b) hypomodified $tRNA^{Gln}_{UmUG}$; (c) normal $tRNA^{Gln}_{UmUG}$.

II and III recognized the UAG termination codon of TMV RNA, whereas the two $tRNA^{Gln}_{CUG}$ species from fractions I and II did not (Fig. 8). It should be noted that the suppressor activity of $tRNA^{Gln}_{UmUG}$ was much weaker than that of *Tetrahymena* $tRNA^{Gln}_{CUG}$ and $tRNA^{Gln}_{UmUA}$, and even weaker than that of tobacco and wheat leaf $tRNA^{Tyr}_{G\psi A}$, which has the anticodon sequence GUA, and is known to be a natural UAG suppressor tRNA in plants (32, 33). Contrary to the case of the *Tetrahymena* suppressor glutamine tRNAs, the nucleotide sequence of mouse liver $tRNA^{Gln}_{UmUG}$ is very similar to that of the major species of mouse liver glutamine tRNA ($tRNA^{Gln}_{CUG}$), the differences in nucleotide sequence between them being at only three residues (the first position of the anticodon and two positions in the CCA stem). Recognition of the UAG codon by $tRNA^{Gln}_{UmUG}$ is facilitated by wobble base-pairs, i.e., at the first and third positions of the anticodon. Wobbling of the guanine residue at the third position of the anticodon with U in the first position of the codon in mRNA has been demonstrated previously in yeast *in vivo* by introduction of multiple copies of a yeast glutamine tRNA gene (29). The major mouse glutamine tRNA

($tRNA^{Gln}_{CUG}$), having CUG as its anticodon, has no UAG readthrough activity, although theoretically it can recognize the UAG codon, if the third position of the anticodon wobbles with U as in the case of $tRNA^{Gln}_{UmUG}$. The major species of glutamine tRNA from *Tetrahymena*, having UmUG as its anticodon sequence, also showed no UAG readthrough activity. The sequences of *Tetrahymena* and mouse liver $tRNA^{Gln}_{UmUG}$s are quite different from each other (see Figs. 4 and 7). Perhaps the conformation of tRNA is another important factor that governs UAG suppressor activity.

VI. A LARGE INCREASE OF $tRNA^{Gln}_{UmUG}$ IN MOUSE CELLS INFECTED WITH Mo-MuLV

A question raised is how a small amount of $tRNA^{Gln}_{UmUG}$ having weak suppressor activity in mouse cells is used efficiently to generate the 180K readthrough product coded by Mo-MuLV messenger (30). One obvious possibility is that the amount of $tRNA^{Gln}_{UmUG}$ is increased in Mo-MuLV-infected cells to allow synthesis of a sufficient amount of the protease. This was in fact shown to be the case. As shown in Fig. 9, glutamine acceptor activity in fraction III in the chromatographic profile of BD-cellulose column chromatography was greatly increased in Mo-MuLV-infected cells. That the increase in glutamine acceptor activity in fraction III is due to an increase in the amount of $tRNA^{Gln}_{UmUG}$ was proved by isolation and nucleotide sequencing of the tRNA present in that fraction and also by its readthrough ability of the UAG codon in TMV RNA.

The high level of $tRNA^{Gln}_{UmUG}$ in NIH-3T3 cells infected with Mo-MuLV is probably due to specific activation of transcription of the tRNA gene for $tRNA^{Gln}_{UmUG}$. It is tempting to speculate that there is a specific transcription factor required for the activation of the suppressor tRNA gene. In the case of the transcription factors TFIIIB and TFIIIC, the latter factor is induced by adenoviral infection and stimulates transcription of many tRNA genes in a nonspecific manner (34,35). The transcription factor for $tRNA^{Gln}_{UmUG}$, however, if it exists, must be different from TFIIIB and TFIIIC, since the increase in the level of $tRNA^{Gln}_{UmUG}$ in Mo-MuLV infection is very specific and the infection has no effect on the level of the major tRNA ($tRNA^{Gln}_{CUG}$).

VII. CONCLUDING REMARKS

Glutamine tRNAs having UAA and/or UAG readthrough ability in *in vitro* translation systems have been isolated from both *Tetrahymena* and mouse cells. In the case of *Tetrahymena*, it seems that all UAA and UAG

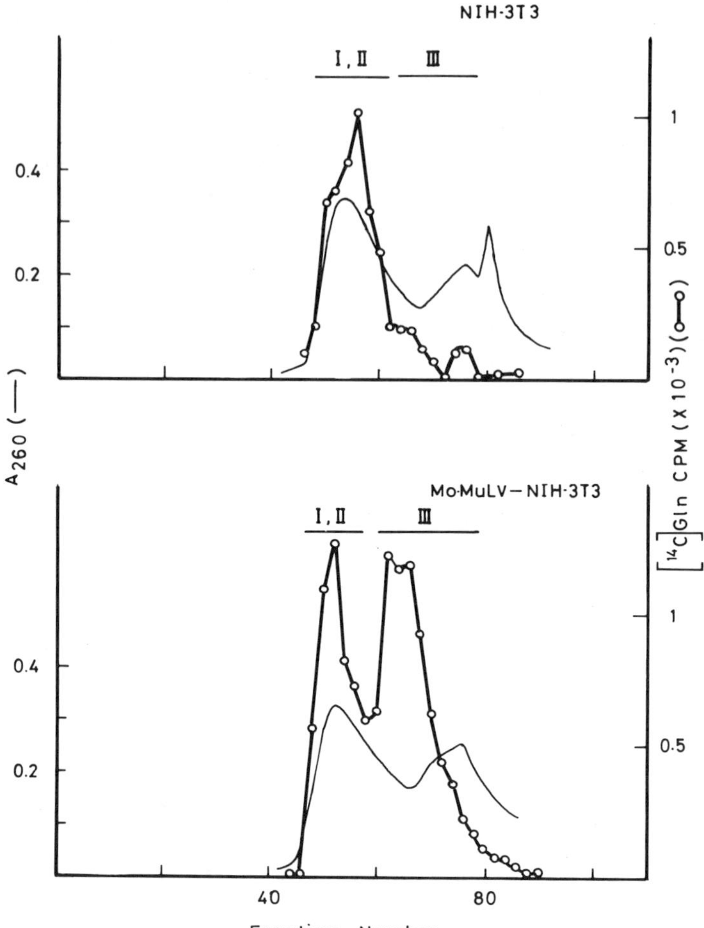

Fig. 9. Chromatographic profile of glutamine acceptor activity of tRNA from NIH-3T3 cells and NIH-3T3 cells infected with Mo-MuLV on a BD-cellulose column (0.5 × 3 cm). Unfactionated tRNA (40 A_{260} units) was loaded on a column and eluted with an 80-ml linear gradient of 0.3–1.2 *M* NaCl in 0.02 *M* NaOAc buffer (pH 6.0). Fractions of 1 ml of effluent were collected. The horizontal bars indicate the fractions.

codons are read as glutamine and are never used as a chain-termination codon. The presence of large amounts of such glutamine tRNAs having strong readthrough activity supports this conclusion. It is striking to find UAG suppressor glutamine tRNA in mouse cells, even though UAG is normally used as a chain-termination codon in this organism. The common finding that glutamine tRNAs can recognize the UAG codon in both ciliate and mammalian cells has tempted us to speculate how the use of the termination codon UAG has evolved in the genetic code. Before ciliates branched off from other eukaryotic cells, an ancestral glutamine tRNA that recognized the UAG codon already existed, and the activity of such a suppressor glutamine tRNA became considerably pronounced during ciliate evolution in order to fix the UAG codon as discussed before. In other eukaryotes such as mouse cells, the suppressor glutamine tRNA remained at a marginal level to function on specific occasions for the regulation of cellular function. Use of nonsense codons as amino acid codons in a specific way was also recently found in the case of UGA suppression, when selenocysteine is inserted in mouse gluthathione peroxidase (36) and *E. coli* formate dehydrogenase (37). Contrary to this hypothesis as to the evolution of mammalian glutamine suppressor tRNA is the fact that the nucleotide sequence of mouse suppressor $tRNA^{Gln}_{UmUG}$ is very similar to that of the major $tRNA^{Gln}_{CUG}$, the difference between the two tRNAs being only three base alterations. This evidence favors the idea that the suppressor $tRNA^{Gln}_{UmUG}$ is derived from the normal glutamine tRNA at more recent stage of animal evolution. If the mouse suppressor glutamine tRNA is derived in a manner quite independent from that of the *Tetrahymena* suppressor glutamine tRNAs, it is puzzling as to why glutamine tRNA is used to read the UAG codon in both organisms, since tRNAs for many other amino acids can be changed to recognize the UAG codon by a single point mutation in their anticodons. To give more insight into the molecular mechanisms of evolution of the genetic code, a search for suppressor tRNAs from other organisms, particularly from those belonging to ancient groups, may be important.

REFERENCES

1. Sprinzl, M., Moll, J., Meissner, F., and Hartmann, T. (1985). *Nucleic Acids Res.* **13,** r1–r49.
2. Kuchino, Y., Ihara, M., Yabusaki, Y., and Nishimura, S. (1982). *Nature (London)* **298,** 684–685.
3. Pang, H., Ihara, M., Kuchino, Y., Nishimura, S., Gupta, R., Woese, C. R., and McCloskey, J. A. (1982). *J. Biol. Chem.* **257,** 3589–3592.
4. Kuchino, Y., Mita, T., and Nishimura, S. (1981). *Nucleic Acids Res.* **9,** 4557–4562.
5. Hall, B. D., Clarkson, S. G., and Tocchini-Valentini, G. (1982). *Cell (Cambridge, Mass.)* **29,** 3–5.
6. Young, L. S., Takahashi, N., and Sprague, K. U. (1986). *Proc. Natl. Acad. Sci. U.S.A.* **83,** 374–378.
7. Goodman, H. M., Olson, M. V., and Hall, B. D. (1977). *Proc. Natl. Acad. Sci. U.S.A.* **74,** 5453–5475.
8. Kaine, B. P., Gupta, R., and Woese, C. R. (1983). *Proc. Natl. Acad. Sci. U.S.A.* **80,** 3309–3312.
9. Kuchino, Y., Hanyu, N., Tashiro, F., and Nishimura, S. (1985). *Proc. Natl. Acad. Sci. U.S.A.* **82,** 4758–4762.

10. Zaug, A. J., Been, M. D., and Cech, T. R. (1986). *Nature* (*London*) **324,** 429–433.
11. Kuchino, Y., Watanabe, S., Harada, F., and Nishimura, S. (1980). *Biochemistry* **19,** 2085–2089.
12. Nishimura, S., and Kuchino, Y. (1983). *In* "Methods of DNA and RNA Sequencing" (S. M. Weissman, ed.), pp. 235–260. Praeger, New York.
13. Helftenbein, E. (1985). *Nucleic Acids Res.* **13,** 415–433.
14. Caron, F., and Meyer, E. (1985). *Nature* (*London*) **314,** 185–188.
15. Preer, J. R., Jr., Preer, L. B., Rudman, B. M., and Barnett, A. J. (1985). *Nature* (*London*) **314,** 188–190.
16. Horowitz, S., and Gorovsky, M. A. (1985). *Proc. Natl. Acad. Sci. U.S.A.* **82,** 2452–2455.
17. Barrell, B. G., Anderson, S., Bakier, A. T., DeBruijn, M. H. L., Chan, E., Coulson, A. R., Drouin, J., Eperon, I. C., Nierlich, D. P., Roe, B. A., Sanger, F., Schreier, P. H., Smith, A. J. H., Staden, R., and Young, I. G. (1980). *Proc. Natl. Acad. Sci. U.S.A.* **77,** 3164–3166.
18. Bonitz, S. G., Berlani, R., Coruzzi, G., Li, M., Marcino, G., Nobrega, F. G., Nobrega, M. P., Thalenfeld, B. E., and Tzagoloff, A. (1980). *Proc. Natl. Acad. Sci. U.S.A.* **77,** 3167–3170.
19. Heckman, J. E., Sarnoff, J., Alzner-DeWeerd, B., Yin, S., and RajBhandary, U. L. (1980). *Proc. Natl. Acad. Sci. U.S.A.* **77,** 3159–3163.
20. deBruijn, M. H. L. (1983). *Nature* (*London*) **304,** 234–241.
21. Yamao, F., Muto, A., Kawauchi, Y., Iwami, M., Iwagami, S., Azumi, Y., and Osawa, S. (1985). *Proc. Natl. Acad. Sci. U.S.A.* **82,** 2306–2309.
22. Hanyu, N., Kuchino, Y., Nishiura, S., and Beier, H. (1986). *EMBO J.* **5,** 1307–1311.
23. Crick, F. H. C. (1966). *J. Mol. Biol.* **19,** 548–555.
24. Pelham, H. R. B. (1978). *Nature* (*London*) **272,** 469–471.
25. Goelet, P., Lomonossoff, G. P., Butler, P. J. G., Akam, M. E., Gait, M., and Karn, J. (1982). *Proc. Natl. Acad. Sci. U.S.A.* **79,** 5818–5822.
26. Efstratiadis, A., Kafatos, F. C., and Maniatis, T. (1977). *Cell* (*Cambridge, Mass.*) **10,** 571–585.
27. Heindell, H. C., Liu, A., Paddock, G. V., Studnicka, G. M., and Salser, W. A. (1978). *Cell* (*Cambridge, Mass.*) **15,** 43–54.
28. Kuchino, Y., Beier, H., Akita, N., and Nishimura, S. (1987). *Proc. Natl. Acad. Sci. U.S.A.* **84,** 2668–2672.
29. Pure, G. A., Robinson, G. W., Naumovski, L., and Friedberg, E. C. (1985). *J. Mol. Biol.* **183,** 31–42.
30. Yoshinaka, Y., Katoh, I., Copeland, T. D., and Oroszlan, S. (1985). *Proc. Natl. Acad. Sci. U.S.A.* **82,** 1618–1622.
31. Yang, J. A., Tai, L. W., Agris, P. E., Gehrke, C. W., and Wong, T. W. (1983). *Nucleic Acids Res.* **11,** 1991–1996.
32. Beier, H., Barciszewska, M., Krupp, G., Mitnacht, R., and Gross, H. J. (1984). *EMBO J.* **3,** 351–356.
33. Beier, H., Barciszewska, M., and Sickinger, H. D. (1984). *EMBO J.* **3,** 1091–1096.
34. Lassar, A. B., Martin, P. L., and Roeder, R. G. (1983). *Science* **222,** 740–748.
35. Yoshinaga, S., Dean, N., Han, M., and Berk, A. J. (1986). *EMBO J.* **5,** 343–354.
36. Chambers, I., Frampton, J., Goldfarb, P., Affara, N., McBain, W., and Harrison, P. R. (1986). *EMBO J.* **5,** 1221–1227.
37. Zinoni, F., Birkmann, A., Stadtman, T. C., and Bock, A. (1986). *Proc. Natl. Acad. Sci. U.S.A.* **83,** 4650–4654.

21

The Purification of Small RNAs by High-Performance Liquid Chromatography

BERNARD S. DUDOCK

Department of Biochemistry
State University of New York at Stony Brook
Stony Brook, New York, 11794

I. INTRODUCTION

RNA has recently provided us with some interesting surprises. An RNA acting as an enzyme (1), an RNA capable of self-splicing (2), and an RNA covalently linked to DNA (3) are a few examples of new and previously unexpected roles for RNA. It is apparent that RNA is a highly complex class of molecules, with a wide variety of cellular roles, from which we still have much to learn. However, we are often limited in our ability to study a specific RNA molecule by the difficulty in obtaining it in pure form. Perhaps this is best illustrated by looking at tRNA, a complex family of RNA molecules with a size range of 75 to about 90 nucleotides. This class of molecules is often quite complex, especially in higher organisms such as mammalian cells. Because of this complexity, although it is relatively easy to isolate a major species of tRNA, it is often extremely difficult to isolate one of the many minor tRNA species. One major reason for this is the limitation of the resolution obtainable in chromatographic isolation procedures. In this chapter we describe a new chromatographic procedure that provides significantly improved resolution of tRNA and other relatively low-molecular-weight RNA species.

II. MATERIALS AND METHODS

The prepacked high-performance liquid chromatography (HPLC) column used in these studies was a 5-μm particle size, 250 × 4.6-mm i.d. W-Porex 5 C-4 column obtained from Phenomenex (Rancho Palos Verdes, California). The chromatograms were run at room temperature (25–27°C) on an LKB Ultro-Chrom GTi Bioseparation System. All buffers were thoroughly degassed and filtered through a 0.45-μm filter prior to use. Columns were run at a flow rate of 1 ml/min, at a starting pressure of about 60 bar. Column fractions were desalted using a Centricon 10 microconcentrator from Amicon used according to the manufacturer's instructions.

Wheat germ tRNA (4) and bovine liver tRNA (5) were prepared as described, as was aminoacyl tRNA synthetase (6). Aminoacylation reactions were performed essentially as previously reported (7) except that the reaction volume was 20 μl and ^{3}H-labeled amino acids and a rat liver extract containing aminoacyl-tRNA synthetases were used. ^{3}H-Labeled amino acids were from ICN. Liquiscint, which was used to count aqueous samples, is a product of National Diagnostics (Somerville, New Jersey). ^{32}P-labeled small nuclear RNA (snRNA) was immunoprecipitated with the SLE Sm antibody from nuclear extracts of L929 cells and was a gift of Gary Zieve and Robert Feeney.

III. RESULTS

When bovine liver tRNA is fractionated on a Porex C-4 column at room temperature with a decreasing ammonium sulfate gradient, a complex pattern of peaks is observed in a 2-hr run (Fig. 1A). About 30 discernible peaks are observed under these conditions. Moreover, by changing the gradient parameters the pattern can be readily altered, allowing for easy optimization of any region of the chromatogram. This is illustrated in Fig. 1, B and C. The quality of resolution of individual bovine liver tRNA isoacceptors achieved on the Porex C-4 column is shown for leucine tRNA (Fig. 2A), alanine tRNA (Fig. 2B), and threonine tRNA (Fig. 2C). In all cases the tRNAs were assayed both directly from the column fractions, i.e., in the presence of ammonium sulfate and potassium phosphate buffer, and after the column fractions had been desalted by Centricon filtration. In every case the pattern was the same in the presence or absence of salt. Thus, it appears that desalting of the column fractions is not necessary prior to aminoacylation.

The quality of the resolution achieved on the Porex C-4 column in most

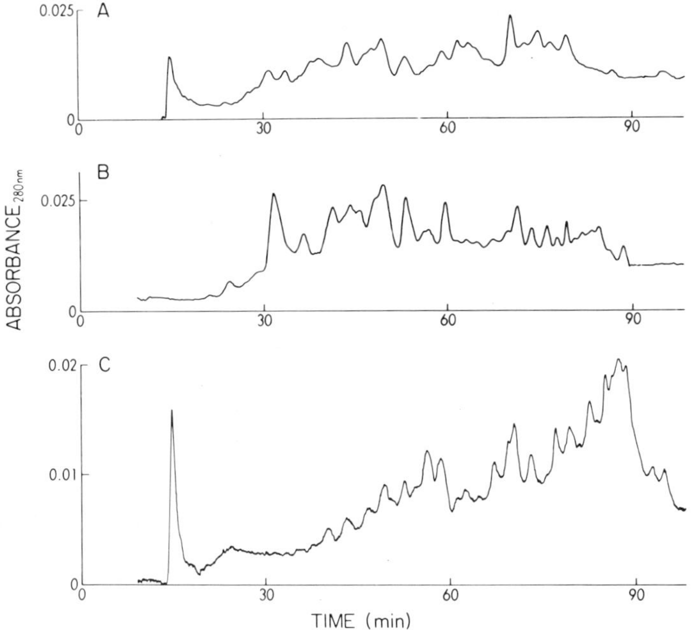

Fig. 1. (A) Crude bovine liver tRNA (0.5 mg) run on a Porex C-4 column. Buffer A is 2 *M* ammonium sulfate, 0.1 *M* potassium phosphate (pH 7.0), 0.75% isopropanol. Buffer B is 0.1 *M* potassium phosphate (pH 7.0), 0.75% isopropanol. Gradient: 10 min at 100% buffer A, 70 min at 55–95% buffer B, 30 min at 95–100% buffer B, 10 min at 100% buffer B. The temperature was 25.5°C, starting pressure was 62 bar, flow rate was 1 ml/min, and the absorbance units (A_{280}) full scale (AUFS) was 0.05. (B) Crude bovine liver tRNA (0.5 mg) was applied. Buffers A and B were the same as above except no isopropanol was present. Gradient: 10 min at 100% buffer A, 70 min at 25–100% buffer B, 20 min at 100% buffer B. Temperature, pressure, flow rate, and AUFS as in A above. (C) Crude bovine liver tRNA (0.3 mg) was applied. Buffers A and B as in A above. Gradient: 10 min at 100% buffer A, 70 min at 50–100% buffer B, 10 min at 100% buffer B. The temperature was 27°C, starting pressure was 55 bar, flow rate was 1 ml/min, and AUFS was 0.02 A_{280} units.

cases was exceptionally good. For example, Fig. 2A shows four distinct major leucine tRNAs and at least 10 minor leucine tRNA species. Bovine liver alanine tRNA also shows a very well resolved pattern (Fig. 2B) consisting of five major species as well as a complex, less well resolved group of minor species. These patterns are quite superior to the corresponding patterns obtained on RPC-5 chromatography (8). Not all tRNAs show highly complex patterns on the Porex C-4 column as seen in the case of threonine (Fig. 2C) where two major and several minor species are shown.

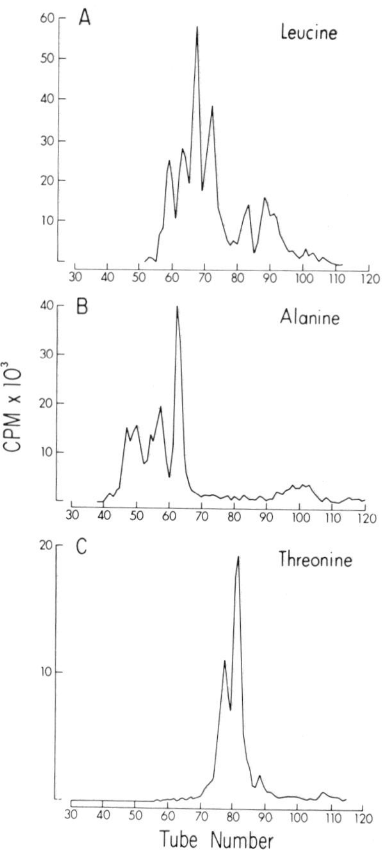

Fig. 2. Aminoacylation of bovine liver tRNA fractionated on a Porex C-4 column. Each fraction contained 1 ml of column effluent and each fraction was assayed. Aminoacylation reactions, performed as described (7), contained 4 μl of column effluent. (A) Leucine tRNA (sp. act. of [^{3}H]leucine = 58 Ci/m*M*), (B) alanine tRNA (sp. act. of [^{3}H]alanine = 50 Ci/m*M*), (C) threonine tRNA (sp. act. of [^{3}H]threonine = 19 Ci/m*M*).

To test the capacity of the Porex C-4 column, a series of samples of partially purified wheat germ tRNA was applied (Fig. 3A–D). As can be seen, this column has a very large capacity to bind tRNA and resolution falls off only gradually with increasing sample load.

It is often desirable to fractionate tRNAs at pH 4.5 so that aminoacylated tRNAs can be studied directly. Figure 4 shows the chromatogram obtained when bovine liver tRNA is fractionated at pH 4.5 with a decreasing ammonium sulfate gradient. Figure 5 shows the leucine tRNA profile

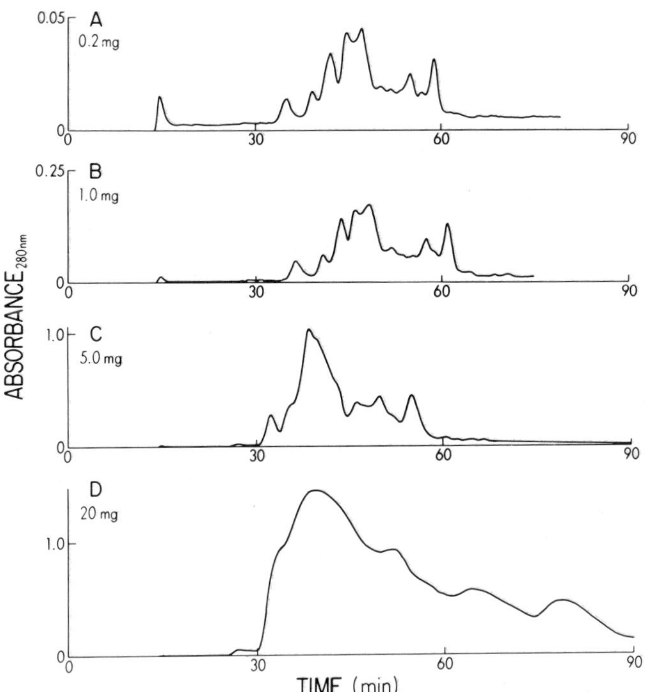

Fig. 3. Chromatography of a partially purified sample of wheat germ tRNA on a Porex C-4 column. The column was run at room temperature, at a flow rate of 1 ml/min, with a starting pressure of 65 bar. Buffers A and B are as in Fig. 1A. (A) 0.2 mg of tRNA; gradient, 10 min at 100% A, 70 min at 55–100% B. (B) 1.0 mg of tRNA; gradient, same as A above. (C) 5.0 mg of tRNA; gradient, same as A above. (D) 20 mg of tRNA; gradient, 10 min at 100% A, 150 min at 55–100% B.

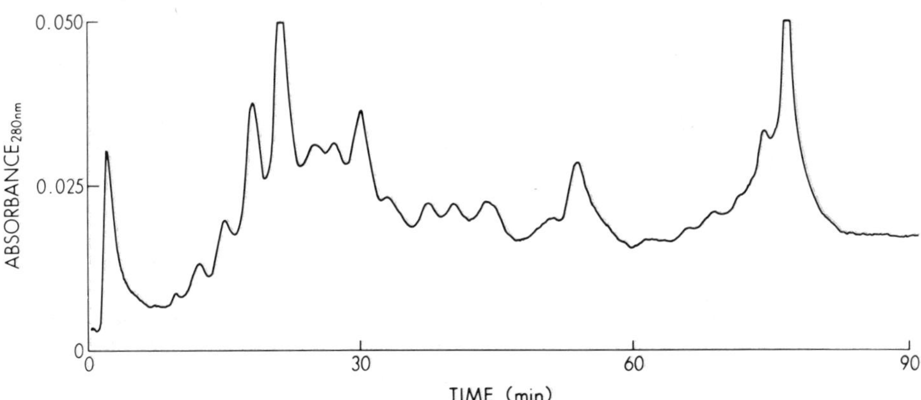

Fig. 4. Chromatography of crude bovine liver tRNA (1.0 mg) on a Porex C-4 column at pH 4.5. Buffer A is 2 *M* ammonium sulfate, 0.2 *M* potassium phosphate (pH 4.5), 0.75% isopropanol. Buffer B is 0.2 *M* potassium phosphate (pH 4.5), 0.75% isopropanol. Gradient: 75 min at 50–100% B, 10 min at 100% B. The column was run at room temperature at a flow rate of 1 ml/min and a starting pressure of 63 bar.

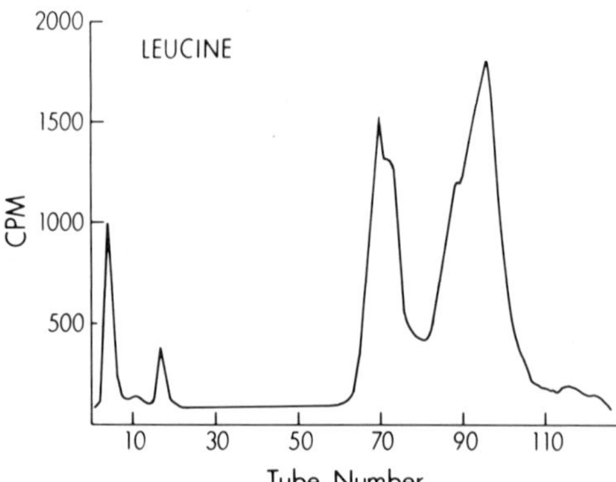

Fig. 5. Bovine liver leucine tRNA was fractionated on a Porex C-4 column at pH 4.5. [^{3}H]Leucine tRNA (140,000 cpm) was added to carrier bovine liver tRNA, and the mixture was chromatographed at pH 4.5 essentially as in Fig. 4. Fractions of 1 ml were collected and aliquots were counted in Liquiscint.

obtained under these conditions. The resolution observed at pH 4.5 is not as good as that obtained at pH 7.0 (compare, for example, the leucine tRNA profile of Fig. 2A to Fig. 5). Attempts to improve the resolution at pH 4.5 are continuing, primarily by altering the nature of the gradient components.

Although the Porex C-4 column showed very good resolution of tRNA, we also wished to determine its applicability to other classes of RNA. Figure 6, A and B, show the resolution obtained when a ^{32}P-labeled sample of snRNA was applied to this column. A highly reproducible pattern is obtained (Fig. 6, A and B, shows two independent column runs). The pattern is quite complex, showing two major snRNAs and many minor components. Indeed, this column may be useful in the isolation and characterization of snRNAs.

IV. DISCUSSION

Although many procedures have been developed for the fractionation of tRNAs, the method that has been the most widely used is RPC-5 (9). Indeed this method has become the standard procedure when the fractionation of tRNAs is to be undertaken. Although this method often gives adequate resolution, it has several disadvantages. RPC-5 is difficult to obtain, and it often provides inadequate resolution for the isolation of minor species of tRNA and sometimes for the major species as well.

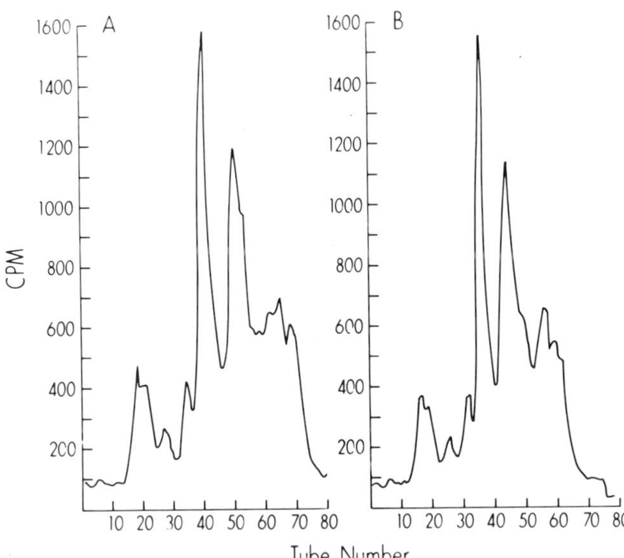

Fig. 6. (A) Chromatography of immunoprecipitated ^{32}P-labeled snRNA from L929 cells on a Porex C-4 column. Buffer A is 1 *M* ammonium sulfate, 0.05 *M* potassium phosphate (pH 7.0), 0.35% isopropanol. Buffer B is deionized water. Gradient: 10 min at 0–90% B, 60 min at 90–100% B, 10 min at 100% B. The column was run at room temperature at a flow rate of 1 ml/min and a starting pressure of 50 bar. Fractions of 1 ml were collected and an aliquot of each fraction was counted in Liquiscint. (B) Same as A above.

The first use of a reverse salt gradient to fractionate tRNA was by Holmes *et al.* (10) who used ammonium sulfate gradients on Sepharose-4B columns to separate *Escherichia coli* tRNAs. More recently Pearson *et al.* (11) *n*-alkylated silicas with short-chain (C_1, C_2, and C_4) chlorosilanes and employed gradients of decreasing ammonium sulfate at 55°C. Bischoff and McLaughlin (12) *n*-alkylated APS-Hypersil WP with short-chain acids (C_2, C_4, C_6, and C_8) and employed a gradient of decreasing ammonium sulfate at 35°C. Zhang *et al.* (13) used simultaneous salt and methanol gradients on a Vydac-C_4 column at 22°C to obtain quite satisfactory separation of *E. coli* tRNAs.

In this chapter we show that a prepacked W-Porex C-4 HPLC column can yield effective fractionation of tRNAs. In several cases, such as for bovine liver leucine and alanine tRNA, the resolution was remarkably good, quite superior to that achieved with RPC-5. Indeed, the leucine tRNA pattern shows 14 chromatographically distinct peaks. If one considers that it is quite likely that additional minor tRNAs are "buried" under the major leucine tRNA peaks, it is clear that the number of different leucine tRNAs in mammalian cells is far higher than previously thought. Other tRNA isoacceptors, such as alanine tRNA, also show this striking degree of complexity, but some, such as threonine tRNA, apparently do not. Extensive use of the Porex C-4 column at pH 7.0 may allow a reasonable estimate of the total number of mammalian tRNA species to be made.

The Porex C-4 column, in addition to its ability to separate tRNA, may also find applicability for the separation of other classes of relatively low-molecular-weight RNAs as well. The highly reproducible and complex pattern achieved with snRNA is an example of this.

The Porex C-4 column used in these studies was used for almost 100 chromatographic runs without noticeable loss of resolution. Occasionally the column was cleaned by passing water or a dilute methanol solution through it. All column runs were performed at a flow rate of 1 ml/min and at a pressure of about 60 bar. With reasonable care, this column should have a very extended lifetime.

The Porex C-4 column has specific advantages as well as disadvantages. The main advantage of this column is the high level of resolution achievable. The column is effective for tRNA, snRNA, and presumably for other relatively low-molecular-weight RNAs as well. In addition it is possible that this column will prove effective in the separation of other complex mixtures, such as DNA restriction fragments. The Porex C-4 column is quite durable, has a relatively large capacity and is commercially available in a ready to use prepacked HPLC format. The main disadvantages are the inferior resolution achieved to date at pH 4.5, and the fact that the column effluent contains ammonium sulfate and potassium phosphate and thus the RNA is not recoverable by simple ethanol precipitation. In the studies reported here, the RNA was recovered by Centricon filtration which, although reasonably efficient, is somewhat costly and more time consuming than ethanol precipitation. Attempts to get the same degree of resolution achieved here, but with salts from which the RNA could be recovered by ethanol precipitation are in progress.

ACKNOWLEDGMENT

We thank Gary Zieve and Robert Feeney for samples of immunoprecipitated [^{32}P]snRNA from L929 cells. We thank Doug McCrory and Vern Kadding of Phenomenex for helpful discussions. We also thank Laura Hirasaka for the aminoacylation assays. This research was supported in part by National Science Foundation Grant PCM8201925 and Grant GM25254 from the National Institutes of Health.

REFERENCES

1. Guerrier-Takada, C., Gardiner, K., Marsh, T., Pace, N., and Altman, S. (1983). *Cell (Cambridge, Mass.)* **35,** 849–857.
2. Cech, T. R., Zaug, A. J., and Grabowski, P. J. (1981). *Cell (Cambridge, Mass.)* **27,** 487–496.
3. Furuichi, T., Inouye, S., and Inouye, M. (1987). *Cell (Cambridge, Mass.)* **48,** 55–62.
4. Dudock, B. S., Katz, G., Taylor E. K., and Holley, R. W. (1969). *Proc. Natl. Acad. Sci. U.S.A.* **62,** 941–945.
5. Roe, B. N. (1974). *Nucleic Acids Res.* **2,** 21–42.
6. Diamond, A., Dudock, B., and Hatfield, D. (1981). *Cell (Cambridge, Mass.)* **25,** 497–506.
7. Reszelbach, R., Greenberg, R., Pirtle, R., Prasad, R., Marcu, K., and Dudock, B. (1977). *Biochim. Biophys. Acta* **475,** 383–392.
8. Hatfield, D., Varricchio, F., Rice, M., and Forget, B. G. (1982). *J. Biol. Chem.* **257,** 3183–3188.
9. Pearson, R. L., Weiss, J. F., and Kelmers, A. D. (1971). *Biochim. Biophys. Acta* **228,** 770–774.
10. Holmes, W. M., Hurd, R. E., Reid, B. R., Rimerman, R. A., and Hatfield, G. W. (1975). *Proc. Natl. Acad. Sci. U.S.A.* **72,** 1068–1071.
11. Pearson, J. D., Mitchell, M., and Regnier, F. E. (1983). *J. Liq. Chromatogr.* **6,** 1441–1457.
12. Bischoff, R., and McLaughlin, L. W. (1984). *J. Chromatogr.* **317,** 251–261.
13. Zhang, S.-B., Bronskill, P. M., Wang, Q.-S., and Wong, J. T.-F. (1986). *J. Chromatogr.* **360,** 282–287.

22

Comparative Studies on the Secondary Structure of the RNAs of Related RNA Coliphages

ANN B. JACOBSON,* MICHAEL ZUKER,† AND AKIKAZU HIRASHIMA‡

**Department of Microbiology*
State University of New York at Stony Brook
Stony Brook, New York 11794
†Division of Biological Sciences
National Research Council
Ottawa, Canada K1A 0R6
‡Yakult Central Institute for Microbiological Research
Tokyo 186, Japan

I. INTRODUCTION

Several years ago we described studies in which extensive and reproducible structural features could be visualized by electron microscopy within single stranded RNA from the bacteriophage MS2 (Jacobson, 1976; Jacobson and Spahr, 1977). The location of individual viral genes within the structure of the viral RNA suggests that these structures may play a role in the regulation of viral gene expression. Our current studies are directed toward examining this question. As a first step we are attempting to identify conserved structural features in phylogentically related RNAs. The novel feature of these studies is the use of electron microscopy as a guide both in computer analysis of conserved structure and in the chemical probing of these same features. Recent studies by electron microscopy have shown that RNA from the phage MS2 can be visualized in two alternate conformations. The observed variation in conformation appears to be limited to specific regions of the molecules (Jacobson *et al.*, 1985). We describe here some recent studies with RNA from the phage MS2 and its relative GA, as well as some new computer tools that are being used in our analyses.

II. RESULTS

A. Comparison of the Sequence of GA with the Sequence of MS2

Our work to date has focused on the comparison of the structure of the RNAs from MS2 and GA using electron microscopy and computer modeling. Information on the biology of MS2 is extensive. A very recent article by van Duin (1986) reviews some of this information as well as some of the results that are obtained by comparing the sequence of MS2 with the sequence of GA. The genetic maps of these two viral strains are shown in Fig. 1. Both MS2 and GA code for four viral genes. These are the maturation or A protein gene, the coat gene, the replicase or β-subunit of the viral polymerase gene, and the lysis gene. The lysis protein overlaps the coat and replicase genes and is read in the +1 reading frame with respect to MS2 coat protein and the −1 reading frame with respect to GA coat protein. The nucleotide sequence of GA RNA is shorter than the se-

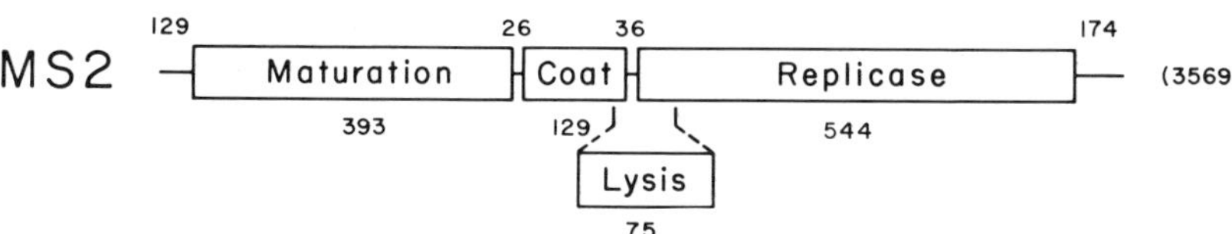

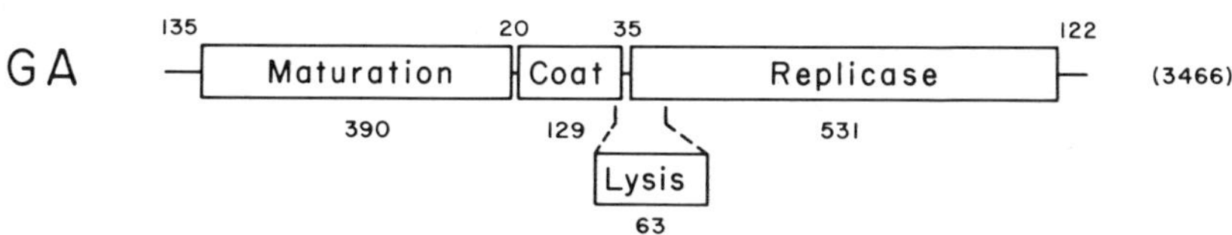

Fig. 1. Genetic maps of MS2 and GA. MS2 is a member of the Group I phage type and GA is a member of the Group II phage group (Furuse *et al.*, 1979; Nishihara, 1969). The lengths of the untranslated regions of each sequence are given in numbers of nucleotides (above the map). The length of the viral proteins is given in numbers of amino acids (below the map).

quence of MS2. This difference in molecular weight is characteristic of Group II bacteriophage strains (Furuse *et al.*, 1979) and is reflected in differences in the molecular weights of both the maturation and replicase proteins, as well as in a number of large deletions in the nontranslated region at the 3′ terminus of the GA RNA relative to MS2 (Inokuchi *et al.*, 1986).

The similarity between the nucleotide sequences of GA and MS2 is shown in Fig. 2 in the form of a dot plot. The results were filtered by requiring a match of 14 nucleotides in a window of 20 nucleotides. Under

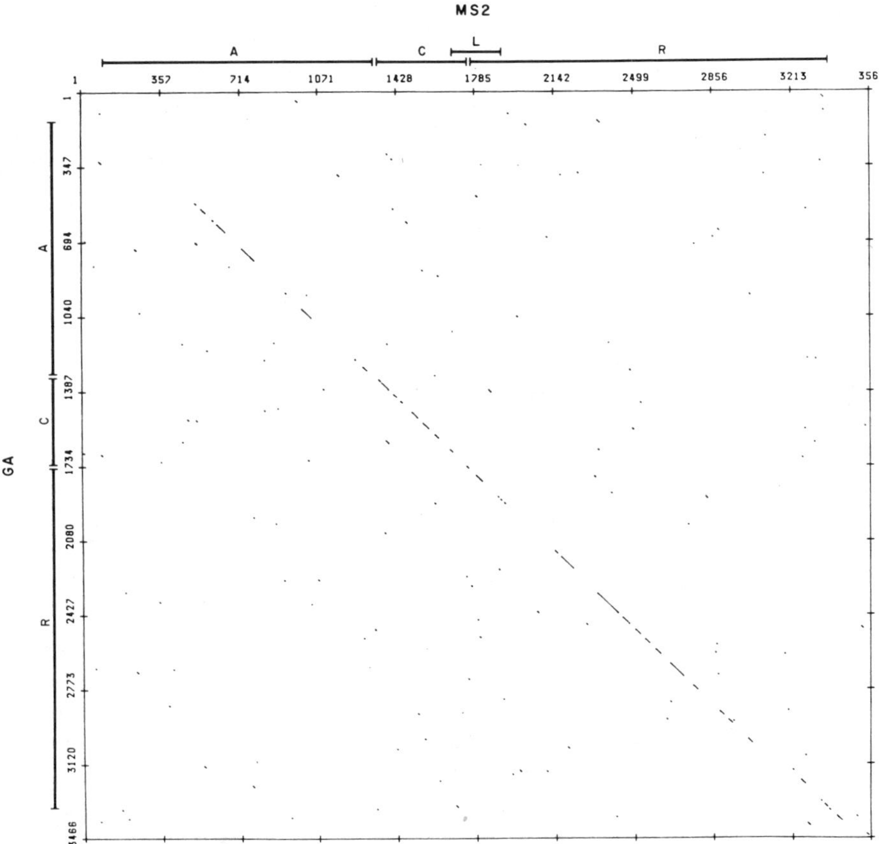

Fig. 2. A dot plot showing the nucleotide sequence similarity between MS2 and GA. The plots were filtered by requiring a match of 14 nucleotides in a window of 20 nucleotides. Plots of random sequences with these same parameter settings showed no significant matches.

these conditions random sequences with the same base composition as GA and MS2 exhibit no significant similarity. The sequences of GA and MS2 differ most from one another within the first 500 nucleotides from the 5′ ends of the sequences. Additional sequence divergence is seen in the last third of the maturation protein as well as in the early part of the replicase gene. In general, the regions of similarity between the two sequences are clustered in short stretches containing many common nucleotides. These are interspersed with regions where substantial sequence divergence has occurred. The region with the strongest sequence homology is located in the center of the replicase gene. This region is conserved in all four RNA coliphage sequences that have been examined to date (manuscript in preparation).

At the amino acid level, the degree of homology between the individual viral genes of GA and MS2 varies. In the viral coat proteins approximately 60% of the amino acids are common between the two sequences, but in the maturation protein only 40% of the amino acids are common between the two sequences. Greater sequence divergence is observed at the nucleotide sequence level, since approximately 50% of the codons exhibit third-position base changes even in the coat genes.

B. Electron Microscopy of GA and MS2 RNAs

The genomic RNAs from MS2 and GA have been examined by electron microscopy using Kleinschmidt spreading techniques in the presence of polycations. Results are presented here on the foldings that are observed in the presence of 0.5 m*M* spermidine. At this concentration of polyamine, extensive regions of the molecules are base-paired, however the folding patterns are still relatively easy to discern. The details of this work have been presented previously (Jacobson *et al.,* 1985). We have shown that under these conditions MS2 RNA folds in two alternate conformations. One of these is a large cloverleaf structure with three large loops, each of which is 600–700 nucleotides in size. The stem of the cloverleaf allows regions of the sequence that are 2500 nucleotides apart to pair. In the second conformation, base-pairing at the base of the central loop and the stem of the cloverleaf is retained. Now, however the intervening regions that form the remaining two large loops in the cloverleaf conformation pair with one another. In addition to long-range contacts, several new small hairpins also appear in this conformation. An example of each of these conformations is shown in Fig. 3.

Like MS2, GA has a well-defined structure that is easy to map by electron microscopy. Three typical molecules and their interpreted drawings are shown in Fig. 4. The molecules were obtained from the same

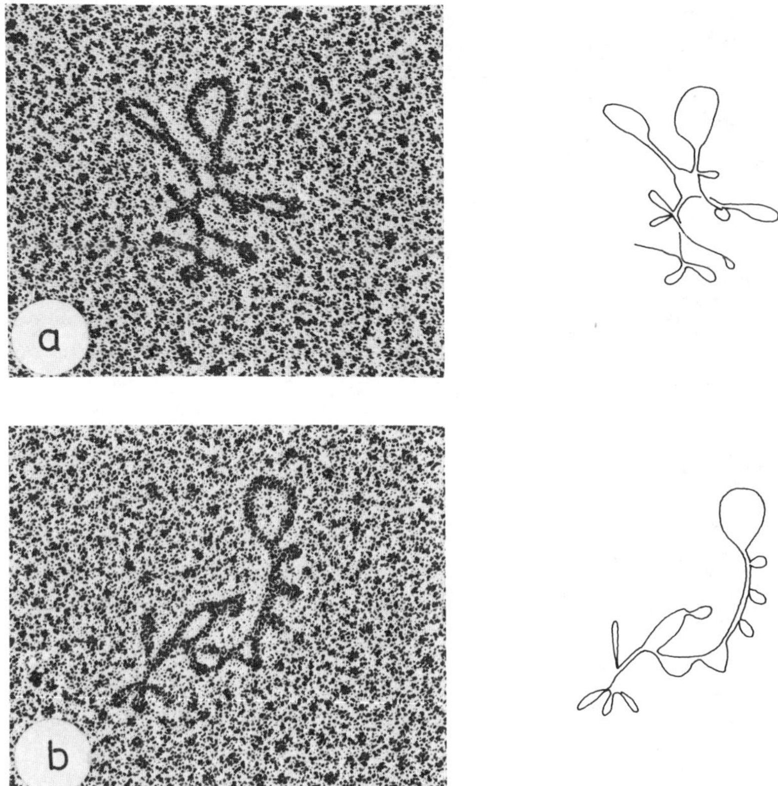

Fig. 3. Electron micrographs of two molecules of MS2 RNA. (a) A typical molecule in the cloverleaf conformation. (b) A molecule in the alternate extended stem conformation. Interpreted drawings are shown adjacent to each panel. The photographs were taken from Kleinschmidt spreadings of the RNA performed in the presence of 0.5 m*M* spermidine and 60% formamide. The molecules were stained with uranyl acetate and rotary-shadowed with platinum to achieve contrast. The length of each molecule is approximately 1.2 μm.

spreading and illustrate some of the variation in folding that is observed within any population of molecules that is examined by electron microscopy. All three molecules are variations of one basic conformation. The molecules have two large loops, as well as characteristic smaller hairpins. The central large loop is similar to the loop structure seen in the same position in MS2 and can be considered homologous. In panels a and c, the central loop appears smaller than its homolog in MS2 due to more extensive base-pairing that occurs within the loop. In these two molecules, the

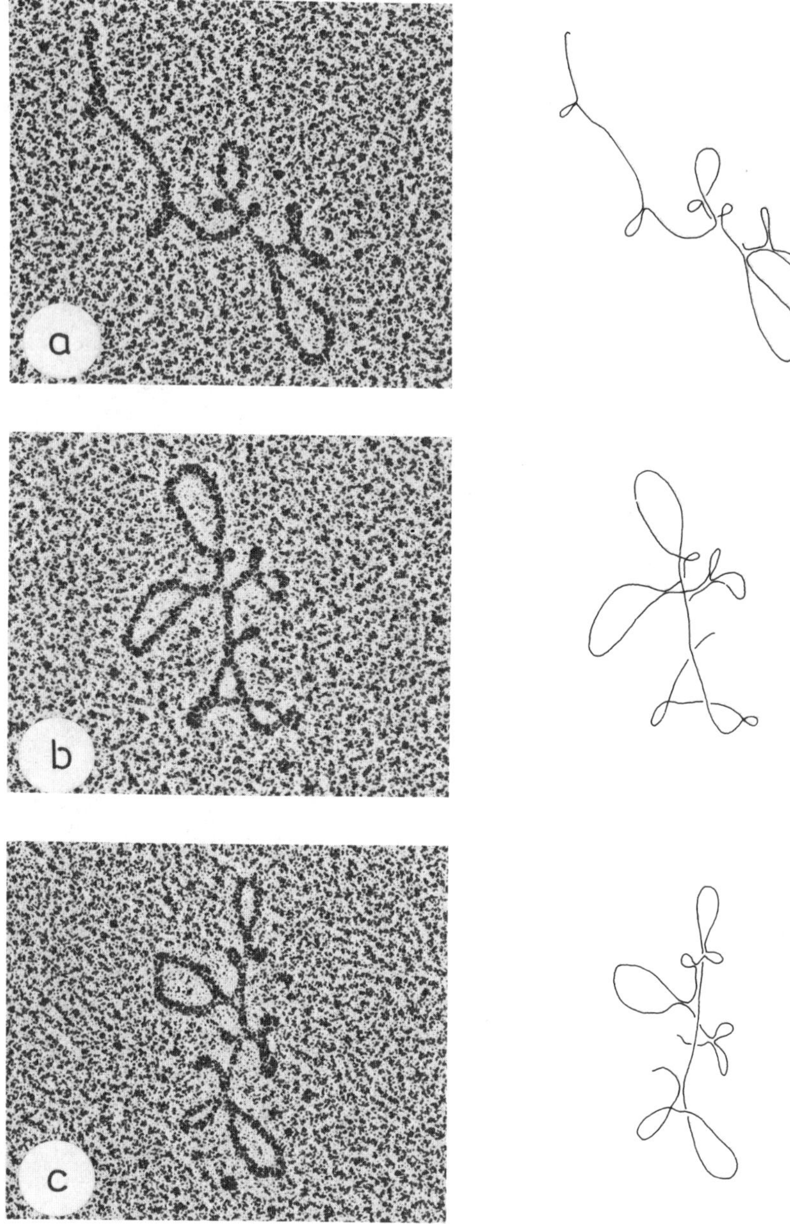

Fig. 4. Electron micrographs of three molecules of GA RNA spread in the presence of 0.5 m*M* spermidine. The experimental protocol is identical with the one described in Fig. 3. The length of each molecule is approximately 1.2 μm.

stem at the base of the central loop appears longer and has two small hairpins at its base, rather than the single hairpin commonly seen in MS2. The determination of the 5′ → 3′ orientation of the structural features that are visualized in GA RNA is still incomplete, and it is unclear whether the second large open loop is located in the 3′ or 5′ arm of the molecule. This loop appears to be larger than either of its counterparts in MS2 and appears to be displaced closer to the end of the molecule. The most striking differences between GA and MS2 are the absence of a large loop in the second arm of the molecule, and the concomitant inability of sequences at the distal end of this arm to base-pair with the distal region of the arm containing the large loop. This difference is likely to correlate with the sequence divergence that is observed at both the 3′ and 5′ ends of GA and MS2. With regard to small hairpins, two features appear to be homologous between GA and MS2. The first is the small hairpin at the base of the central loop. The second is composed of two small hairpins at the end of the arm with the large open loop. This group appears similar to, but slightly larger than, a similar cluster at the 3′ end of MS2. It should be noted, however, that structures of this type are observed near the terminus at both the 3′ and 5′ ends of MS2 RNA, and the apparent homology with the 3′ end needs to be confirmed experimentally.

The positions of the initiation codons for three of the four viral genes are shown within the folded molecules visualized by electron microscopy in Fig. 5. The initiation region of the viral coat gene is located at the base of the large central loop, and the replicase initiation region appears to be located near the apex of this loop in both RNAs. The initiation region for the lysis gene lies immediately to the left (see Fig. 1) of the replicase gene and is thus also located near the apex of the central open loop. It is intriguing that both the lysis and replicase regions lie within relatively exposed regions of both molecules, and the coat initiation site, which is the preferred ribosome-binding site in native RNA, lies in a complex region that has many base pairs. As discussed above, the 5′ → 3′ orientation of the structural features in GA is still uncertain. The assignment of the positions of the coat and replicase genes would not be altered markedly by a reversal in orientation: the initiation codon for the maturation protein, however, would be located on the other arm of the molecule.

The positions of the genes that are given in Fig. 5 were determined by assuming colinearity of the structural and genetic maps. Currently we are developing methods to test the truth of this assumption and to improve the precision with which we are able to assign nucleotide sequences to structural features that are visualized in the RNA by electron microscopy. In these experiments the RNA is cleaved with RNase H after it has been hybridized with synthetic oligodeoxynucleotides. The amount of cleavage

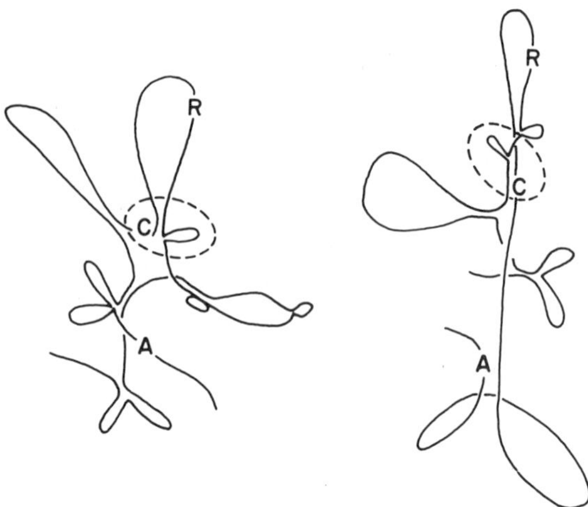

Fig. 5. Interpreted drawings of folded molecules of GA and MS2 showing the location of the initiation codons of three of the four viral genes. The lysis protein initiation region lies to the left of the replicase initiation region in the central loop and was omitted from the figure for the sake of clarity. As discussed in the text, the $5' \rightarrow 3'$ orientation of the individual structural features is still uncertain. Due to the symmetry of the structural features, this uncertainty does not affect the assigned position of the coat or replicase initiation regions greatly. However, the start of the A (maturation) protein could be at either end of the molecule. Dashed circles represent regions that are modeled in Fig. 9.

is monitored by gel electrophoresis and the structure of the cleaved molecules is examined by electron microscopy. Figure 6 shows a gel of MS2 RNA that was cleaved with RNase H after hybridization with an oligodeoxynucleotide that was complementary to nucleotides 1738–1752. (The initation codon for the viral replicase gene is located at nucleotide 1761.) All of the RNA molecules were cleaved under the experimental conditions used. Typical electron micrographs of the cleaved molecules are shown in Fig. 7. The cleavage site is located in the middle of the central large loop and is indicated by arrows in the interpreted drawings adjacent to each molecule. A histogram summarizing the analysis of 20 cleaved molecules that are in the cloverleaf conformation is shown in Fig. 8. A histogram of 28 native RNA molecules in the cloverleaf conformation is also shown. Comparing the two figures, it can be seen that the frequency and position of the large loops is similar in the two preparations. However, the peaks corresponding to the large loops located in the 3′ and 5′ arms of the molecules appear broader in cleaved than in native RNA, reflecting a greater degree of heterogeneity of the loop structures observed in the electron micrographs in this region of the RNA. The vertical

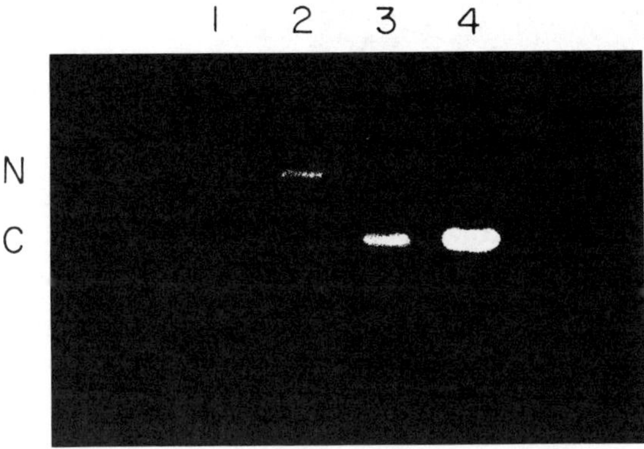

Fig. 6. Agarose gel electrophoresis of MS2 RNA cleaved with RNase H and a synthetic oligodeoxynucleotide. The reaction conditions were similar to those described by Kastelein and van Duin (see Kastelein *et al.*, 1983). The reactions were performed in 12.5 μl of buffer containing 5 m*M* Tris (pH 8.0), 1.5 m*M* magnesium chloride, and 0.5 m*M* dithiothreitol (DTT). MS2 RNA was incubated with a synthetic oligonucleotide complementary to nucleotides 1738–1752 of MS2 in the presence of RNase H at 37°C for 1 hr. The cleaved RNA was denatured with formamide at 50°C for 3 min and applied to a 1% horizontal agarose gel. (Lane 1) 2.4 μg RNA, 1 μg oligo; (lane 2) 2.4 μg RNA, 1 unit RNase H; (lane 3) 2.4 μg RNA, 1 μg oligo, 1 unit RNase H; (lane 4) 4.8 μg RNA, 1 μg oligo, 1 unit RNase H.

line that bisects the plot through the central loop indicates the position of the cleavage site close to the center of the loop. The small hairpins are less frequent in the cleaved RNA than in native molecules. In general, the details of structure in the cleaved molecules appear to be more variable than in uncleaved RNAs under similar experimental conditions. The reasons for these differences need to be explored.

The structure of the cleaved central loop in Fig. 7c is of interest, since the strand on one side of the cleavage site appears to be much shorter than the strand on the other. This is due to the formation of a second small loop in one strand of the loop. The small loop is observed at low frequency both in native and in cleaved MS2 RNA. It may correspond to the second small loop seen in the stem of the central loop of GA RNA (see above).

C. Computer Modeling on the Stem of the Central Loop

We have attempted to deduce the detailed nucleotide sequence at the stem of the central loop of GA and MS2. These studies are motivated by the observations that (1) these regions are very stable by electron microscopy, (2) that the loop is conserved both in GA and in MS2, and (3) that the ribosome-binding site for the coat gene is located within this stem. It

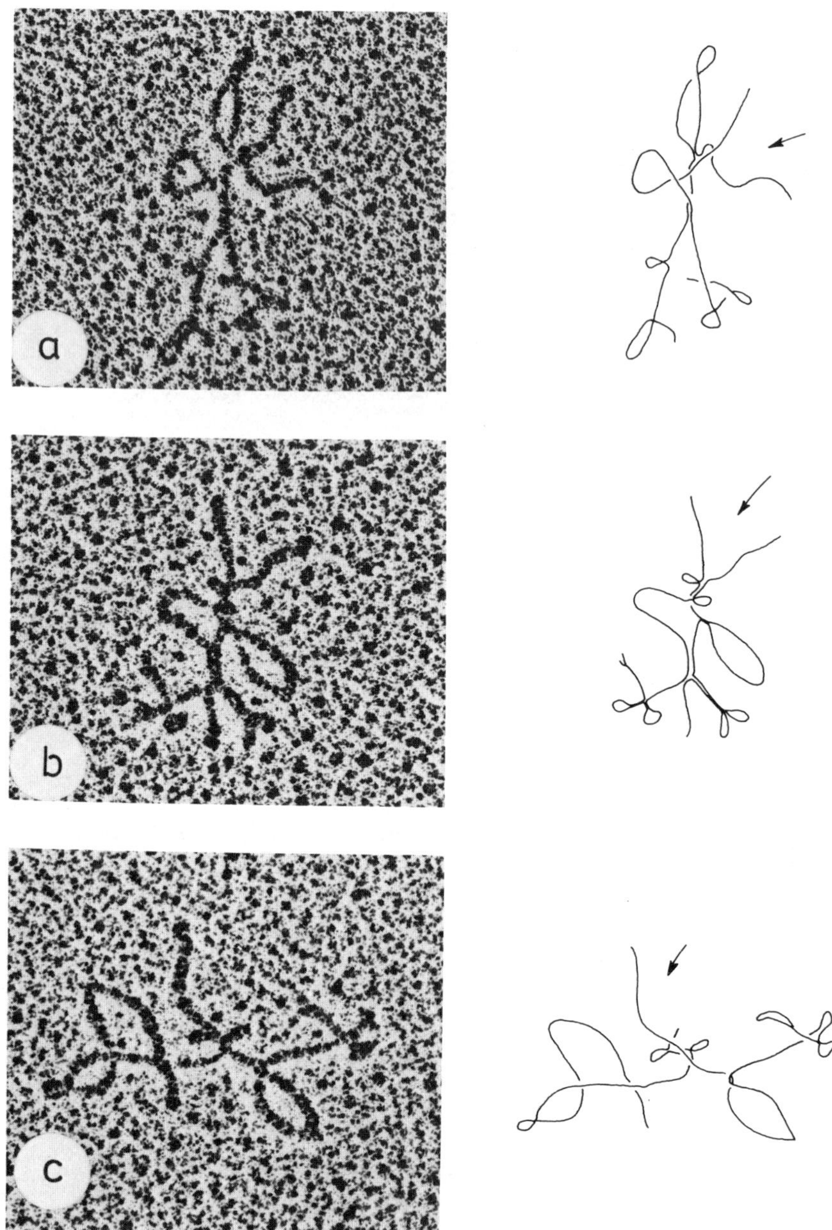

Fig. 7. Electron micrographs of MS2 RNA that were cleaved in the region of the central open loop using a synthetic oligodeoxynucleotide complementary to nucleotides 1738–1752 of MS2 RNA and RNase H. The arrow in the interpreted drawing indicates the cleavage site in each molecule. Material from lane 4 of Fig. 6 was taken for electron microscopy prior to application to the gel. Samples from the reaction mixture were diluted 100-fold and spread under our standard spreading conditions (Jacobson *et al.*, 1985) in the presence of 0.5 m*M* spermidine.

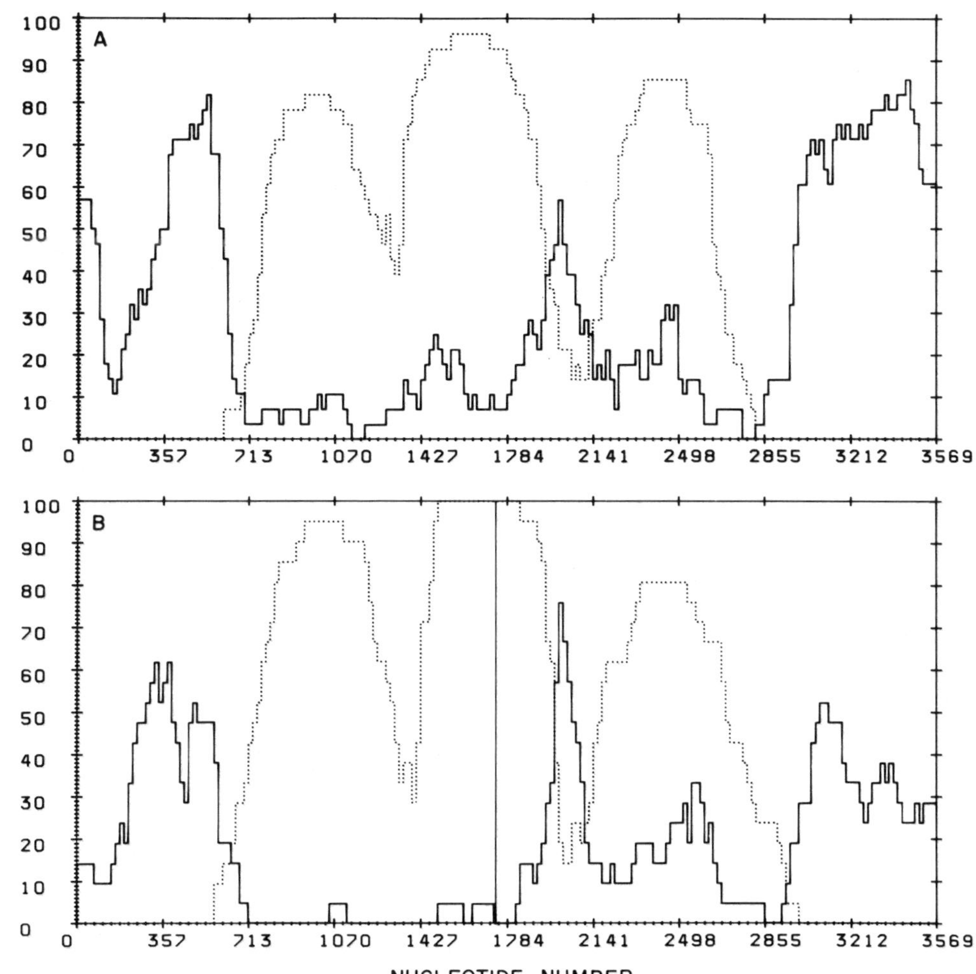

Fig. 8. Histograms showing the frequency of structural features along the length of native and cleaved MS2 RNA. (A) A plot of 28 molecules of native MS2 RNA spread in 0.5 m*M* spermidine. (B) A plot of 20 molecules of MS2 RNA cleaved with RNase H as described in Fig. 6 and spread in 0.5 m*M* spermidine. The individual molecules were normalized against the known cleavage position at nucleotides 1738–1752 prior to calculation of the histograms. (—) Small hairpin loops that range in size between 100 and 200 nucleotides; (···) large open loops that range in size from 600–700 nucleotides. The presence of these loops leads to the characteristic cloverleaf appearance of MS2 RNA.

seems likely that structure in this region of the RNA will be shown to have functional significance. The regions that are being modeled are indicated by dashed circles in Fig. 5.

The first modeling studies were performed with MS2 before the sequence of GA was known. The addition of the comparative analysis with GA has improved the quality of the models substantially, although many details still remain to be resolved. The method that has been followed in these analyses is first to identify all possible structures in both sequences that are compatible with electron microscopic data. (Due to the uncertainty associated with electron microscopic mapping procedures, structures are accepted if they lie within 500 nucleotides of the measured positions of the loops in the electron micrographs.) The structures from both sequences are compared, and common structures are identified on the basis of nucleotide sequence alignment. As discussed below, the models presented here for GA and MS2 differ from one another in their detailed structure. These differences are due to sequence divergence between the two viral RNAs and the fact that only a few base-paired helices have been conserved exactly in the common structures that form in these regions of the RNAs.

The best models that we have been able to obtain for the stem of the central loops in GA and MS2 are shown in Fig. 9. The analysis of structure in this region of MS2 RNA has been discussed previously (Jacobson *et al.*, 1985). Briefly, three possible structures were identified. Each contained a long stem corresponding to the stem of the central loop and a large hairpin that branches from the stem and corresponds to the small base-paired loop that is identified in the electron micrographs in this region of the RNA. The structures all had the large hairpin branch, but differed in the nucleotides forming the 5′ complement of the main stem. The hairpin that branches from the main stem has been retained in the current model (nucleotides 2001–2125) and can be seen in Fig. 9. It consists of a long stem with three short hairpins that form a triple-branch loop at its apex. the surprising feature of this hairpin was its behavior during computer modeling. In general, structural features vary during modeling, depending on the sequence environment in which they are located, due to the large number of alternate structures that are similar energetically. In contrast, the structure of the triple-branched hairpin remained invariant, independent of the external sequence environment that was present during the computer modeling studies.

Computer modeling in the structure of the stem of the central loop of GA RNA began with a search for a structure that might be similar to the triple-branch hairpin in MS2. Surprisingly, a structure with morphology almost identical with the structure seen in MS2 was identified at positions

1979–2118 of the GA sequence. Like the homologous hairpin in MS2, the folding of this hairpin remained invarient regardless of the external sequence environment in which the foldings were performed.

The search for long-range contacts corresponding to the stem of the central loop in GA was performed as in MS2. Only one possible structure was identified in the GA sequence. The nucleotide sequence of this structure aligned well with one of the three possible structures that had been identified for MS2. In addition to the main stem and triple-branched hairpin, the structures in Fig. 9 each have two small hairpins in the initiation region of the coat gene. In the case of GA these hairpins are included in the best structure that can be calculated for this region. For MS2, the best structure has a larger, single hairpin in the coat initiation region. In this hairpin, the coat initiation codon is buried in the stem of the hairpin. The double hairpin that is shown for MS2 is based on a comparative analysis of the sequence of GA and MS2 in the coat initiation region. The structure shown differs by 5 kcal from the optimal structure that can be obtained for this region and is well within the experimental error of the calculation.

The alignment of the sequences that were used to construct the models is shown in Fig. 10. The exact sequences comprised in the structures shown in Fig. 9 are indicated by square brackets and the sequences corresponding to the triple-branched hairpin are underscored. The sequence of the triple-branched hairpin in GA is slightly longer than the sequence of the triple-branched hairpin of MS2. Although the regions corresponding to the triple-branched hairpin align well with one another, no significant homology is observed between the two sequences of these hairpins.

Using a new computer program that we have developed to identify conserved helices between related sequences that have compensatory base changes, we have identified three conserved helices between GA and MS2 in the stem of the central loop. Each helix has 5 base pairs and each has one compensatory base change. The positions of the three helices are shown in Fig. 9. Two are indicated by dashed lines, and the third is indicated by the boxes labeled a1, a2, and a′. Note that two of the helices lie adjacent to one another in the upper stem region of the model. For one of them, the helix may form in two alternate ways, since regions a1 and a2 can both pair with a′. One compensatory base change is observed in each form of the helix. Pairing between a2 and a′ is shown in Fig. 9A and leads to the formation of a small hairpin that branches from the main stem. In Fig. 9B a1 pairs with a′ and the a1–a′ helix forms part of the structure of the main stem.

Recently we have begun to explore the folding potential of the sequences in the stem of the central loop using a new algorithm that allows us to examine alternate and suboptimal foldings within a preset range

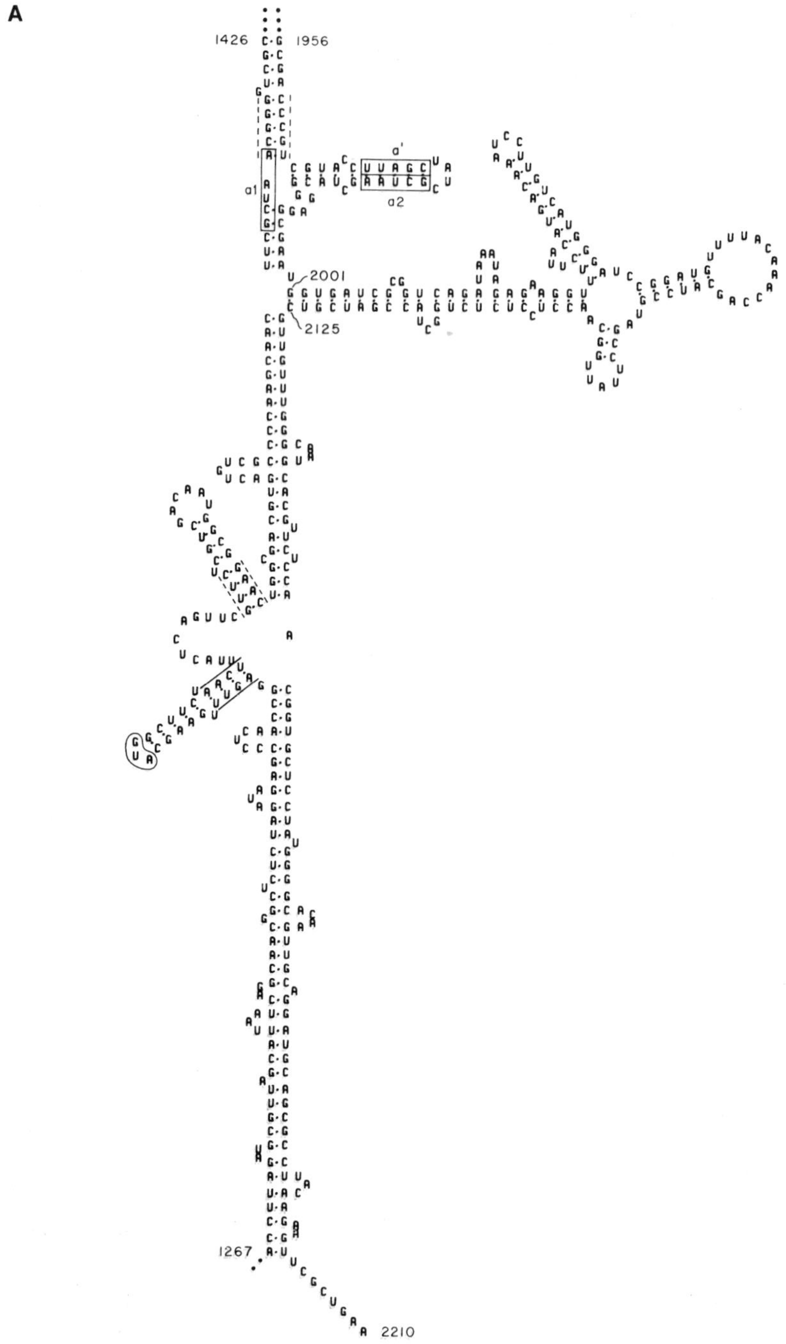

Fig. 9. Computer-generated secondary structure models of the nucleotide sequences located at the base of the central open loops in GA and MS2 RNAs. (A) MS2, calculated free

energy −184.7 kcal. (B) GA Calculated free energy −177.6 kcal. The structure models were generated using the algorithm of Zuker (1987) and Zuker and Stiegler (1981) and the energy parameters compiled by Salser (1977).

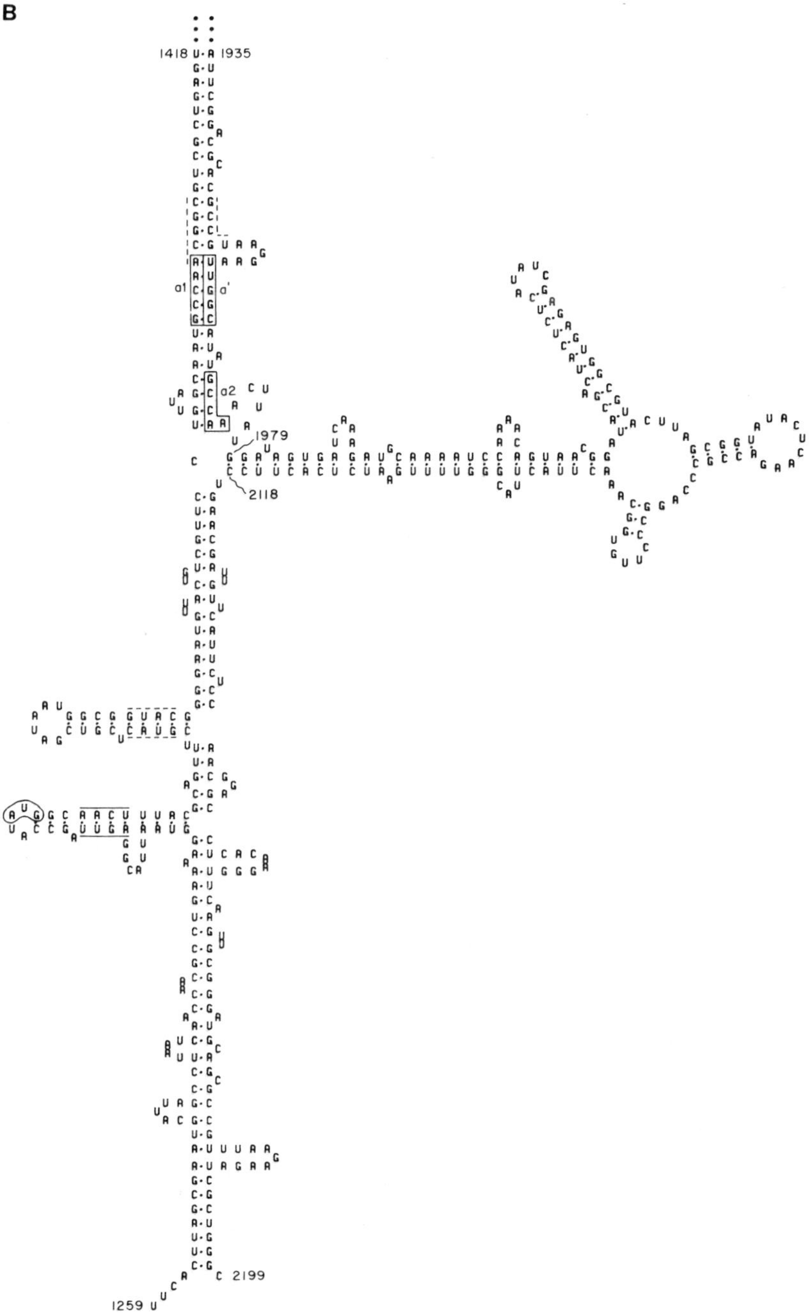

```
AT[ACCTTAG- --ATGCGTTA GCATTAATCA GGCAACGGCT CTCTAGATAG AGCCCTCAAC CGGAGTTTGA AGCATGGCTT CTAACTTTAC TCAG-TTCGT 1360
 *  *****    ** ****   ** *******   ***** **      ** *   *   * *  ******* *    *****     ********  *** *****
[TTCACTTAGC GAATGCATTA GCCTTAATCA ACCAACGCCT -----GAAAA GGTAATTA-- CGGAGTTAGC CATATGGC-- --AACTTTAC GCAGTTTCGT 1347

TCTCGTCGAC AATGGCGGAA CTGGCGACGT GACTGTCGCC CCAAGCAACT TCGCTAACGG GGTCGC]TGAA TGGATCAGCT CTAACTCGCG TTCAC      1455
 ********  ******** * * **  * **  *******  **    * *     ** *****  ********  *** *      *********     *
ACTCGTCGAT AATGGCGGTA CGGGGAATGT TACTGTCGTT CCTGTTAGCA ATGCCAACGG CGTCGCTGAG ]TGGCTTTCTA ATAACTCGCG CAGTC      1442


TACAGCAATT GCTTACTTAA GGGACGAATT GCTCACAAAG CATCCGACCT TAGGTTCTGG TAATGACGAG [GCGACCCGTC GTACCTTAGC TATCGCTAAG 1985
*   * * ** ** *** *   * ** ****   ** ** ** *****  *   * **     * ****   *    **   ****       ** **     ** **
TGATGGACTT GCCTACCTTC GTGATGAATG TCTAACTAAA CATCCTTCAC TTGGAGACAG TA[ATTCGGAC GCACGCCGTA AGGAATTGGC ATATGCCAAA 1973

CTACGGGAGG CGAATGGTGA TCGCGGTCAG ATAAATAGAG AAGGTTTCTT ACATGACAAA TC-CTTGTCA TGGGATCCGG A-----TGTT TTACAAACCA 2079
**   **       ** **** **   *     * **    *  *  **  *   * * *** *  ** * * **  * *   **  *      * *  ***  *  *
CTTATGG--- ---ATAGTGA TCAAAGATGC AAAATCCAAA ACAGTAACGG ATACGACTAC TCTCATATCG AGAGTGGCGT ACTTAGCGGT ATACTCAAGA 2067

GCATCCGTAG CCTTATTGGC AACCTCCTCT CTGGCTACCG ATCGTCGTTG TTTGGGCAAT GCACGTTCTC CAACGGTGCT CCTATGGGGC ACAAGTTGCA 2179
 *  **  *  **** *     *** * **   * ** *     ***    **   *     * * *  * ***** ****** **   *    ***    ********
CCGCCCAGGC CCTTGTGGCA AACTTACTTA CGGGTTTTGA ATCTCACTTC CTGAACGATT GTTCATTCTC CAACGGAGCC TCACAAGGGT TCAAGTTGCG 2167

GGATGCAGCG CCTTACAAGA AGTTCGCTGA A]C                                                                          2211
********** ** *  **** ** ******   *
GGATGCAGCG CCGTTTAAGA AGATCGCTGG GC]                                                                          2199
```

Fig. 10. Alignment of the sequences comprising the stem of the central loop in GA and MS2. The alignment was performed using a Needleman Wunsch algorithm. The parameters setting used were: match = 1, mismatch (transversion) = 0, mismatch (transition) = 0.01, indel = $-(2 + .01n)$, where n is the number of nucleotides in the indel. The initiation codon for the coat protein gene is boxed. The regions corresponding to the triple-branched hairpins are underscored, and the nucleotide sequences corresponding to the structural models in Fig. 7 are indicated by square brackets.

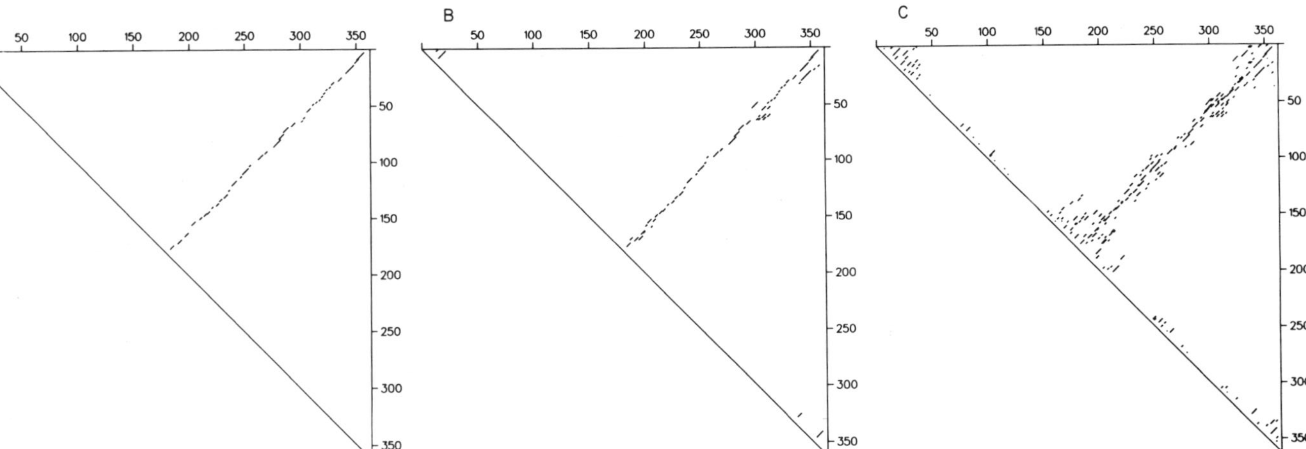

Fig. 11. Computer-generated energy dot plots showing the distribution of alternate and suboptimal structures that lie within 15 kcal of the optimum for the nucleotide sequence of PSTV RNA. The free energy calculated for the optimum folding is −190.1 kcal. (A) The optimal folding. (B) Base pairs located in alternate and suboptimal structures that lie within 5 kcal from the optimal folding. (C) Base pairs located in structures that lie within 15 kcal from the optimal folding.

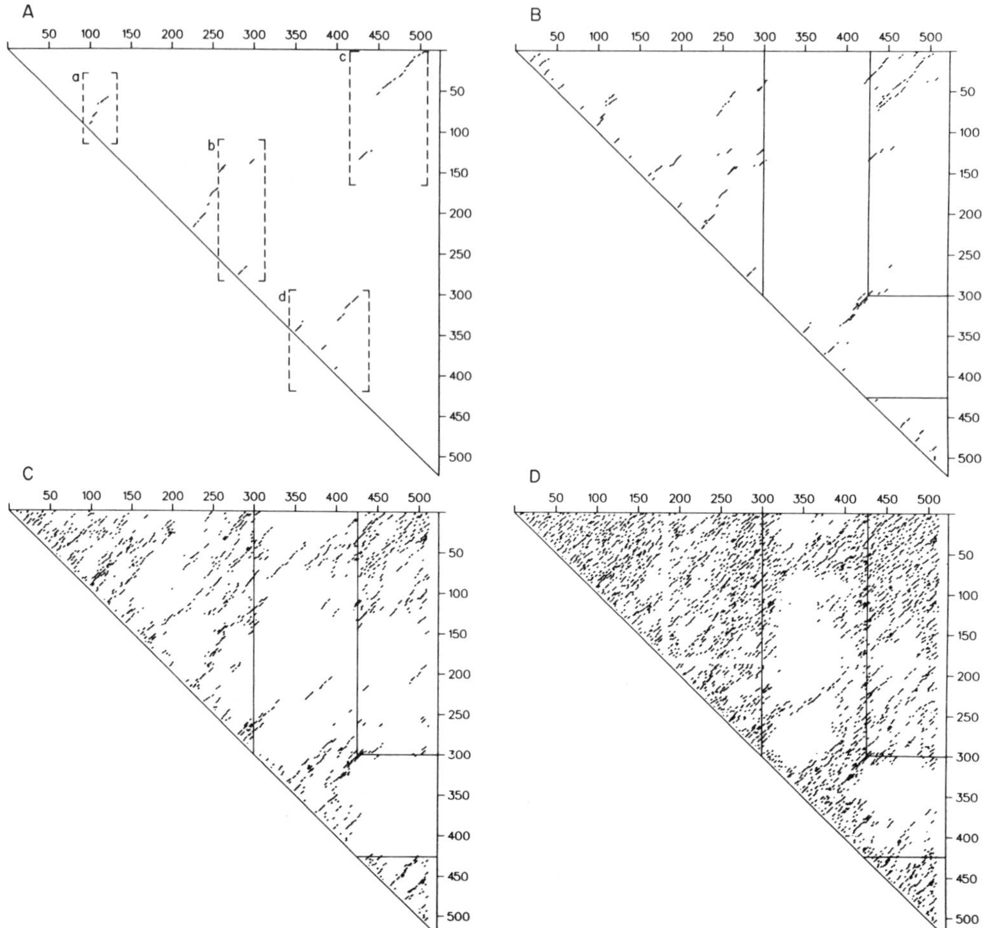

Fig. 12. Computer-generated energy dot plots for the stem of the central loop of MS2. The plots shown the distribution of alternate and suboptimal structures that lie within 15 kcal of the optimum. The sequences used in these foldings correspond to nucleotide positions 1265–1455 and 1886–2211. The free energy calculated for the optimum folding of this sequence is −206.0 kcal. (A) Base pairs within structures in the optimal folding. The annotation of this plot identifies the structural features associated with each region of the plot. (a) Helices located in the coat initiation region; (b) helices located in the upper stem; (c) helices located in the lower stem; (d) helices located in the triple-branched loop. (B, C, D) Base pairs within structures that lie within 5, 10, or 15 kcal from the optimum structure.

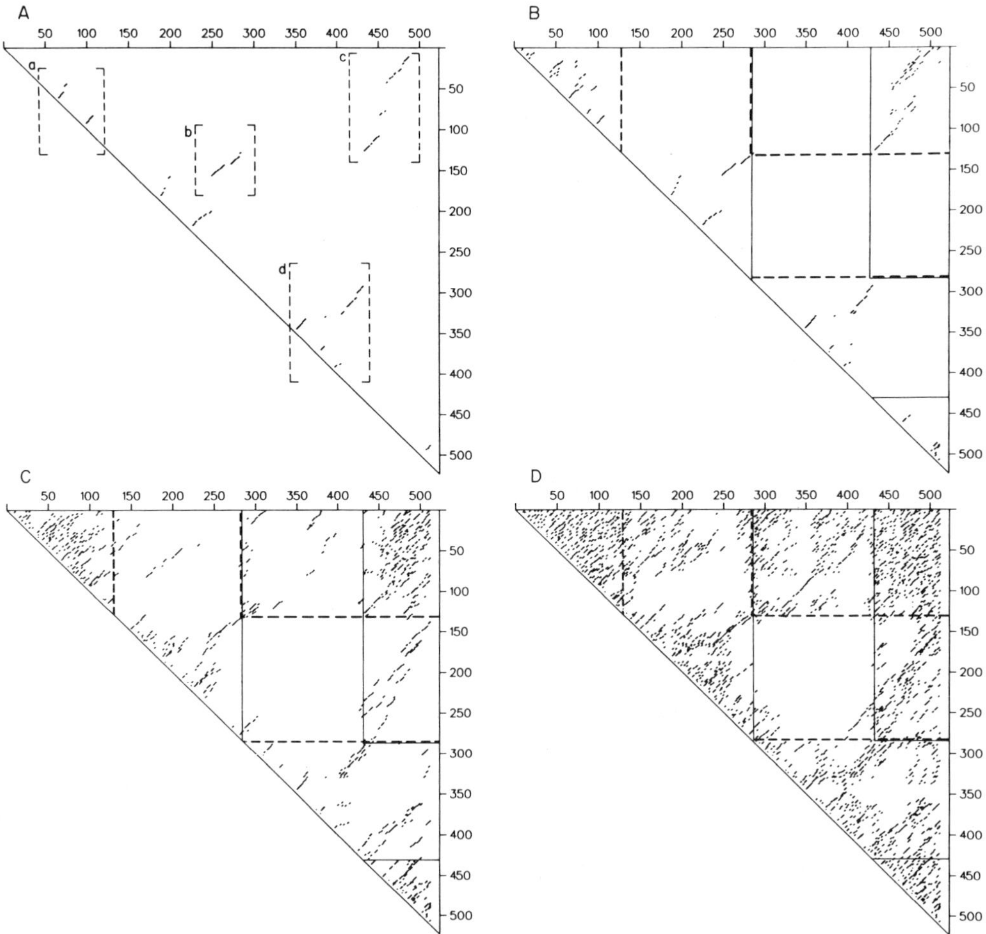

Fig. 13. Computer-generated energy dot plots for the stem of the central loop of GA. The plots show the distribution of alternate and suboptimal structures that lie within 15 kcal of the optimum. The sequences used in these foldings correspond to nucleotide positions 1259–1442 and 1874–2199. The free energy that is calculated for the optimal folding is −191.0 kcal. The annotation in A corresponds to the annotation in Fig. 12A. The parameter settings of A, B, C, and D are identical with the settings used to create Fig. 12.

from the best folded structure (Zuker, 1987). The results of some of these calculations are shown in Fig. 11–13 and are presented in the form of dot plots where each dot represents a base pair found in one of the optimal or suboptimal structures that lie within a certain free-energy range from the best computed structure. The plots will be referred to as energy dot plots.

Base pairs in the same row or column of the plot represent base pairs associated with helices located in alternate structures that include a common base. Results are shown for calculations ranging from 0 to 15 kcal from the optimal structure.

In Fig. 11 we show the foldings obtained for the nucleotide sequence of PSTV (potato spindle tuber viroid). The structure of the RNA is well studied: it is known to form a stable, double stranded, unbifurcated hairpin (Steger *et al.,* 1984). This structure is one of the few RNA structures that can be computed correctly with the original Zuker–Stiegler RNA folding algorithm using the energy rules that were compiled by Salser (1977; Zuker and Stiegler, 1981). Figure 11A shows the energy dot plot representation of the best folding that is obtained by computer for this sequence. The hairpin is represented as a discontinuous diagonal line from the upper righthand corner to the base of the triangle that forms the bounds of the figure. The discontinuities in the diagonal line represent sites of bulge and interior loops in the stem of the hairpin. Branches in the structure would be observed as additional diagonal lines at the base of the figure and are not seen.

Similar plots are shown for GA and MS2 (Fig. 12A and 13A) in the region of the central loop. The structures in this region are more complex than the viroid hairpin and are similar to, but not identical with, the two-dimensional structures drawn in Fig. 9. In each instance a long central stem is interrupted by several branches. The central stem is divided into upper (b) and lower (c) regions. The triple-branch hairpin (d) is at the lower right of the plots. In the coat initiation region (a) of MS2 only one hairpin is found in the optimal folding. The sequences that were used for the energy dot plots of MS2 and GA are the same sequences used that were used for the alignments shown in Fig. 10. The relative insertion regions (indicated by dashed lines) were deleted from the appropriate sequence and the fused sequences were renumbered as shown in the plots. The deletion of the inserted regions allows us to compare structures by superimposing the plots.

Plots showing the folded structures of PSTV RNA that lie within 5 and 15 kcal of the optimum are shown in Fig. 11, B and C. The striking feature of these plots is the absence of alternate structures. In contrast, in the central stem of MS2 and GA (C and D of Fig. 12 and 13), a large number of suboptimal structures can be identified that lie within 15 kcal of the optimum structure.

Although the overall impression that is obtained from examination of the plots of the central stem sequences of GA and MS2 is that many alternate conformations can be calculated in these regions of the sequence, a careful examination of the plots suggests that specific patterns can be discerned as well. In MS2 extensive clear areas are observed in the region corresponding to nucleotides 300–425 (nucleotides 2001–2125 in the original sequence). The region is identified in the figures by two truncated rectangles that extend from the borders of the plots toward the center. Examining Fig. 12D in detail, it can be seen that substantial local variation is observed in the region corresponding to the triple-branch hairpin itself, and this variability can also be observed in Fig. 12C, where structures that lie within 10 kcal of the optimal structure are shown. Few base-paired alternatives can be observed, however, between nucleotides 1 and 299, or 426 and 511 and the region from 300 to 425. By analogy with the results of our analysis of the sequence of PSTV, we believe that clear areas in the plots indicate regions of the sequence that are well defined and likely to form stable structures. As discussed above, computer modeling studies have shown that the triple-branched hairpin of MS2 was extremely resistant to alterations in sequence context during computer simulation studies. The new plots confirm this observation.

In addition to identifying stable regions in the sequence, the energy plots also allow us to identify regions that exhibit extensive structural variation. In MS2 these regions correspond to the lower stem and coat initiation regions. In variable regions of this type, computer calculations probably do not provide a reliable indication of the structure of the RNA since all the structures observed lie within the error limits of the energy rules that are used in the calculations.

The interpretation of the plots corresponding to the central stem of GA is more difficult (Fig. 13). The overall impression obtained from these plots is that the clear areas are more extensive than they are in MS2. We believe that the results are best interpreted as indicating the presence of two unusually stable structures in this region of the sequence. The first of these corresponds to nucleotides 130–280 and corresponds to the upper stem region of the structure (It also includes the excision region of the sequence). The second, from nucleotides 284 to 430 corresponds to the triple-branch loop that is also seen in MS2. More alternate structures are seen for the sequences corresponding to the triple-branch loop in GA than in MS2. A striking feature of these plots is the fact that the central region of the plot is devoid of alternate structure, indicating that pairing between nucleotides 130 and 280 (the upper stem region) and nucleotides 284 and 430 (the triple-branched loop) is very unlikely.

The observation that regions in the stem of the central loop are stable by computer modeling is intriguing in view of the fact that these structures are also stable when examined by electron microscopy. It seems likely that the sequences in these stable regions of secondary structure will be shown to have important functional significance in the regulation of viral gene expression, and might define nucleation sites that determine the overall folding of the whole mRNA molecule.

III. DISCUSSION

The analysis of folding in large mRNA molecules is an intriguing problem. The genetic code of these sequences is more complex than that of structural RNAs, since the molecules contain information defining both protein and nucleic acid structure. The ability to obtain detailed structural information on the structure of coliphage RNAs by electron microscopy provides a unique handle with which to explore these problems.

Our results suggest that the central loop is an important structure because it is conserved between GA and MS2 despite substantial differences in the nucleotide composition of these RNAs, and because, in MS2, where we observe alternate conformations of the RNA, the central loop is conserved in both structures. These results suggest that this structural feature may play an important role in the regulation of viral gene expression. Unfortunately, little is known concerning mechanisms by which large global structural features in an RNA molecule might play a role in the regulation of gene expression. Early studies by Vollenweider and Koller (see Vollenweider *et al.*, 1976) by electron microscopy have shown that the viral polymerase of Qβ can be cross-linked with glutaraldehyde to a loop structure in Qβ RNA that is reminiscent of the loop structure that we see in GA and MS2. The binding of polymerase in this region of the RNA is though to inhibit ribosome binding in the coat initiation region and to result in translational shut-off (Kolakofsky and Weissman, 1971). Possibly, structures associated with the stem of the central loop facilitate replicase and ribosome recognition and/or binding to the appropriate nucleotide sequences. Very recent studies by Boni *et al.* (1986) show that the sequence 2030–2056 in the replicase region of MS2 is protected by the ribosomal protein S1, when 30 S ribosomes are bound in the coat initiation region. These regults provide the first evidence that nucleotides associated with the stem of the central loop interact with ribosomes during the initiation of coat protein synthesis.

The presence of long-range folding has been proposed to explain translational coupling of the coat and replicase genes of MS2 (Min Jou *et al.*, 1972). In this model nucleotides 1409–1423 of the viral coat gene base-pair with nucleotides 1738–1769 of the replicase initiation region. Translation of the early region of the coat gene would disrupt this structure and allow ribosomes to bind to the replicase initiation region. Recent studies by Berkhout and van Duin (1985) show that when the nucleotide sequences from positions 1380 to 1432 are deleted from MS2, translational coupling of the coat and replicase gene is no longer observed. Although these results provide unequivocal evidence that an internal sequence at the beginning of the coat gene affects regulation of the expression of the replicase gene, it is more difficult to establish that these nucleotides are paired with the initiation region for the replicase gene. In our modeling studies we find that the sequences 1410–1423 pair best with nucleotides 1959–1999, more than 200 nucleotides downstream from the replicase initation region. Our model derives from electron microscopic visualization of structure, computer modeling, as well as an analysis of conserved helices with compensatory base changes for both folding alternatives. Nonetheless, it still needs to be verified experimentally, both by comparison with additional related RNA coliphage sequences, as well as by using structure-specific biochemical probes. Should the model we propose for the stem of the central loop prove to be correct, an alternate model would be required to explain the coupled translation of the coat and replicase genes.

The comparative analysis that we present here for the stem of the central loop is still tentative since we only consider two nucleotide sequences, and only observe single base changes in the common helices that we identify. Studies with 16 S ribosomal RNAs suggest that multiple related sequences need to be analyzed and at least two independent compensatory base changes should be identified for each helix before it can be considered likely (Gutell *et al.,* 1985). We are currently planning further studies with several additional closely related RNA coliphage strains (Furuse *et al.,* 1979; Furuse, 1982; Inokuchi *et al.,* 1982; Nishihara, 1969). Nonetheless, the results by electron microscopy show that homologous structural features can be identified between GA and MS2 RNA despite the large amount of divergence that has occurred between the two nucleotide sequences. The results are encouraging and suggest that the phylogenetic method will be useful in the analysis of messenger as well as structural RNAs.

ACKNOWLEDGMENT

We thank Masayori Inouye for encouraging the comparative aspect of this study. David Kleinman participated in the analysis of the electron micrographs. Lawrence Chan and Cher Win Lin wrote computer programs that were used in the modeling of RNA folding as well as in the analysis of electron micrographs. The research was supported in part by National Institutes of Health Grants AI15273 and GM38425 to A.B.J.

REFERENCES

Berkhout, B., and van Duin, J. (1985). Mechanism of translational coupling between coat protein and replicase genes of RNA bacteriophage MS2. *Nucleic Acids Res.* **13,** 6955–69675.

Boni, I. V., Isaeva, D. M., and Budowsky, E. I. (1986). Ribosomal protein binds to the internal region of the replicase gene within the complex of *E. coli* 30S ribosomal subunit with MS2 phage RNA. *Bioorg. Khim.* **12,** 293–296.

Furuse, K. (1982). Phylogenetic studies on RNA coliphage RNAs. *J. Keio Med. Soc.* **59,** 265–274.

Furuse, K., Hirashima, A., Harigai, H., Ando, A., Watanabe, K., Kurosawa, K., Inokuchi, Y., and Wantanabe, I. (1979). Grouping of RNA coliphages based on analysis of the sizes of their RNAs and proteins. *Virology* **97,** 328–341.

Gutell, R. R., Weiser, B., Woese, C. R., and Noller, H. F. (1985). Comparative anatomy of 16-S-like ribosomal RNA. *Prog. Nucleic Acid Res. Mol. Biol.* **32,** 155–216.

Inokuchi, Y., Hirashima, A., and Watanabe, I. (1982). Comparison of the nucleotide sequences at the 3′-terminal region of RNAs from RNA coliphages. *J. Mol. Biol.* **158,** 711–730.

Inokuchi, Y., Takahashi, R., Hirose, T., Inayama, S., Jacobson, A. B., and Hirashima, A. (1986). The complete nucleotide sequence of RNA coliphage GA. *J. Biochem.* (*Tokyo*) **99,** 1169–1180.

Jacobson, A. B. (1976). Studies on the secondary structure of single-stranded RNA from the bacteriophage MS2. *Proc. Natl. Acad. Sci. U.S.A.* **73,** 307–311.

Jacobson, A. B., and Spahr, P. F. (1977). Studies on the secondary structure of single-stranded RNA from the bacteriophage MS2 II. Analysis of the RNase IV cleavage products. *J. Mol. Biol.* **115,** 279–294.

Jacobson, A. B., Kumar, H., and Zuker, M. (1985). The effect of spermidine on the conformation of MS2 RNA: Electron microscopy and computer modeling. *J. Mol. Biol.* **181,** 517–531.

Kastelein, R. A., Berkhout, G., Overbeek, G. P., and van Duin, J. (1983). Effect of the sequences upstream from the ribosome-binding site on the yield of protein from the cloned gene for phage MS2 coat protein. *Gene* **23,** 245–254.

Kolakofsky, D., and Weissman, C. (1971). Qβ replicase as a repressor of Qβ RNA-directed protein synthesis. *Biochem. Biophys. Acta* **246,** 596–599.

Min Jou, W., Haegeman, G., Ysebaert, M., and Fiers, W. (1972). Nucleotide sequence of the gene coding for the bacteriophage MS2 coat protein. *Nature* (*London*) **237,** 82–88.

Nishihara, T. (1969). Chemical studies on RNA phages. 1. Chemical properties of various RNA phages. *J. Keio Med. Soc.* **46,** 351–361.

Salser, W. (1977). Globin messenger-RNA sequences—analysis of base-pairing and evolutionary implications. *Cold Spring Harbor Symp. Quant. Biol.* **42,** 985–1002.

Steger, G., Hofmann, H., Förtsch, J., Gross, H. J., Randles, J. W., Sänger, H. L., and Riesner, D. (1984). Conformational transitions in viroids and virusoids: Comparison of results from energy minimization algorithm and from experimental data. *J. Biomol. Struct. Dyn.* **2,** 543–571.

van Duin, J. (1988). The single stranded RNA bacteriophages. *In* "Bacteriophages" (R. Calendar, ed.), in press. Plenum, New York.

Vollenweider, H. J., Koller, T., Weber, H., and Weissman, C. (1976). Physical mapping of Qβ replicase binding sites on Qβ RNA. *J. Mol. Biol.* **101,** 367–377.

Zuker, M. (1987a). Manuscript in preparation.

Zuker, M. (1987b). The use of dynamic programming algorithms in RNA secondary structure prediction. *In* "Mathematical Methods for DNA Sequences" (M. S. Waterman, ed.), in press. CRC Press, Boca Raton, Florida.

Zuker, M., and Stiegler, P. (1981). Optimal computer folding of large RNA sequences using thermodynamics and auxiliary information. *Nucleic Acids Res.* **9,** 133–148.

VI

RNA in Regulation and Repression

23

Autogenous Regulation of Transcription of the *crp* Operon by a Divergent RNA Transcript

KEINOSUKE OKAMOTO*
AND MARTIN FREUNDLICH†

**Department of Microbiology*
School of Medicine
Fujita-Gakuen University
Aichi 470-11, Japan
†Department of Biochemistry
State University of New York at Stony Brook
Stony Brook, New York 11794

I. INTRODUCTION

The classic operon model of Jacob and Monod (1961) suggested that the activity of structural genes was controlled by a trans-acting product made by regulatory genes. Subsequent studies in numerous prokaryotic systems have shown that the products of these regulatory genes are proteins that bind to specific sites on the DNA to influence gene transcription (Gilbert and Mueller-Hill, 1966; Ptashne, 1967). The cyclic AMP receptor protein (CRP) is an example of a DNA-binding protein that acts to regulate the expression of numerous genes in *Escherichia coli* including many operons involved in the metabolism of secondary carbon sources (Ullmann and Danchin, 1983). Recently, it has been shown that cyclic AMP-CRP bind to the *crp* promoter region and strongly inhibit transcription of the *crp* gene (Aiba, 1983). These and other data (Cossart and Gicquel-Sanzey, 1985) suggest that *crp* expression is negatively controlled by autogenous regulation.

In this chapter we present evidence that the autogenous control of the

Molecular Biology of RNA
New Perspectives

crp operon is accomplished by a previously unrecognized mechanism for the regulation of gene expression. This control involves the inhibition of *crp* transcription by a trans-acting RNA. This regulatory RNA is produced by cyclic AMP-CRP activation of a divergent promoter located within the *crp* promoter region.

II. ACTIVATION OF A DIVERGENT PROMOTER BY CYCLIC AMP-CRP IS REQUIRED FOR *crp* AUTOREGULATION

A. Cyclic AMP-CRP Activates an RNA Transcribed in the Opposite Direction to the *crp* Gene

Aiba (1983) showed that cyclic AMP-CRP strongly inhibits transcription from the *crp* promoter *in vitro*. We have found that in addition to blocking *crp* transcription, cyclic AMP-CRP activates a new RNA species (Okamoto and Freundlich, 1986). This new RNA is essentially not made without added cyclic AMP-CRP (Fig. 1). A very strong correlation can be seen between the appearance of this transcript and the inhibition of *crp* transcription (Fig. 2). Using a variety of templates of different lengths at the 5′ and 3′ ends, we showed that the new RNA was probably not transcribed by the *crp* promoter but by a new promoter which initiates transcription in the opposite direction to the *crp* gene (Okamoto and Freundlich, 1986). An examination of the nucleotide sequence on the opposite strand downstream from the initiation of the *crp* transcript (Fig. 3) shows a sequence beginning at +39 that is similar to sequences found in many *E. coli* promoters (Hawley and McClure, 1983). Although the spacing between the proposed −10 and −35 regions in this promoter is larger (21 bp) than the consensus (17 bp), five known promoters have this same separation (Hawley and McClure, 1983). Taken together, these data suggest that the cyclic AMP-CRP-dependent RNA initiates close to, but in the reverse direction to, *crp* mRNA. This was confirmed by S1 mapping experiments which showed that transcription from the divergent promoter begins 3 nucleotides upstream and on the opposite strand from the initiation of the *crp* transcript (Okamoto and Freundlich, 1986).

B. A Functional Divergent Promoter Is Necessary for *crp* Autoregulation

The dependence of the divergent promoter on cyclic AMP-CRP and its proximity to the *crp* promoter suggested that the divergent promoter might be a major element in the mechanism of *crp* autoregulation. We tested this possibility by altering the divergent promoter by the insertion

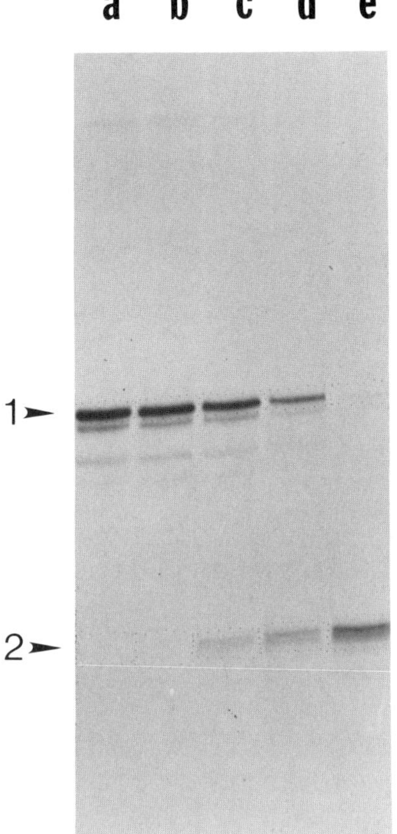

Fig. 1. Effect of cyclic AMP-CRP on *in vitro* transcription. Experiments were carried out as described previously (Okamoto and Freundlich, 1986). Added CRP (μg/ml): 0 (a), 0.75 (b), 2 (c), 4 (d), 7.5 (e). Arrowheads: 1, *crp* mRNA; 2, new RNA.

of a *Bam*HI linker at the *Rsa*I site at +24 (Fig. 3). There is no overlap in this region with either the *crp* promoter or with the downstream CRP binding site. However, the insertion increases the distance between the proposed −35 and −10 regions of the divergent promoter by 8 base pairs which should prevent the functioning of this promoter (Hawley and McClure, 1983). The fragment with the insertion and the unaltered fragment were used as templates for *in vitro* transcription. In contrast to the normal fragment, the addition of cyclic AMP-CRP did not activate the divergent transcript when the altered template was transcribed, and no inhibition of *crp* transcription was found (Okamoto and Freundlich, 1986). A DNase

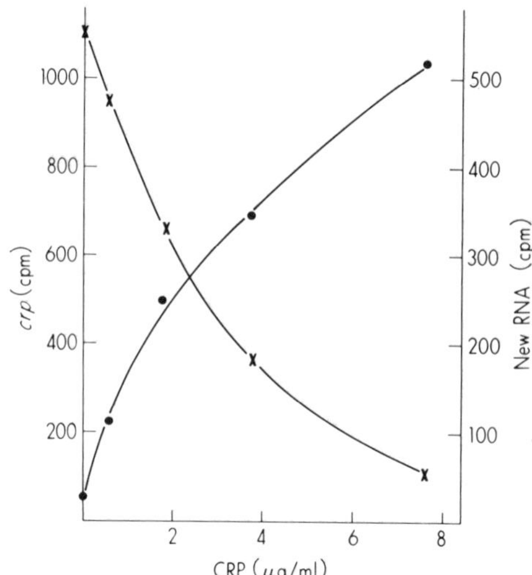

Fig. 2. Correlation between inhibition of *crp* mRNA and appearance of the new RNA. The RNA bands described in Fig. 1 were cut out of the gel and counted in a liquid scintillation counter. The amount of radioactivity in crp mRNA (×) is plotted against that of the new RNA (●).

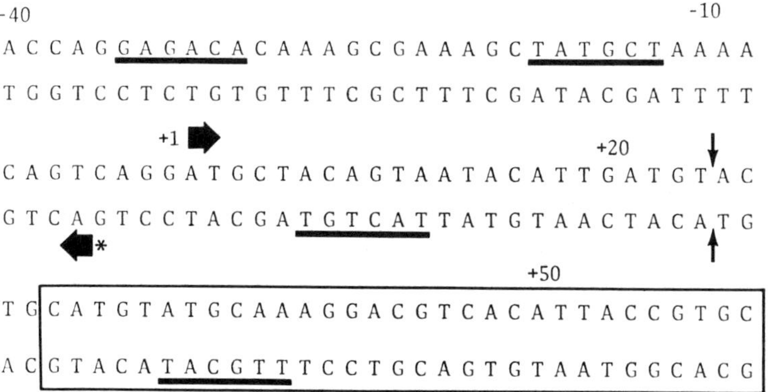

Fig. 3. Nucleotide sequence of the *crp* promoter region. The nucleotide sequence is taken from Aiba (1983). The nucleotides are numbered from the proposed start of initiation of *crp* mRNA at position 1 (Aiba, 1983). The proposed start of transcription of the anti-sense RNA is indicated by an asterisk. The suggested −35 and −10 regions for the *crp* promoter (Aiba, 1983) and for the divergent promoter are underlined. The CRP binding site that is necessary for negative autoregulation is boxed. The *Rsa*I cutting site is indicated by vertical arrows.

footprint (Galas and Schmitz, 1978) of CRP binding showed essentially the same protection pattern in the normal and altered fragment, suggesting that CRP interactions were not changed by the 8-base-pair insertion (Okamoto and Freundlich, 1986). In addition, experiments to be described later show that even though *crp* expression in the altered fragment is no longer inhibited by cyclic AMP-CRP, the *crp* promoter still retains the capacity for autoregulation (Fig. 6). Taken together, these results indicate that the divergent promoter has an essential role in the autoregulation of the *crp* operon.

III. AUTOREGULATION OF *crp* IS MEDIATED BY DIVERGENT RNA

A. Divergent RNA Inhibits *crp* Transcription *in Vitro*

Since the *crp* and divergent promoters overlap, activation of the divergent promoter by cyclic AMP-CRP could inhibit *crp* transcription by a polymerase competition mechanism (Herbert *et al.*, 1986). Alternatively, a product of divergent promoter function could be responsible for the inhibition. The latter hypothesis was tested by purifying a 92-nucleotide divergent RNA made when a *Mlu*I–*Hin*dIII *crp* promoter fragment was transcribed in the presence of cyclic AMP-CRP (Okamoto and Freundlich, 1986). The synthesis of *crp* mRNA was strongly inhibited by the addition of the divergent RNA to the *in vitro* transcription reaction (Fig. 4). A *bla* promoter fragment, which was transcribed in the same reaction as the *crp* fragment, was essentially not effected by the divergent RNA (Fig. 4). The data in Fig. 5 shows that *crp* transcription was inhibited 50% by 100 ng/ml of divergent RNA. This concentration corresponds to two divergent RNA molecules per template molecule. We find that divergent RNA as small as the first 16 nucleotides effectively blocks *crp* transcription (T. Inoue and M. Freundlich, unpublished data).

We previously inactivated the divergent promoter by inserting a *Bam*HI linker at +24 (Okamoto and Freundlich, 1986). The binding of CRP and *crp* promoter function were maintained in the altered fragment but cyclic AMP-CRP no longer inhibited *crp* transcription. However, the altered fragment has retained the capacity for autoregulation since added divergent RNA blocks transcription from this fragment to the same extent as the normal template (Fig. 6).

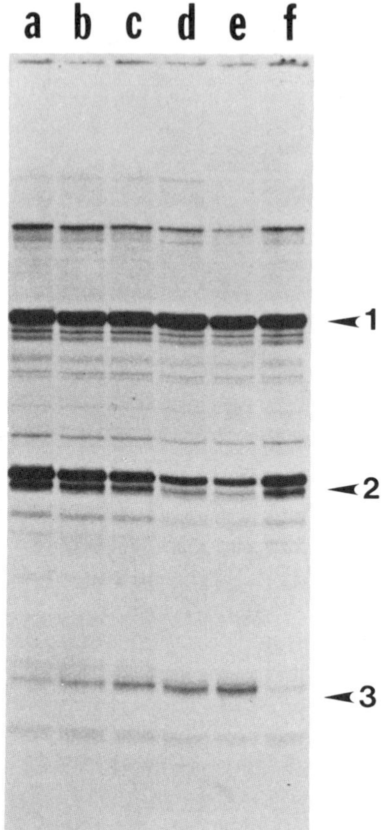

Fig. 4. Effect of added divergent RNA on *crp* transcription. A 92-nucleotide divergent RNA was added to the standard transcription assay. A *bla* promoter fragment was used as a control. Divergent RNA added (ng/ml): 0 (a,f); 42 (b); 84 (c); 126 (d); 150 (e). Arrows: 1, *amp* RNA; 2, *crp* RNA; 3, added divergent RNA.

B. Autoregulation of *crp in Vivo*

To assess the physiological significance of the *in vitro* results we measured the amount of CRP made in cells grown with and without cyclic AMP. These experiments used the isogenic strains MF3000 and MF3001, each of which contains a plasmid that carries the *crp* structural gene. In

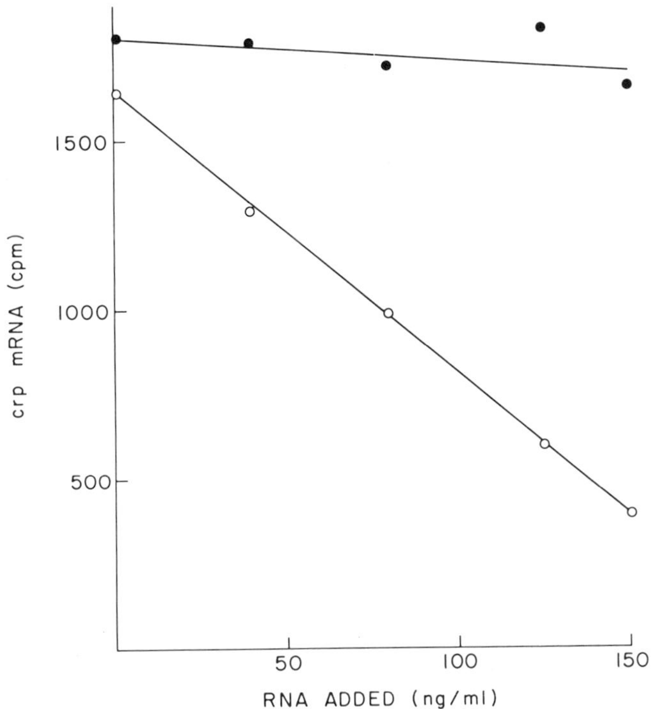

Fig. 5. Inhibition of *crp* transcription by divergent RNA. The RNA bands described in Fig. 4 were cut out of the gel and counted in a liquid scintillation counter. The amount of radioactivity in counts per min (cpm) is plotted against the amount of added divergent RNA. Symbols: (●) *bla* RNA; (○) *crp* RNA.

MF3000 the cloned *crp* gene is under control of its own promoter, while in MF3001 *crp* expression is mediated by a plasmid promoter. The data in Fig. 7 show a strong reduction in the protein band corresponding to CRP when MF3000 was grown with increasing amounts of cyclic AMP. Quantification of these data revealed a 6.5-fold decrease in CRP in the culture grown with the highest level of cyclic AMP (Fig. 8). Significantly, strain MF3001 showed no reduction in CRP levels under these same conditions. Evidence that the mechanism for *crp* autoregulation *in vivo* is similar to that found *in vitro* was obtained previously using *in vivo* S1 mapping experiments. These experiments showed that negative autoregulation of *crp* mRNA and positive activation of the divergent transcript by cyclic AMP-CRP take place *in vivo* (Okamoto and Freundlich, 1986). The importance of divergent RNA for *in vivo* autoregulation was investigated by construction of a plasmid containing the *crp* gene expressed through the

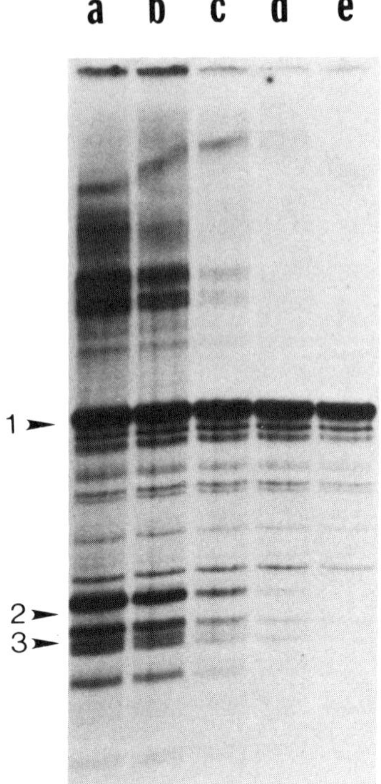

Fig. 6. Effect of linker mutagenesis in the divergent promoter on divergent RNA inhibition of *crp* transcription. The altered, normal and *bla* promoter fragment (0.1 pmole each) were used as templates in the standard transcription assay. Divergent RNA added (ng/ml): 0 (a); 30 (b); 60 (c); 120 (d); 240 (e). Arrowheads: 1, *bla* RNA; 2, *crp* RNA from the altered fragment; 3, *crp* RNA from the normal fragment.

crp promoter fragment that has an inactive divergent promoter. This fragment, described in Section II.B., produces no detectable divergent RNA. No autoregulation of CRP production was found in cells containing the plasmid with the altered *crp* promoter region (K. Okamoto and M. Freundlich, unpublished data). In contrast, cells containing a control plasmid, constructed with a normal divergent promoter, showed strong autoregulation.

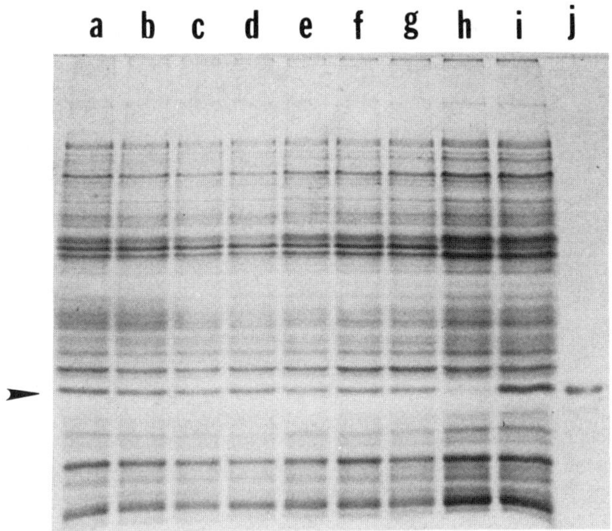

Fig. 7. *In vivo* autoregulation of *crp*. Cells were grown for 14 hr in Luria broth and the indicated amounts of cyclic AMP. Total proteins were extracted and analyzed as described by Anderson *et al.* (1973). Strain MF3000 (pHA5/pp47) contains a plasmid with *crp* under control of its own promoter. Strain MF3001 (pHA7/pp47) contains a plasmid with *crp* under plasmid promoter control (Aiba *et al.*, 1982). Strain MF3001 grown in the presence of cyclic AMP (m*M*): 0 (a); 0.5 (b); 1 (c); 2 (d). Strain MF3000 grown in the presence of cyclic AMP (m*M*): 0 (e); 0.5 (f); 1 (g); 2 (h). Addition of purified CRP (2 μg) to the protein extract used in lane h (i) and purified CRP (2 μg) added alone to the gel (j). The arrow indicates the CRP band.

IV. A MODEL FOR *crp* AUTOREGULATION

Autoregulation of the *crp* gene by a divergent RNA appears to be the first example of the control of transcription by a *trans*-acting RNA. The molecular events involved in this process are not apparent but an examination of the sequence at the 5′ ends of the divergent RNA and *crp* mRNA suggest a possible mechanism. There is very strong homology between nucleotides 2–15 in the divergent RNA and nucleotides 2–11 in *crp* mRNA (Fig. 9A). The next 11 nucleotides of *crp* mRNA (bases 12–22) are AU rich. The binding of the initial nucleotides of the divergent RNA to the homologous 5′ segment of *crp* mRNA would result in a strong RNA–RNA hybrid. This structure would resemble a rho-independent transcription terminator, except for an extended series of uridine residues immediately distal to the RNA duplex (Rosenberg and Court, 1979; Platt, 1981).

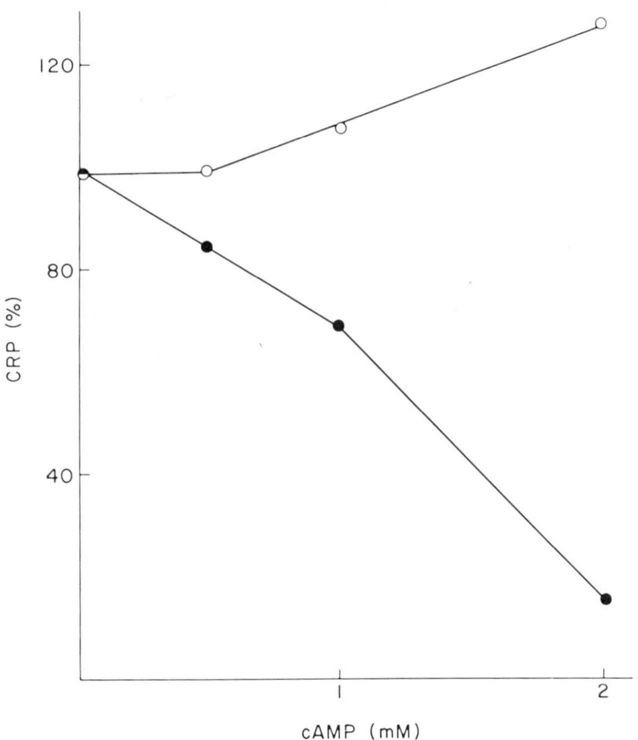

Fig. 8. Autoregulation of *crp in vivo*. The gel in Fig. 7 was analyzed in a Shimadzu scanning densitometer. The values were corrected for the total protein in each lane on the gel. Symbols: (○) extracts from MF3001; (●) extracts from MF3000. The apparent increase in CRP synthesis in strain MF3001 with increasing concentrations of cyclic AMP may reflect increased stability of CRP when saturated with cyclic AMP.

TABLE I

Comparison of the *crp* Terminator with Known Terminators

Terminator	$-\Delta G$	Stem length	A/T length
Known[a]	18	8 ± 2	10
crp	14.7	10	11

[a] The properties of 30 known rho-independent terminators, as described by Brendel and Trifonov (1984), are compared with the properties of the proposed *crp* regulatory terminator.

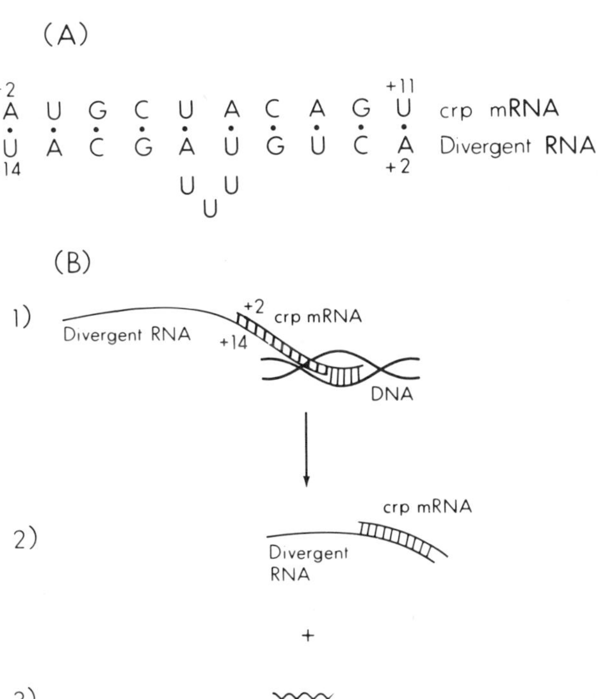

Fig. 9. Model for divergent RNA inhibition of *crp* transcription. (A) The proposed RNA–RNA duplex between the 5′ end of *crp* mRNA and the initial 14 nucleotides of the divergent RNA. The properties of this hybrid are consistent with those present in the initial segment of rho independent terminators (Table I; Brendel and Trifonov, 1984). (B) Mechanism for inhibition of *crp* transcription. The structure of the proposed rho-independent terminator is shown (B1). This structure consists of the RNA–RNA hybrid followed by an AU-rich RNA–DNA hybrid of 11 bp. We propose that the RNA–RNA hybrid causes the polymerase to pause, and as suggested for rho-independent terminators (Platt, 1981; von Hippel *et al.*, 1984), the instability of the AU-rich RNA–DNA hybrid leads to release of the transcript (B2) and collapse of the DNA bubble (B3).

TABLE II

Effect of Temperature on *crp* Autoregulation

	Percent inhibition			
Condition[a]	37°C	30°C	25°C	20°C
cAMP-CRP	81	75	29	21
Divergent RNA	75	66	48	32
In vivo + cAMP	71	—	11	—

[a] The *in vitro* experiments (inhibition of *crp* transcription by cyclic AMP-CRP or by divergent RNA) were done as described in Figs. 1 and 4. Autoregulation *in vivo* was done as described in Fig. 7.

However, recent evidence indicates that an extended run of uridines is not essential for all rho-independent terminators (Brendel and Trifonov, 1984). The data in Table I suggest that the proposed structure has very similar properties to the 30 rho-independent terminators described by Brendel and Trifonov (1984). Autoregulation of *crp* would therefore be initiated by the activation of the divergent promoter by cyclic AMP-CRP, followed by the synthesis of divergent RNA. This would lead to formation of an RNA–RNA duplex (Fig. 9B-1) followed by RNA release (Fig. 9B-2), DNA–DNA reannealing (Fig. 9B-3), and very early termination of *crp* transcription. Consistent with this model are results obtained on *crp* autoregulation at low temperature. We find that the *in vitro* inhibition of *crp* transcription by cyclic AMP-CRP or by divergent RNA, and *crp* autoregulation *in vivo,* is strongly reduced at temperatures below 30°C (Table 2). It is known that rho-independent termination is markedly decreased at low temperature (Farnham and Platt, 1980).

V. CONCLUSION

We have shown that the negative autoregulation of the *crp* operon is due to the activation by cyclic AMP-CRP of a divergent promoter located within the *crp* promoter region. The data indicate that this control is manifested primarily by inhibition of *crp* transcription by the divergent RNA. We suggest that this transcriptional inhibition by a trans-acting RNA constitutes a previously unrecognized mechanism for the regulation of gene expression. The molecular events concerned with the blocking of *crp* transcription by the divergent RNA are not known. However, the strong homology between the 5′ end of *crp* mRNA and the divergent RNA suggest that the mechanism may involve a hybrid formed by the initial nucleotides of the two RNAs. We are currently examining this possibility. In addition, we are investigating the feasibility of blocking transcription in other systems by adding RNAs that are complementary to the 5′ end of the mRNA of selected genes.

ACKNOWLEDGMENTS

This work was supported by National Institutes of Health Grants GM17152 and GM36339.

REFERENCES

Aiba, H. (1983). *Cell (Cambridge, Mass.)* **32,** 141–149.
Aiba, H., Fujimoto, S., and Ozaki, N. (1982). *Nucleic Acids Res.* **10,** 1345–1361.
Anderson, W. W., Braun, P. R., and Gesteland, R. F. (1973). *J. Virol.* **12,** 241–252.
Brendel, V., and Trifonov, E. N. (1984). *Nucleic Acids Res.* **12,** 4411–4427.
Cossart, P., and Gicquel-Sanzey, B. (1985). *J. Bacteriol.* **161,** 454–457.
Farnham, P. J., and Platt, T. (1980). *Cell (Cambridge, Mass.)* **20,** 739–748.
Galas, D. J., and Schmitz, A. (1978). *Nucleic Acids Res.* **5,** 3157–3170.
Gilbert, W., and Mueller-Hill, B. (1966). *Proc. Natl. Acad. Sci. U.S.A.* **56,** 1891–1898.
Hawley, D., and McClure, W. (1983). *Nucleic Acids Res.* **11,** 2237–2255.
Herbert, M., Kolb, A., and Buc, H. (1986). *Proc. Natl. Acad. Sci. U.S.A.* **83,** 2807–2811.
Jacob, F., and Monod, J. (1961). *J. Mol. Biol.* **3,** 318–356.
Okamoto, K., and Freundlich, M. (1986). *Proc. Natl. Acad. Sci. U.S.A.* **83,** 5000–5004.
Platt, T. (1981). *Cell (Cambridge, Mass.)* **24,** 10–23.
Ptashne, M. (1967). *Proc. Natl. Acad. Sci. U.S.A.* **57,** 306–313.
Rosenberg, M., and Court, D. (1979). *Annu. Rev. Genet.* **13,** 319–353.
Ullmann, A., and Danchin, A. (1983). *Adv. Cyclic Nucleotide Res.* **15,** 1–53.
von Hippel, P. H., Bear, D. G., Morgan, W. D., and McSwiggen, J. A. (1984). *Annu. Rev. Biochem.* **53,** 389–446.

24

The Role of Translational Regulation in Growth Rate–Dependent and Stringent Control of the Synthesis of Ribosomal Proteins in *Escherichia coli*

J. R. COLE AND M. NOMURA

Department of Biological Chemistry
University of California, Irvine
Irvine, California 92717

It has been generally accepted that the regulation of the synthesis of most proteins in prokaryotes takes place at the level of transcription. Although the presence of translational regulation has been firmly established for the synthesis of ribosomal proteins (r-proteins) in *Escherichia coli* (for a review, see Nomura *et al.,* 1984), it has been frequently argued that even in this case the major regulation takes place at the transcription step and that the translational feedback mechanisms demonstrated both *in vitro* and *in vivo* play only a minor role, perhaps a role in fine adjustment in balancing the synthesis rates or a regulatory role under cetain non-steady-state conditions. Here we summarize some recent experiments which show that the two classical regulatory phenomena of r-protein synthesis, growth-rate dependence (Maaløe and Kjeldgaard, 1966; Dennis, 1974) and stringent control (Dennis and Nomura, 1974) are abolished in the absence of translational regulation, and hence, translational feedback regulation plays an important role in cell physiology.

I. SECONDARY STRUCTURE OF THE L1 TARGET SITE ON L11 mRNA

The L11 r-protein operon consists of the genes for r-proteins L11 and L1. L1 is the translational repressor protein for this operon and regulates its own synthesis and the synthesis of L11 by interacting with a target site on L11 operon mRNA (Yates *et al.*, 1980; Dean and Nomura, 1980). The target site for L1 has been previously defined *in vivo* by deletion analysis as well as site-directed mutagenesis, and several mutations that disrupt translational regulation were created on a plasmid molecule (Baughman and Nomura, 1983, 1984). One such mutation called MN2 involves a two-base substitution in the leader region of the mRNA (Fig. 1) and is used in the experiments described in the following sections. The importance of this stem structure of the mRNA for regulation was demonstrated by combining MN2 with another mutation that restored the stem structure, and by showing restoration of regulation by L1 in this double mutant (Baughman and Nomura, 1984). More recent work has involved using an *in vivo* system that allows isolation of mutants with altered regulation. The mutants were isolated as those producing high levels of an L11-β-galactosidase fusion protein from a L11–*lacZ* hybrid gene in the presence of excess L1. It was found that the majority had mutational alterations in the stem structure of the L1 target site mentioned above (Thomas and M. Nomura, 1987).

II. GROWTH RATE–DEPENDENT CONTROL OF RIBOSOMAL PROTEIN SYNTHESIS

Previous experiments have shown that r-protein promoters are not involved in the growth rate control of r-protein synthesis. This conclusion was obtained from analyses of the expression of the *lacZ* gene (or *galK* gene) fused to the *spc,* S10, or α r-protein operon promoters (Miura *et al.*, 1981; Gourse *et al.*, 1986).

Recent experiments have directly tested for the involvement of translational regulation in growth rate control. The MN2 mutation, discussed above, was transferred to the chromosomal L11 operon, resulting in a strain that is identical to a control strain in all respects except that the L11 operon is no longer subject to translational control (Fig. 2).

Measurements of the differential synthesis rates of L11 and L1 (relative to the rate of total protein synthesis) at three different growth rates

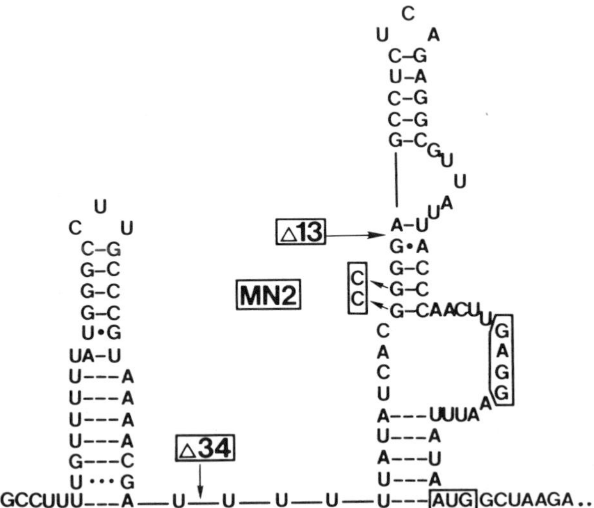

Fig. 1. Secondary structure model of the L1 target site on the L11 mRNA. The secondary structure of the L1 target site on the mRNA is based on the results of structure-specific nuclease treatment and computer analysis of the L11 mRNA (K. Kearney and M. Nomura, in press) as well as site-directed mutagenesis experiments (Baughman and Nomura, 1984). The sequence shown represents the first 103 nucleotides of the L11 mRNA (Post *et al.*, 1979). Base pairs indicated by a single hyphen or dot are those that have clear experimental support, while base pairs indicated by triple hyphens or dots are less strongly supported. Δ 34 is the deletion end point of an mRNA that is still regulated *in vitro* and *in vivo* by L1, while Δ 13 is that of an mRNA that is no longer subject to L1 regulation (Baughman and Nomura, 1983). MN2 is a two-base substitution mutation constructed by *in vitro* mutagenesis that disrupts the proposed two GC base pairs and was found to abolish the regulation by L1 both *in vitro* and *in vivo* (Baughman and Nomura, 1984). The presumptive Shine–Dalgarno sequence (GAGG) and the AUG start codon for the L11 cistron are also indicated.

showed that there is no growth rate dependency in the mutant. Other r-proteins tested (L3 and L6) in the mutant, as well as all these r-proteins in the control strain, showed growth-rate dependency as previously observed. Therefore, we conclude that translational control is responsible for the normal growth rate–dependent control of L11 and L1 synthesis (Cole and Nomura, 1986a). From these data, it appears highly likely that the growth rate–dependent control of most r proteins is a consequence of the growth rate–dependent control of rRNA synthesis with r-protein synthesis indirectly linked to rRNA synthesis through translational regulation.

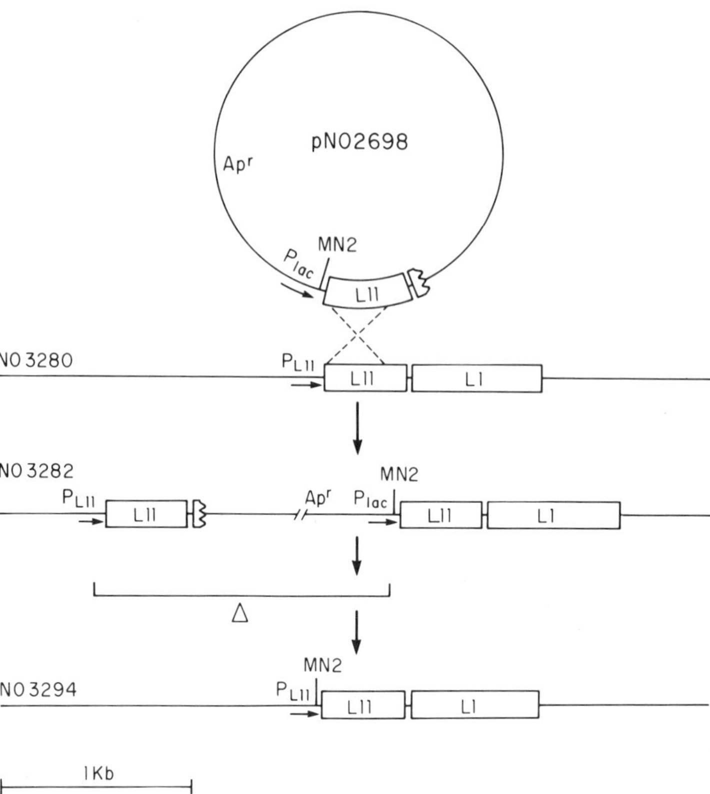

Fig. 2. Construction of a strain (NO3294) carrying the L1 target mutation, MN2, in the chromosomal L11 operon. The MN2 mutation is the mutation carried by the plasmid pNO2597 described by Baughman and Nomura (1984) and is shown in Fig. 1. Details of the method used for construction are described in Cole and Nomura (1986a). Briefly, pNO2698 (a derivative of pNO2597) carrying the MN2 mutation was inserted into the chromosome of a *polA* strain (NO3280, which is *rif*^{r}) by homologous recombination to give strain NO3282. Strain NO3282 thus contains two copies of the L11 gene but only one complete copy of the L1 gene, and this single L1 gene is under the *lac* promoter control. A culture of NO3282 was then enriched for those cells that had lost the integrated *amp*^{r} gene and *lac* promoter by homologous recombination by growing in the absence of both the *lac* operon inducer and ampicillin. Phage P1 grown on this revertant pool was used to transduce a *polA*$^{+}$ strain to *rif*^{r}. The *rif*^{r} transductants were screened for cotransduction of the MN2 mutation that would lead to overproduction of L1 and L11, and one of the transductants carrying the MN2 mutation (NO3294) was obtained. The presence of the MN2 mutation on the chromosome was demonstrated directly by Southern analysis using a synthetic oligodeoxynucleotide with the mutant sequence as a radioactive probe.

III. STRINGENT CONTROL OF RIBOSOMAL PROTEIN SYNTHESIS

Stringent control was initially defined for rRNA (and tRNA) synthesis; during amino acid starvation the synthesis rate of stable RNA is depressed in *relA*$^+$ strains but not in *relA*$^-$ strains (Stent and Brenner, 1961). Subsequently it was found that during amino acid starvation the differential synthesis rate of r-proteins decreases in *relA*$^+$ strains and increases in *relA*$^-$ strains, and, hence, r-protein synthesis is also under stringent control (Dennis and Nomura, 1974). This change in the differential synthesis rate was found to be accompanied by a change in the steady-state amount of r-protein mRNA (Dennis and Nomura, 1975). Furthermore, based on the results of pulse-labeling experiments, it was concluded that stringent control of r-protein synthesis occurs at the level of transcription (Maher and Dennis, 1977). However, since translational regulation has been found to be associated with a change in the stability of r-protein mRNA (Fallon *et al.*, 1979), this apparent change in r-protein mRNA synthesis rates could be due to a decrease in mRNA stability caused by translational regulation instead of an actual change in mRNA synthesis rates, with rRNA synthesis directly under stringent control and r-protein synthesis indirectly linked to rRNA synthesis through translational regulation.

We used the same MN2 chromosomal mutation described above to test for the involvement of translational regulation in the stringent control of L11 and L1 synthesis (Cole and Nomura, 1986a). Four isogenic strains were constructed. Two of them were *relA*$^+$ and *relA*$^-$ strains with the MN2 mutation which disrupts the regulation at the target site by L1. It was found that in the presence of the MN2 mutation, the synthesis rates of L11 and L1 decrease to the same extent as the total protein synthesis rate during amino acid starvation in both *relA*$^+$ and *relA*$^-$ strains, and consequently are not under stringent control. The synthesis of other r-proteins not regulated by L1 showed the typical stringent response as is the case for all r-proteins examined in the control strains with the wild-type target site (Table I). Thus, stringent control of the synthesis of L11 and L1 does not act directly on the transcription of r-protein genes, and like growth rate control it is a consequence of translational regulation, which links r-protein synthesis to rRNA synthesis.

It has been previously noted that many r-protein operons share a similar GC-rich nucleotide sequence around the transcription start site (Post *et al.*, 1978), and it has been proposed that this common structural feature might be important in stringent control (Travers, 1980). The promoter for the L11 operon (Post *et al.*, 1979) also shares this common sequence feature. Based on the conclusion on the mechanism of stringent control in

TABLE I

Relative Differential Synthesis Rates of r-Proteins during Amino Acid Starvation (+valine/−valine)[a]

r-proteins	Strains with normal regulation		Strains with MN2 mutation	
	relA$^+$	*relaA*$^-$	*relA*$^+$	*relaA*$^-$
L1	0.45	2.44	<u>0.80</u>	<u>0.75</u>
L11	0.53	3.29	<u>1.04</u>	<u>0.93</u>
L3	0.38	1.70	0.30	2.07
L5	0.55	2.34	0.49	2.27
L6	0.43	1.21	0.42	1.55
S14	0.62	2.25	0.56	2.39

[a] Four isogenic strains (see text) were grown in a synthetic medium. Each culture was divided into two parts. One part was treated with valine (100 μg/ml) for 15 min to cause isoleucine starvation. The other part served as control. Cells were pulse-labeled with [^{3}H]lysine and differential synthesis rates (relative to total protein synthesis rates) were measured for each of the r-proteins shown. The values in the table are the values for the valine-treated cultures divided by the corresponding values for the control culture. The underlined values indicate the breakdown of stringent control of L1 and L11 synthesis in the presence of the MN2 mutation. The data are taken from Cole and Nomura (1986a).

the L11 operon described above, it appears that the importance of this feature in r-protein operons will have to be sought in some other connection.

In contrast to the conclusion obtained with the L11 operon described above, stringent control of r-protein synthesis was previously thought to act at the level of transcription. This conclusion was based on the analysis of radioactive RNA labeled for a short time under conditions of partial inhibition of aminoacyl-tRNA synthesis (Maher and Dennis, 1977). Since it is now well established that an increase in translational repression causes a decrease in the half-life of r-protein mRNA (Fallon *et al.*, 1979; Singer and Nomura, 1985; Cole and Nomura, 1986b), the apparent change in r-protein mRNA synthesis rate observed in the earlier experiments may have been due to a large decrease in mRNA half-life caused by translational repression. However, the results of more recent experiments by Freedman and co-workers (1985) on the S10 r-protein operon appear to contradict this hypothesis. Using several operon fusions, they obtained results that indicate that the promoter for the S10 operon is responsible for the stringent control of r-protein synthesis in this operon. It is possible that the apparent discrepancy between our conclusion and that obtained for the S10 operon indicates a basic difference in regulation between the S10 operon and other r-protein operons, as noted previously by these workers (Zengel and Lindahl, 1985; Zengel *et al.,* 1984). It should be noted, however, that differential synthesis rates of most of r-proteins, including the r-proteins in the S10 operon, *increase* in *relA*$^-$ strains (see Table I and Dennis and Nomura, 1974) and this feature is difficult to explain on the basis of the model that guanosine tetraphosphate (ppGpp) accumulated in *relA*$^+$ strains inhibits transcription from r-protein promoters. In *relA*$^-$ strains, such accumulation of ppGpp does not take place and r-protein synthesis rates should be decreased by amino acid starvation to the same extent as the total protein synthesis rate, as was seen with the synthesis rate of L11 and L1 in the presence of the MN2 mutation shown in Table I. The increase in differential synthesis rate of r-proteins in *relA*$^-$ strains during stringent conditions is more easily explained by the model of stringent control as a result of translational regulation. In *relA*$^-$ strains, total protein synthesis decreases during amino acid starvation, but rRNA synthesis continues, resulting in derepression of r-protein translation and hence increase in their differential synthesis rates.

IV. CONCLUDING REMARKS

Translational regulation plays a dominant role in the known regulatory phenomena of r-protein synthesis. Since r-proteins comprise as much as a quarter of total cellular protein in *E. coli* cells growing in rich media, translational regulation controls a significant portion of total cellular protein synthesis. In addition, there are also several other systems known to involve translational regulation in prokaryotes (reviewed in Campbell *et al.,* 1983, and other more recent work, e.g., Butler *et al.,* 1986). Since translational regulation may affect mRNA stability, as is the case for r-protein regulation, and since many central regulatory loops are probably interconnected, other important regulatory mechanisms involving translation may have been overlooked. With the powerful new technologies for manipulating genes, we expect that more examples of translational regulation, as well as important new regulatory loops operating *in vivo,* will be recognized.

As for the regulation of ribosome biosynthesis, it is now almost certain that it is the synthesis of rRNA that is rate-limiting under most exponential growth conditions, and hence, understanding its regulation is crucial for understanding the regulation of ribosome synthesis and cellular growth. Elsewhere we have extensively discussed a ribosome feedback regulation model which proposes products of rRNA operons regulate rRNA and tRNA synthesis by negative feedback, and we have used this model to explain how the growth rate-dependent control of rRNA synthesis is achieved (Jinks-Robertson *et al.*, 1983; Nomura *et al.*, 1984; Gourse *et al.*, 1985, 1986). However, the molecular mechanisms involved in this feedback inhibition, as well as those involved in stringent control of rRNA synthesis, are still not clear and their elucidation requires further studies.

ACKNOWLEDGMENTS

The work described here was supported by grants from the National Science Foundation (DMB 8543776) and National Institutes of Health (GM 35949).

REFERENCES

Baughman, G., and Nomura, M. (1983). Localization of the target site for translational regulation of the L11 operon and direct evidence for translational coupling in *Escherichia coli*. *Cell* (*Cambridge, Mass.*) **34,** 979–988.

Baughman, G., and Nomura, M. (1984). Translational regulation of the L11 ribosomal protein operon of *Escherichia coli:* analysis of the mRNA target site using oligonucleotide-directed mutagenesis. *Proc. Natl. Acad. Sci. U.S.A.* **81,** 5389–5393.

Butler, J. S., Springer, M., Dondon, J., and Grunberg-Manago, M. (1986). Posttranscriptional autoregulation of *Escherichia coli* threonyl tRNA synthetase expression *in vivo*. *J. Bacteriol.* **165,** 198–203.

Campbell, K. M., Stormo, G. D., and Gold, L. (1983). Protein-mediated translational repression. *In* "Gene Function in Prokaryotes" (J. Beckwith, J. Davies, and J. A. Gallant, eds.), pp. 185–210. Cold Spring Harbor Lab., Cold Spring Harbor, New York.

Cole, J. R., and Nomura, M. (1986a). Translational regulation is responsible for growth rate–dependent and stringent control of the synthesis of ribosomal proteins L11 and L1 in *Escherichia coli*. *Proc. Natl. Acad. Sci. U.S.A.* **83,** 4129–4133.

Cole, J. R., and Nomura, M. (1986b). Changes in the half-life of ribosomal protein messenger RNA caused by translational repression. *J. Mol. Biol.* **188,** 383–392.

Dean, D., and Nomura, M. (1980). Feedback regulation of ribosomal protein gene expression in *Escherichia coli*. *Proc. Natl. Acad. Sci. U.S.A.* **77,** 3590–3594.

Dennis, P. P. (1974). *In vivo* stability, maturation and relative differential synthesis rates of individual ribosomal proteins in *Escherichia coli* B/r. *J. Mol. Biol.* **88,** 25–41.

Dennis, P. P., and Nomura, M. (1974). Stringent control of ribosomal protein gene expression in *Escherichia coli*. *Proc. Natl. Acad. Sci. U.S.A.* **71,** 3819–3823.

Dennis, P. P., and Nomura, M. (1975). Stringent control of the transcriptional activities of ribosomal protein genes in *E. coli. Nature* (*London*) **255,** 460–465.

Fallon, A. M., Jinks, C. S., Strycharz, G. D., and Nomura, M. (1979). Regulation of ribosomal protein synthesis in *Escherichia coli* by selective mRNA inactivation. *Proc. Natl. Acad. Sci. U.S.A.* **76,** 3411–3415.

Freedman, L. P., Zengel, J. M., and Lindahl, L. (1985). Genetic dissection of stringent control and nutritional shift-up response of the *Escherichia coli* S10 ribosomal protein operon. *J. Mol. Biol.* **185,** 701–712.

Gourse, R. L., Takebe, Y., Sharrock, R. A., and Nomura, M. (1985). Feedback regulation of rRNA and tRNA synthesis and accumulation of free ribosomes after conditional expression of rRNA genes. *Proc. Natl. Acad. Sci. U.S.A.* **82,** 1069–1073.

Gourse, R. L., de Boer, H. A., and Nomura, M. (1986). DNA determinants of rRNA synthesis in *E. coli:* Growth rate dependent regulation, feedback inhibition, upstream activation, antitermination. *Cell* (*Cambridge, Mass.*) **44,** 197–205.

Jinks-Robertson, S., Gourse, R. L., and Nomura, M. (1983). Expression of rRNA and tRNA genes in *Escherichia coli:* Evidence for feedback regulation by products of rRNA operons. *Cell* (*Cambridge, Mass.*) **33,** 865–876.

Maaløe, O., and Kjeldgaard, N. O. (1966). "Control of Macromolecular Synthesis." Benjamin, New York.

Maher, D. L., and Dennis, P. P. (1977). *In vivo* transcription of *E. coli* genes for rRNA, ribosomal proteins and subunits of RNA polymerase : Influence of the stringent control system. *Mol. Gen. Genet.* **155,** 203–211.

Miura, A., Kruger, J. H., Itoh, S., de Boer, H. A., and Nomura, M. (1981). Growth-rate-dependent regulation of ribosome synthesis in *E. coli:* Expression of the *lacZ* and *galK* genes fused to ribosomal promoters. *Cell* (*Cambridge, Mass.*) **25,** 773–782.

Nomura, M., Gourse, R., and Baughman, G. (1984). Regulation of the synthesis of ribosomes and ribosomal components. *Annu. Rev. Biochem.* **53,** 75–117.

Post, L. E., Arfsten, A. E., Reusser, F., and Nomura, M. (1978). DNA sequences of promoter regions for the *str* and *spc* ribosomal protein operons in *E. coli. Cell* (*Cambridge, Mass.*) **15,** 215–229.

Post, L. E., Strycharz, G. D., Nomura, M., Lewis, H., and Dennis, P. P. (1979). Nucleotide sequence of the ribosomal protein gene cluster adjacent to the gene for RNA polymerase subunit β in *Escherichia coli. Proc. Natl. Acad. Sci. U.S.A.* **76,** 1697–1701.

Singer, P., and Nomura, M. (1985). Stability of ribosomal protein mRNA and translational feedback regulation in *Escherichia coli. Mol. Gen. Genet.* **199,** 543–546.

Stent, G., and Brenner, S. (1961). A genetic locus for the regulation of ribonucleic acid synthesis. *Proc. Natl. Acad. Sci. U.S.A.* **47,** 2005–2014.

Thomas, M. S., and Nomura, M. (1987). Translational regulation of the L11 ribosomal protein operon of *Escherichia coli:* Mutations that define the target site for repression by L1. *Nucleic Acids Res.* **15,** 3085–3096.

Travers, A. A. (1980). Promoter sequence for stringent control of bacterial ribonucleic acid synthesis. *J. Bacteriol.* **141,** 973–976.

Yates, J. L., Arfsten, A. E., and Nomura, M. (1980). *In vitro* expression of *Escherichia coli* ribosomal protein genes; autogenous inhibition of translation. *Proc. Natl. Acad. Sci. U.S.A.* **77,** 1837–1841.

Zengel, J. M., and Lindahl, L. (1985). Transcriptional control of the S10 ribosomal protein operon of *Escherichia coli* after a shift to higher temperature. *J. Bacteriol.* **163,** 140–147.

Zengel, J. M., Archer, R. H., Freedman, L. P., and Lindahl, L. (1984). Role of attenuation in growth rate-dependent regulation of the S10 r-protein operon of *E. coli. EMBO J.* **3,** 1561–1565.

25

Sequence and Structural Elements Associated with the Degradation of Apolipoprotein II Messenger RNA

GREGORY S. SHELNESS,[1] ROBERTA BINDER, SHENG-PING L. HWANG, CLINTON MACDONALD, DAVID A. GORDON, AND DAVID L. WILLIAMS

Department of Pharmacological Sciences
Health Sciences Center
State University of New York at Stony Brook
Stony Brook, New York 11794

I. INTRODUCTION

Differences in mRNA stability can be an important determinant of the characteristic steady-state mRNA levels achieved by a cell (Darnell, 1982; Brawerman, 1987). In addition there are cases in which rates of specific mRNA turnover are modulated by hormones and other physiological effectors such as estrogen (Brock and Shapiro, 1983), prolactin (Guyette *et al.*, 1979), growth hormone (Laverriere *et al.*, 1983), interferon (Dani *et al.*, 1985), cyclic AMP (Mangiarotti *et al.*, 1983) and factors associated with viral infection (Wilson and Darnell, 1981), cellular differentiation (Aviv *et al.*, 1976), and regulation of DNA synthesis (Sittman *et al.*, 1983). Particularly well-characterized examples of regulation at the level of mRNA stability occur in cells responsive to estrogenic steroids.

[1] Present address: Laboratory of Cell Biology, Rockefeller University, New York, New York 10021-6399.

Molecular Biology of RNA
New Perspectives

The ability of estrogens to regulate both rates of transcription and turnover of specific mRNAs was first observed in studies of the estrogen-mediated accumulation of ovalbumin mRNA in chick oviduct (Palmiter and Carey, 1974; Palmiter *et al.*, 1977; Schutz *et al.*, 1977). In estrogenized hens, ovalbumin comprises approximately 50% of total protein synthesis in the oviduct tubular gland cell. Upon removal of hormone, ovalbumin synthesis declines to approximately 5% of total protein synthesis within 20 hr and is nearly undetectable after 40 hr (Palmiter and Carey, 1974). Examination of the decay rate of ovalbumin mRNA activity during this period revealed complex, non-first-order kinetics in which the $t_{1/2}$ decreased approximately 10-fold from 24 hr at steady state to 2–3 hr within 20 hr after hormone withdrawal. Similar effects of estrogen on the $t_{1/2}$ of other oviduct mRNAs have also been observed (Hynes *et al.*, 1979).

Since these studies, similar results have been obtained when the induction and decay kinetics of other estrogen-regulated mRNAs have been examined. Brock and Shapiro (1983) have shown that the $t_{1/2}$ of estrogen-induced vitellogenin mRNA at steady state in isolated *Xenopus* liver explants is 500 hr; however, when hormone is removed from the culture medium, vitellogenin mRNA decays with a $t_{1/2}$ of only 16 hr. Cytoplasmic mRNA destined for degradation after hormone withdrawal can be rapidly restabilized by addition of estrogen, suggesting a mechanism involving a rapidly reversible modification of the mRNA (Brock and Shapiro, 1983). Similar effects of estrogen have been observed in the avian liver for both vitellogenin mRNA and the mRNA for apolipoprotein II (apo II) of very-low-density lipoprotein (Wiskocil *et al.*, 1980).

Central to the regulation of mRNA stability is the question of mRNA structure. At some level, specific mRNAs whose cellular concentrations are subject to control at the level of turnover must contain sequences, structures, and/or factors selectively recognized by the appropriate regulatory system. Although no evidence for this prediction currently exists, there are numerous cases in which mRNA–protein interactions and/or modulation of mRNA structure, contribute to both transcriptional and posttranscriptional forms of gene regulation. Examples include translational polarity observed in polycistronic phage mRNAs (Gold *et al.*, 1981), attentuation in *Escherichia coli* amino acid biosynthetic operons (Kolter and Yanofsky, 1982), retro-regulation of bacteriophage mRNA activities (Gottesman *et al.*, 1982), and regulation of mRNA translational efficiency (Wood *et al.*, 1984; Pelletier and Sonenberg, 1985).

In an effort to understand the relationship between native mRNA structure and modulation of specific mRNA stability by estrogens, we have initiated studies of the estrogen-induced mRNA for apo II of very-low-density lipoprotein. This yolk protein mRNA is induced to 10% of total liver mRNA by estrogen and possesses a longer $t_{1/2}$ in the presence of hormone (15 hr) than during hormone withdrawal (1.5 hr) (Wiskocil *et al.*, 1980; Gordon *et al.*, 1987). Apo II mRNA is a particularly attractive candidate for analysis of the relationship between mRNA structure and mRNA stability because of its high abundance and its compact size (ca. 660 bases) (Wieringa *et al.*, 1981).

The initial goal of this analysis was to determine whether the multiple forms of 5′ noncoding sequence in apo II mRNA (Shelness and Williams, 1984) influence mRNA stability. Subsequently the secondary structure of protein-free apo II mRNA was analyzed using enzymatic probes and reverse transcriptase—a methodology that could be extended to the analysis of mRNA structure and mRNA–protein interactions within ribonucleoprotein particles. Finally, sites of degradation by endogenous nucleases were examined. Evidence is presented for an endogenous activity that cleaves apo II mRNA *in vitro* at hypersensitive sites in the 3′ noncoding region. This activity may have a particularly high affinity for AU-rich sequences which map to internal and bulged loop structures in apo II mRNA.

II. EFFECT OF 5′ NONCODING SEQUENCES ON apo II mRNA STABILITY

Previous studies on the primary structure of apo II mRNA revealed heterogeneity in the 5′ untranslated region. The six different species of mRNA identified (Table I) arise due to three sites of transcription initiation and differential joining of the first two exons in the pre-mRNA (Shelness and Williams, 1984). To test whether this sequence variation in the 5′ untranslated region can alter cytoplasmic stability, a quantitative primer extension assay was used to assess the relative abundance of each mRNA species during various stages of the hepatic response to estrogen and following hormone withdrawal. This analysis revealed no significant change in the relative abundance of each mRNA species. Table I shows data from one experiment in which the various forms of apo II mRNA were monitored after hormone stimulation was abruptly terminated. These results indicate that the first 17 nucleotides (relative to the −11 cap site position, Table I) do not influence mRNA stability. Similarly, the trinucleotide insertion at the exon I–exon II boundary failed to alter apo II mRNA stability. If determinants of stability are present in the 5′ region of the mRNA, as has been shown for some prokaryotic mRNAs (Gorski *et al.*, 1985; Belasco *et al.*, 1986), more extensive alterations must be required to disrupt their function.

TABLE I

Six Forms of apo II mRNA[a]

```
  -11              +1        +7                                            *    *
1A. AGATGAGCATCAACCTCAGCTTCAGCCTGGGAGAGAGAAAGCAGGACAGCAGGTCTCTTGGTGT
1B.    AGATGAGCATCAACCTCAGCTTCAGCCTGGGAGAGAGAAAGCAGGACAGGTCTCTTGGTGT
2A.                  AACCTCAGCTTCAGCCTGGGAGAGAGAAAGCAGGACAGCAGGTCTCTTGGTGT
2B.                        AACCTCAGCTTCAGCCTGGGAGAGAGAAAGCAGGACAGGTCTCTTGGTGT
3A.                              AGCTTCAGCCTGGGAGAGAGAAAGCAGGACAGCAGGTCTCTTGGTGT
3B.                                    AGCTTCAGCCTGGGAGAGAGAAAGCAGGACAGGTCTCTTGGTGT
```

mRNA form	Hormone treatments		
	E-14d	E-14d, 1h Tam	E-14d, 8h Tam
1A	8.9	8.3	7.7
1B	3.9	4.0	3.9
2A	61	60	61
2B	21	23	21
3A	3.8	4.1	3.8
3B	1.5	1.5	1.6
Totals: mRNA per cell	19,400	9,910	4,670

[a] Forms 1, 2, and 3, arise from different sites of transcription initiation (−11, +1, and +7 respectively). Forms A and B arise from alternate splicing at the duplicated 5′-CAG-3′ (*) at the first intron–exon border (Shelness and Williams, 1984). The relative abundance of each species of mRNA is indicated at 14 days of estrogen (E) treatment and at 1 hr and 8 hr after acute hormone withdrawal initiated by administration of the antiestrogen tamoxifen (Tam) citrate. apo II mRNA remaining at the indicated times is given as mRNA molecules per cell.

III. USE OF ENZYMATIC PROBES AND REVERSE TRANSCRIPTASE TO ANALYZE apo II mRNA SECONDARY STRUCTURE

Methods employing structure-specific ribonucleases (Vournakis *et al.*, 1981) and chemical probes (Peattie and Gilbert, 1980) have been developed for studying the secondary structure of RNA in solution. This approach requires the purification of the RNA of interest, labeling of its 5′ or 3′ terminus, and treatment with base and structure-specific reagents under conditions of single-hit kinetics. Sites of cleavage are detected by denaturing gel electrophoresis and autoradiography such that the electrophoretic mobilities of the resultant fragments define the distance between the labeled terminus and the site of cleavage in the RNA. If a sufficient number of structure probes are employed, extensive regions of the RNA can be mapped to either single-stranded, double-stranded, or inaccessible domains. These data are then used, usually in conjunction with a computer modeling program, to generate a secondary structure model of the RNA.

Although this method has been used to considerable advantage in the analysis of ribosomal, transfer, phage, and catalytic RNAs, it has serious limitations when applied to mRNAs. These include difficulty in purifying a specific mRNA, heterogeneity at the ends of the mRNA due to the poly(A) tail or multiple transcription initiation sites, and the presence of 5′ cap structures in eukaryotic mRNAs which must be removed before end-labeling. Although some of these problems may be overcome in the analysis of naked mRNA (Vournakis *et al.*, 1981; Vary and Vournakis, 1984), they are probably intractable when applied to the analysis of mRNA structure and mRNA–protein interactions within a messenger ribonucleoprotein (mRNP) particle. An approach that circumvents many of these problems is the use of primer extension to determine sites of nuclease cleavage (Qu *et al.*, 1983). As outlined in Fig. 1, this approach permits the analysis of specific mRNAs within a complex population and is applicable to the analysis of mRNA structure in a RNP particle.

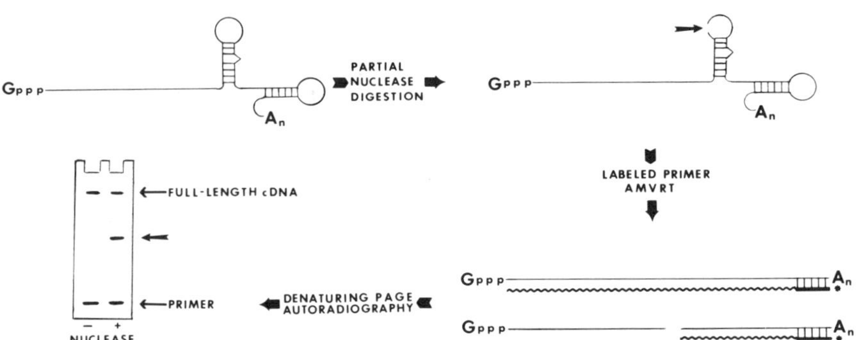

Fig. 1. Locating sites of nuclease cleavage by primer extension. Total RNA containing the mRNA of interest is partially digested with a structure-specific nuclease which, in this case, places a single cleavage (bold arrow) in a hairpin loop structure. This site is detected by annealing an end-labeled (*) primer (bold line) to the nuclease-cleaved RNA followed by extension (wavy line) with AMV reverse transcriptase (AMVRT) in the presence of the four deoxynucleoside triphosphates. Following heat denaturation, the labeled cDNA products are analyzed by denaturing gel electrophoresis and autoradiography. The cDNA product present only in the + lane arises from the interaction of the mRNA with the structure specific nuclease. The site of this cleavage in the mRNA is determined by accurately sizing the nuclease-dependent cDNA. (From Shelness and Williams, 1985.)

A. Evaluation of the Accuracy of Primer Extension

Annealing of labeled DNA primers to mRNA or pre-mRNA, followed by extension of the primer with avian myeloblastosis virus (AMV) reverse transcriptase has been used extensively to map sites of transcription initiation and hence the 5′ ends of mRNAs (Ghosh *et al.*, 1978). In extending this technique to the analysis of mRNA structure, we examined the ability of reverse transcriptase to make cDNAs whose lengths accurately reflected the lengths of their corresponding templates. For this purpose total RNA from estrogen-induced roosters was treated with the guanosine-specific ribonuclease T1 under denaturing conditions (Donis-Keller *et al.*, 1977) so that cleavages would occur independently of secondary structure. The T1 and mock-treated RNA preparations were used as templates for primer extension, and the labeled cDNA products were analyzed by electrophoresis on a denaturing polyacrylamide gel followed by autoradiography. If primer extension accurately locates sites of T1 cleavage, T1-dependent primer extension products should migrate one ladder position faster than G bands in a homologous sequencing ladder (RNase T1 cleaves at the 3′ side of G residues).

As shown in Fig. 2, primer extension of mock-treated (−) RNA gives rise to a background of incomplete cDNA products, most of which map to C residues that occur in CA and CU dinucleotides (Shelness and Williams, 1985). RNase T1 digestion results in numerous cDNAs not present in the mock-treated (−) lane. Most of these cDNAs migrate at the expected positions; however, there also exists a number of T1-dependent primer extension products that comigrate exactly with the G bands of the sequencing ladder. This result suggests the production of cDNAs one nucleotide longer than the mRNA template. The degree of conversion to this "plus one" form of cDNA varies between no conversion (coordinate 516) to nearly 100% conversion (coordinate 500). We confirmed by direct sequencing of the apo II mRNA population by the dideoxy method that the basis for this primer extension artifact is not due to apo II mRNA sequence variants.

Interestingly, the generation of terminal cDNA doublets during primer extension of capped mRNA has been reported (Proudfoot *et al.*, 1980; Luse *et al.*, 1981; Lee and Roeder, 1981; Benyajati *et al.*, 1983). This phenomenon has usually been attributed to the termination of some transcripts at the 5′ penultimate nucleotide of the mRNA, possibly due to steric hindrance from the cap structure (Lee and Roeder, 1981; Benyajati *et al.*, 1983). The results reported here, however, suggest that the faster migrating band of the doublet is the full-length extension product while the slower-migrating band represents a cDNA with a mobility corresponding to one nucleotide longer than the template. In fact, Gupta and

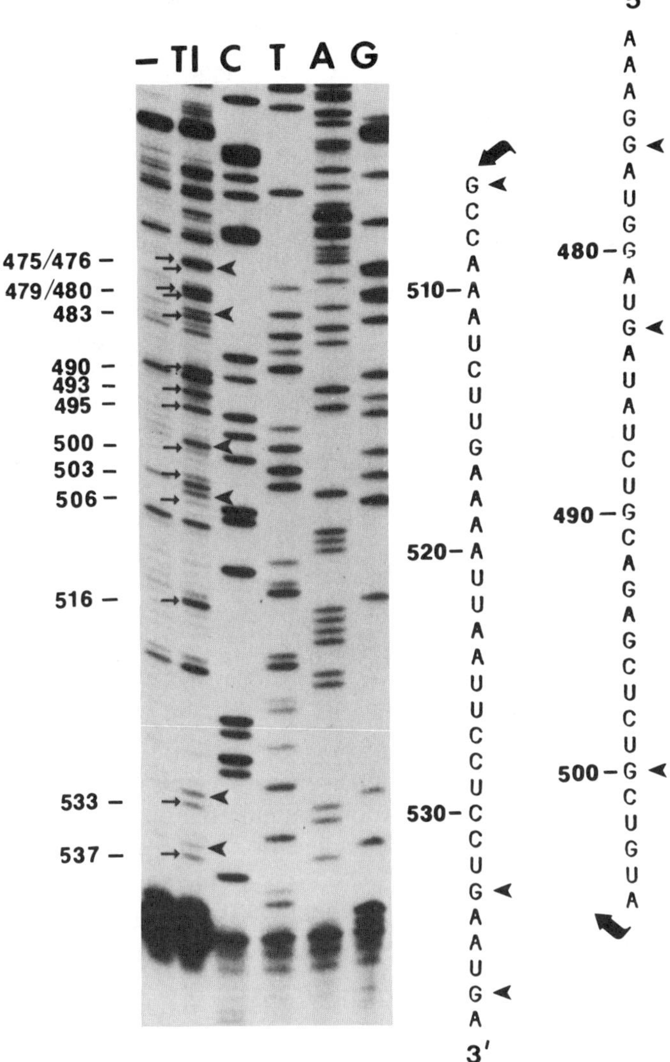

Fig. 2. Evaluating the accuracy of primer extension. T1 and mock (–) treated RNA were used as templates for primer extension. The end-labeled primer used in this and all subsequent figures is located near the 3′ poly(A) addition site of apo II mRNA (coordinates 543–573). Following primer extension, the labeled cDNA products were electrophoresed with a homologous sequence ladder (C, T, A, and G) generated by annealing the same primer used for primer extension to a sense-strand apo II M13 cDNA clone, followed by dideoxy sequencing. The small arrows to the left of the T1 lane and their corresponding translation coordinates listed to the left of the (–) lane indicate the theoretical positions of primer extension products due to cleavage at the 3′ side of G residues. Arrowheads to the right of the T1 lane indicate the T1-dependent cDNAs which migrate at a ladder position corresponding to one nucleotide longer than the T1-cleaved mRNA template. The positions of these aberrant cDNAs are also indicated by arrowheads on the apo II mRNA sequence. Note that in this and all subsequent figures in which a sequencing ladder appears, the labels used to identify each lane correspond to the template (i.e., the mRNA) sequence. (From Shelness and Williams, 1985.)

Kingsbury (1984) have reported that primer extension of the mRNA for NS protein of vesicular stomatitis virus yields both the full-length cDNA and a band corresponding to one nucleotide longer than the known 5′ terminus. While they suggest that reverse transcriptase may incorporate a nucleotide directed by the cap, our results with 5′-hydroxylated RNA suggest a reaction that is template independent. This phenomenon is therefore a general property of AMV reverse transcriptase (or a contaminating activity) and should be considered when interpreting results of primer extension analysis.

The identification of an apparent template-independent nucleotide addition by reverse transcriptase or a contaminating activity does not result in inaccuracy of mapping T1 cleavages since this nuclease possesses a strict base specificity. When non-base-specific probes are used, however, the possibility exists that with some low frequency cleavages will be assigned one nucleotide to the 5′ side of the actual site in the RNA.

B. Identification of Nuclease Cleavage Sites in apo II mRNA

Total cellular RNA was purified from rooster liver as described (Protter *et al.*, 1982) using guanidine hydrochloride as a denaturant, and was renatured by incubation for 5 min at 37°C in a buffer (25 m*M* Tris-HCl, pH 7.5, 200 m*M* NaCl, 5 m*M* $MgCl_2$) approximating physiological salt concentrations. Each structure-specific nuclease was used at a series of concentrations sufficient to produce digestions ranging from mild (less than 5% of mRNA cleaved) to almost complete digestion of the mRNA (Shelness and Williams, 1985). Use of multiple RNase concentrations permits the identification of cleavages that occur at enzyme concentrations at which most of the mRNA is full-length. These sites are considered to be primary cleavages and are distinguished from secondary cleavages that appear only after extensive degradation of the mRNA. Examples of this analysis with T1 RNase (panel A) or cobra venom (CV) RNase (panel B) are shown in Fig. 3. In comparison to the primer extension products seen with mock-treated RNA (lane −), numerous nuclease-dependent cDNAs were produced as the concentration of nuclease was increased. Cleavages designated as primary (for example, nucleotides 500 and 516) occur at the lowest concentration of RNase used. These cleavages can be contrasted with the site indicated by the unlabeled arrow between 500 and 516 which appears only after the bulk of full-length message has been degraded. By repeating this analysis with each structure-specific nuclease, primary cleavage sites were identified for T1 and S1 nucleases (specific for unpaired nucleotides) and for CV nuclease (specific for base-paired nucleo-

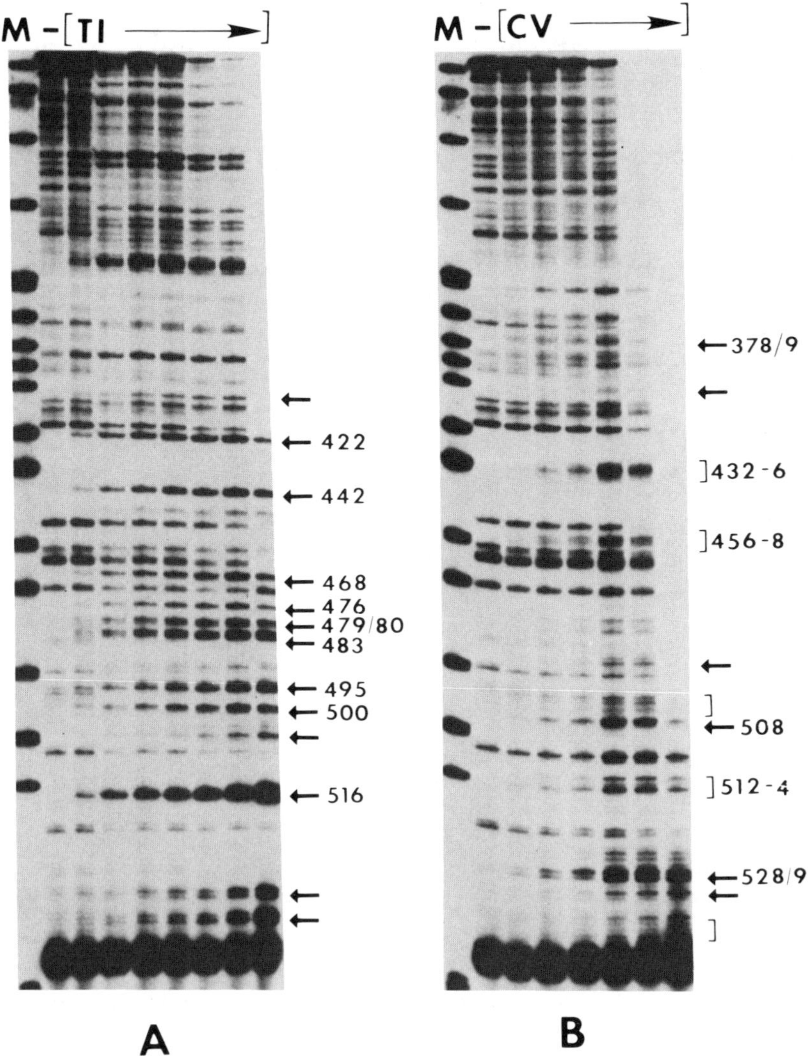

Fig. 3. Digestion of apo II mRNA by increasing concentrations of CV and T1 RNases. RNA samples digested with increasing concentrations of CV and T1 RNases (bracketed lanes) were primer-extended with a labeled cDNA primer. cDNAs not present in the mock-treated (−) RNA samples are indicated with arrows or brackets. The arrows and brackets labeled with translation coordinates arise due to primary cleavage events based on the criteria described in the text. Note that the brackets do not necessarily mean that each nucleotide within the bracketed region is actually cleaved. Lane M contains end-labeled MspI-digested pBR322. (From Shelness and Williams, 1985.)

tides) (Vournakis *et al.*, 1981). Primary cleavages along the entire length of the mRNA were identified by primer extensions with four different primers separated by intervals of 150–200 nucleotides.

C. Fine Mapping of Primary Nuclease Cleavage Sites

To map the exact sites of primary cleavage in apo II mRNA as identified by concentration curve analysis, single samples of nuclease-cleaved RNA were primer-extended and coelectrophoresed with a homologous sequencing ladder. An example of this analysis in which T1, S1, and CV nuclease cleavage sites were mapped in one region of apo II mRNA is shown in Fig. 4. Bands identified as primary cleavages by concentration curve analysis are indicated with arrowheads or brackets next to the appropriate gel lane. These are in turn labeled with the translation coordinates of the sites within the mRNA. These sites are also indicated on the mRNA sequence shown to the right of the gel lanes.

In addition to mapping sites of nuclease cleavage, the analysis shown in Fig. 4 allows side-by-side comparison of cleavages produced by enzymes with different specificities. This analysis, when extended to the entire length of apo II mRNA, shows no overlap between sites of primary cleavage by single-strand- (T1 and S1) and double-strand-specific (CV) nucleases. This result suggests that the secondary structure of apo II mRNA is probably discrete and does not consist of many stable alternate forms or multiple forms in rapid equilibrium. In addition, several different renaturation and digestion protocols were used to determine if the nuclease cleavage profiles obtained under standard buffer conditions reflected an inherent structure or whether the structure was readily affected by the environment to which the RNA was exposed. Alternate conditions included slowly cooling the RNA to either 25°C or 37°C after heat (70°C) denaturation in the absense of monovalent and divalent ions, followed by the standard equilibration and digestion protocol, as well as equilibration and digestion at 25°C in the absence of NaCl. In each case the nuclease cleavage profiles remained the same. These data argue that the mapping results obtained under standard conditions reflect a predominant secondary structure that reforms after denaturation and is stable to moderate changes in thermal and ionic conditions.

D. Computer Modeling of apo II mRNA

The locations of primary cleavage sites were used in conjunction with the RNA folding program of Zucker and Stiegler (1981) to formulate structure maps of apo II mRNA. Since T1 cleavages were mapped with single

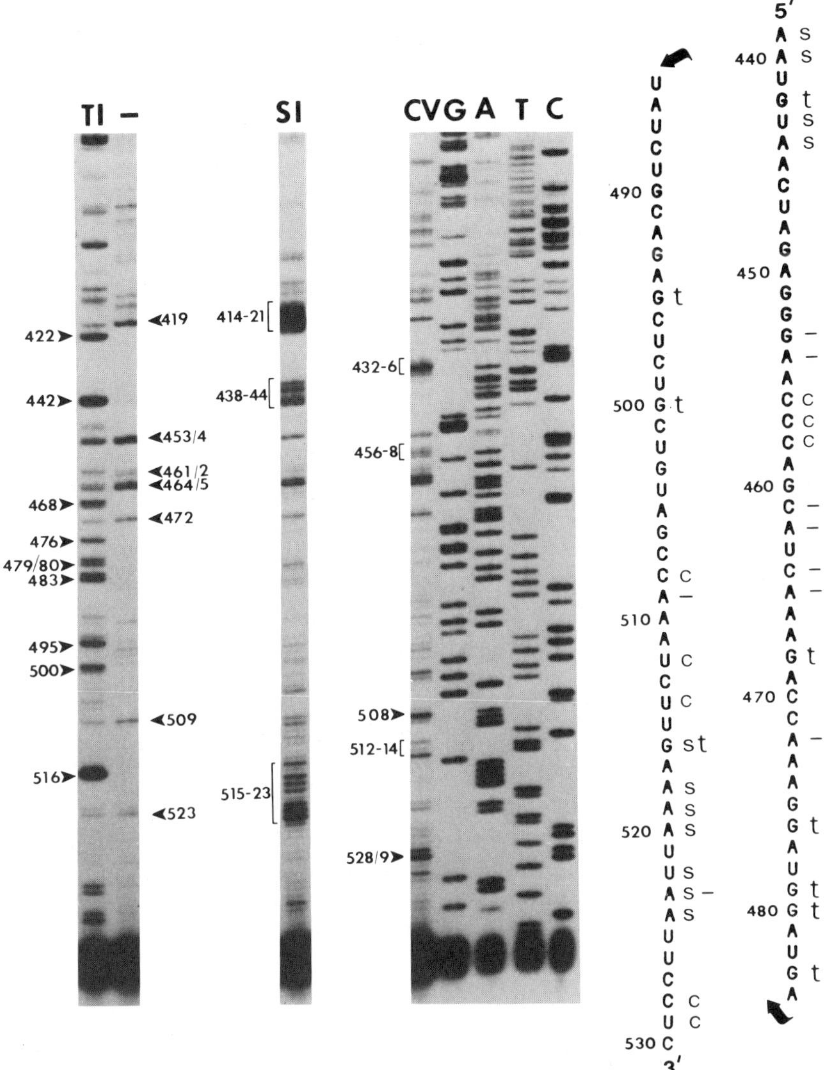

Fig. 4. Fine mapping of primary nuclease cleavage sites. Single samples of mock treated (−) RNA or RNA samples digested with T1, CV, and S1 nucleases were used for primer extension. The cDNA products were sized by electrophoresis with a homologous dideoxy sequencing ladder (lanes G, A, T, C). The translation coordinate(s) for each primary cleavage are indicated next to the appropriate gel lane and next to the apo II sequence to the right of the figure (t,T1; s,S1; c,CV; −, background incomplete cDNAs). (From Shelness and Williams, 1985.)

nucleotide accuracy, T1 cleavage data were used as constraints in generating the secondary structure models. Consistency of these structures with CV and S1 data was confirmed manually. A structure generated for the entire apo II mRNA is shown in Fig. 5. This structure was formulated with no restrictions on the linear distance (along the mRNA sequence) over which base pairs can form. Under these conditions long-range interactions are permitted, and sequences near the 5′ and 3′ ends are base-paired. While the T1 data are in agreement with this structure, only hairpin structures A, B, and C are in nearly complete agreement with CV and S1 cleavage results.

Long-range base-pairing interactions between the translated and untranslated regions of the mRNA may occur in the naked RNA but are probably not favored *in vivo* when the coding region is loaded with ribosomes. A perhaps more realistic way to view apo II mRNA is to consider secondary structures formed by shorter-range interactions that would be compatible with translation. In particular the 3′ untranslated region of the mRNA was considered potentially important since 3′ sequences of some mRNAs are sufficient to specify cytoplasmic stability (Wickens and Stephenson, 1984; Shaw and Kamen, 1986). When the extensive 3′ untranslated domain of apo II mRNA (coordinates 319–579) was folded independently of the rest of the molecule (Fig. 6), the hairpin structures A and B first identified in the model shown in Fig. 5 were retained. Hairpin A is, with one potential exception, in complete agreement with the nuclease cleavage data. This exception, the apparent cleavage of the bulged U residue at coordinate 358 by CV RNase, may be a "plus one" primer extension artifact of cleavage of the base-paired A at position 359. CV cleavages that are inconsistent with the structure of hairpin B occur at coordinates 514, 528, and 529. It should be noted, however, that CV nuclease is sensitive to long-range and tertiary interactions (Lockard and Kumar, 1981) which cannot be deduced from models based solely on secondary structure predictions.

IV. ANALYSIS OF apo II mRNA IN POLYSOMES

Of ultimate interest is the structure of the apo II mRNA within a polysome or free mRNP. To extend the analysis to the level of mRNP, mRNA extracted from gradient fractionated polysomes corresponding to a ribosome density of four or more per mRNA was examined. Comparison of cDNAs generated by primer extension of polysomal and guanidine-extracted RNA revealed several prominent incomplete cDNAs formed only with polysomal RNA. These cleavages, presumably the result of endoge-

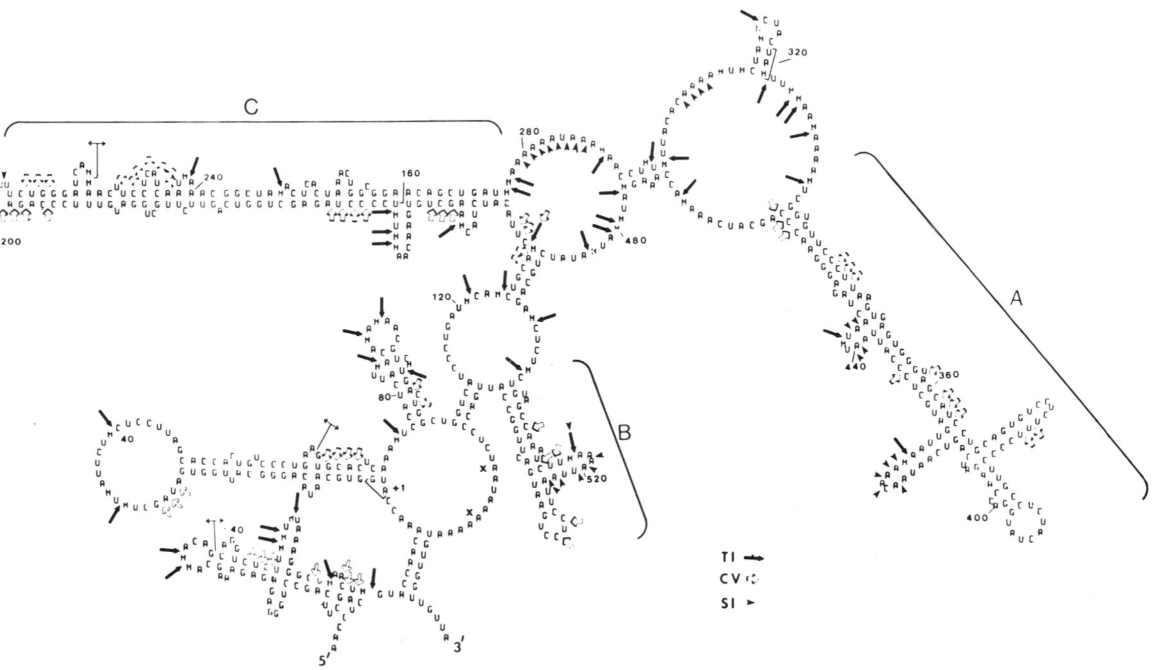

Fig. 5. Secondary-structure model of apo II mRNA. A computer model was formulated for the major form of apo II mRNA (apo II-2A, Table I) using T1 RNase cleavage data and no restrictions on the distance over which base pairs can form. The bracketed regions indicate major hairpin structures discussed in the text. The initiation and termination codons are bracketed and splice junctions are identified by double-headed arrows. The first and last nucleotides of the polyadenylation sequence, AAUAAA, are marked (×). Sites of structure-specific nuclease cleavage are identified with the indicated symbols. In addition, an "H" is used to represent G residues which are single-stranded, as judged by susceptibility to T1 RNase.

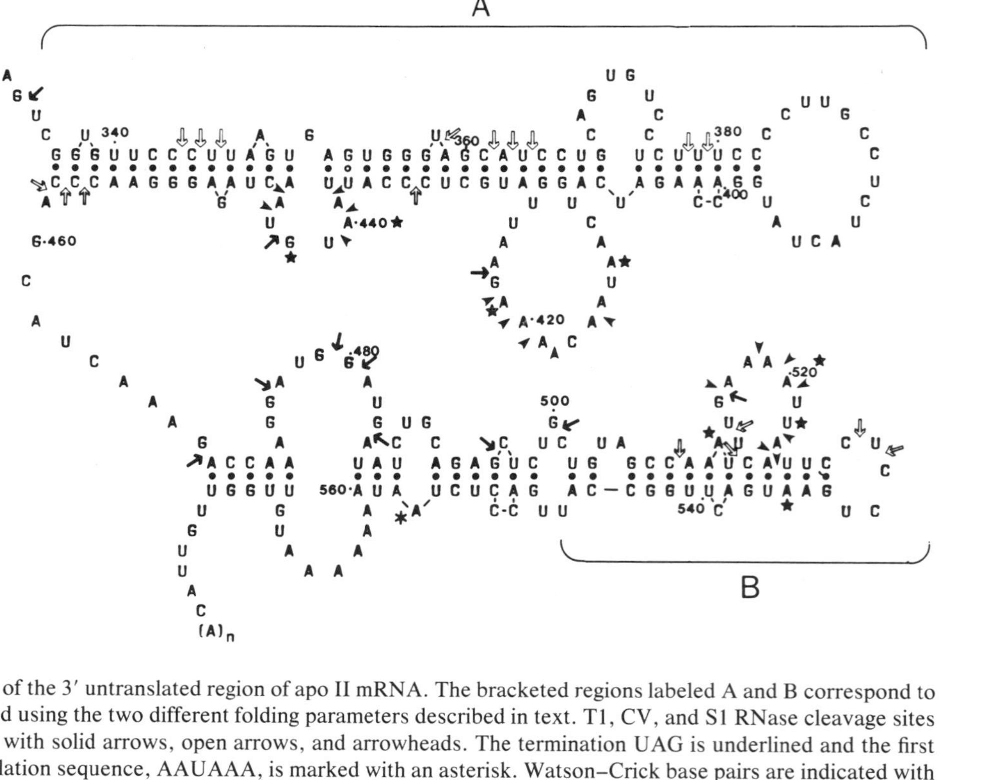

Fig. 6. Structure model of the 3′ untranslated region of apo II mRNA. The bracketed regions labeled A and B correspond to structures that are conserved using the two different folding parameters described in text. T1, CV, and S1 RNase cleavage sites are identified, respectively, with solid arrows, open arrows, and arrowheads. The termination UAG is underlined and the first nucleotide of the polyadenylation sequence, AAUAAA, is marked with an asterisk. Watson–Crick base pairs are indicated with closed circles. The single GU base pair between residues 353 and 437 is indicated with an open circle. The hypersensitive sites of cleavage in apo II polysomal mRNA are indicated by the stars on the nucleotide to the 5′ side of the cleavage. (From Shelness and Williams, 1985.)

nous nuclease activity during polysome isolation, mapped to eight sites in three interior and bulged loop regions within hairpins A and B (stars, Fig. 6). S1 nuclease mapping with a 3′-end-labeled probe confirmed the locations of these cleavages. The S1 analysis as well as primer extension with a probe-primer located immediately 5′ of nucleotide 415 also showed that the cleavages were essentially restricted to the 3′ untranslated region of the mRNA.

In addition to their specificity for potentially looped regions of the mRNA, these polysome-specific cleavages demonstrate a strong bias for AU dinucleotides. Five of the eight cleavages occur between A and U residues of the trinucleotide 5′-AAU-3′. Several of the sites show a similar sequence of surrounding nucleotides, but a consensus sequence for all of the cleavage sites is not apparent. However, the predominance of the AAU sequence and the location of seven cleavage sites in potential loop regions suggests that a base- and structure- specific ribonuclease may be involved. This prediction is further supported by the fact that one interior loop (centered at nucleotide 479) and another region (nucleotides 321–335) that are highly susceptible to T1 ribonuclease in the naked mRNA are not cleaved in polysomal apo II mRNA. Hence, if one can extrapolate from the results with naked mRNA, cleavage by this endogenous nuclease is not related simply to accessibility. However, further studies will be required to test directly which regions of the mRNA in polysomes are actually accessible to nucleases.

In an attempt to learn more about the endogenous activity responsible for the polysome-specific cleavages, we monitored apo II mRNA in tissue homogenates. When liver was homogenized in the same buffer used for polysome isolation (Palmiter, 1974) but lacking the nonionic detergent Triton X-100, apo II mRNA was very stable and showed only nonspecific degradation when incubated at 25°C for up to 1 hr. The specific cleavages seen in polysomal RNA were not observed. However, when EDTA was added to the homogenate, apo II mRNA was rapidly and extensively degraded. This result is illustrated by the primer extension analysis shown in Fig. 7 (lane 3). Note that the major cDNAs correspond to the cleavage sites identified in polysomal apo II mRNA (Fig. 6). The absence of cDNAs extending beyond nucleotide 440 indicates that essentially all apo II mRNA molecules had been cleaved at these hypersensitive sites.

To characterize this process further, the effects of dilution on *in vitro* cleavage of the AU-rich sites was examined. Since the dilution buffer contained heparin, the heparin-to-protein ratio was increased by homogenate dilution. The results of this analysis (Fig. 7) showed that threefold dilution (lane 4) prevents extensive degradation as judged by the presence of full-length cDNA. Further dilution (lanes 5–7), however, did not in-

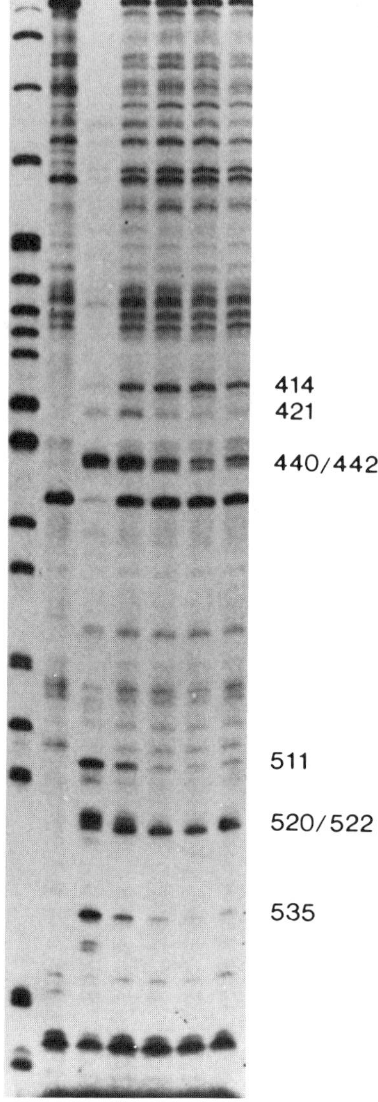

Fig. 7. Hypersensitive sites for apo II mRNA degradation. Liver was homogenized in 25 m*M* Tris (pH 7.5), 25 m*M* NaCl, 5 m*M* $MgCl_2$, 1 mg/ml heparin, 20 m*M* EDTA and was undiluted (lane 3) or diluted with homogenization buffer 3 (lane 4), 9 (lane 5), 27 (lane 6), or 81 (lane 7) fold before incubation for 10 min at 25°C. Total RNA was then purified by the guanidine HCl procedure (Protter *et al.*, 1982). RNA samples were primer-extended with an end-labeled primer complementary to nucleotides 563–579, and the resultant cDNAs were analyzed by denaturing gel electrophoresis. Lane 1 shows an end-labeled *Msp*I digest of pBR322, and lane 2 shows the primer extension products with undigested apo II mRNA purified directly from liver by guanidine-HCl extraction.

crease the proportion of full-length cDNA, suggesting that nonspecific nucleases were inhibited adequately under these conditions. In contrast to the nonspecific degradation, a threefold dilution (lane 4) had little effect on cleavage at the eight hypersensitive sites (nucleotides 414, 421, 440, 442, 511, 520, 522, and 535) in the 3′ noncoding region. Cleavage at three sites (511, 520, 535) was inhibited by a ninefold dilution (lane 5), while the others were unaffected or only partially reduced even by an 81-fold dilution (lane 7). These results suggest that the AU cleavage activity is resistant to heparin or is present in vast excess compared to nucleases that degrade the mRNA in a nonspecific fashion. Another possibility is that this activity has a high affinity for its AU-rich target sites in the mRNA. Consistent with this interpretation is the fact that some of the hypersensitive sites are more resistant to dilution and heparin inhibition than others. Unless two separate activities are responsible for the observed cleavages, simple competitive inhibition by heparin would not likely be site-selective. In addition, this activity is directed to only a small percentage of the total number of AU sites within the mRNA, again suggesting a high affinity for certain mRNA structures and/or mRNA-associated factors present in the 3′ noncoding region.

Interestingly, other experimental evidence supports a role for AU-rich 3′ untranslated sequences in mRNA degradation. For example, Shaw and Kamen (1986) have shown that AU-rich segments found in the 3′ untranslated region of certain transiently expressed mRNAs can confer cytoplasmic instability to an otherwise stable mRNA. In addition Albrecht *et al.* (1984) have reported that truncation of rabbit β-globin mRNA, presumably due to the action of an endogenous reticulocyte nuclease, occurs in the 3′ noncoding region of the mRNA at sites containing the dinucleotide AU. Experiments are currently underway in this laboratory to establish the relationship between the AU- and structure-specific degradative activity described above and the estrogen-regulated turnover of apo II mRNA observed *in vivo*.

V. SUMMARY AND PERSPECTIVE

A method has been developed for analyzing the structure of mRNA that can be extended to mRNA structure and mRNA–protein interactions within an mRNP. Analysis of protein-free mRNA has identified a stable structural element in the 3′ untranslated domain of apo II mRNA. Interior and bulged loops present in the 3′ noncoding region of the naked mRNA contain AAU sequences that are hypersensitive targets for cleavage by an endogenous nuclease activity *in vitro*. Cleavage of these sites during polysome isolation does not lead to loss of the 3′ fragment from higher-order polysomes. This result may indicate that this region is involved in a stable structure in polysomes that is similar to the major stem-and-loop structure proposed for the naked mRNA. The highly specific behavior of this nuclease activity suggests that it possesses an affinity for sequence or structural elements of apo II mRNA or perhaps for an mRNA-associated factor. This nuclease activity and the hypersensitive sites in apo II mRNA are candidates to be evaluated for potential roles in the *in vivo* degradation of apo II mRNA and the modulation of this process by estrogenic hormones. Our present results suggest that the ability to monitor and manipulate endogenous degradative activity directed toward specific mRNAs *in vitro* might be a valuable means of elucidating mechanisms associated with constitutive and regulated forms of mRNA catabolism.

ACKNOWLEDGMENT

This research was supported by a grant from the National Institutes of Health (AM 18171). G. S. S. and D. A. G. were supported by Pharmacological Sciences training Grant GM07518.

REFERENCES

Albrecht, G., Krowczynska, A., and Brawerman, G. (1984). *J. Mol. Biol.* **178,** 881–896.

Aviv, H., Voloch, Z., Bastos, R., and Levy, S. (1976). *Cell* (*Cambridge, Mass.*) **8,** 495–503.

Belasco, J. G., Nilsson, G., von Gabain, A., and Cohen, S. N. (1986). *Cell* (*Cambridge, Mass.*) **46,** 245–251.

Benyajati, C., Spoerel, N., Haymerle, H., and Ashburner, M. (1983). *Cell* (*Cambridge, Mass.*) **33,** 125–133.

Brawerman, G. (1987). *Cell* (*Cambridge, Mass.*) **48,** 5–6.

Brock, M. L., and Shapiro, D. J. (1983). *Cell* (*Cambridge, Mass.*) **34,** 207–214.

Dani, C., Mechti, N., Piechaczyk, M., Lebleu, B., Jeanteur, P., and Blanchard, J. M. (1985). *Proc. Natl. Acad. Sci. U.S.A.* **82,** 4896–4899.

Darnell, J. E. (1982). *Nature* (*London*) **297,** 365–371.

Donis-Keller, H., Maxam, A. M., and Gilbert, W. (1977). *Nucleic Acids Res.* **4,** 2527–2538.

Ghosh, P. K., Reddy, V. B., Swinscoe, J., Lebowitz, P., and Weissman, S. (1978). *J. Mol. Biol.* **126,** 813–846.

Gold, L., Pribnow, D., Schneider, T., Shinedling, S., Singer, B. S., and Stormo, G. (1981). *Annu. Rev. Microbiol.* **35,** 365–403.

Gordon, D. A., Shelness, G. S., and Williams, D. L. (1987). In preparation.

Gorski, K., Rock, J.-M., Prentki, P., and Krisch, H. M. (1985). *Cell* (*Cambridge, Mass.*) **43,** 461–469.

Gottesman, M., Oppenheim, A., and Court, D. (1982). *Cell* (*Cambridge, Mass.*) **29,** 727–728.

Gupta, K. C., and Kingsbury, D. W. (1984). *Nucleic Acids Res.* **12,** 3829–3841.

Guyette, W. A., Matusik, R. J., and Rosen, J. M. (1979). *Cell (Cambridge, Mass.)* **17,** 1013–1023.

Hynes, N. E., Groner, B., Sippel, A. E., Jeep, S., Wurtz, T., Nguyen-Huu, M. C., Giesecke, K., and Schutz, G. (1979). *Biochemistry* **18,** 616–624.

Kolter, R., and Yanofsky, C. (1982). *Annu. Rev. Genet.* **16,** 113–134.

Laverriere, J. N., Morin, A., Tixier-Vidal, A., Truong, A. T., Gourdji, D., and Martial, J. A. (1983). *EMBO J.* **2,** 1493–1499.

Lee, O. C., and Roeder, R. G. (1981). *Mol. Cell. Biol.* **1,** 635–651.

Lockard, R. E., and Kumar, A. (1981). *Nucleic Acids Res.* **9,** 5125–5140.

Luse, D. S., Haynes, J. R., VanLeeuween, D., Schon, E. A., Cleary, M. L., Shapiro, S. G., Lingrel, J. B., and Roeder, R. G. (1981). *Nucleic Acids Res.* **9,** 4339–4354.

Mangiarotti, G., Ceccarelli, A., and Lodish, H. F. (1983). *Nature (London)* **301,** 616–18.

Palmiter, R. D. (1974). *Biochemistry* **13,** 3606–3615.

Palmiter, R. D., and Carey, N. H. (1974). *Proc. Natl. Acad. Sci. U.S.A.* **71,** 2357–2361.

Palmiter, R. D., Mulvihill, E. R., McKnight, G. S., and Senear, A. W. (1977). *Cold Spring Harbor Symp. Quant, Biol.* **42,** 639–647.

Peattie, D. A., and Gilbert, W. (1980). *Proc. Natl. Acad. Sci. U.S.A.* **77,** 4679–4682.

Pelletier, J., and Sonenberg, N. (1985). *Cell (Cambridge, Mass.)* **40,** 515–526.

Protter, A. A., Wang, S.-Y., Shelness, G. S., Ostapchuk, P., and Williams, D. L. (1982). *Nucleic Acids Res.* **10,** 4935–4950.

Proudfoot, N. J., Shander, M. H. M., Manley, J. L., Gefter, M. L., and Maniatis, T. (1980). *Science* **209,** 1329–1336.

Qu, L. H., Michot, B., and Bachellerie, J.-P. (1983). *Nucleic Acids Res.* **11,** 5903–5920.

Schutz, G., Nguyen-Huu, M. C., Giesecke, K., Hynes, N. E., Groner, B., Wurtz, T., and Sippel, A. E. (1977). *Cold Spring Harbor Symp. Quant. Biol.* **17,** 617–624.

Shaw, G., and Kamen, R. (1986). *Cell (Cambridge, Mass.)* **46,** 659–667.

Shelness, G. S., and Williams, D. L. (1984). *J. Biol. Chem.* **259,** 9929–9935.

Shelness, G. S., and Williams, D. L. (1985). *J. Biol. Chem.* **260,** 8637–8646.

Sittman, D. B., Graves, R. A., and Marzluff, W. F. (1983). *Proc. Natl. Acad. Sci. U.S.A.* **80,** 1849–1853.

Vary, C. P. H., and Vournakis, J. N. (1984). *J. Biol. Chem.* **259,** 3299–3307.

Vournakis, J. N., Celantano, J., Finn, M., Lockard, R. E., Mitra, T., Pavlakis, G., Troutt, A., van den Berg, M., and Wurst, R. M. (1981). *In* "Gene Amplification and Analysis" (J. G. Chirikjian and T. S. Papas, eds.), Vol. 2, pp. 268–294. Elsevier/North-Holland, New York.

Wickens, M., and Stephenson, P. (1984). *Science* **226,** 1045–1051.

Wieringa, B., Ab, G., and Gruber, M. (1981). *Nucleic Acids Res.* **9,** 489–499.

Wilson, M. C., and Darnell, J. E. (1981). *J. Mol. Biol.* **148,** 231–251.

Wiskocil, R., Bensky, P., Dower, W., Goldberger, R. F., Gordon, J. I., and Deeley, R. G. (1980). *Proc. Natl. Acad. Sci. U.S.A.* **77,** 4474–4478.

Wood, C. R., Boss, M. A., Patel, T. P., and Emtage, J. S. (1984). *Nucleic Acids Res.* **12,** 3937–3950.

Zucker, M., and Stiegler, P. (1981). *Nucleic Acids Res.* **9,** 133–148.

26

A New Immune System against Viral Infection Using Antisense RNA: micRNA-Immune System

**AKIKAZU HIRASHIMA,*
MASATOSHI TAKAHASHI,*
AND MASAYORI INOUYE†**

**Yakult Central Institute for Microbiological Research*
Tokyo 186, Japan
†Department of Biochemistry
Robert Wood Johnson Medical School at Rutgers
University of Medicine and Dentistry of New Jersey
Piscataway, New Jersey 08854

I. INTRODUCTION

It has been a long time since Jacob and Monod established the fact that repressors consisting of proteins play a central role in regulating gene expression (Jacob and Monod, 1961). On the other hand, the finding of RNA repressors is much more recent (see reviews by Green *et al.*, 1986; Pines and Inouye, 1986). In contrast to protein repressors which bind directly to the target genes to block transcription (the first step of gene expression), RNA repressors inhibit the last step of gene expression by binding to mRNAs to form double-stranded RNAs. As a result, translation of the mRNAs by ribosomes is blocked. RNA repressors are able to bind the target mRNAs, because they consist of single-stranded RNAs which are complementary to the mRNAs. Consequently they are called antisense RNA or micRNA (*m*RNA-*i*nterfering *c*omplementary RNA). Needless to say, production of an antisense RNA can be easily achieved by transcribing a gene in the opposite direction to the regular mRNA

Molecular Biology of RNA
New Perspectives

transcription. In fact, many cases have been reported to date, where expression of a specific gene is regulated by the transcription of the gene in both orientations. The gene giving rise to an antisense RNA against *ompF* mRNA has been shown to be located on the *Escherichia coli* chromosome completely independently from *ompF,* the gene for a major outer membrane protein, OmpF (Mizuno *et al.,* 1984). The gene is designated *micF* and is mapped at 47 min, 26 min away from *ompF* (which maps at 21 min) on the *E. coli* chromosome. The micRNA was detected by Northern blot analysis and its promoter was identified by S1 nuclease mapping. Based on DNA sequence analysis, the *micF* RNA transcript was defined as a 174-nucleotide transcript. The *micF* gene was found to be located immediately upstream of the *ompC*. Recently, Andersen *et al.* (1987) successfully isolated, from *E. coli* cells, *micF* RNAs that were unambiguously identified as the *micF* gene products. In this work, two species of *micF* RNA were found: The most abundant form consisted of 93 nucleotides (4.5 S), whereas much smaller amounts of an 174-nucleotide RNA (6 S) were detected. In consistent agreement with this result, we have also found that the *micF* gene has two tandem promoters that direct the synthesis of these two transcripts, and that the 4.5 S promoter was approximately 25 times stronger than the 6 S promoter (Andersen *et al.,* 1987). It is important to note that both the 6 S and 4.5 S RNAs contain sequences complementary to the *ompF* mRNA, including Shine–Dalgarno and translation-initiation sequence. In addition, they both have very stable stem-and-loop structures at their 3′ ends.

The finding of such natural RNA repressors prompted us to develop a system designed to regulate bacterial gene expression artificially with antisense RNA (Coleman *et al.,* 1984). In addition to facilitating the controlled expression of various bacterial genes, the artificial micRNA system has provided insight into the mechanism of antisense RNA regulation and the characteristics of effective antisense RNAs. An artificial *mic* gene can be created easily by positioning a DNA fragment coding for a portion of an mRNA between a suitable promoter and a transcription terminator in the reverse orientation. The *mic* transcripts from such a construct would be complementary to the target mRNA over the region covered by the cloned DNA fragment. To achieve artificial micRNA regulation a *mic* cloning vector, pJDC406, was constructed as shown in Fig. 1. This vector has been used to efficiently construct a number of antisense genes in *E. coli* (Coleman *et al.,* 1984, 1985). pJDC406 contains the pIN-II (Nakamura and Inouye, 1982) promoter system, which consists of the highly expressed lipoprotein (*lpp*) promoter (Nakamura and Inouye, 1979) and the *lac* promoter-operator region in tandem. It also contains the *lpp* transcription terminator sequence separated by unique *Xba*I and *Eco*RI sites.

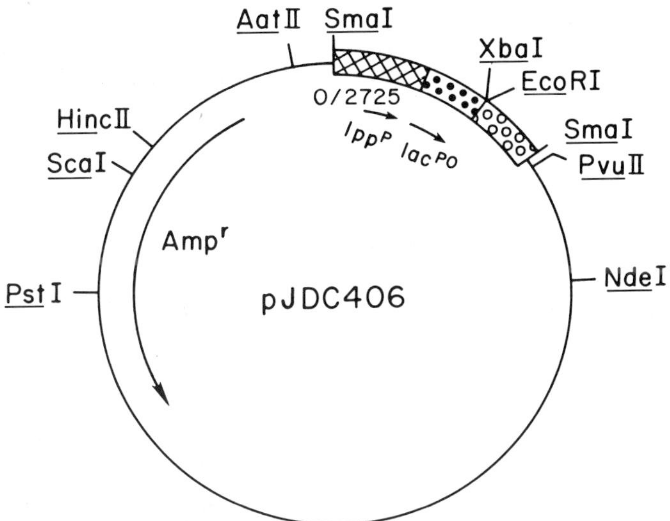

Fig. 1. micRNA cloning vector, pJDC406. Restriction map of *mic* vector pJDC406 (Coleman *et al.*, 1985). Numbers refer to size in kilobase pairs. Crosshatches represent the *lpp* promoter, filled circles represent the *lac*UV5 promoter-operator, and the open circles represent the *lpp*-independent transcription terminator. pJDC406 was constructed from pJDC402 (Coleman *et al.*, 1984) by deleting 0.3 kb downstream of the *lpp* terminator and changing both *Hin*fI sites flanking the *mic* transcription unit to *Sma*I sites with the use of *Sma*I linkers (CCCGGG). (Reprinted by permission from *Nature* **315,** 601–603. Copyright © 1985 Macmillan Magazines Limited.)

This configuration ensures high-level transcription of DNA fragments inserted into the *Xba*I or *Eco*RI site in the presence of *lac* inducers such as IPTG (isopropyl-β-D-thiogalactopyranoside). Another feature incorporated into the *mic* vector is the presence of sequences with the potential to form stem-and-loop structures on either side of the *Xba*I site. These structures flank the antisense sequence of the artificial micRNAs and were fashioned after the predicted structure of the natural *micF* RNA (Mizuno *et al.*, 1983, 1984).

II. micRNA MUTAGENESIS

Using the vector described above, a number of artificially manipulated antisense RNA genes have been created against various *E. coli* genes. For example, a small fragment (112 nucleotides) was excised from the *lpp* gene and inserted into the unique *Xba*I site of the micRNA cloning vector in reverse orientation with respect to its own promoter (Coleman *et al.*,

1984). This fragment includes the Shine–Dalgarno sequence and the coding region for the first 29 amino acid residues of prolipoprotein. In the presence of IPTG, transcription of the resulting *mic(lpp)* gene yields antisense RNA that is complementary to the ribosome binding site region of the *lpp* mRNA. The synthesis of lipoprotein, the most abundant protein in *E. coli,* was markedly inhibited, and within 5 min after induction of this artificial *mic(lpp)* RNA, a 16-fold reduction in lipoprotein synthesis was observed. Similarly, an almost complete inhibition of the production of OmpC, another major outer membrane protein was achieved using a *mic(ompC)* construct covering most of the *ompC* mRNA leader, the Shine–Dalgarno sequence, and 32 amino acid residues of the coding sequence (Coleman *et al.,* 1984). From these studies we are able to summarize the mechanism of micRNA-mediated repression as follows: (1) The induction of micRNA production blocks the expression of the target gene very rapidly, within a time period less than the half-life of the mRNA. (2) The micRNA reduces the amount of target mRNA in the cell. This is probably due to preferential digestion of double-stranded RNA formed between target mRNA and micRNA by ribonuclease III, a ribonuclease specific for double-stranded RNA. (3) There is a clear gene dosage effect. (4) micRNAs complementary to regions of the mRNA known to interact with ribosomes are the most effective.

III. micRNA-IMMUNE SYSTEM

In addition to micRNA mutagenesis described above, another exciting application of the artificial micRNA system involves the development of an immune system against viral infection. Since artificial micRNA is able to block expression of a specific gene, it is certainly feasible to use artificial micRNA to inhibit viral gene expression. As shown in Fig. 2, a viral particle that infects a cell releases the viral genome into the cell. The genome that can be either RNA or DNA is duplicated in the cell and also transcribed into mRNAs. The viral mRNAs are then translated into various viral proteins. The viral genome and the viral proteins are eventually assembled into new viral particles. However, if the cell is capable of producing micRNA against viral mRNA, the micRNA is able to block translation of the viral mRNA. As a result, production of mature viral particles is prevented. Such a cell carrying a gene to produce micRNA against viral mRNA thus becomes immune against that virus.

We have attempted to examine whether such a micRNA-immune system is possible. As a model system, we have chosen RNA coliphage SP. This RNA phage is specific for *E. coli* F^+ and its genome consists of a single-stranded positive RNA of approximately 4000 nucleotides (see Fig.

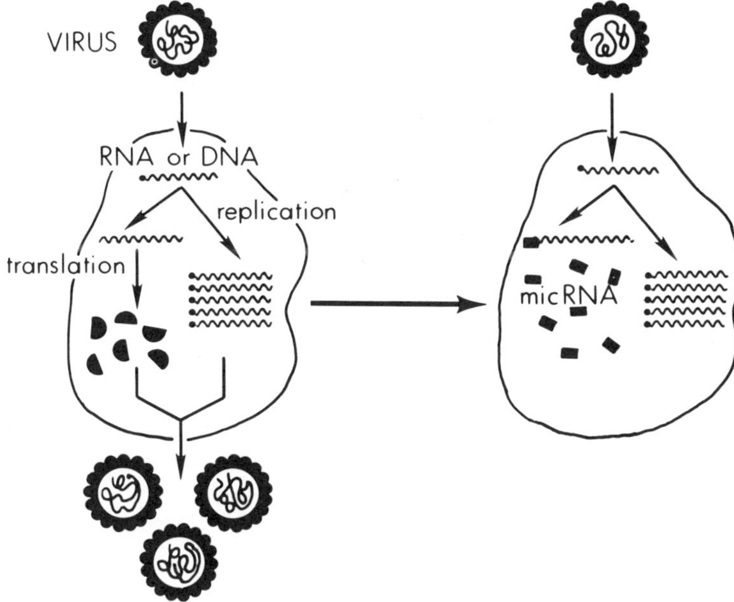

Fig. 2. micRNA immune system.

3). cDNA for the entire genome, which contains three genes, has been cloned, and the entire nucleotide sequence of the genome has been determined (Y. Inokuchi and A. Hirashima, unpublished results).

We first constructed the micRNA-immune system against the second (coat protein) and the third (replicase) genes of phage SP (Coleman *et al.*, 1985). The basic design of all constructions places the plasmid-carried genes under the control of the *lpp* plus *lac* promoters using the micRNA cloning vector, pJDC406 (Fig. 1). The following plasmids were constructed; pMIC-A, a plasmid carrying a gene that produces a 247-base-pair micRNA complementary to the translation initiation region of the mRNA for coat protein; pMIC-B, which inducibly produces an 159-base-pair micRNA complementary to the translation initiation region of the mRNA for replicase; and pMIC-C, which produces a micRNA complementary to the 518-base-pair 3′-end of the phage RNA including the 411-base-pair sequence from the carboxy-terminal coding region of the replicase gene and the 107-base-pair sequence from the 3′-noncoding region (see Fig. 3a).

These clones exerted inhibitory effects against phage SP proliferation upon induction with IPTG: 69% inhibition for pMIC-A, 42% for pMIC-B, and 40% for pMIC-C (Coleman *et al.*, 1985; see also Table I). The first

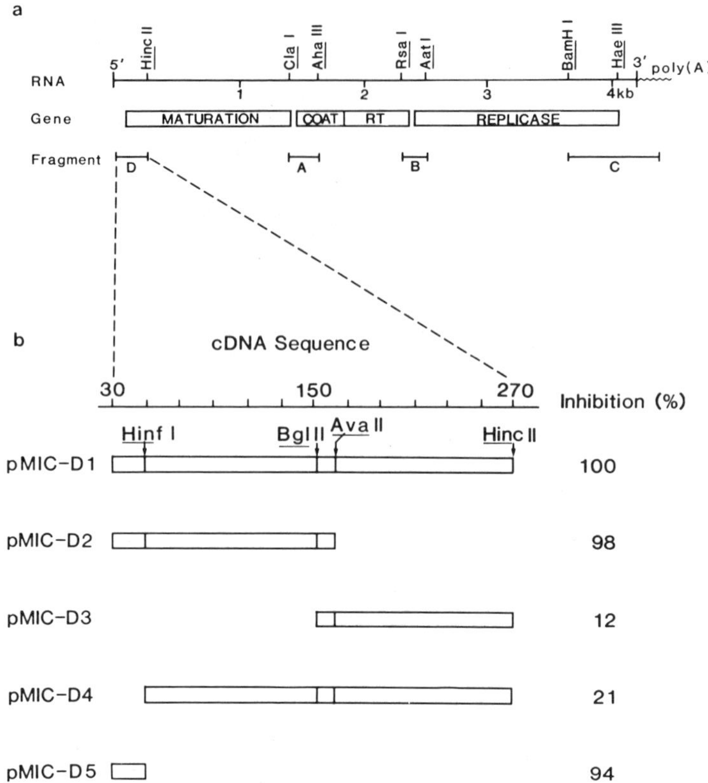

Fig. 3. (a) Diagrammatic representation of bacteriophage SP. Numbers refer to the distance (in kilobases) from the 5′ end and poly (A) was joined to the RNA. Only the relevant restriction sites are shown. RT, Coat readthrough protein. Fragments A, B, C, and D were used to construct the micRNA plasmids pMIC-A, -B, -C, and -D, respectively. These fragments were inserted into the unique *Xba*I site of pJDC406 (Fig. 1). (b) Construction of various micRNAs from the D region (see above). pMIC-D2 and pMIC-D5 were constructed in the same way as pMIC-D1 except that the cDNA clone was digested with *Mst*I plus *Ava*II for pMIC-D2 and with *Mst*I plus *Hin*fI for pMIC-D5 followed by Klenow polymerase I treatment to create the blunt end of the fragment. In the cases of pMIC-D3 and pMIC-D4, the 123-base-pair *Bgl*II–*Hinc*II fragment and 221-base-pair *Hin*fI–*Hinc*II fragment from the cDNA clone were cloned into pJDC406, respectively. Nucleotide numbers represent those from the 5′ end of phage SP. The exact nucleotide sequence is shown in Fig. 4. Inhibition of phage production with these micRNAs was examined using *E. coli* JA221/F′*lacI*^{q} pJDC406 as control. The results were expressed as percentage inhibition of the control titer. All the experiments were carried out in the presence of 2 m*M* IPTG. (From Hirashima *et al.*, 1986.)

gene of phage SP, encoding the maturation protein, has been cloned recently and the gene product, while known to be a very minor component of phage particles (one molecule per phage), is an essential factor for the binding of phage to its host cell (Weber and Konigsberg, 1975). It was of interest therefore to construct micRNA-immune systems using this region and to examine whether micRNAs against genes for minor viral proteins are more effective than those against major species such as coat protein and replicase (Hirashima *et al.*, 1986). The DNA sequence of cDNA at the 5′ end of phage SP is shown in Fig. 4. For this region, we used various restriction fragments, which were inserted into pJDC406 (Fig. 1): One clone carrying a single copy of the DNA fragment inserted into pJDC406 in an antisense orientation to the promoter of pJDC406 was designated pMIC-D1 (see also Fig. 3b). *E. coli* cells transformed with pMIC-D1 are thus able to produce the mic-D1 RNA (complementary to the region of phage SP from position 31 to 270; see Fig. 4) upon the addition of IPTG. The *mic*-D1 RNA covers the 24-base 5′-noncoding region including the Shine–Dalgarno sequence for ribosome binding and 216 bases of the 5′ end of the coding region for maturation protein including the initiation codon AUG.

As shown in Table I, the immunity to phage SP conferred by pMIC-D1 was much more effective than not only pMIC-A, pMIC-B, and pMIC-C but also pMIC-A:B:C (Table I) (Hirashima *et al.*, 1986). When the D1 fragment was inserted in the sense orientation [pMIC-D1 (sense)], there was no immune activity (Table I). In the case of pMIC-D1, the number of infective centers was approximately five times lower than that in the case of pJDC406. In addition, the final phage production at 90 min in the case of pMIC-D1 was approximately 2% of that in the case of pJDC406. Adsorption efficiencies of phages were identical for both control cells and

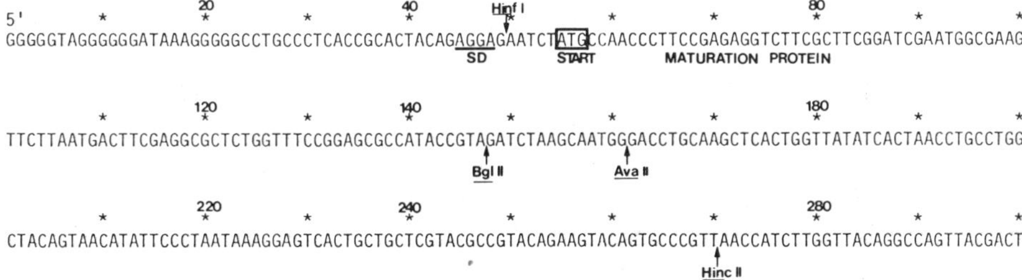

Fig. 4. The nucleotide sequence of the cDNA corresponding to the 5′-terminal region of bacteriophage SP. SD, Shine–Dalgarno sequence. ATG in a box represents the initiation codon for the maturation protein. (From Hirashima *et al.*, 1986.)

TABLE I

Effects of Various micRNA Genes on Plaque Formation of Bacteriophage SP[a,b]

Plasmid	Gene	% Inhibition IPTG (2 m*M*) +	−
pMIC-A	Coat protein	69	32
pMIC-B	Replicase	42	22
pMIC-C	3′-Terminal region	40	4
pMIC-D1	Maturation protein	98	54
pMIC-D1 (sense)	Maturation protein in the sense orientation	2	0
pMIC-A:B:C	Coat protein plus replicase plus 3′-terminal region	91	44

[a] From Hirashima *et al.*, 1986.

[b] Phage titers were measured with use of *E. coli* JA221/F′*lacI*[q] harboring various micRNA clones in the presence or absence of IPTG (Coleman *et al.*, 1985). Percent inhibition was calculated on the basis of the titer using the cells carrying pJDC406, a micRNA cloning vector.

cells carrying pMIC-D1. Therefore, the lower numbers for infective centers in the case of cells carrying pMIC-D1 is due to the function of antisense RNA produced by the plasmid.

To demonstrate which part of the pMIC-D1 RNA is essential for the immune system, we next utilized three unique restriction sites to examine the immune activity of various regions of the *mic*-D1 RNA (see Fig. 3b and Fig. 4). The DNA fragments tested are shown in Fig. 3b: pMIC-D2 and pMIC-D5 were constructed in the same way as pMIC-D1 except that the cDNA clone was digested with *Mst*I plus *Ava*II for pMIC-D2 and with *Mst*I plus *Hin*fI for pMIC-D5, instead of *Mst*I plus *Hin*cII. Therefore pMIC-D2 lost the 109-base-pair DNA fragment from position 162 (*Ava*II site) to position 270 (*Hin*cII site) in the coding region of the maturation gene in comparison with pMIC-D1 (see Fig. 3b and Fig. 4). Similarly, pMIC-D5 lost the 251-base-pair fragment from position 50 (*Hin*fI site) to position 270 (*Hin*cII site) leaving only the 19-base-pair fragment from position 31 to position 49 (see Fig. 3b and Fig. 4). In the cases of pMIC-D3 and pMIC-D4, the 123-base-pair *Bgl*II–*Hin*cII fragment and the 151-base-pair *Hin*fI–*Hin*cII fragment from the cDNA clone were cloned into pJDC406, respectively (see Fig. 3b and Fig. 4).

The micRNA-immune effects of these plasmids on phage SP production in the presence of IPTG are shown in Fig. 3b (Hirashima *et al.*, 1986). As is evident in the case of pMIC-D2, the mic immunity was hardly affected by removing approximately half of the coding region from pMIC-D1. On the other hand, the removal of the 5′ end half of the pMIC-D1 sequence resulted in significant decrease of the mic-immune effect as can be seen with pMIC-D3 (from 100 to 12%; see Fig. 3b). The complete recovery of the coding region of pMIC-D1 by adding the 98-base-pair *Hin*fI-*Bgl*II fragment to pMIC-D3 (which includes the initiation codon, AUG) improved the immune effect only to a small extent, from 12 to 21% (Fig. 3b). These results suggest that the 19-base-pair sequence from position 31 to position 49 (*Hin*fI site) encompassing the Shine–Dalgarno sequence plays an essential role in this micRNA-immune system. Strong evidence for this conclusion was obtained with pMIC-D5, which contains only the 19-base-pair fragment from phage SP: This plasmid is able to inhibit phage production almost as well as pMIC-D1 (Fig. 3b).

It has been shown that there are other positive, single-stranded RNA coliphages such as Qβ and GA, which are in the same class of RNA phages as phage SP (Furuse *et al.,* 1979). The nucleotide sequences of all these phages have been determined recently and show a similar gene arrangement (Y. Inokuchi and A. Hirashima, unpublished results). Therefore, we examined whether the present micRNA-immune system against phage SP also exerts immune effects against phages Qβ and GA.

We found that the immune system of cells harboring pMIC-D1 was highly specific for phage SP, showing little effect on phage Qβ and no effect on phage GA. However, as a part of the antisense RNA covering the coding region for maturation protein was removed, the immune effect on phage Qβ as well as GA increased dramatically without decreasing the immune effect on phage SP. The cells harboring pMIC-D2 became approximately 40% immune to phage Qβ and approximately 50% immune to phage GA. It was particularly surprising that pMIC-D5, the 19-base micRNA encompassing the Shine–Delgarno sequence, was no longer specific for phage SP (94%) and exerted potent immune effects on both phages Qβ (78%) and GA (98%).

IV. CONCLUSION

In micRNA mutagenesis, we have shown that micRNA function is most effective when it covers the sequence translation initiation region of the target mRNA (Coleman *et al.,* 1984). This was also true in micRNA immune system. However, it is not well understood at present why the specificity of the micRNA-immune system is broadened to related RNA phages when the sequence of the micRNA complementary to the coding region is shortened; inclusion of the coding region apparently prevents mic-immune activity from inhibiting the related RNA phages. It is also

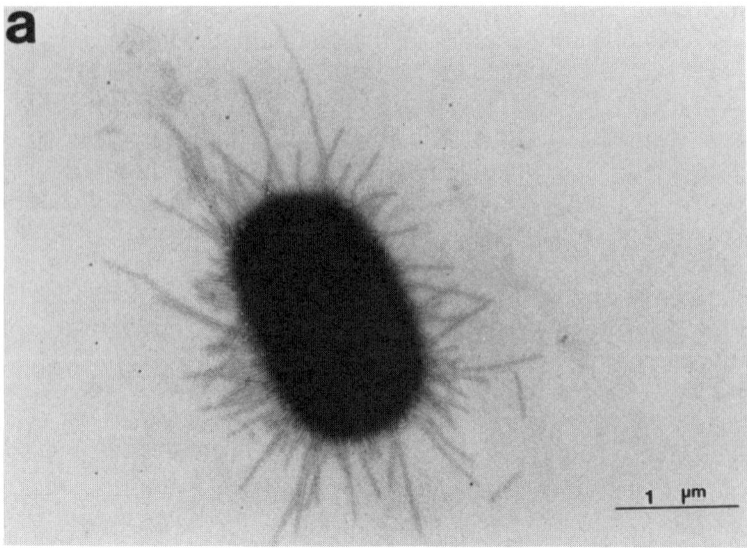

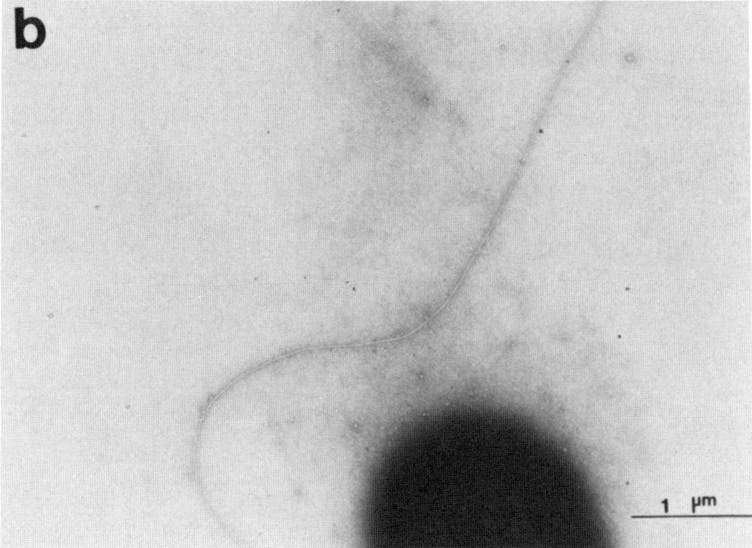

Fig. 5. Electron micrographs of *E. coli* JA221. (a) The cells harboring pMIC-a (see the text) were incubated in L broth containing 2 m*M* IPTG at 37°C for 2 hr. (b) The cells harboring pMIC-D1 were incubated as above.

surprising that the micRNA of only 19 bases in length (mic-D5 RNA) which is complementary to the region of the Shine–Dalgarno sequence of the maturation gene of phage SP is able to exert mic-immune activity against both phage Qβ and phage GA. The sequence similarities between the phages at these regions are only approximately 50% (Hirashima *et al.*, 1986). Therefore mic-immune activity of mic-D5 RNA against phages Qβ and GA may be due to another element within the RNA, which is able to block the proliferation of related RNA phages.

In this regard, it is interesting to note that when a micRNA immune plasmid (pMIC-a) was constructed with the 5′-end fragment of the SP genome from residues 1 to 30 (Fig. 4), the immunity conferred by this *mic* clone was not as effective as pMIC-D5. However, when cells harboring pMIC-a were incubated in the presence of IPTG, cells became covered with many short pili (Fig. 5a), whereas cells harboring pMIC-D1 have only one long F-pilus per cell (Fig. 5b). This result indicates that the micRNA produced from pMIC-a appears to have an effect on F-pilus production of the host F^+ cell. This result further raises an intriguing possibility to make cells immune against viral infection by modifying viral (or phage) receptors on the cell surface using micRNA mutagenesis.

The micRNA immune system presents an exciting possibility for a novel immune system against viral infection. The ability to inhibit the production of not only a specific virus but also a family of viruses is intriguing. Although many pathogenic viruses in higher systems are in fact single-stranded positive RNA viruses (i.e., various plant viruses, poliovirus, foot-and-mouth disease virus, mouse hepatitis virus, etc.) like phage SP, it would also be of interest to develop a micRNA-immune system against DNA viruses. It has been shown that micRNA mutagenesis is effective in higher systems as well (see reviews by Green *et al.*, 1986; Pines and Inouye, 1986). Furthermore, synthetic oligonucleotides complementary to viral mRNAs have been shown to have antiviral activity in tissue culture cells (Zamecnik and Stephenson, 1978; Stephenson and Zamecnik, 1978; Zamecnik *et al.*, 1986). The development of an inheritable micRNA immune system in animals may have significant potential in rendering resistance against specific viruses.

REFERENCES

Andersen, J., Delihas, N., Ikenaka, K., Green, P. J., Pines, O., Ilercil, O., and Inouye, M. (1987). *Nucleic Acids Res.* **15,** 2089–2101.

Coleman, J., Green, P. J., and Inouye, M. (1984). *Cell (Cambridge, Mass.)* **37,** 429–436.

Coleman, J., Hirashima, A., Inokuchi, Y., Green, P. J., and Inouye, M. (1985). *Nature (London)* **315,** 601–603.

Furuse, K., Hirashima, A., Harigai, H., Ando, A., Watanabe, K., Kurosawa, K., Inokuchi, Y., and Watanabe, I. (1979). *Virology* **97,** 328–341.
Green, P. J., Pines, O., and Inouye, M. (1986). *Annu. Rev. Biochem.* **55,** 569–598.
Hirashima, A., Sawaki, S., Inokuchi, Y., and Inouye, M. (1986). *Proc. Natl. Acad. Sci. U.S.A.* **83,** 7726–7730.
Jacob, F., and Monod, J. (1961). *J. Mol. Biol.* **3,** 318–356.
Mizuno, T., Chou, M.-Y., and Inouye, M. (1983). *Proc. Jpn. Acad. Sci.* **59,** 335–338.
Mizuno, T., Chou, M.-Y., and Inouye, M. (1984). *Proc. Natl. Acad. Sci. U.S.A.* **81,** 1966–1970.
Nakamura, K., and Inouye, M. (1979). *Cell (Cambridge, Mass.)* **18,** 1109–1117.
Nakamura, K., and Inouye, M. (1982). *EMBO J.* **1,** 771–775.
Pines, O., and Inouye, M. (1986). *Trends Genet.* **2,** 284–287.
Stephenson, M. L., and Zamecnik, P. C. (1978). *Proc. Natl. Acad. Sci. U.S.A.* **75,** 285–288.
Weber, K., and Konigsberg, W. (1975). *In* "RNA Phages" (N. Zinder, ed.), p. 51. Cold Spring Harbor Lab., Cold Spring Harbor, New York.
Zamecnik, P. C., and Stephenson, M. L. (1978). *Proc. Natl. Acad. Sci. U.S.A.* **75,** 280–284.
Zamecnik, P. C., Goodchild, J., Taguchi, Y., and Sarin, P. S. (1986). *Proc. Natl. Acad. Sci. U.S.A.* **83,** 4143–4146.

27

Regulation of IS*10* Transposase Expression by RNA/RNA Pairing

N. KLECKNER,* J. D. KITTLE,*
AND R. W. SIMONS†

**Department of Biochemistry and Molecular Biology*
Harvard University
Cambridge, Massachusetts 02138
†Department of Microbiology and the Molecular Biology Institute
University of California at Los Angeles
Los Angeles, California 90024

I. INTRODUCTION

The tetracycline resistance transposon Tn*10* has at its ends inverted repeats of a 1329-base-pair insertion sequence, IS*10* (Fig. 1). IS*10*-Right is a fully functional transposition module that is capable of acting as an independent unit. Transposase encoded by IS*10*-Right acts at the ends of IS*10* to promote IS*10* transposition and at the ends of Tn*10* to promote Tn*10* transposition. IS*10*-Left does not make functional transposase (Foster *et al.*, 1981). A detailed picture of IS*10*-Right is also shown in Fig. 1. Of particular importance are the transposase gene and two promoters: pIN, the promoter for the transposase gene, and pOUT, a strong promoter that opposes pIN and directs transcription outwards toward (and occasionally across) the end of IS*10* (Halling *et al.*, 1982; Simons *et al.*, 1983).

Transposition of Tn*10* and of IS*10* and occurrence of Tn*10*-promoted chromosomal rearrangements are subject to regulation at several levels. The regulation discussed below is that generally referred to as "multicopy

Molecular Biology of RNA
New Perspectives

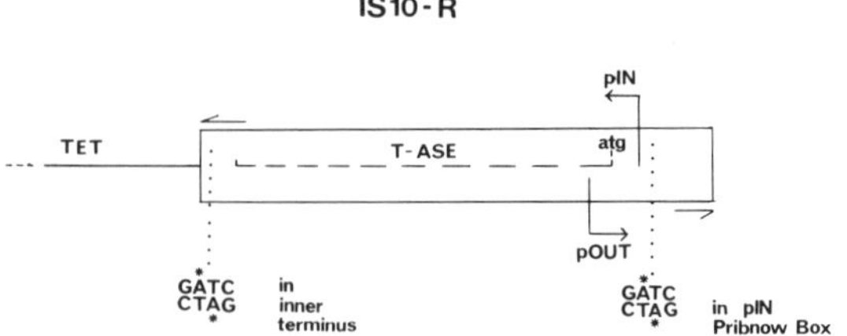

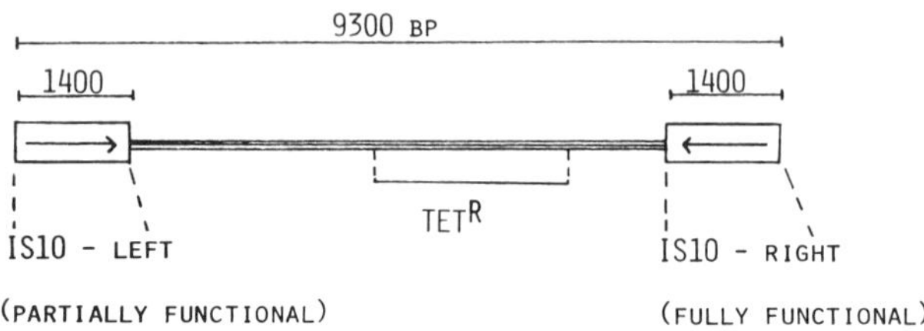

Fig. 1. Structures of Tn*10* and IS*10*-Right.

inhibition." The activity of a single marked IS*10* or Tn*10* element in the chromosome is decreased 10- to 20-fold by introduction of a multicopy plasmid containing IS*10* (Simons and Kleckner, 1983). One might naively have expected transposition of the marked element to increase, owing to an increased level of transposase provided by the extra IS*10* copies. Such an increase does not occur because, to a first approximation, IS*10* transposase works only on the DNA molecule encoding it; that is, in cis (Morisato *et al.,* 1983). Transposition of the marked element actually *decreases* as a result of the action of a trans-acting negative regulator encoded by IS*10* whose effectiveness increases with concentration. This negative regulator is the product of IS*10*'s pOUT promoter, an antisense RNA, hereafter referred to as RNA-OUT. RNA-OUT inhibits transposition by pairing with the transposase gene mRNA, hereafter referred to as RNA-IN, over the 36-base-pair region of complementarity between the two RNAs. Such pairing inhibits expression of the transposase gene at some step subsequent to transcription. The original and simplest proposal is that pairing of RNA-OUT to RNA-IN prevents binding of the ribosome to the translation start region (Simons and Kleckner, 1983).

II. BIOLOGICAL ROLE OF MULTICOPY INHIBITION

The most likely role for RNA-OUT regulation is to decrease the rate of IS*10* transposition per element as the number of elements in the cell increases. In the absence of such control, accumulation of more and more elements would result in an exponential increase in the rate and number of transposition events, and eventually to death of the cell line.

In support of this view, it has recently been possible to demonstrate rigorously that inhibition by RNA-OUT plays no role in the expression of transposase from a single-copy IS*10* element. Mutations exist that abolish transcription from pOUT, decrease the stability of RNA-OUT *in vivo,* or decrease the ability of RNA-OUT to pair with RNA-IN *in vitro* (all described below) without affecting any other known genetic determinants in IS*10*. None of these mutations confers any phenotype on a chromosomal Tn*10* element present in single copy, while all result in increased transposase expression and transposition on plasmid-borne Tn*10* elements (A. Toyofuku, J. Matsunaga, and R. W. Simons, unpublished).

The exact relationship between the extent of multicopy inhibition and the number of Tn*10*/IS*10* copies per cell has not yet been determined. Available data do demonstrate directly that the amount of transposition per copy does decrease as the number of copies of the element increases from about 10 to about 250 (A. Toyofuku, J. Matsunaga, and R. W. Simons, unpublished).

III. MOLECULAR MECHANISM OF MULTICOPY INHIBITION

The original postulate, that RNA-OUT directly inhibits binding of ribosomes to the transposase gene mRNA, remains the simplest explanation for multicopy inhibition.

A second reasonable model would be that cleavage of the paired duplex region by RNase III, which is known to cleave duplex RNAs, might separate the ribosome binding segment of the transposase gene from the distal portion of the coding region. This possibility seems unlikely because RNA-OUT inhibition is normal in strains carrying the RNase III allele *rnc*105 (C. Masada and R. W. Simons, unpublished).

Other, more *ad hoc,* models are also possible. For example, pairing could destabilize the 5′ end of RNA-IN due to the action of enzyme(s) other than RNase III, or duplex RNA might be sequestered somehow in the cell.

It is difficult to distinguish rigorously among the various models. Since prokaryotic transcription and translation are tightly coupled, a decrease in the amount of RNA-IN may occur either directly from degradation or indirectly from inhibition of translation. In several systems where the effects of antisense RNA on gene expression have been analyzed, there is evidence for a reduction in the steady-state level of the target RNA (Light and Molin, 1983; Mizuno *et al.,* 1984). The question remains whether these reductions are primary or secondary effects of antisense inhibition.

RNA-OUT does affect slightly the level of functional transposase mRNA. The strongest evidence for post-transcriptional regulation is the observation that RNA-OUT strongly inhibits expression of β-galactosidase from transposase–*lacZ* protein fusions in which *lacZ* expression is directed by IS*10* signals for both transcription and translation, but only weakly inhibits expression for β-galactosidase from analogous operon fusions in which transcription of *lacZ* is promoted by pIN but translation begins at the normal *lacZ* ribosome binding site. However, the weak reduction in operon fusion expression is real and reproducible. It is increased and decreased by all of the genetic alterations that increase or decrease the effectiveness of RNA-OUT (see below). Experiments in progress are consistent with the idea that the effect on operon fusions is an indirect consequence of translational "polarity"; the small reduction can be quantitatively mimicked by blocking translation with frameshift mutations in the transposase gene sequences upstream of *lacZ* (C. Jain and N. Kleckner, unpublished).

IV. MECHANISM OF PAIRING BETWEEN RNA-IN AND RNA-OUT *IN VITRO*

RNA-IN and RNA-OUT transcripts synthesized *in vitro* from purified DNA restriction fragment templates can pair with one another with a rate constant of about 3×10^5 moles/liter per second. One such *in vitro* pairing experiment is described in Fig. 2. Inspection of the sequences of RNA-IN and RNA-OUT and analysis of the effects of single base alterations in one or both RNAs has led to a specific molecular model for the mechanism of pairing between these two RNA molecules which is outlined below (Kittle and Kleckner, 1987).

RNA-IN does not contain any obvious secondary structure, at least within the 5′ terminal 350 nucleotides used in these *in vitro* experiments. RNA-OUT, in contrast, is likely to form a structure close to that in Fig. 3. The RNA-OUT molecule that occurs *in vivo* probably consists of bases 116 to ~47, that is, a single stem-and-loop structure with short 5′ and 3′

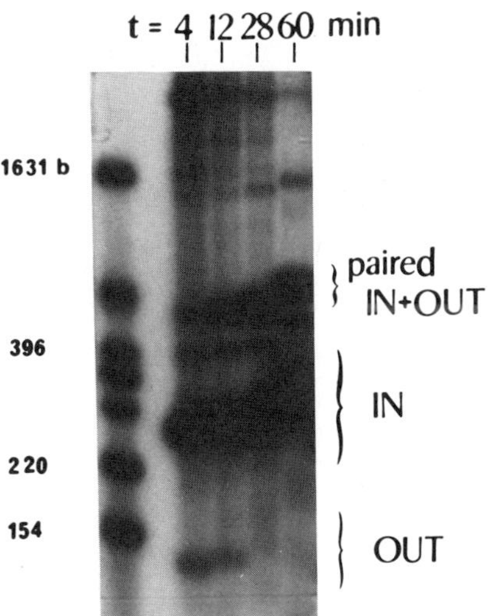

Fig. 2. Pairing between RNA-IN and RNA-OUT *in vitro*. DNA restriction fragment template was preincubated with buffer and RNA polymerase. At $t = 0$ a single round of transcription was initiated by addition of the four nucleotide triphosphates plus heparin. Under the conditions used, transcription is complete within 1 min. The reaction mixture was incubated further, and at indicated times samples were withdrawn and loaded directly onto a nondenaturing polyacrylamide gel, which was running slowly but continuously throughout the experiment to insure rapid migration of samples into the gel. In this particular experiment, a single DNA fragment containing the end of a pIN up-promoter mutant was used as template; as a result, RNA-IN is in excess over RNA-OUT. As the pairing reaction proceeds, RNA-OUT disappears and a new species corresponding to the paired RNA-IN/RNA-OUT hybrid appears. The exact conditions for *in vitro* transcription are described in Simons *et al.* (1983).

tails (Lee and Schmidt, 1985; C. Case and R. W. Simons, unpublished). In any case, portions of RNA-OUT 3′ to this structure are irrelevant to pairing *in vitro;* RNA-OUT molecules made from templates truncated distal to the primary stem-and-loop at base pair 35 pair with RNA-IN at the same rate as an RNA-OUT molecule whose 3′ end is at base pair 1 of IS*10*.

The RNA-OUT structure can be roughly divided into two domains, the bottom tightly paired stem domain (−19 kcal mole), and the top loop domain (<+2 kcal mole) in which some bases may or may not ever be paired. The 5′ end of RNA-IN begins at the sequences complementary to the top of the loop and the pairing region extends through the loop domain and down through the stem as indicated in Fig. 3.

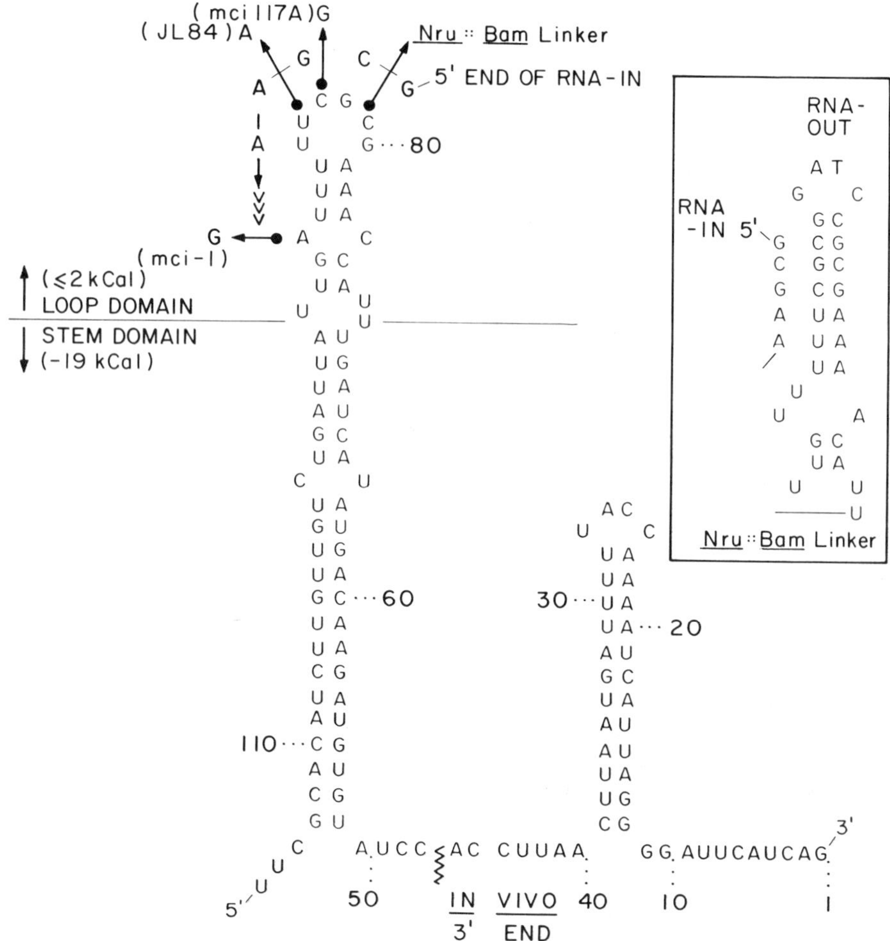

Fig. 3. RNA-OUT, RNA-IN, and point mutations in the postulated region of nucleation between the 5′ end of RNA-IN and the loop domain of RNA-OUT. Bases are numbered from the outside end of IS*10* inward. Also indicated is the 3′ end of the RNA-OUT species isolated *in vivo* (Lee and Schmidt, 1985).

The relationship of the RNA-OUT and RNA-IN sequences suggested that the pairing reaction might begin by a rate-limiting nucleation step in which the 5′ end of RNA-IN interacts with the loop domain of RNA-OUT. This first step, which is required for the two molecules to remain in stable juxtaposition, would then be followed by pairing of RNA-IN with RNA-OUT sequences in the stem region by a non-rate-limiting strand displacement reaction.

Analysis of base changes in RNA-IN and RNA-OUT fully supports this working model.

A. Role of the Stem Domain

Two lines of evidence indicate that the stem domain of RNA-OUT poses no barrier to the rate of the pairing reaction. First, a truncated RNA-OUT molecule (base pair 116 to base pair 82, Fig. 3), which contains no base-paired stem, pairs with RNA-IN at the same rate as normal RNA-OUT. Second, none of the eight existing mutations that change bases in the stem alter the ability of mutant RNA-OUT to pair with wild-type RNA-OUT, despite their disparate effects on the structure and stability of RNA-OUT and of the IN/OUT hybrid. Three of these mutations increase and five decrease the (calculated) stability of the stem. Furthermore, two of the eight are in the RNA-OUT strand that pairs with RNA-IN and six are in the displaced strand; both of the mutations in the pairing strand decrease the stability of the resulting IN/OUT hybrid molecule.

In view of these observations, we suggest that the stem region of RNA-OUT probably exists to allow the RNA-OUT molecule to be stable *in vivo*. Preliminary experiments place its *in vivo* half-life at roughly 20 min, and mutations that weaken the stem make RNA-OUT unstable (R. W. Simons, unpublished, see below).

A further interesting feature of the stem region is that the several base-pair mismatches and GU base pairs present in the stem may serve to keep the strand displacement reaction proceeding irreversibly in the correct direction. Each IN/OUT base pair formed will be perfect, owing to the perfect complementarity of the two molecules; and thus in several instances the perfect IN/OUT base pair will replace an imperfect OUT/OUT stem base pair. Each such replacement will impose a barrier to strand displacement of the newly formed IN/OUT hybrid by repairing of the OUT/OUT stem.

B. Critical Role of Loop Domain and the 5′ End of RNA-IN

In sharp contrast to the situation with RNA-OUT, every base alteration in the loop domain of RNA-OUT and two extensions of the 5′ end of RNA-IN drastically decrease the rate of pairing. For example, insertion of a *Bam*HI linker within the loop region of RNA-OUT does not alter the homology between wild-type RNA-IN and the altered RNA-OUT, but should result in strong pairing to close up the loop domain (Fig. 3). This mutation completely abolishes pairing of altered RNA-OUT with wild-type RNA-IN. A similar but more subtle example is provided by the single base mutation *mci*1, which should allow formation of a critical GC pair and dramatic stabilization of a closed loop structure (Fig. 3). And indeed, this mutation abolishes the ability of mutant RNA-OUT to pair with wild-type RNA-IN. That the effect of this mutation is actually due to closure of the loop domain is suggested by the fact that its *in vitro* pairing defect is eliminated if the mutant RNA-OUT is truncated at the top of the loop to eliminate one-half of the loop domain. Furthermore, this mutation does not exert its effect (solely) by destabilizing the homology between RNA-OUT and RNA-IN, because RNA-OUT containing the *mci*1 mutation also fails to pair with the homologous *mci*1-containing RNA-IN.

Two other mutations in the loop domain, the transition mutations *mci*117A and *JL84* (Fig. 3), also critically alter pairing *in vitro*. Both mutations drastically disrupt pairing in the two heterologous IN/OUT combinations: mutant RNA-OUT pairing with wild-type RNA-IN, and mutant RNA-IN pairing with wild-type RNA-OUT. However, for both of these mutations, the decreases can be attributed directly to a decrease in stability of the IN/OUT hybrid molecule due to disruption of homology. To a first approximation, neither mutation affects the rate of IN/OUT pairing in the homologous mutant combinations where the two interacting RNA species contain the same (compensating) base alterations. These experiments provide critical evidence for the role of RNA homology in general as well as for the particular details of the proposed model.

Finally, two RNA-INs with altered 5′ ends also fail to pair with wild-type RNA-OUT: a pIN transcript produced by a single base mutant DR33 is several bases longer than the normal pIN transcript; and the transcript produced by insertion of a *ptac* promoter close to the normal location of pIN results in a transcript containing several extra bases of IS*10* sequence plus some extra heterologous material at the very 5′ end.

Taken together, these data provide strong support for the proposed *in vitro* pairing model. Analysis of the *in vivo* effects of loop mutations *mci*117A and *JL84* should provide additional rigorous evidence for the role of RNA/RNA pairing *in vivo* and for (or against) the applicability of the *in vitro* model to RNA pairing *in vivo*.

V. *IN VIVO* PHENOTYPES OF MUTATIONS IN RNA-OUT

All of the mutations discussed in the previous section abolish the effect of mutant RNA-OUT to inhibit expression of transposase from wild-type RNA-IN *in vivo*. For those mutations that inhibit RNA/RNA pairing *in vitro,* this phenotype is presumably a direct consequence of the failure of the pairing reaction. *In vivo* RNA analysis of one such pairing mutant confirms that RNA-OUT is present in normal amounts and thus is presumably simply ineffective (R. W. Simons, unpublished).

Mutations that do not alter pairing *in vitro* but nonetheless abolish inhibition *in vivo* can be further divided into two classes. Members of one class decrease the calculated stability of the RNA-OUT stem and the corresponding RNA-OUT molecules made *in vitro* exhibit a lower than normal mobility on nondenaturing polyacrylamide gels, consistent with an actual destabilization of the stem structure. It seems probable that these mutations cause RNA-OUT to be unstable *in vivo*. *In vivo* RNA analysis of one such mutant confirms that no RNA-OUT is present. Members of the second class actually increase the stability of the RNA-OUT stem and do not affect *in vitro* RNA-OUT gel mobility. Instead, both of these mutations cause a dramatic increase in the rate of transcription from the pIN promoter, and exert their effects by providing extra plasmid-encoded RNA-IN transcripts which titrate away the plasmid-encoded RNA-OUT, which is otherwise normal in its stability and pairing properties. Titration by one of these mutations, HH104, was described during the original identification and documentation of IS*10* antisense RNA regulation (Simons and Kleckner, 1983).

REFERENCES

Foster, T., Davis, M. A., Takeshita, K., Roberts, D. E., and Kleckner, N. (1981). *Cell (Cambridge, Mass.)* **23.** 201–213.

Halling, S. M., Simons, R. W., Way, J. C., Walsh, R. B., and Kleckner, N. (1982). *Proc. Natl. Acad. Sci. U.S.A.* **79,** 2608–2612.

Kittle, J. D., and Kleckner, N. (1987). In preparation.

Lee, Y., and Schmidt, F. J. (1985). *J. Bacteriol.* **164,** 556–562.

Light, J., and Molin, S. (1983). *EMBO J.* **2,** 93–98.

Mizuno, T., Chou, M.-Y., and Inouye, M. (1984). *Proc. Natl. Acad. Sci. U.S.A.* **81,** 1966–1970.

Morisato, D., Way, J. C., Kim, H.-J., and Kleckner, N. (1983). *Cell (Cambridge, Mass.)* **32,** 799–807.

Simons, R. W., and Kleckner, N. (1983). *Cell (Cambridge, Mass.)* **34,** 683–691.

Simons, R. W., Hoopes, B., McClure, W., and Kleckner, N. (1983). *Cell (Cambridge, Mass.)* **34,** 673–682.

28

Characterization and Functional Analysis of the Factors Required for Transcription of the Adenovirus Major Late Promoter

DANNY REINBERG,[1] OSVALDO FLORES,[1] AND LEONARD BUCKBINDER[1]

Department of Biochemistry
State University of New York at Stony Brook
Stony Brook, New York 11794–5215

I. INTRODUCTION

The initiation of transcription by RNA polymerase II is the first of several steps in the regulation of gene expression. An understanding of site selection by RNA polymerase II will be crucial for elucidating some of the mechanisms that regulate differential gene expression.

Class II promoters, those transcribed by RNA polymerase II, are composed of several sequence elements usually located upstream of the start site of transcription (CAP site). An element found in most class II promoters is located at or near position −28 and contains a consensus sequence TATAAA or ATAAA. This so-called "TATA box" element, in conjunction with the CAP site, appears to position the start site of transcription (Breathnach and Chambon, 1981). One or several additional upstream elements with respect to the TATA box are required for optimal

[1] Present address: Department of Biochemistry, Robert Wood Johnson Medical School at Rutgers, University of Medicine and Dentistry of New Jersey, Piscataway, New Jersey 08854–5635.

Molecular Biology of RNA
New Perspectives

expression. There appear to be several subclasses of these upstream elements. Some, like the CCAT, GGGCG, or GGCCACGTGACCAA homologies are found in several different genes. Others, like the heat shock consensus sequences, metallothionein metal response element, or the several histone-specific sequences appear to be gene specific (Dynan and Tjian, 1985). Another class of genes, in which the TATA box element is absent from the promoter, has now been identified (Dynan and Tjian, 1985). There are a number of genes in this category and a common alternate transcriptional control sequence has not been detected.

Cell-free transcription systems have demonstrated that the complex multisubunit RNA polymerase II recognizes and accurately transcribes purified class II promoters only when supplemented with additional factors present in crude cellular extracts (Weil *et al.*, 1979). The fractionation of cellular extracts has led to the identification of two sets of transcription factors: a group that (like RNA polymerase II) appears to be required by most class II genes and to operate through common promoter elements, and thus defined as general transcription factors, and a group that recognizes only the gene-specific elements in different promoters, and thus are defined as gene-specific factors (for review, see Dynan and Tjian, 1985).

In this chapter we will describe our studies on the factors required for transcription of the adenovirus major late promoter.

II. FRACTIONATION AND FUNCTIONAL ANALYSIS OF THE FACTORS REQUIRED FOR TRANSCRIPTION OF THE ADENOVIRUS MAJOR LATE PROMOTER

A. Fractionation of the General Transcription Factors

In our efforts to understand the mechanism of initiation of transcription from class II promoters, we have attempted to purify and characterize mechanistically the human factors required for transcription of a model promoter, the adenovirus major late promoter (Ad-MLP). However, to focus the analysis on those transcription factors that would probably be shared by other class II promoters, the Ad-MLP construct used in the initial studies contained only the minimum sequences required for transcription initiation *in vitro,* that is, those around the TATA box and the CAP site (Hu and Manley, 1981).

In the case of the human HeLa cells, chromatographic fractionation of extracts has led to the resolution of five fractions (designated TFIIA, -B, -C, -D, and -E) that were required for transcription of the Ad-MLP in relatively crude reconstituted systems (Matsui *et al.*, 1980; Dignam *et al.*,

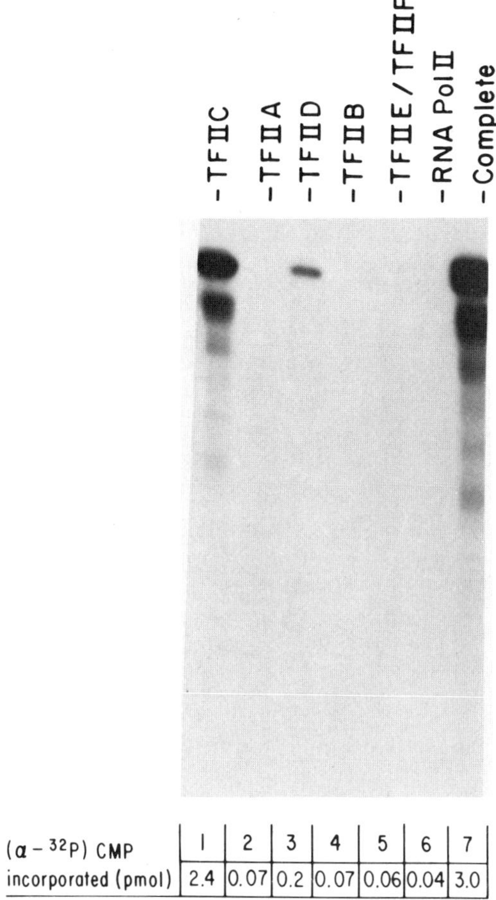

(α - ³²P) CMP	1	2	3	4	5	6	7
incorporated (pmol)	2.4	0.07	0.2	0.07	0.06	0.04	3.0

Fig. 1. Factor requirements for transcription of the Ad-MLP. Reaction mixtures (40 μl) were as previously described (Reinberg and Roeder, 1986a) and contained, except as indicated: TFIIC (single-stranded DNA agarose fraction, 0.09 μg), TFIIA (DEAE-5 PW fraction, 0.06 μg), TFIIB (Sephacryl AcA44 fraction, 0.23 μg), TFIIE/TFIIF (Sephacryl AcA44 fraction, 2.1 μg), TFIID (Sephacryl S200 fraction, 0.78 μg), RNA polymerase II (43 ng, 146,000 units/mg protein), and a DNA template (pML) in which the Ad-MLP was ligated (at position +11) to a synthetic DNA/fragment with no cytidylic residues downstream to the promoter (Swadogo and Roeder, 1985a). Transcription of this template produced a 400-nucleotide transcript resistant to ribonuclease T1. After 30 min of incubation at 30°C, half of the reaction was precipitated with trichloroacetic acid and the other half was analyzed on a 4% polyacrylamide 7 *M* urea gel.

1984). The active component in one required fraction, TFIIC, was identified as poly(ADP-ribose) polymerase and was found to suppress random transcription *in vitro* by binding to nicks on the template. This activity was dispensable when the other factors were purified more extensively (Slatery *et al.*, 1983). The further fractionation of the A, B, E, and D fractions had led to the resolution of five transcription factors (TFIIA, TFIIB, TFIID, TFIIE, and TFIIF) that were required, in addition to RNA polymerase II, for transcription of the Ad-MLP. Omission of any of the above factors or RNA polymerase II resulted in the absence of detectable levels of transcription (Fig. 1 and Fig. 3, below).

All the activities required for transcription of the adenovirus major late promoter have been isolated from uninfected HeLa cell nuclear extracts as previously described (Reinberg and Roeder, 1986a; Reinberg *et al.*, 1986) and summarized in Fig. 2. The IIA protein has been purified 2200-fold with 8% recovery of activity. The initial three steps of the purification are as previously described (Reinberg *et al.*, 1986) and we have included three additional steps, which involve fractionations on Bio-Gel TSK phenyl-5-PW, Bio-Gel HPHT, and DEAE-5PW columns (Fig. 2). TFIIA activity eluted from high-salt gel filtration (1.0 *M* KCl) on Sephacryl AcA44 or HPLC-GF-250 with an apparent molecular weight of 84,000. Our most purified TFIIA preparation still contained about 12 major polypeptides and we have not been able to identify, thus far, the polypeptide(s) that coeluted with the activity. Samuels and Sharp (1986) have purified the IIA factor from calf thymus and indicated that the activity

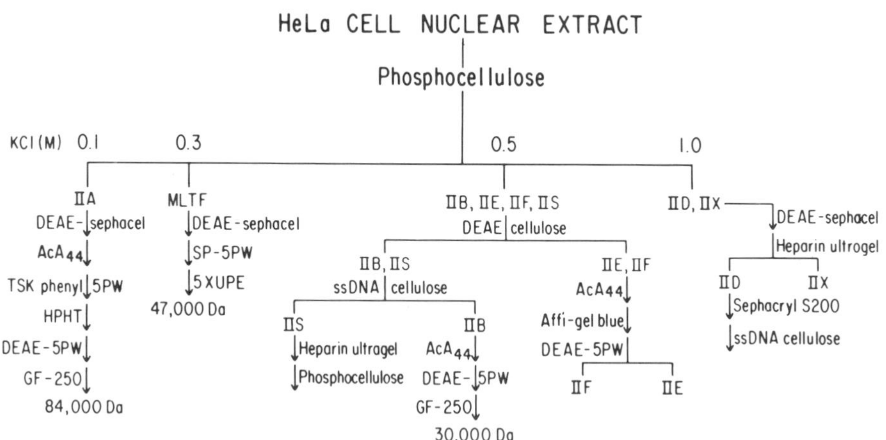

Fig. 2. A schematic representation of the separation of the factors required for transcription of the Ad-MLP.

was contained in three polypeptides species with molecular weights of 19,600, 19,100, and 12,800. The native protein eluted from gel filtration with an apparent molecular weight of 25,600–35,000. We do not understand the different molecular weight values obtained for the native proteins; however, it may reflect some difference in the physical properties between the human- and the calf-derived factors. Previous studies (Egly *et al.,* 1984) had indicated that TFIIA may be actin. Our most purified preparation, while not homogenous, is free of actin. In addition, we have not been able to replace TFIIA activity by any of the commercially available actin preparations. Our results, together with those obtained by Samuels and Sharp (1986), suggest that TFIIA is not actin.

TFIIB protein has been purified to near homogeneity (Reinberg and Roeder, 1986a). The activity appeared to be contained in a single polypeptide that eluted from a high-salt gel filtration column (Sephacryl AcA44 or HPLC-GF-250) with an apparent molecular weight of 29,500 to 33,000. It has been suggested that the HeLa cell-derived TFIIB corresponded to an activity isolated from Ehrlich ascites tumor cells that was found to stimulate RNA polymerase II in a random transcription reaction (Sekimizu *et al.,* 1979). The purified TFIIB did not stimulate purified RNA polymerase II, and it was separated from such a stimulatory activity (IIS, Reinberg and Roeder, 1986b) by chromatography on a single-strand DNA agarose column (Reinberg and Roeder, 1986a). The IIS protein has been purified and mechanistically characterized (Reinberg and Roeder, 1986b). These studies indicated that the HeLa IIS is an elongation factor that stimulated the efficiency by which RNA polymerase II passed through pausing sites. A similar activity has been purified from calf thymus (Rappaport *et al.,* 1986).

The TFIIE activity has been purified over 3200-fold from the nuclear extract as outlined in Fig. 2 and described elsewhere (Flores and Reinberg, 1987). Upon fractionation of the Affi-Gel blue protein fraction (see Fig. 2) on a DEAE-5PW column, the activity split into two fractions, TFIIE and TFIIF, both of which were required to restore maximal levels of transcription (Fig. 3). Addition of either TFIIE or TFIIF to an *in vitro* reconstituted system composed of factors IIA, IIB, IID, RNA polymerase II and the Ad-MLP produced low levels of transcription (Fig. 3). This low level of activity may suggest either cross-contamination of TFIIE and TFIIF or that much higher levels of a single factor (TFIIE or TFIIF) are required to reach saturation when only TFIIE or TFIIF are added to the reaction. TFIIE and TFIIF were required for transcription from circular or linear templates (Fig. 3A,B).

A

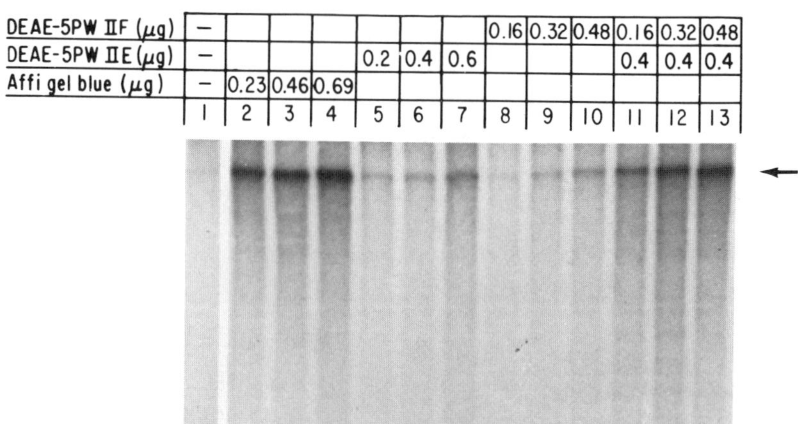

CMP incorporated (pmol)	0.05	0.5	0.6	0.7	0.15	0.15	0.24	0.12	0.16	0.2	0.4	0.55	0.7

B

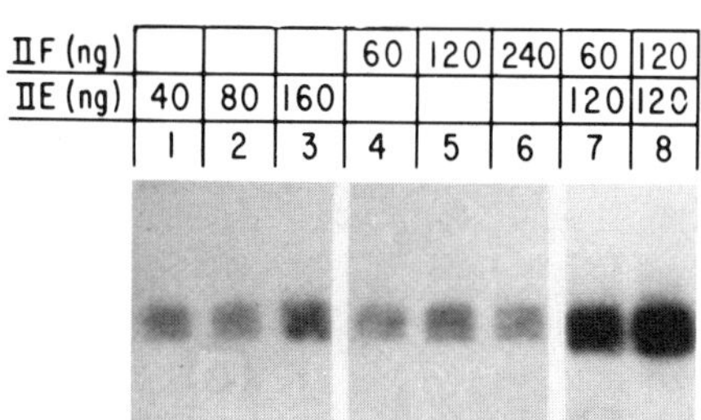

CMP incorporated (pmol)	0.2	0.2	0.4	0.15	0.25	0.2	0.8	1.2

Fig. 3. Optimal transcription of the Ad-MLP requires factors IIE and IIF. Reaction mixtures (40 μl) were as previously indicated and contained: TFIIC (single-stranded DNA agarose, 0.09 μg), TFIIA (DEAE-Sephacel fraction, 2.1 μg), TIIB (Sephacryl AcA 44 fraction, 0.2 μg), TFIID (DEAE-Sephacel fraction, 2.4 μg), RNA polymerase II (43 ng, 146,000 units/mg protein), TFIIE, and/or TFIIF as indicated in the figure. (A) Contains 0.5 μg of *Bgl*II-cut pD139 DNA (Reinberg and Roeder, 1986a). (B) Contains 0.3 μg of circular pML DNA (Sawadogo and Roeder, 1985b).

B. Functional Analysis of the General Transcription Factors

The various factors described above were required for the overall transcription reaction (Figs. 2 and 3); however, it was shown, using nonspecific inhibitors (heparin or Sarkosyl) which affected different steps in the transcription reaction, that factors IIA, IIB, IIE/IIF, IID, and RNA polymerase II participated in the formation of a preinitiation complex (Reinberg and Roeder, 1986a; Reinberg *et al.*, 1986).

It has been documented that heparin is a very effective inhibitor of RNA polymerases. Only those RNA polymerase molecules that are in transcription complexes or in tight association with the DNA will be relatively resistant to heparin. Thus, heparin was used to separate the preinitiation step from the initiation/elongation step. The formation of a heparin-resistant complex, capable of elongation upon addition of the ribonucleoside triphosphates, required factors IIA, IIB, IID, IIE/IIF, and RNA polymerase II (Reinberg *et al.*, 1986). Transcription occurred if, and only if, all factors were added prior to heparin, clearly indicating factor associations that sequestered them from heparin inactivation. However, because a free (unbound) factor might be bound by heparin, and thus not function, it was not possible to conclude that complexes containing less than the full set of factors were not heparin resistant (Reinberg and Roeder, 1986a).

The addition of increasing Sarkosyl concentrations defined three functional steps exhibiting different sensitivities to Sarkosyl during the initiation of transcription from the Ad-MLP (Hawley and Roeder, 1985): (1) a template commitment step occurred in the presence of 0.015% Sarkosyl, (2) conversion of the committed complex to a rapid-start complex, a step blocked by 0.015% Sarkosyl, and (3) a step that involved conversion of the rapid-start complex to an initiated complex (Hawley and Roeder, 1986). When Sarkosyl, at a concentration of 0.02%, was used to analyze the factor(s) required for the formation of a committed complex, TFIIA and TFIID were shown to be required to render the reaction resistant to low concentrations of Sarkosyl (Reinberg *et al.*, 1986). The yield of the reaction was higher if RNA polymerase II was added together with TFIIA and TFIID. A committed, 0.02% Sarkosyl-resistant complex could not be formed by incubation of the TFIIB and TFIIE/TFIIF either alone or in combination with each other and RNA polymerase II (Reinberg *et al.*, 1986). These results suggested that TFIID and TFIIA were required at an early stage during the formation of a complex and that this association was followed by the binding of RNA polymerase II. In addition, these results are in agreement with the observation that the TFIID protein fraction contained an activity that specifically bound to the TATA sequence (Nakajimi *et al.*, 1987; Sawadogo and Roeder, 1985b) and with previous studies that indicated, by template competition experiments, that two crude protein fractions, equivalent to TFIIA and TFIID were required for the formation of a stable committed complex (Davidson *et al.*, 1983; Fire *et al.*, 1984).

Kinetic analyses combined with order of addition experiments have suggested that TFIIA acted first, followed by TFIID, during the initiation of transcription (Reinberg *et al.*, 1986). This result was surprising since purified TFIID binds to the TATA sequence and it has not been possible to observe any specific DNA-binding activity associated with TFIIA, as monitored by DNase I protection experiments. However it is possible that TFIIA produces a more stable association of TFIID to the TATA sequence. To determine if TFIIA and/or TFIID may serve as an entry site for RNA polymerase II in the transcription cycle, we analyzed binding of the polymerase to the promoter by DNase I protection experiments. It was only possible to footprint RNA polymerase II on the major late promoter when TFIID was added to the reaction (Reinberg, 1987).

Although TFIIB and TFIIE/TFIIF were required to produce a heparin resistant-initiation competent complex (Reinberg and Roeder, 1986a), it was not possible to detect any specific binding of these factors to promoter sequences when analyzed by DNase I protection experiments. When the possibility that an association of these proteins with the promoter was mediated through interaction with other factors that either directly or indirectly recognized specific promoter sequences was studied, it was possible to demonstrate, by glycerol gradient sedimentation, an interaction between TFIIE and/or TFIIF and RNA polymerase II (Reinberg and Roeder, 1986a).

Similar studies with TFIIB failed to demonstrate clearly an association with the polymerase. When TFIIB was incubated with RNA polymerase II and sedimented, the recovered TFIIB activity was low and present in a diffuse area of the gradient extending from the position of RNA polymerase II to the top of the gradient, at the position where the purified TFIIB sedimented (D. Reinberg and R. G. Roeder, unpublished observations). However, when TFIIB and TFIIE/TFIIF were incubated with RNA polymerase II and sedimented through a glycerol gradient, the peak fraction containing polymerase activity also contained TFIIE and/or TFIIF and low but significant levels of TFIIB activity. When these fractions were then analyzed in a complementation assay dependent upon TFIIB, TFIIE/TFIIF, and RNA polymerase II, no activity was observed (D. Reinberg and R.G. Roeder, unpublished observations). These results were explained as if another factor, beside TFIIB and TFIIE, was required for transcription and this presumptive factor contaminated both the TFIIB and TFIIE fractions (Reinberg and Roeder, 1986a). In lieu of the isolation of TFIIF from the TFIIE protein fraction, the possibility of an interaction between TFIIB and RNA polymerase II is currently being investigated.

The above experiments (interaction of TFIIB, TFIIE/TFIIF, and polymerase) also suggested the possibility of an interaction between TFIIB with TFIIE and/or TFIIF. When this was analyzed, by glycerol gradient sedimentation, it was possible to demonstrate an association between TFIIE and/or TFIIF with TFIIB (Reinberg and Roeder, 1986a).

The results discussed above indicated that TFIIB and TFIIE/TFIIF participated in the formation of a stable preinitiation complex. This complex was deduced on the basis of its formation in the absence of nucleoside triphosphates and its resistance to (ability to initiate in the presence of) heparin concentrations which blocked the action of a free factor. This result strongly suggests that IIB and IIE/IIF are initiation factors, or that if they are elongation factors their function in elongation may be dependent upon entry into the transcription cycle through interaction with the preinitiation complex, more certainly via RNA polymerase II. The observation that TFIID could be footprinted at the promoter (Nakajimi *et al.*, 1987) in addition to the results obtained with Sarkosyl, namely that factors IIA and IID were required to produce a committed complex, strongly indicate that TFIIA and TFIID are initiation factors.

The above results along with the kinetic analyses combined with order of addition experiments (Reinberg *et al.*, 1986) suggested that there is a sequential assembly of the factors onto the DNA (Fig. 4). An ordered assembly of factors to the DNA was also suggested by the studies of Fire *et al.* (1984) and Hawley and Roeder (1985).

C. Studies on the Ad-MLP Upstream Promoter Factor

The above six general transcription factors (IIA, IIB, IID, IIE, IIF, IIS) and RNA polymerase II are the minimum proteins required for transcription of the Ad-MLP and operate via minimum promoter sequences, the TATA box and the CAP site. When the Ad-MLP containing sequences that extended from nucleotides −400 to +546 were used in the reaction, two other activities that affected transcription were detected. The requirement for one of the activities, TFIIX, has been correlated with sequences downstream of the CAP site (Reinberg *et al.*, 1986). The exact role of this activity in transcription is not known.

The initial studies of Hen *et al.*, (1982) and Jove and Manley (1984)

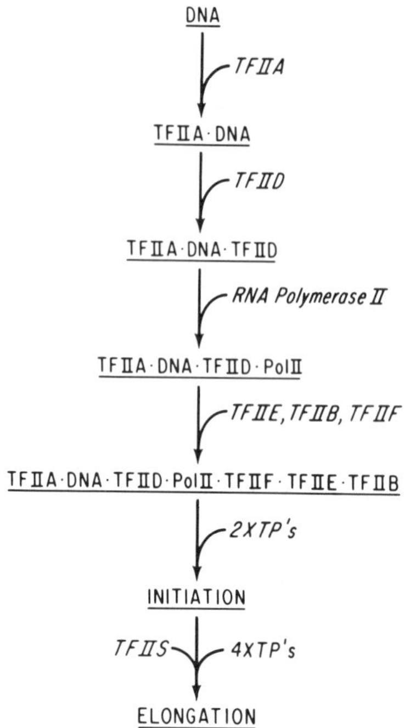

Fig. 4. Proposed model for ordered assembly of the transcription factors on the DNA. For details, see text and Reinberg *et al.* (1986).

defined an upstream element for the Ad-MLP contained within positions −51 to −66. Subsequent studies by others resulted in the isolation of a protein fraction (USF, Sawadogo and Roeder, 1985b; MLTF, Carthew *et al.*, 1985; UEF, Miyamoto *et al.*, 1985) that specifically interacted with this region. Addition of this protein fraction to a partially reconstituted system resulted in a 10- to 20-fold stimulation of transcription (Sawadogo and Roeder, 1985b). Using a reconstituted system for transcription of the Ad-MLP as well as a DNA-binding assay, we have purified MLTF from uninfected HeLa cells and 293 cells. The factor was purified to near homogeneity in four steps (Fig. 2 and Buckbinder *et al.*, 1986). One of these steps involved affinity chromatography using a column that contained multiple copies of the MLTF binding site (5× UPE, Fig. 2). The isolated DNA fragment containing these sites was end-labeled with biotinylated nucleotides and the labeled fragments were retained on a Streptavidin agarose column as initially described by Kasher *et al.* (1986). This step allowed us to purify the protein over 20,000-fold. The activity coincided with a single polypeptide of 47,000 daltons as judged by SDS–

polyacrylamide gel electrophoresis, gel filtration on a HPLC-GF-250 column, and glycerol gradient sedimentation. The purified protein was found to interact specifically with the upstream element of the major late promoter as evidenced by its ability to: (1) affect the mobility of DNA fragment only when they contain the MLTF target site in a gel retardation assay (Fig. 5A) and (2) compete for the binding of MLTF with a DNA that contained the Ad-MLP (pBal E) but not by a DNA that contained a point mutation at −56 (pTA-56) previously shown to abolish binding of MLTF (Shi *et al.*, 1986) nor by pBR322 DNA (Fig. 5B).

To address more directly the functional role of this protein, the effect of the MLTF in transcription was examined. The experiment described in Fig. 6 demonstrated that purified MLTF stimulated the overall transcription reaction 12-fold in the presence of the purified general factors, RNA polymerase II and the MLP. The recognition by the MLTF of an upstream promoter element suggested that the effect of this factor may be mediated through the preinitiation complex. To analyze if the MLTF contributed to the stability and/or rate of complex formation, the following reaction was performed. Transcription complexes were formed at the Ad-MLP as previously described (Reinberg *et al.*, 1986) in the presence and absence of the MLTF. At specific times of preincubation, as indicated in Fig. 7, aliquots of the reaction were withdrawn and Sarkosyl was added to block further complex formation. The reaction was then incubated for an additional minute, at which time ribonucleoside triphosphates were added and elongation allowed to proceed for 30 min. As shown in Fig. 7, the addition of MLTF did not stimulate the rate of complex formation; rather it increased the amount of complex capable of transcription. Since transcription was limited to only one round by the addition of Sarkosyl (Hawley and Roeder, 1985), the amount of transcript obtained is a direct measurement of the amount of competent complex formed; under these conditions only a three-fold stimulation was observed. Preliminary results indicated that the MLTF directly stimulated the formation of a committed complex (Buckbinder *et al.*, 1986). Previous studies have suggested that the Ad major late upstream promoter factor interacted cooperatively with the TATA binding protein (TFIID) to render a more stable protein : promoter sequence interaction (Sawadogo and Roeder, 1985b). Since the MLTF, together with TFIID and TFIIA, were involved in the production of a stable complex (committed complex) and, as previously indicated this complex was recognized by RNA polymerase II and also stabilized by the polymerase in reactions carried out in the absence of MLTF (Reinberg *et al.*, 1986), it is thus pertinent to postulate that the MLTF permits multiple RNA polymerase molecules to enter the transcription cycle. Reactions carried out in the absence of the MLTF will generate a stable committed

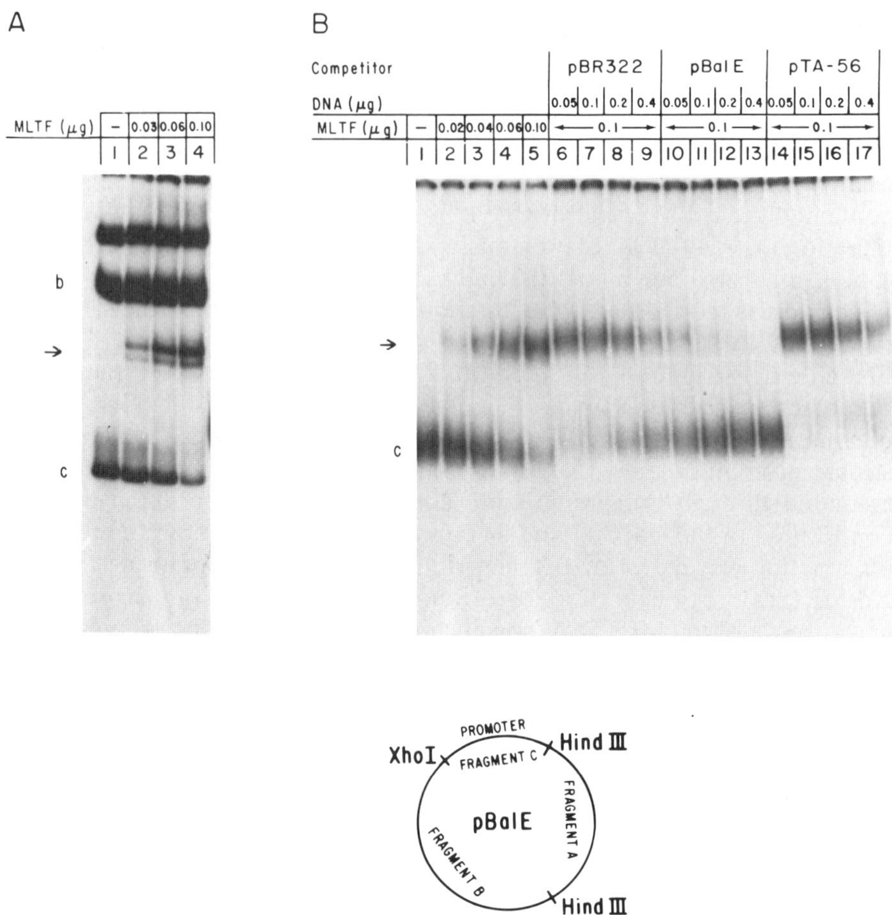

Fig. 5. Analysis of the specificity of the MLTF binding to DNA by a gel retardation assay. (A) Reaction mixtures (20 μl) containing 20 m*M* Hepes-NaOH (pH 7.9), 10 m*M* Tris-HCl (pH 7.9), 60 m*M* KCl, 0.6 m*M* dithiothreitol, 10% glycerol, 1% polyethyleneglycol 8000, 1 μg poly(dA:dT), 0.6 m*M* EDTA, 50 ng of 3′-end-labeled *Xho*I–*Hin*dIII double-digested pBal-E DNA, and different amounts of MLTF (SP-5PW fraction) as indicated in the figure. Reactions were incubated at 30°C for 1 hr, then the different size DNA fragments resulting from digesting pBal-E DNA with *Xho*I and *Hin*dIII (indicated by a, b, and c) as well as the DNA fragment containing MLTF (indicated by an arrow) were separated by electrophoresis on a 4% polyacrylamide gel. (B) Reaction conditions were as described in A. The isolated 456-bp fragment (15 ng) containing the Ad-MLP DNA was incubated with different amounts of MLTF (lanes 1–5). Lanes 6–17 show the effect of different competitor DNAs, as indicated on the figure, on the MLTF binding to the DNA.

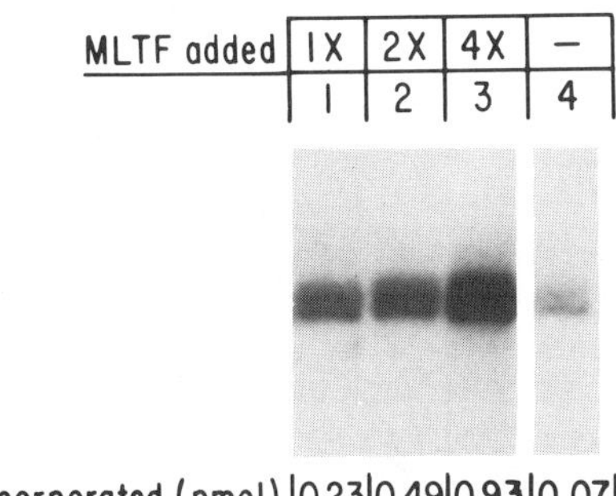

Fig. 6. Effect of the MLTF on transcription of the Ad-MLP. Reaction mixtures (40 μl) were as previously described and contained TFIIC (single-stranded DNA agarose fraction, 0.06 μg), TFIIA (Sephacryl AcA44 fraction, 0.7 μg), TFIIB (Sephacryl AcA44 fraction, 0.18 μg), TFIIE/TFIIF (Sephacryl AcA44 fraction, 1.6 μg), TFIID (phosphocelluose fraction, 2.9 μg), RNA polymerase II (43 ng, 146,000 units/mg protein), pML DNA (0.1 μg), and different amounts of MLTF (5× UPE fraction). The protein concentration of MLTF added to the reaction was indeterminate: this fraction contained 200 μg/ml albumin and no protein above the exogeneous added was detected. Reaction mixtures were incubated at 30°C for 45 min and the products separated on a 4% polyacrylamide 7 *M* urea gel. The amount of [α^{32}P]CMP incorporated into the transcript was determined by cutting the band of the gel and counted.

complex only when polymerase is an integral part of it. Thus, when the polymerase leaves the promoter during RNA elongation, the complex dissociates. Our preliminary results indicated that reactions carried out in the presence of the MLTF undergo multiple rounds of transcription. The results of Sawadogo and Roeder (1985b, see Fig. 3) also suggested multiple round of transcription.

The Ad-MLP CAP site is separated by 210 nucleotides from the CAP site of the IVa2 promoter, a non-TATA sequence containing promoter which is transcribed from the opposite DNA strand (Natarajan *et al.*, 1984). Two DNA elements have been defined as positively regulating this promoter; a proximal element centered at position −40 relative to the IVa2 initiation site, and a distal element that overlaps with the MLTF binding site (Natarajan *et al.*, 1985). To learn more about the mechanism of action of the MLTF, we have partially reconstituted transcription from the IVa2 promoter. Our preliminary results suggested that the addition of the MLTF to a IVa2 reconstituted transcription system icreased IVa2 transcription by fourfold (S. Lobos and D. Reinberg, unpublished observations). This suggested that the MLTF binding site operated bidirectionally. Furthermore, Weinmann and co-workers have inverted the major late upstream promoter element and obtained wild-type levels of transcription from the MLP (personal communication), indicating that the

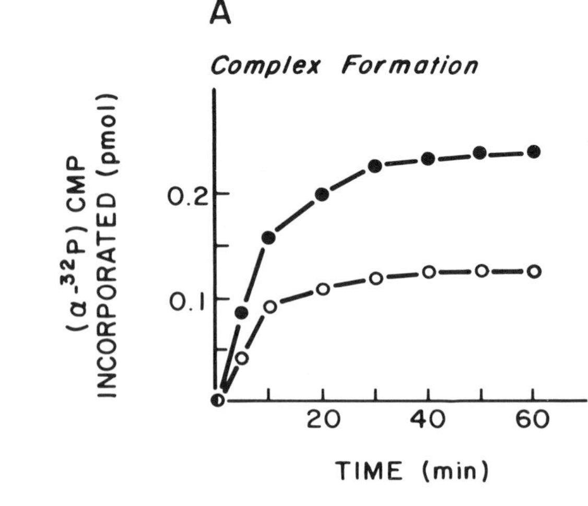

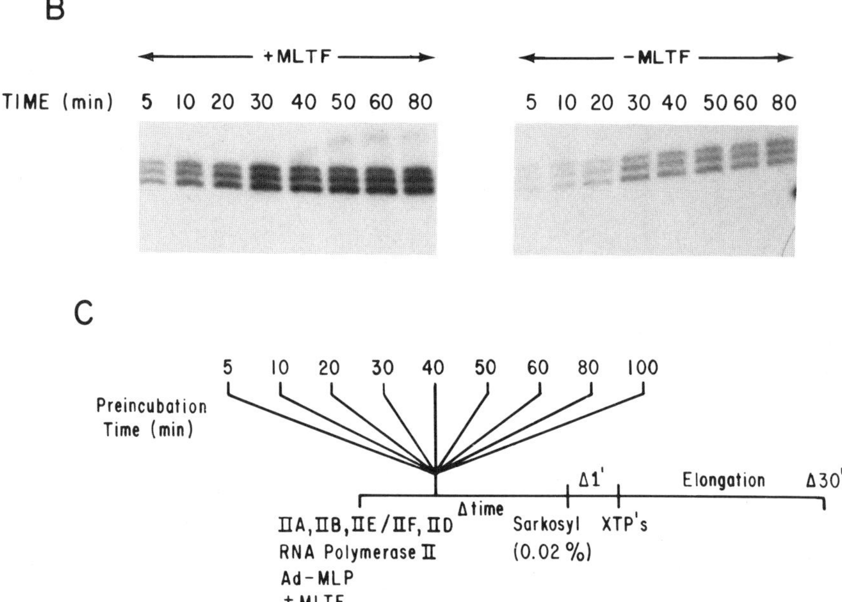

Fig. 7. Effect of the MLTF in the formation of a preinitiation complex at the Ad-MLP. Two 400-μl reaction mixtures that contained TFIIC (single-stranded DNA agarose fraction, 0.6 μg), TFIIA (Sephacryl AcA44 fraction, 7 μg), TFIIB (Sephacryl AcA44 fraction, 1.8 μg), TFIIE/TFIIF (Sephacryl AcA44 fraction, 16 μg), TFIID (Sephacryl S200 fraction, 5.3 μg), RNA polymerase II (0.4 μg, 146,000 units/mg protein), pML DNA (1.0 μg), and MLTF (5× UPE fraction) were incubated at 30°C. After different periods of incubation, as indicated in the figure (C), an aliquot (40 μl) was removed and Sarkosyl (to 0.02%) was added and the reaction further incubated for 1 min. Then the ribonucleoside triphosphates were added and the reaction elongated for 30 min. Products were separated on a 4% polyacrylamide–7 *M* urea gel (B) and the bands cut out of the gel, counted, and plotted (A).

MLTF binding site can operate bidirectionally. Interestingly enough, the partially reconstituted IVa2 transcription system was independent of TFIID; however, another protein fraction (PPF) was required for IVa2 transcription (S. Lobos and D. Reinberg, unpublished observations). This preliminary result suggested that the MLTF may function via different factors (TFIID for the MLP and PPF for the IVa2 promoter). Another possibility is that the effect of the MLTF in transcription is mediated via another component of the transcription machinery that is common for both promoters and operates via a committed complex, such as, for example, TFIIA. The results of Sawadogo and Roeder (1985b) suggested that TFIID directly affected the binding of the Ad-MLP upstream factor which seems to argue against an effect of the MLTF on transcription via a factor such as TFIIA; however, their transcription system was independent of exogenously added TFIIA. The possibility exists that TFIIA was present as a contaminant in one of their fractions required for transcription. The copurification of TFIIA with TFIID has been documented (Reinberg *et al.*, 1986). Furthermore, it has been suggested that TFIIA activity was required for stable binding of the TATA binding protein to the promoter (Davidson *et al.*, 1983; Fire *et al.*, 1984; Reinberg *et al.*, 1986) and, therefore, if this is a general function of TFIIA and TFIIA was present in the TFIID fraction used in the studies of Sawadogo and Roeder (1985b), then it is possible that the effect of the TFIID fraction on the upstream promoter factor was mediated by TFIIA. This remains to be investigated further.

III. CONCLUDING REMARKS

Initiation of transcription from class II promoters is a complex process. We have demonstrated that at least five transcription factors (IIA, IIB, IID, IIE, and IIF), in addition to RNA polymerase II, were required for transcription. These factors operated via minimum promoter sequences (TATA box and CAP site) and participated in the formation of a preinitiation complex. This complex was defined by its ability to start transcription, upon addition of the nucleoside triphosphates, in the presence of heparin concentration that inhibited the action of a free factor (Reinberg and Roeder, 1986a; Reinberg *et al.*, 1986). Thus far, every TATA sequence containing promoter (mouse β-globin, rabbit α-globin, adenovirus-encoded major late, EIV, IX, E1a, and E1b) that has been analyzed in the *in vitro* transcription system exhibited a requirement for these factors; thus, they were defined as general transcription factors (Reinberg and Roeder, 1986a; Nakajimi *et al.*, 1987).

While the above factors were sufficient for transcription of the Ad-MLP and other TATA sequence containing promoters *in vitro,* other factors that recognized upstream promoter elements (relative to the TATA box) were necessary for transcription *in vivo* and in some cases stimulated transcription *in vitro*. Different upstream promoter elements have been identified in several genes; some appeared to be gene specific and/or to confer cell specificity (for review, see Dynan and Tjian, 1985). The simplest case was observed using the Ad-MLP in which a single element, upstream of the TATA box, appeared to regulate the promoter (Hen *et al.,* 1982). The studies of Sawadogo and Roeder (1985a) as well as our studies indicated that the protein that recognized this element stimulated transcription *in vitro*. We have extended these studies and demonstrated that the MLTF affected the amount of preinitiation complex formed. Similar results were obtained when studying the role of the upstream element present in the fibroin promoter (Tsuda *et al.,* 1986). The exact mechanism by which the MLTF stimulated the amount of preinitiation complex formed at the adenovirus major late promoter is unknown. It could be that the MLTF interacted with another transcription factor and this interaction resulted in a more stable association of the factors to the promoter. It is also possible that the MLTF by binding to the major late upstream promoter element affected the secondary structure of the promoter and induced some changes on the DNA that are more favorably recognized by the other transcription factors.

The Ad-MLP contains inverted repeat sequences which reside between −26 to −12 and −7 to +8 (Ziff and Evans, 1978) and direct repeats flanking the TATA box (Mace *et al.,* 1983). Hu and Manley (1986) have shown that the formation of a nuclease S1-sensitive structure at the Ad-MLP (Larsen and Weintraub, 1982) was probably the result of strand slippage and involved the direct repeats. They have proposed that in the "slipped" structure the TATA box element existed in a single-stranded form and the two strands were physically removed from each other in separate loops. Under this configuration, the inactive one, the TATA-binding protein was unable to bind to the DNA and therefore to promote the formation of a preinitiation complex.

Previous studies (Knezetic and Luse, 1986) demonstrated that the presence of nucleosomes on a DNA template that contained the Ad-MLP prevented initiation of transcription. Also, the studies of Tsuda *et al.* (1986) demonstrated that the addition of histone proteins to an *in vitro* transcription system containing the fibroin promoter inhibited transcription; however, the inhibition observed was lower if the promoter contained upstream sequences and was preincubated with an extract depleted of histone proteins. It is thus possible that one role of the adenovirus major late upstream promoter factor is to induce and maintain a structure on the DNA that is more favorably recognized by the TATA-binding protein or equivalent. It may well be that *in vivo,* where the DNA is associated with proteins and undergoing several processes, the formation of an active structure at the Ad-MLP will be prevented by nucleosome assembly. Thus, it is possible that nucleosome assembly induces the "slipped" structure while the major late upstream promoter factor prevents the formation of such a structure.

ACKNOWLEDGMENTS

We wish to thank Dr. Lynne D. Vales and members of the laboratory for helpful discussion while the work was in progress. We also thank Dr. R. Weinmann for communicating unpublished results and Dr. L. D. Vales for critical reading of the manuscript.

This work was supported by National Institutes of Health Grant GM 37120 and an institutional grant from the American Cancer Society to the State University of New York at Stony Brook.

REFERENCES

Breathnach, R., and Chambon, P. (1981). *Annu. Rev. Biochem.* **50,** 349–393.

Buckbinder, L., Flores, O., Lee, R., Weinmann, R., and Reinberg, D. (1986). Submitted for publication.

Carthew, R. W., Chodosh, L. A., and Sharp, P. A. (1985). *Cell (Cambridge, Mass.)* **43,** 439–488.

Davidson, B. L., Egly, J. M., Mulvihill, E. R., and Chambon, P. (1983). *Nature (London)* **301,** 680–686.

Dignam, J. D., Martin, P. L., Shastry, B. S., and Roeder, R. G. (1984). In "Methods in Euzymology" (W. B. Jakoby, ed.), Vol. 104, pp. 582–598. Academic Press, New York.

Dynan, W. S., and Tjian, R. (1985). *Nature (London)* **316,** 774–778.

Egly, J. M., Miyamoto, N. G., Moncollin, V., and Chambon, P. (1984). *EMBO J.* **3,** 2363–2371.

Fire, A., Samuels, M., and Sharp, P. A. (1984). *J. Biol. Chem.* **259,** 2509–2516.

Flores, O., and Reinberg, D. (1987). In preparation.

Hawley, D., and Roeder, R. G. (1985). *J. Biol. Chem.* **260,** 8163–8142.

Hen, R., Sassone-Corsi, P., Corden, J., Gaub, M. P., and Chambon, P. (1982). *Proc. Natl. Acad. Sci. U.S.A.* **79,** 7132–7136.

Hu, S., and Manley, J. L. (1981). *Proc. Natl. Acad. Sci. U.S.A.* **78,** 820–824.

Hu, Y. T., and Manley, J. L. (1986). *Cell (Cambridge, Mass.)* **48,** 743–751.

Jove, R., and Manley, J. L. (1984). *J. Biol. Chem.* **289,** 8513–8521.

Kasher, M. S., Pintel, D., and Ward, D. C. (1986). *Mol. Cell. Biol.* **6,** 3117–3127.

Knezetic, J. A., and Luse, D. S. (1986). *Cell (Cambridge, Mass.)* **45,** 98–104.

Larsen, A., and Weintraub, H. (1982). *Cell (Cambridge, Mass.)* **29,** 609–622.

Mace, H. A. F., Pelham, H. R. B., and Travers, A. A. (1983). *Nature (London)* **304,** 555–557.

Matsui, T., Segall, J., Weil, A. P., and Roeder, R. G. (1980). *J. Biol. Chem.* **285,** 11992–11996.
Miyamoto, N. B., Moncollin, V., Egly, J. M., and Chambon, P. (1985). *EMBO J.* **4,** 3563–3570.
Nakajimis, Horikoshi, M., and Roeder, R. G. (1987). In preparation.
Natarajan, V., Madden, M. J., and Salzman, N. P. (1984). *Proc. Natl. Acad. Sci. U.S.A.* **81,** 6290–6294.
Natarajan, V., Madden, M. J., and Salzman, N. P. (1985). *J. Virol.* **55,** 10–15.
Rappaport, J., Reinberg, D., Zandomeni, R., and Weinmann, R. (1986). *J. Biol. Chem.* (in press).
Reinberg, D. (1987). Submitted for publication.
Reinberg, D., and Roeder, R. G. (1986a). *J. Biol. Chem.* (in press).
Reinberg, D., and Roeder, R. G. (1986b) *J. Biol. Chem.* (in press).
Reinberg, D., Horikoshi, M., and Roeder, R. G. (1986). *J. Biol. Chem.* (in press).
Samuels, M., and Sharp, P. A. (1986). *J. Biol. Chem.* **261,** 2003–2013.
Sawadogo, M., and Roeder, R. G. (1985a). *Proc. Natl. Acad. Sci. U.S.A.* **2,** 4394–4398.
Sawadogo, M., and Roeder, R. G. (1985b). *Cell (Cambridge, Mass.)* **43,** 165–175.
Sekimizu, K., Nakamishi, Y., Mizuno, D., and Natori, S. (1979). *Biochemistry* **18,** 1582–1588.
Shi, X. P., Lee, R., and Weinmann, R. (1986). *Nucleic Acids Res.* **14,** 3429–3744.
Slatery, E., Dignam, J. D., Matsui, T., and Roeder, R. G. (1983). *J. Biol. Chem.* **258,** 5955–5959.
Tsuda, M., Hirose, S., and Suzuki, Y. (1986). *Mol. Cell. Biol.* **6,** 3928–3933.
Weil, A. P., Luse, D. S., Segall, J., and Roeder, R. G. (1979) *Cell. (Cambridge, Mass.)* **18,** 469–484.
Ziff, E., and Evans, R. M. (1978). *Cell (Cambridge, Mass.)* **15,** 1463–1475.

Index

M

R

U

V

Y